Microbiology of
Marine Food Products

MICROBIOLOGY OF MARINE FOOD PRODUCTS

Edited by

Donn R. Ward

Department of Food Science
North Carolina State University

Cameron Hackney

Department of Food Science and Technology
Virginia Polytechnic Institute and State University

Springer Science+Business Media, LLC

An AVI Book
(AVI is an imprint of Van Nostrand Reinhold)

Copyright © 1991 by Springer Science+Business Media New York
Originally published by Van Nostrand Reinhold in 1991
Softcover reprint of the hardcover 1st edition 1991
Library of Congress Catalog Number 90-37357
ISBN 978-0-442-23346-4

Library of Congress Cataloging-in-Publication Data
Microbiology of marine food products / edited by Donn R. Ward and
  Cameron R. Hackney.
     p.   cm.
  "An AVI book."
  Includes bibliographical references and index.
  ISBN 978-0-442-23346-4   ISBN 978-1-4615-3926-1 (eBook)
  DOI 10.1007/978-1-4615-3926-1
  1. Seafood—Microbiology. 2. Seafood—Safety measures.   I. Ward,
Donn R., 1949–   . II. Hackney, Cameron Ray.
QR118.M53   1991   90-37357
576'.163—dc20   CIP

Contents

Preface

In recent years, consumption of seafood products has risen dramatically. However, along with the growth in consumption, there has been growing enthusiasm for efforts to improve the quality and the perceived safety of seafoods. This has culminated in the debate on "Mandatory Seafood Inspection." While *quality* and *safety* are the principal issues behind the inspection debate, microbiology is one of the principal sciences associated with quality and safety.

All food commodities have their own distinctive microbiology. However, of all the food commodities, seafoods have one of the most, if not *the* most, diverse and complex microbiologies. Unlike meat or poultry products where only a few species are represented in each group, the term seafoods encompasses hundreds of genera and species. Concerns associated with mollusks are often quite different from those associated with finfish or crustaceans. Other factors contributing to the microbiological complexity are the range of environmental habitat (freshwater to saltwater; tropical waters to arctic waters; pelagic swimmers to sessile bottom dwellers) and processing practices (iced fresh products to commercially sterile canned products; hand labor to mechanized processes).

This book provides a comprehensive examination of microbiological quality and safety concerns of seafood from harvest through processing. Many of the chapters are the most comprehensive reviews to date. A concerted effort has been made to incorporate discussions on topics that are both timely and timeless. Examples of the former include "U.S. Seafood Inspection and HACCP" and "Vibrionacae," whereas examples of the latter include "Irradiation" and "Packaging." There are also topics which are not necessarily "microbiological" in nature, but are important to any meaningful discussion on seafood safety; these include: "Parasites," "Natural Toxins," "Thermal Processing," and "Depuration."

This book views seafood microbiology from the perspectives of environmental influences and processing influences. It will hopefully serve as a stimulus for future research in these two areas, and bring the questions of quality and safety into a proper perspective.

Contributors

Linda S. Andrews, M.S.
Department of Food Science
Louisiana State University
Baton Rouge, LA 70803

Tuu-Jyi Chai, Ph.D.
Seafood Science Program
Horn Point Environmental Laboratories
University of Maryland
Cambridge, MD 21613

Ralph R. Cockey
Seafood Science Program
Horn Point Environmental Laboratories
University of Maryland
Cambridge, MD 21613

Mary T. Cole, M.S.
Department of Biology
Nicholls State University
Thibodaux, LA 70310

David W. Cook, Ph.D.
Gulf Coast Research Laboratory
Ocean Springs, MS 39564-7000

Thomas L. Deardorff, Ph.D.
Department of Structural and Cellular
  Biology
College of Medicine
University of South Alabama
Mobile, AL 36688

George J. Flick, Ph.D.
Department of Food Science
  and Technology
Virginia Polytechnic Institute
  and State University
Blacksburg, VA 24061

E. Spencer Garrett, M.S.
National Marine Fisheries Service
National Seafood Inspection Laboratory
Pascagoula, MS 39568-1207

Robert M. Grodner, Ph.D.
Department of Food Science
Louisiana State University
Baton Rouge, LA 70803

Cameron R. Hackney, Ph.D.
Department of Food Science
  and Technology
Virginia Polytechnic Institute
  and State University
Blacksburg, VA 24061

Sherwood Hall, Ph.D.
Food and Drug Administration
Washington, DC 20204

Martha Hudak-Roos
National Marine Fisheries Service
National Seafood Inspection Laboratory
Pascagoula, MS 39568-1207

Steven C. Ingham, Ph.D.
Applied Microbiology and Food Science
University of Saskatchewan
Saskatoon, Saskatchewan S7N OWO
Canada

Howard Kator, Ph.D.
School of Marine Science
Virginia Institute of Marine Science
College of William and Mary
Gloucester Point, VA 23062

Marilyn B. Kilgen, Ph.D.
Department of Biology
Nicholls State University
Thibodaux, LA 70310

John E. Kvenburg, Ph.D.
Division of Microbiology
Food and Drug Administration
Washington, DC 20204

Lucina E. Lampila, Ph.D.
Seafood Experiment Station
Virginia Polytechnic Institute
   and State University
Hampton, VA 23669

Brian K. Mayer, M.S.
Campbell Soup Company
Camden, NJ 08103-1799

Russell J. Miget, Ph.D.
Texas A&M Marine Advisory Service
Port Aransas, TX 78373

Robin M. Overstreet, Ph.D.
Gulf Coast Research Laboratory
Ocean Springs, MS 39564-7000

Alvin P. Rainosek, Ph.D.
University of South Alabama
Mobile, AL 36688

Martha W. Rhodes, M.S.
School of Marine Science
Virginia Institute of Marine Science
College of William and Mary
Gloucester Point, VA 23062

Gary P. Richards
National Marine Fisheries Service
Southeast Fisheries Center
Charleston Laboratory
Charleston, SC 29412-0607

Thomas E. Rippen, M.S.
Seafood Experiment Station
Virginia Polytechnic Institute
   and State University
Hampton, VA 23669

Gary E. Rodrick, Ph.D.
Department of Food Science
   and Human Nutrition
University of Florida
Gainesville, FL 32611

Jayne E. Stratton
Department of Food Science
   and Technology
   and Food Processing Center
University of Nebraska
Lincoln, NE 68583-0919

Steve L. Taylor, Ph.D.
Department of Food Science
   and Technology
   and Food Processing Center
University of Nebraska—Lincoln
Lincoln, NB 68583

Donn R. Ward, Ph.D.
Department of Food Science
North Carolina State University
Raleigh, NC 27695-7624

Part 1

1

Microbiology of Finfish and Finfish Processing

Brian K. Mayer and Donn R. Ward

Finfish are generally regarded as being much more perishable than other high-protein muscle foods. This high degree of perishability is due primarily to the large concentration of nonprotein nitrogenous compounds present in fish muscle. These compounds, which include free amino acids and volatile nitrogen bases such as ammonia, trimethylamine, creatine, taurine, the betaines, uric acid, anserine, carnosine, and histamine are utilized actively by bacteria during spoilage (Jay 1986). Another factor that contributes to the perishability of finfish is the temperature of the water from which they are harvested. The bacterial flora of cold-water fish species are not inhibited as effectively by refrigeration as are the normal flora of fish harvested from warm tropical waters. When handled properly, tropical fish species are generally less prone to rapid spoilage and exhibit a longer refrigerated shelf life than cold-water species (Disney 1976; Poulter et al. 1981; Sumner et al. 1984). This broad generalization, however, may have to be reconsidered. A review by Lima dos Santos (1981) on the storage of tropical fish on ice contends that there are too many variables for direct comparisons, and that there are instances of cold-water fish such as halibut and grenadiers having a shelf life, on ice, for up to 3 weeks.

It is generally accepted that the internal flesh of live, healthy fish is sterile. The natural bacterial flora reside mainly in the outer slime layer of the skin, on the gills, and in the intestines of feeding fish. Bacterial numbers range from 10^2 to 10^6 colony forming units (cfu) per square centimeter on the skin, on the gills from 10^3 to 10^5 per gram, and in the intestine from very few in nonfeeding fish to 10^7 or greater in feeding fish (Liston et al. 1976). These initial microflora are directly related to the environment, whereas the total microbial load is subject to seasonal variation (Liston 1956; Shewan 1961). Shewan (1977) indicated that warm-water fish seem to have a more mesophilic, Gram-positive microflora (micrococci, bacilli, coryneforms), whereas cold-water fish harbor a predominately Gram-negative psychrophilic population (*Moraxella, Acinetobacter, Pseudomonas, Flavobacterium,* and *Vibrio*). Regardless of the differences in the initial microflora, the spoilage patterns of finfish during iced storage are usually quite similar and are caused by *Pseudomonas* spp. and *Alteromonas putrefaciens* (Barile et al. 1985b). This chapter will discuss the factors that affect the growth and invasion of these spoilage bacteria during harvesting and processing and ultimately affect quality.

HARVESTING AND ONBOARD HANDLING

The initial quality and microbial load of fresh finfish is affected by the method of harvesting. It is at this point that quality maintenance must begin. Abusive handling at harvest will be detrimental to subsequent quality and shelf life at the retail level.

A wide variety of fishing gear and methods are employed to harvest finfish commercially. These include traps and barriers, hook and line techniques, and various types of nets (Alverson 1976). Although little quantitative data are available to compare the microbial load of freshly landed fish by different harvesting methods, Shewan (1949) demonstrated that trawled fish usually carry microbial loads that are 10–100 times greater than those of line-caught fish. This increase is attributed to dragging along the ocean bottom, which stirs up the mud and contaminates the fish, and to compaction of the fish, which causes gut contents to be expressed. When trawling for fish, it is generally accepted that longer tows will result in lower quality (Costakes et al. 1982). During periods of heavy fishing the cod end of the net becomes very full and the resulting catch, which may have been dead for hours, is bruised and crushed from compression. Furthermore, larger catches take longer to stow properly. The fish are therefore subjected to the physical abuse of sliding on the boat deck and exposure to ambient temperature and sunlight. In this case fishing for quantity adversely affects quality.

The amount of stress that the fish endures during capture, just prior to death, has also been shown to affect postharvest quality. Fish that are normally very active, such as tuna and mackerel, may become excited and die in a frenzied state when harvested by purse-seining. Inherent physiological features of tuna make them unique from other fish species. They are among the fastest swimming fish, with burst speeds for yellowfin (*T. albacores*) reported as 21 body lengths per second (Walters and Fierstine 1964). Tuna have very high metabolic rates and some species have the ability to adjust their body temperature. When tuna are captured in a highly stressed state, the buildup of lactic acid in the muscle combined with elevated muscle temperature results in a serious flesh defect known as burnt flesh (Goodrick 1987). The flesh is no longer bright red and the flavor is acidic with a metallic aftertaste. In this state the tuna is still acceptable for canning but unacceptable for the highly lucrative Japanese sashimi market. For this reason, longlining is the desired harvesting method for minimizing stress and maintaining postharvest quality of tuna.

The effect of stress on postharvest quality has been documented with other fish species as well. Salmon, harvested by gill netting, die after an exhausting struggle. As a result rigor mortis sets in quickly and earlier signs of deterioration occur during icing (Dassow 1976). Harvesting by hook and line, where the fish is quickly brought aboard the vessel and killed, minimizes stress and its associated quality deterioration. This concept of "clean kill" is well known in the slaughter of poultry and livestock.

A study on the harvesting of Atlantic cod indicated that the extent of struggling significantly affected protein and caloric content of the flesh by reducing moisture (Botta et al. 1987a). A concurrent study by the same researchers indicated that the method of catching Atlantic cod was more significant than the time of season in its effect on sensory quality (Botta et al. 1987b). Studies by Herborg and Villadsen (1975) indicated that the quality and shelf life of rainbow trout were adversely affected by physical stress. They also stated that the bacterial infection level in the fish muscle increased with stress.

Because it is not always possible to be selective in the harvesting methods utilized for many commercial fish species, it is critical that the fish be handled in a quality conscious manner as soon as they are landed on the vessel. Handling of fish with gaff hooks, picks, or forks should be avoided or at least limited to the head region. Breaks in the skin and unsightly holes in the flesh quickly introduce spoilage bacteria and accelerate quality deterioration. Likewise, stepping or standing on the fish should be avoided as this may result in bruising.

Unlike the red meat and poultry industry, where animals in good physiological condition are brought live to the processing facility, finfish are usually harvested in remote locations and must be stored for several hours or days aboard the vessel prior to processing.

In 1973, the Food and Agricultural Organization published its *Code of Practice for Fresh Fish* (FAO 1973), the primary focus being the handling of fish at sea. Section 3.1.1 states: "Fish are extremely perishable food, and should be handled at all times with great care and in such a way as to inhibit the growth of microorganisms. Fish quality deteriorates rapidly and potential keeping time is shortened if they are not handled and stored properly. Much of the fish landed for human consumption is subjected to fairly rough handling treatment. Fish should not be exposed to direct sunlight or to the drying effect of wind, but should be carefully cleaned and cooled down to the temperature of melting ice, 0°C, as quickly as possible. Any careless treatment or delay in reducing the temperature of fish will have a marked effect on their potential keeping time."

Note that one of the suggestions in the quote is that the fish should be carefully "cleaned." The word "cleaned" is not defined, but several states require that certain ground fish be bled, gutted, and gilled on board the fishing vessel. Shewan (1961) points out that the main advantage to gutting is to prevent autolytic spoilage, rather than bacterial decomposition. Indeed, FAO (1973) states that "bad gutting might be worse than no gutting" inasmuch as it can facilitate entry of bacteria into the flesh. Stansby and Lemon (1941) reported that gutting of fresh mackerel could increase rather than diminish bacterial numbers. In a more recent study on board a New England trawler, Samuels et al. (1984) found that gutted cod and haddock actually had lower psychrotrophic counts than did cod immediately after catch, although the differences were not significant. Furthermore, they found that the practice of washing the gutted fish in circulating seawater prior to ice storage had no effect on bacterial numbers, although this practice did remove mud and other visible extraneous matter.

Work by Ravesi et al. (1985) on spiny dogfish demonstrated that microbial numbers on fillets were a function of the level of handling and processing prior to removal of the fillet. In other words, dogfish stored whole had lower counts than gutted fish, which in turn had lower counts than those headed and gutted. Interestingly, however, the researchers recommended gutting if the fishing trip is longer than 2 days, because it results in increased shelf life of the fillets, increases in bacterial counts notwithstanding. On the other hand, if the trips are short (less than 2 days) the fishermen's time would be best spent on adequately icing, and thus rapidly chilling, the whole dogfish. The researchers reported that heading and gutting was not helpful, because removing the head creates another cut surface for potential bacterial contamination and does not benefit shelf life.

A study by Scott et al. (1986) compared microbiological and sensory assessment of whole and headed and gutted orange roughy during iced storage. Prior to storage the whole fish and the headed and gutted fish were both washed with seawater. Microbiological results indicated that there were no significant differences between the bacterial counts of the two groups. The shelf life of whole orange roughy stored in ice, as determined by sensory analysis, was between 11 and 13 days with only a slight increase resulting from heading and gutting. The authors concluded that the slight increase in the shelf life of the headed and gutted fish was due to reduced autolysis rather than to reduced microbial activity. This autolytic spoilage by digestive enzymes is more prevalent in fish that are actively feeding at the time of catch (Dassow 1976).

A number of sources recommend that fish be bled, as well as gutted, prior to storage aboard the fishing vessel (Costakes et al. 1982; Strom and Lien 1984; Valdimarsson et al. 1984). The Canadian grading standards for Atlantic groundfish specify that the color of the flesh should be characteristic of bled fish (DFO 1983). There is some question, however, whether the bleeding of fish at harvest actually benefits quality in all situations. A cooperative project by the Maine Groundfish Association and the Maine Department of Marine Resources concluded that the bleeding of live-caught cod, before gilling and gutting, appeared to have no significant effect on quality or shelf life when compared to identically harvested but unbled cod samples from the same tow (Moser 1986). In this study samples were subjected to color measurement with a color difference meter, sensory panel evaluations, torrymeter readings, surface bacterial load tests, and determination of trimethylene and hypoxanthine (spoilage indicators) concentrations. In all cases no significant differences were found between the bled and unbled fillets. The author pointed out that Maine boats tend to fish deeper waters and/or make longer tows than boats in the New Bedford fishery, where prior information on bleeding was generated. The Maine boats bring up a combination of live, stunned, and dead fish, whereas in the New Bedford fishery more fish are live caught.

A study by Botta et al. (1986) revealed that the bleeding of northern cod, which were caught by Canadian Otter trawlers, was beneficial to quality only if conducted within 1–2 hours of the fish being brought aboard. These observations were based on either a one- or two-step bleeding procedure. The authors stressed that the benefits of bleeding may vary with location of catch, and time of year.

Mayer et al. (1986) reported that onboard processing (bleeding, gutting, and gilling) did not extend the shelf life of bluefish, inasmuch as fillets from these fish were very similar in sensory scores to the fillets cut from fish boxed and stored whole. However, the authors pointed out that because the whole bluefish were only on ice 1–2 days before processing, there may not have been sufficient time lapse to see any quality benefit from onboard processing. This agrees with the suggestions of Ravesi et al. (1985).

It is obvious from the conflicting results in the literature that the universal application of bleeding and gutting procedures should not be applied in all fish harvesting operations. Each fishery will have to make its own decision based on the species and method of harvest. Whatever on-board processing is conducted it must not subject the fish to lengthy delays in icing. The use of ergonomic design principles has recently been suggested for improving on-board fish handling and processing procedures (Rodman 1987).

The FAO (1973) report recommended that fish should be cooled down to the temperature of melting ice, 32°F (0°C), as quickly as possible. The rationale for this is apparent. Barile et al. (1985a) found that the shelf life of Faughn's mackerel, on ice, was reduced by 1 day for each hour delay in icing or exposure to ambient temperatures of 82.4–86°F (28–30°C). Moreover, at the point of rejection by a trained taste panel, the fish samples had the following aerobic plate counts at 68°F (20°C): fish iced immediately, 10^9 organisms/g; fish iced after a 3-hour delay, 10^7; fish iced after a 6-hour delay, 10^6; and fish iced after 9- and 12-hour delays, 10^5. There was also a distinct change in the spoilage microflora. The principal spoilage organisms after 0–6 hours of delay in icing were *Pseudomonas* spp. and *Alteromonas putrefaciens,* whereas prolonged delay (9–12 hours) resulted in a final spoilage flora of *Bacillus* spp., *Aeromonas hydrophila,* and *Pseudomonas* spp. These fish were purse-seined off the west coast of the Philippines.

Interestingly, work done on trench sardines (Chinivasagam and Vidanapathirana 1986) indicated little difference between the quality of the fish iced immediately after landing and after a 5-hour delay. Other studies (DeSilva 1978; Jayaweera 1980) have also reported no detrimental effects from delayed icing, for specific fish, for up to 12 hours. Chinivasagam and Vidanapathirana offered two possibilities as to why delaying icing is apparently not detrimental. One is the low incidence of *Pseudomonas* organisms on tropical fish, and the second is the observation of Poulter et al. (1981) that when warm-water fish are kept at high temperatures, the onset of rigor is slow and of longer duration, thus delaying bacterial growth. Nonetheless, the work in which Barile et al. (1985a) found that shelf life was reduced by 1 day for every hour delay in icing was done on a tropical species.

Most large, commercial fishing vessels in the mid-Atlantic and New England regions store their catch on ice in the ship's hold. An exception to this would be smaller vessels which leave the docks early in the morning and return later in the day. These vessels, commonly referred to as day boats, store fish in the holds, but often without ice. Samuels et al. (1984) reported significant increases in the psychrotrophic counts on the surface of noniced bluefish, stored in the hold of these vessels, from the time of catch to the arrival at the dock. However, there were no significant increases on fish caught at the same time but held in iced coolers.

Delays in the icing of fresh fish are of particular concern with fish of the Scombridae family (tuna, mackerel, bonita), due to the possibility of histamine poisoning. Other fish have been implicated, however, including mahi-mahi and bluefish. Histamine poisoning is a chemical intoxication resulting from the ingestion of foods that contain unusually high levels of histamine. Histamine is formed by the decarboxylation of the amino acid histidine by the microbial enzyme histidine decarboxylase. Histidine is found at high levels in the above fish. Because the production of histamine is a function of microbial enzymatic activity, the simplest method of prevention is rapid chilling of harvested fish and maintenance of low temperature through consumption. Chen et al. (1987) reported that *Morganella morganii, Klebsiella pneumoniae,* and *Hafnia alvei* are the only histamine-producing bacteria that have actually been isolated from fish causing scombroid or histamine poisoning. There are, however, numerous fish isolates that have been reported to display histidine decarboxylase activity. These include *Pseudomonas putrefaciens, Aeromonas hydrophilia, Proteus vulgaris, Clostridium perfringens, Vibrio alginolyticus,* and *Enterobacter aerogenes* (Middlebrooks et al. 1988).

Besides the traditional method of stowing fish in bulk on ice in the hold of the vessel, fisherman have also utilized short-shelving and boxing at sea. When fish are stowed in bulk they are often stacked 5–6 feet deep in ice. The fish near the bottom are therefore under a tremendous amount of weight. This pressure causes poorer quality, decreased shelf life, and lower yields. The benefit of short-shelving is the reduction of this pressure on the fish. Short-shelving consists of layering fish and ice between shelves that help support the weight. Shelving requires the use of removeable shelves that can be easily cleaned, such as corrugated aluminum. The shelves should be placed 18–22 inches apart. When economically feasible, boxing at sea can offer an excellent method of maintaining quality. Plastic boxes of various sizes and design are available and in use at present. Boxing of fish at sea is done, at least to some extent, in almost every fishing nation in the world. Although boxing is still a relatively new concept in North America, it has been a common practice in countries such as Norway and Iceland. In addition to reduced compression, boxing provides less abusive handling and quicker unloading at dockside. According to a study by the New England Fisheries Development Foundation (Costakes et al. 1982) boxing at sea provided a 2- to 6-day increase in shelf life and a 7–15% increase in landed weight over traditionally bulk stowed fish. In general the longer the trip, the more dramatic the difference in quality.

Although the use of ice is beneficial in the preservation of fish quality, it is not without its disadvantages. Holston and Slavin (1965) suggested that some of the undesirable attributes of ice are a tendency to injure and bruise the flesh and a leaching of flavor components and nutritionally desirable minerals, as well as water-soluble proteins. Also, icing of fish can be very labor intensive and expensive, in view of the large catches that are typical of some fisheries (Lee and Kolbe 1982; Reppond et al. 1985). Consequently, systems using some form of mechanically refrigerated seawater (RSW) have begun to replace ice in certain fisheries, especially in the Pacific Northwest (Lee and Kolbe 1982).

RSW and CO_2-modified refrigerated seawater (MRSW) have been shown to delay spoilage in shrimp and fish, compared to storage on ice (Bullard and Collins 1978; Reppond et al. 1979; Reppond and Collins 1983). Roach et al. (1967) stressed the importance of proper sanitation, especially on RSW installations: "While with iced fish only a part of a load may spoil, with fish held in tanks of RSW even under proper conditions there is a danger that all, or nearly all, the fish will be rejected if spoilage occurs." The implications of this statement are underscored in work performed by Lee and Kolbe (1982). In their study, aerobic plate counts (APC) [at 77°F (25°C)] of RSW from a commercial vessel were found to be 7.7×10^4 organisms/ml on day 1, rising to 1.5×10^6/ml on day 8. The authors reported that the recirculating seawater seemed to be quickly contaminated by inadequately cleaned surfaces of the fish hold and by debris remaining in the pipes and on the wooden bin boards. The authors' contention seemed to be further supported by the finding that ceiling scrapings of the fish hold yielded a microbial count of 4.5×10^8 organisms/g, even after the hold had been cleaned and sanitized.

Bronstein et al. (1985) examined the storage of dressed salmon in refrigerated freshwater, diluted seawater, seawater, and ice. Although differences between storage media were not consistently statistically significant, bacterial growth in fish held on ice appeared to be more rapid than on fish held in chilled-water systems. The authors suggest that the differences may be due to the more aerobic conditions present during iced storage, which allow the normal aerobic spoilage flora to grow more rapidly.

Several studies have reported the MRSW system, in which CO_2 is injected to control bacterial growth by reducing the pH of the brine (Nelson and Barnett 1973), to be superior to both ice and RSW in slowing product deterioration (Barnett et al. 1971; Bullard and Collins 1978; Lemon and Regier 1977; Longard and Regier 1974). Tomlinson et al. (1974), however, determined that although the addition of CO_2 to RSW did help to restrict bacterial growth it was detrimental to certain sensory attributes of the stored fish. In their study eviserated salmon and lingcod were stored in RSW and MRSW. The authors stated that the addition of CO_2 promoted salt uptake, increased the susceptibility of the fish to rancidity during frozen storage, and in the case of salmon, promoted the flesh color to fade. Reppond and Collins (1983) found that the quality of Pacific cod was acceptable to 6 days in ice and 9 days in MRSW, but they also stated that the absorption of salt during MRSW storage may present a problem.

FISHING VESSEL SANITATION

Within the food industry, sanitation of food-contact surfaces and food handling equipment is generally given high priority. On board fishing vessels, however, sanitation as practiced by most food industries is not always practical. Because of the naturally occurring high bacterial loads on fish, counts on board the vessels are quite high. Samuels et al. (1984) reported the $\log_{10}$ psychrotrophic counts/inch^2 of fish-contact surfaces to be: hand gaffs, 8.2 organisms; decks, 7.3; fish hold, 6.5; and shovels/pushes, 7.7. The researchers went on to report that although the sanitary environment on board commercial fishing vessels could be improved, it was not shown to be detrimental to a well-iced catch. This suggestion is in agreement with the work performed by Huss et al. (1974). While strict sanitary control may not benefit the quality and shelf life of well-iced whole fish, it is still important to avoid cross contamination of the stowed fish with bilge oil or dirty ice from previous trips. If on board processing is performed, however, an unsanitary environment will contribute to reduced quality and shelf life by exposing the previous sterile flesh to high numbers of spoilage bacteria. Fish holds should be constructed or so altered to make them conducive to sanitary practices. Wood, for example, cannot be sanitized properly.

PROCESSING

Given the large diversity of commercially exploited species and the products that can be derived from them, the term "processing" can encompass operations that range from the very simple to the complex. Moreover, when it comes to processing, as with most food industries, evidence indicates that the sanitary conditions of the seafood processing plant do correlate well with the microbial quality of the finished product (Phillips and Peeler 1972; Wentz et al. 1985).

Work by Samuels et al. (1984) in finfish processing plants indicated that in some instances processing lines seem to reach a contamination "saturation point." Fish from the boats quickly contaminated the processing equipment (sorting tables, weigh scales, scalers, totes, knives, cutting boards) with significant levels of psychrotrophic spoilage bacteria (1.0^5 cfu/inch^2). The authors theorized that establishing an effective in-plant sanitation program could possibly lead to improved product quality and shelf life. Effective sanitation would, however, be contingent on decontaminating or reducing microbial loads on fish before they enter the processing lines. To accomplish this they used a high-pressure wash on the whole fish to remove surface slime and its associated microflora. This treatment reduced psychrotrophic counts by 99%. Moreover, taste panel evaluations judged flounder fillets processed under the recommended procedures to have a shelf life of 11–12 days, as opposed to 7–8 days for fillets processed in the traditional manner.

One area of concern, with respect to microbial contamination of finfish during processing, is the use of wash tanks. Although wash tanks are effective in removing blood and physical debris, they can be a source of microbial contamination. A study by Mayer et al. (1986) revealed that when similar lots of fish were processed at nearby processing plants, the microbial load differed greatly depending on the use of wash tanks or high pressure washing. APC [68°F (20°C)] of dressed Atlantic mackerel, which were passed through a wash tank, were more than 4.0×10^5 cfu/inch^2 higher than high-pressure washed fish. Similar results were also obtained with dressed sea bass and dressed porgy. Additional studies indicated that finfish fillet quality can be best maintained by high-pressure washing whole fish, prior to filleting, rather than high-pressure washing the fillets themselves. The physical appearance of the fillets, especially those from soft fleshed fish, is easily abused by the high-pressure spray.

The effectiveness of high-pressure washing with surfactants on reducing surface microflora was also tested by Mayer et al. (1986). High-pressure washing whole croaker, with a solution of 0.1% cetylpyridinium chloride, provided a 1.08 log cfu/inch^2 reduction over high-pressure washing with tap water alone. By scaling the fish, prior to high-pressure washing with 0.1% cetylpyridinium chloride, the APC was reduced an additional 0.88 log (average APC 2.43 log cfu/inch^2).

Kosak and Toledo (1981) also investigated the effect of microbial decontamination on the storage stability of finfish. Chlorine dip was used for microbiological decontamination of the fish prior to packaging. The authors reported a doubling of shelf life by using the combined effects of a 3.5-minute dip in a 1,000 μg/ml free chlorine solution followed by packaging under vacuum or in polyethylene bags. However, Samuels et al. (1984) found that dipping in a 200 ppm hypochlorite solution for 4 minutes followed by overwrapping in an oxygen-permeable film was only marginally advantageous over dipping in water. Mayer et al. (1986) reported that a spray application of 2,000 ppm of free chlorine solution reduced the aerobic plate count of gray sea trout from 4.18 to 2.98 log cfu/g. The total reduction, even at this high concentration, was only 1.20 log units. It appears that chlorine application is only marginally effective at reducing the spoilage microflora on fish surfaces.

The effect of ozone on the iced storage life of fresh gutted Atlantic cod was investigated by Ravesi et al. (1987). The ozone was incorporated in either the ice, rinse water, or chilled seawater. The authors reported that the shelf life, as assessed by sensory, chemical, and microbiological tests, was not sufficiently extended by any of the treatments.

Because of the unique nature of seafood and seafood processing operations, it can be a difficult task to maintain extremely rigid standards with respect to microbial numbers on processing-plant equipment. One of the plants reported by Samuels et al. (1984) to be operating under good sanitary conditions had psychrotrophic concentrations on equipment surfaces ranging from 1.3×10^1 to 2.5×10^7 organisms/inch^2. Because of gross contamination resulting from contact with these heavily contaminated surfaces, there has been research regarding the benefits of delayed processing. Shaw et al. (1984), working with cod, found that maximum overall shelf life could be attained by delayed filleting, thereby reducing the time stored as fillets. The data indicated a 2-log difference in total APC [69.8°F (21°C)] from fillets cut from 1-day postmortem fish. The $\log_{10}$ counts were 7.3 and 5.3 cfu/g, respectively. (These fillets were from cod that had been eviscerated at sea.)

Townley and Lanier (1981) looked at the possibility of delayed evisceration. Working with Atlantic croaker and gray trout, they reported no advantages to delayed processing. In fact, fish eviscerated immediately on landing maintained class 1 quality 7–10 days longer than uneviscerated fish or iced fish eviscerated 3 days after harvest. Microbial populations were reported to remain lower in eviscerated fish, for both species, throughout a 2-week storage period [32–33.8°F (0–1°C) with top icing].

Mayer et al. (1986) investigated the effect of delayed processing of boxed at sea bluefish. Bluefish that were processed immediately had a 10-day shelf life [tray-packed fillets, 33°F (0.55°C)] whereas bluefish that were stored whole on ice for a period of 4 and 7 days before processing lasted approximately 11 and 13 days from day of catch. The authors stated that the increase in shelf life would be dependent on whether the fish were actively feeding prior to harvest. Actively feeding fish have accelerated quality deterioration when held uneviscerated.

PRESERVATIVES

Given the perishable nature of seafood, a satisfactory method for extending the shelf life of chilled products that ensures quality and a continuity of supply with a minimum of waste has been the ultimate goal. Research with certain preservatives to aid in this effort has met with varying degrees of success.

Sorbic acid and its potassium salt have been the subject of significant research activity. Fey and Regenstein (1982) reported potassium sorbate dips to be of minimal value in extending the shelf life of red hake and salmon. Chung and Lee (1981) indicated that inhibition of microbial growth by potassium sorbate was concentration dependent. With English sole, the inhibitory effects at 0.1% concentration were barely noticeable, whereas those at 1.0% concentration were quite pronounced. A delay in the microbial lag phase was reported as 1 day and 6 days, respectively. The researchers reported that the typical spoilage pattern was unchanged, with *Pseudomonas* spp. being the predominant spoilage flora for treated and untreated samples.

Shaw et al. (1983) tested the effectiveness of 3% potassium sorbate applications on the shelf life of Atlantic cod. Their results indicated that sorbate dips were effective in extending the shelf life of cod fillets, but were ineffective for gutted fish. Growth of *A. putrefaciens* on sorbate-dipped fillets was significantly suppressed in comparison with the control, although total aerobes were not reduced.

Work by Statham et al. (1985) on the Australian fish morwong indicated that the most effective treatment for extending the shelf life of fillets stored at 39.2°F (4°C) was to dip them in a combined solution of 1.2% potassium sorbate and 10% polyphosphate, followed by packaging in 100% CO_2. Miller and Brown (1984) reported that dipping in a combination of 1% potassium sorbate plus 5 ppm chlortetracycline, followed by vacuum packaging and storage at 35.6°F (2°C), was best at retaining the fresh properties of rockfish fillets after 14 days of storage.

Yet another attempt at chemical preservation of finfish was reported by Wesley (1982). Whole flounder and flounder fillets were given enzyme treatment consisting of glucose oxidase, catalase, and glucose solution applied in three forms: as a dip, as enzymatic ice, and immobilized in algin blankets. Glucose oxidase is an enzyme that catalyzes the oxidation of glucose to gluconic acid with molecular oxygen being reduced to hydrogen peroxide. Commercial enzyme preparations include catalase, to break down the hydrogen peroxide that forms, and glucose, to ensure the presence of sufficient carbohydrate to initiate the reaction. It is the generation of gluconic acid, with the resultant pH drop on the fish surface, that is credited with the inhibition of spoilage microorganisms. The results indicated that untreated controls were acceptable for 8 days, whereas treated samples remained acceptable for 15 days. Shaw et al. (1986) reported on their work with glucose oxidase dips and gluconic acid dips on the preservation of cod fillets. Their results indicated only marginal differences in the acceptability of either treatment over that of the controls. Consequently, the researchers felt that these treatment methodologies have little potential.

With the present negative attitude that consumers have against the use of chemical preservatives in food products, it is unlikely that their use will become widespread for fresh finfish. If fishermen and processors are conscientious about the factors previously mentioned, which affect quality and shelf life of fresh finfish, the use of chemical preservatives to retard microbial growth should be unnecessary. It is important to remember that every stage of handling from harvest to consumption affects quality. On the fishing vessels fishermen must avoid abusive handling. In the processing plants good sanitary conditions and strict temperature must be maintained. It will take a combined effort to provide the quality finfish that the consumer demands.

REFERENCES

Alverson, D. L. 1976. Fishing gear and methods. In: Stansby, M. E. (ed.) *Industrial Fishery Technology*. Robert E. Krieger, New York, pp. 45–64.

Barile, L. E., Milla, A. D., Reilly, A., and Villadsen, A. 1985a. Spoilage patterns of mackerel (*Rastrelliger faudhni* Matsui). 1. Delays in icing. *ASEAN Food J.* **1**:70.

Barile, L. E., Milla, A. D., Reilly, A., and Villadsen, A. 1985b. Spoilage patterns of mackerel (*Rastrelliger faudhni* Matsui). 2. Mesophilic and psychrophilic spoilage. *ASEAN Food J.* **1**:121.

Barnett, H. J., Nelson, R. W., Hunter, P. J., Bauer, S., and Groniger, H. 1971. Studies on the use of carbon dioxide dissolved in refrigerated brine for the preservation of whole fish. *Fish. Bull.* **69**:433.

Botta, J. R., Squires, B. E., and Johnson, J. 1986. Effect of bleeding/gutting procedures on the sensory quality of fresh raw Atlantic cod (*Gadus morhua*). *Can. Inst. Food. Technol. J.* **19**:186.

Botta, J. R., Kennedy, K., and Squires, B. E. 1987a. Effect of method of catching and time of season on the composition of Atlantic cod (*Gadus Morhua*). *J. Food Sci.* **52**:922.

Botta, J. R., Bonnell, G., and Squires, B. E. 1987b. Effect of method of catching and time of season on sensory quality of fresh raw Atlantic cod (*Gadus Morhua*). *J. Food Sci.* **52**:928.

Bronstein, M. N., Price, R. J., Strange, E. M., Melvin, E. F., Dewees, C. M., and Wyatt, B. B. 1985. Storage of dressed chinook salmon, *Oncorhynchus tshawytscha*, in refrigerated freshwater, diluted seawater, seawater, and in ice. *Mar. Fish. Rev.* **47**:68.

Bullard, F. A., and Collins, J. 1978. Physical and chemical changes of pink shrimp, *Pandalus borealis*, held in carbon dioxide modified refrigerated seawater compared to pink shrimp held in ice. *Fish. Bull.* **76**:73.

Chen, C.-M., Marshall, M. R., Koburger, J. A., Otwell, W. S., and Wei, C. I. 1987. Determination of minimal temperatures for histamine production by five bacteria. Tropical and Subtropical Fisheries Technological Society of the Americas. In: *Proceedings from the first joint conference with Atlantic Fisheries Technological Society*, 1988. W. T. Otwell (compiler), Florida Sea Grant, University of Gainesville. pp. 365–376.

Chinivasagam, H. N., and Vidanapathirana, G. S. 1986. Quality changes and bacterial flora associated with trench sardines (*Amblygaster sirm*) under delayed icing conditions. FAO Fisheries Rept. 317. Food and Agriculture Organization of the United Nations, Rome, Italy.

Chung, Y., and Lee, J. S. 1981. Inhibition of microbial growth in English sole (*Parophrys retulus*). *J. Food Protect.* **44**:66.

Costakes, J., Connors, E., and Paquette, G. 1982. Quality at sea—recommendations for on-board quality improvement procedures. New Bedford Seafood Producers Association and New England Fisheries Development Foundation.

Dassow, J. A. 1976. Handling fresh fish. In: Stansby, M. E. (ed.). *Industrial Fishery Technology*. Robert E. Krieger, New York, pp. 45–64.

DeSilva, G. T. K. 1978. Storage life of Soora puraw *(Selas leptolepsis)* with delayed icing of 9 and 12 hours. Cited in Chinivasagam and Vidanapathirana (1986).

DFO. 1983. Grading standard for fresh and frozen Atlantic groundfish products. Minister of Supply and Services, Department of Fisheries and Oceans, Ottawa, Ontario. Cat. No. sF 38-2/1-1983F.

Disney, J. G. 1976. The spoilage of fish in the tropics. In: Cobb, B. F. and Stockton, A. B. (eds.). *Proceedings of the First Annual Tropical and Subtropical Fisheries Technology Conference, Vol. 1;* Texas A&M University Sea Grant, College Station, Texas, p. 23.

FAO. 1973. Code of practice for fresh fish. FAO Fisheries Circ. C318. Food and Agriculture Organization of the United Nations, Rome, Italy.

Fey, M. S., and Regenstein, J. M. 1982. Extending the shelf-life of fresh wet red hake and salmon using CO_2–O_2 modified atmosphere and potassium sorbate ice at 1°C. *J. Food Sci.* **47**:1048.

Goodrick, B. 1987. Postharvest quality of tuna meat, a question of technique. *Food Technology in Australia.* **39**:343.

Herborg, L., and Villadsen, A. 1975. Bacterial infection/invasion in fish flesh. *J. Fd Technol.* **10**:507.

Holston, J., and Slavin, J. W. 1965. Technological problems in the preservation of fish, iced fish requiring more knowledge from fundamental research. In: Kreuzer, R. (ed.). *Technology of Fish Utilization;* Fishing News (Books), London, p. 41.

Huss, H. H., Dalsgaard, D., Hansen, L., Ladefoged, H., Pedersen., and Zitten, L. 1974. The influence of hygiene in catch handling on the storage life of iced cod and plaice. *J. Food Technol.* **9**:213.

Jay, J. M. 1986. *Modern Food Microbiology*. Van Nostrand Reinhold, New York, p. 227.

Jayaweera, V. 1980. Storage life of silverbelly (*Leiognathus* spp.) with delayed icing. *Bull. Fish. Res. Stat., Sri Lanka* **30**:53.

Kosak, P. H., and Toledo, R. T. 1981. Effects of microbial decontamination on the storage stability of fresh fish. *J. Food Sci.* **46**:1012.

Lee, J. S., and Kolbe, E. 1982. Microbiological profile of Pacific shrimp, *Pandalus jordani,* stowed under refrigerated seawater spray. *Marine Fish. Rev.* **44**:12.

Lemon, D. W., and Regier, L. W. 1977. Holding of Atlantic mackerel *(Scomber scombrus)* in refrigerated sea water. *J. Fish. Res. Board Can.* **34**:439.

Lima dos Santos, C. A. M. 1981. The storage of tropical fish in ice—a review. *Trop. Sci.* **23**:97.

Liston, J. 1956. Quantitative variations in the bacterial flora of flatfish. *J. Gen. Microbiol.* **15**:305–314.

Liston, J., M. E. Stansby, and H. S. Olcott. 1976. Bacteriological and chemical basis for deteriorative changes. In: Stansby M. E. (ed.). *Industrial Fishery Technology*. Robert E. Krieger, New York, pp. 345–358.

Longard, A. A., and Regier, L. W. 1974. Color and some compositional changes in ocean perch *(Sebastes marinus)* held in refrigerated sea water with and without carbon dioxide. *J. Fish. Res. Board Can.* **31**:456.

Mayer, B., Samuels, R., Flick, G., Hackney, C., Rippen, T., Soul, D., Sanders, L., Coale, C., DuPaul, W., Grulich., R., Cahill, P., Smith, C., and Clusman, A. 1986. A seafood quality program for the Mid-Atlantic region, Part II. A report submitted to the

Mid-Atlantic Fisheries Development Foundation. Virginia Polytechnic Institute and State University, Sea Grant, Blacksburg.

Middlebrooks, B. L., Toom, P. M., Douglas, W. L., Harrison, R. E., and McDowell, S. 1988. Effects of storage time and temperature on the microflora and amine development in Spanish mackerel *(Scomberomorus maculatus)*. *J. Food Sci.* **53**:1024.

Miller, S. A., and Brown, W. D. 1984. Effectiveness of chlortetracycline in combination with potassium sorbate or tetrasodium ethylenediaminetetraacetate for preservation of vacuum packed rockfish fillets. *J. Food Sci.* **49**:188.

Moser, M. D. 1986. Maine groundfish association vessel quality handling project. A report submitted to the New England Fisheries Development Foundation, Boston, Mass. S. K. Grant #NA84EAH00023. National Marine Fisheries Service.

Nelson, R. W., and Barnett, H. J. 1973. Fish preservation in refrigerated seawater modified with carbon dioxide. *Proc. 13th Intl. Cong. Refrig.* **3**:57.

Phillips, F. A., and Peeler, J. T. 1972. Bacteriological survey of the blue crab industry. *Appl. Microbiol.* **24**:958.

Poulter, R. G., Curran, C. A., and Disney, J. G. 1981. Chilled storage of tropical and temperate water fish—differences and similarities. Advances in refrigerated treatment of fish, especially underutilized species. *Bull. Int. Inst. Refrig.* **49**:111.

Ravesi, E. M., Licciardello, J. J., Tuhkunen, B. E., and Lundstrom, R. C. 1985. The effects of handling or processing treatments on storage characteristics of fresh spiny dogfish, *Squalus acanthias*. *Marine Fish. Rev.* **47**:48.

Ravesi, E. M., Licciardello, J. J., and Racicot, L. D. 1987. Ozone treatments of fresh Atlantic cod, *Gadus morhua*. *Marine Fish. Rev.* **49**:37.

Reppond, K. D., and Collins, J. 1983. Pacific cod *(Gadus macrocephalus):* change in sensory and chemical properties when held in ice and CO_2 modified refrigerated seawater. *J. Food Sci.* **48**:1552.

Reppond, K. D., Bullard, F. A., and Collins, J. 1979. Walleye pollock *(Theragra chalcogramma):* physical, chemical and sensory changes when held in ice and in carbon dioxide modified refrigerated seawater. *Fish. Bull.* **77**:481.

Reppond, K. D., Collins, J., and Markey, D. 1985. Walleye pollock *(Theragra chalcogramma):* changes in quality when held in ice, slush-ice, refrigerated seawater, and CO_2-modified refrigerated seawater then stored as block of fillets at $-18°C$. *J. Food Sci.* **50**:985.

Roach, S. W., Tarr, H. L. A., and Tomlinson, N. 1967. Chilling and freezing salmon and tuna in r frigerated sea water. *Bull. 160 Fish. Res. Bull. Can.*

Rodman, W. K. 1987. On board fish handling systems for offshore wetfish trawlers "work smarter, not harder." Tropical and Subtropical Fisheries Technological Society of the Americas. In: *Proceedings from the first joint conference with Atlantic Fisheries Technological Society*, 1988. W. T. Otwell (compiler), Florida Sea Grant, University of Gainesville, pp. 332–342.

Samuels, R. D., DeFeo, A., Flick, G. J., Ward, D. R., Rippen, T., Riggens, J. K., Coale, C., and Smith, C. 1984. Demonstration of a quality maintenance program for fresh fish products. A report submitted to Mid-Atlantic Fisheries Development Foundation. Virginia Polytechnic Institute and State University, Sea Grant, VPI-SG-84-04R, Blacksburg.

Scott, D. N., Fletcher, G. C., Hogg, M. G., and Ryder, J. M. 1986. Comparison of whole and headed and gutted orange roughy stored in ice: sensory, microbiology and chemical assessment. *J. Food Sci.* **51**:79.

Shaw, S. J., Bligh, E. G., and Woyewoda, A. D. 1983. Effect of potassium sorbate application on shelf life of Atlantic cod *(Gadus morhua)*. *Can. Inst. Food Sci. Technol. J.* **16**:237.

Shaw, S. J., Bligh, E. G., and Woyewoda, A. D. 1984. Effect of delayed filleting on quality of cod fish. *J. Food Sci.* **49**:979.

Shaw, S. J., Bligh, E. G., and Woyewoda, A. D. 1986. Spoilage pattern of Atlantic cod fillets treated with glucose oxidase/gluconic acid. *Can. Inst. Food Sci. Technol. J.* **19**:3.

Shewan, J. M. 1961. The microbiology of sea-water fish. In: Borgstrom, G. (ed.) *Fish as Food, Vol. 1.* Academic Press, New York, pp. 487–560.

Shewan, J. M. 1977. The bacteriology of fresh and spoiling fish and the biochemical changes induced by biochemical action. In: *Proceedings of the Conference of Handling, Processing, and Marketing of Tropical Fish.* Tropical Products Institute, London, p. 51.

Stansby, M. E., and Lemon, J. M. 1941. Studies on the handling of fresh mackerel *(Scomber scombus).* Res. Rept. 1, U.S. Fish and Wildlife Service, Washington, D.C.

Statham, J. A., Bremner, H. A., and Quarmby, A. R. 1985. Storage of morwong *(Nemadactylus macropterus* Bloch and Schneider) in combinations of polyphosphate, potassium sorbate and carbon dioxide at 4°C. *J. Food Sci.* **50**:1580.

Strom, T., and Lien, K. 1984. Fish handling on board Norwegian fishing vessels. In: Moller A. (ed.) *Fifty Years of Fisheries of Fisheries Research in Iceland.* Proceedings of a Jubilee seminar, 1984, p. 15.

Sumner, J. L., Gorczyca, E., Cohen, D., and Brady, P. 1984. Do fish from tropical waters spoil less rapidly in ice than fish from temperate waters? *Food Technol. Aust.* **36**:328.

Tomlinson, N., Geiger, S. E., Boyd, J. W., Southcott, B. A., Gibbard, G. A., and Roach, S. W. 1974. Comparison between refrigerated sea water (with or without added carbon dioxide) and ice as storage media for fish to be subsequently frozen. *I.I.F.I.I.R.—Commissions B2, D3 Tokyo. p. 163.*

Townley, R. R., and Lanier, T. C. 1981. Effect of early evisceration on the keeping quality of Atlantic croaker *(Micropogon undulatus)* and grey trout *(Cynoscion regalis)* as determined by subjective and objective methodology. *J. Food Sci.* **46**:863.

Valdimarsson, G., Matthiasson, A., and Stefansson, G. 1984. The effect of onboard bleeding and gutting on the quality of fresh, quick frozen, and salted products. In: Moller A. (ed.) *Fifty Years of Fisheries of Fisheries Research in Iceland.* Proceedings of a Jubilee seminar, 1984, p. 61.

Walters, V., and Fierstine, H. L. 1964. Measurement of swimming speeds of Yellowfin tuna and Wahoo. *Nature (London).* **202**:208.

Wentz, B. A., Duran, A. P., Swartzentruber, A., Schwab, A. H., McClure, F. D., Archer, D., and Read, R. B., Jr. 1985. Microbiological quality of crabmeat during processing. *J. Food Protect.* **48**:44.

Wesley, P. 1982. Glucose oxidase treatment prolongs shelf-life of fresh seafood. *Food Dev.* Jan., p. 36.

2

Microbiology of Bivalve Molluscan Shellfish

David W. Cook

The majority of all seafood-related illnesses in the United States are associated with the consumption of bivalve molluscan shellfish (Anon. 1988). This disproportionate incidence of disease is related to the biology of the animals, the quality of the waters in which they grow, the handling techniques after harvest, and the fact that they are frequently consumed raw. This chapter will consider all these factors from a microbiological standpoint.

Most of the information on the microbiology of bivalve molluscan shellfish has been developed relative to public health aspects of the shellfish industry. This information has been interpreted into state and federal regulations that govern that industry. To grasp an understanding of the shellfish industry and the necessity for regulations, consult the most recent edition of the National Shellfish Sanitation Program, Manual of Operations, Parts I and II (Food and Drug Administration 1989a, b).

BIOLOGY OF THE BIVALVES

Oysters, clams, and mussels are soft-bodied animals that are enclosed in a rigid, bilaterally symmetrical shell of two parts. Thus, they are termed bivalve molluscan shellfish or "bivalves" for short.

Oysters and mussels are sessile, dwelling on the bottom or attached to structures in the water column. Clams are benthic and live burrowed in the mud, but maintain contact with the water by means of a syphon tube. With the exception of surf clams and ocean quahogs, most bivalve resources of commercial importance grow in shallow, near-shore estuarine waters.

Shellfish contributed to the food supply of native Americans and were a staple in the diet of early coastal residents because they could easily be harvested year round. Though many species of bivalves are edible, only a few are sufficiently abundant to make them commercially important in the United States (Table 2.1).

Bivalves are filter feeders and pass large volumes of water across their gills to obtain oxygen and food. Particulate matter from the water, including microor-

Table 2.1   Commercially Important Bivalve Molluscan Shellfish in the U.S.

Common Name	Scientific Name
Clams	
Butter clam	*Saxidomus giganteus*
Gaper or horse clam	*Tresus capax*
	Tresus nutalli
Geoduck	*Panope generosa*
Hard clam or Quahog	*Mercenaria campechiensis*
	Mercenaria mercenaria
Mahogany or ocean Quahog	*Arctica islandica*
Manila or Japanese littleneck clam	*Venerupis (Tapes) japonica*
Pacific littleneck clam	*Protothaca staminea*
Razor clam	*Siliqua patula*
Soft shell clam	*Mya arenaria*
Surf clam	*Spisula solidissima*
Mussels	
Blue or bay mussel	*Mytilus edulis*
California sea mussel	*Mytilus californianus*
Oysters	
Eastern (American) oyster	*Crassostrea virginica*
Pacific oyster	*Crassostrea gigas*
Olympia oyster	*Ostrea lurida*

ganisms, is trapped in mucus on the gills and transported toward the mouth by ciliary action. As the mucus passes the labial palps, particles are sorted and nonfood items are rejected as pseudofeces. The remaining items entrapped in the mucus enter the mouth. The food passes through a short esophagus to the stomach where it is mixed with digestive enzymes by the rotating action of the crystalline style. Small food particles are transported into the blind tubules of the digestive diverticula where they are absorbed by the cells lining the tubules, or are ingested by phagocytic cells that migrate into the tubules.

Particles that do not enter the digestive diverticula are passed out of the stomach into the midgut and are eventually discharged through the anus. This process requires <2 hours in actively feeding adult oysters (Galtsoff 1964). Many microorganisms ingested by oysters survive the digestive process (Cook 1984; Rowse and Fleet 1982).

PUBLIC HEALTH CONCERNS

Public health concerns related to consumption of shellfish are prompted by the fact that bivalves are frequently consumed raw. Further, the whole animal is consumed rather than just the muscle tissue, as in the case of raw fish and raw crustaceans.

Microbiological Pollution

Because bivalves are filter feeders, they may accumulate pathogenic microorganisms from waters impacted by sewage pollution. On ingestion by humans, bivalves may then become a vector for disease transmission. Between 1900 and 1986 in the United States there were 12,376 documented cases of bivalve shellfish-borne disease, not including cases of paralytic shellfish poisoning. It has been estimated that this represents only 5–10% of the actual number of cases occurring. Of the documented cases, 26% were typhoid; 11% infectious hepatitis; 11% Norwalk virus; 2% vibrio; 7% unspecified; and 43% gastroenteritis, food poisoning, diarrhea, etc. Specific etiologic agents isolated in cases are listed in Table 2.2 (Rippey and Verber 1986).

The National Shellfish Sanitation Program, which is administered by the U.S. Food and Drug Administration and carried out by state shellfish regulatory authorities, works to ensure that only shellfish harvested from approved growing areas reach commercial markets. Growing areas are classified for their microbiological acceptability on the basis of a sanitary survey of the shoreline to detect potential pollution sources and bacteriological analysis of water samples taken from the area.

Table 2.2 Microorganisms Causing Illnesses in Humans Associated with Molluscan Shellfish Consumption

Bacterial	Viral
Salmonella typhi	Hepatitis A
Vibrio cholera 01	Norwalk virus
Vibrio cholera non 01	Other viral agents
Salmonella spp.	
Shigella spp.	
Vibrio vulnificus	
Vibrio parahaemolyticus	
Vibrio fluvialis	
Vibrio mimicus	
Vibrio hollisae	
Vibrio furnissii	
Escherichia coli	
Bacillus cereus	
Staphylococcus sp.	
Campylobacter sp.	
Pleisomonas shigelloides	
Aeromonas hydrophila	

Fecal coliform bacteria are generally used as indicators of domestic pollution in shellfish growing waters, although total coliforms may be used for that purpose. Areas that pass a sanitary survey must maintain a median fecal coliform most probable number (MPN) of less than 14/100 mL of water to obtain an approved area classification (Food and Drug Administration 1989a). Fecal coliform levels in shellfish meats are not used for the purpose of classifying shellfish growing waters.

Following harvest, two microbiological parameters are applied to gauge the acceptability of shellfish meats. Wholesale market level meats should have a 95°F (35°C) standard plate count (SPC) of <500,000/g and a fecal coliform level not in excess of 230 MPN/100 g. Counts above these levels indicate that the shellfish may have come from an improperly classified area, been processed under unsanitary conditions, been subjected to temperature abuse during storage, and/or experienced excessive storage time (Food and Drug Administration 1989b).

Microbiological procedures for testing samples of water from shellfish growing areas and for testing shellfish meat samples have been standardized and published (American Public Health Association 1970; Greenberg and Hunt 1985). MPN techniques are used for measuring fecal coliform levels in water and shellfish meat samples rather than membrane filter techniques because of problems with samples clogging the filters.

Vibrios

Several *Vibrio* species known to be pathogenic for humans have been isolated from bivalves. These bacteria are part of the natural estuarine microflora and may be accumulated by shellfish during feeding. With the possible exception of *V. cholerae* 01, the presence of these vibrios appears unrelated to pollution. A number of deaths caused by *V. vulnificus* have been linked to the consumption of raw oysters (Ronk 1988). These deaths and the large number of illnesses caused by other vibrios have raised concern about the safety of shellfish harvested from approved growing waters.

Biotoxins

Some species of phytoplankton produce toxins. When ingested as food by bivalves, the toxins accumulate in their tissues. The bivalves are apparently unaffected by the toxins. However, when consumed by humans, these toxins cause a disease known as paralytic shellfish poisoning. This will be discussed in Chapter 12 *(this volume)*.

Chemical Pollution

Finally, it should be noted that bivalves growing in waters polluted with industrial and agricultural chemicals may concentrate and accumulate those chemicals in their tissues. Consumption of shellfish so contaminated presents an additional public health problem. Though this topic is beyond the scope of this text, it is an area of significant concern to the shellfish regulatory agencies and the shellfish industry.

Purification of Bivalves

Bivalves can be cleansed of most microbiological contaminants by placing them in water that is free of the microorganisms to be removed and under conditions where the shellfish will actively feed. This may be accomplished by transferring the animals harvested from moderately polluted waters into approved shellfish growing waters. This process is known as "relaying" and requires about 15 days for the bivalves to cleanse, provided the water temperature remains above 50°F (10°C) and the salinity remains at an acceptable level for the species (Cook and Ellender 1986; Richards 1988).

A second approach is "controlled purification" or "depuration" in which the bivalves are brought into a facility where they are maintained in tanks of clean water with the appropriate salinity and temperature. Under these conditions, the bivalves will cleanse in 48–72 hours (Fleet 1978; Richards 1988).

Relaying and depuration are intended to cleanse bivalves of pathogenic microorganisms that they may accumulate in moderately polluted waters. Neither process is approved for removing chemical pollutants or biotoxins from the animals. Moreover, there are questions concerning the removal of some viruses. Chapter 16 (this volume) provides a more in-depth discussion of purification of bivalves.

MICROFLORA OF BIVALVES AT HARVEST

The microflora of bivalves represent both microorganisms that are commensals of the bivalves and microorganisms that are being filtered from the water and ingested as food. The numbers and types of microorganisms contained in the water depend on its salinity, temperature, nutrient concentration, and pollution level. In an estuary, these factors are variable and complicate a study of bivalve microflora.

Bivalve molluscs contain a commensal microflora that has not been cultured on laboratory media. *Cristispira pectineus*, a spirochete, frequently colonizes the crystalline style of the Eastern oyster (Tall and Nauman 1981) and the Pacific oyster (Bernard 1970). Spirochetes of the genus *Saprospira* have been observed microscopically in the crystalline style, stomach, and intestine of Eastern oysters. The function of the spirochetes in the style and digestive process of the oyster is unknown. These spirochetes have no pathological significance in humans (Dimitroff 1926).

Various culture procedures have been used to enumerate the cultureable microflora in bivalves. The SPC is used by regulatory agencies to estimate shellfish meat quality (Greenberg and Hunt 1985). This pour plate procedure employs plate count agar prepared with distilled water and an incubation period of 48 hours at 95°F (35°C). Plate counts higher than those obtained with the SPC are achieved when the medium is prepared with sea water or 3% NaCl and when the plates are incubated for 3–5 days at 68–77°F (20–25°C) (Colwell and Liston 1960; Cox 1965). Thus, studies of bivalve microflora that employ only the SPC procedure may greatly underestimate the microflora.

Bivalves at harvest have a SPC of 10^3–10^5 bacteria/g. These levels are usually 1–2 log units greater than the numbers found in the water from which they are harvested (Vasconcelos et al. 1969). The concentration of bacteria by the shellfish during feeding may be partially responsible for this; however, Colwell and Liston (1960) suggested that bivalves have a commensal microflora other than the spirochetes. This view was supported by Kueh and Chan (1985), who showed that various sections of the Pacific oyster's digestive system contained different levels and species composition of bacteria. Garland et al. (1982) failed to demonstrate any microorganisms attached to the external or gut surfaces of the oysters; therefore, it is assumed that the commensal microflora is selected by physicochemical conditions within the gut.

The microflora of bivalves at harvest consists primarily of Gram-negative rods. *Pseudomonas* and *Vibrio* species usually dominate the microflora of the mantel fluid in Pacific oysters (Colwell and Liston 1960) and Eastern oysters from Chesapeake Bay (Lovelace et al. 1968) as well as the whole animal microflora of Eastern oysters (Murchelano and Brown 1968) and Pacific oysters, clams, and mussels (Kueh and Chan 1985; Vasconcelos and Lee 1972). Pigmented Gram-negative rods of the *Flavobacterium/Cytophaga* group are occasionally numerous in both the Eastern and Pacific oyster, comprising about 30% of the microflora. Other isolates from oysters include *Acinetobacter*, coryneforms, *Achromobacter*, *Alcaligenes*, *Micrococcus*, and *Bacillus*. In Eastern oysters from the Gulf of Mexico, the microflora are dominated by *Vibrio*, *Aeromonas*, *Moraxella*, and *Pseudomonas* species (Vanderzant et al. 1973).

Among the species of *Vibrionaceae* naturally occurring in estuarine waters are several that are pathogenic for humans. These include *Vibrio parahaemolyticus*, *V. vulnificus*, *V. cholera* non-01, *V. mimicus*, *V. hollisae*, *V. fluvialis*, and

Aeromonas hydrophila. These bacteria are filtered from the water by shellfish during feeding and become part of the natural bivalve microflora.

Yeasts are present in bivalve microflora but usually in low numbers. *Rhodotorula rubra* and *Trichosporon* sp. were the most frequently encountered yeasts in Eastern oysters and hard clams in Florida (Hood 1983). At harvest, *R. rubra* was found in 86% of the oysters and 50% of the clams at levels of $3/g$ and $0.8/g$, respectively. Human-associated yeasts represented by *Candida* and *Torulopsis* species have been found at low levels in Eastern oysters, quahogs, and blue mussels (Buck et al. 1977). Although these findings demonstrate the presence of pathogenic yeasts in bivalves, no clear association of human mycosis with shellfish consumption has been found.

Seasonal changes have been noted in the numbers of microorganisms in bivalves. These changes are affected by the microflora in the water which changes seasonally and the feeding rate of the bivalves which is temperature dependent. In northern areas where the water temperature drops to near 32°F (0°C), bivalves enter a state of hibernation and stop feeding. Under these conditions, the bivalve microflora are reduced greatly (Hunter and Harrison 1928).

Estuarine waters, where the majority of commercially important bivalves grow, are frequently the receiving waters for both treated and untreated human wastes. Those waters also receive fecal material from wild and domestic animals in runoff water following rainfall. Thus, any microorganism excreted in fecal material has the potential for finding its way into bivalve-growing waters. Coliforms, fecal coliforms, and *E. coli* are used to indicate the presence of fecal pollution because they are usually present at concentrations greater than those of pathogens. Shellfish growing areas that pass a sanitary survey may contain low levels of these indicator bacteria and thus may contain other microorganisms associated with fecal material. Enteric pathogens that have been isolated from bivalves harvested from approved waters include enteroviruses (Ellender et al. 1980a, b; Fugate et al. 1975; Goyal et al. 1979; Vaughn et al. 1980; Wait 1983), salmonellae (Andrews et al. 1975; Fraiser and Koberger 1984; Sobsey 1980), *Yersinia enterocolitica* (Peixotto 1979), *Pleisomonas shigelloides* (Koburger and Miller 1984), and *Aeromonas hydrophila* (Abeyta 1986; Cook and Ruple 1989; Hood et al. 1983, 1984).

In summary, the microflora of the bivalve at the time of harvest are comprised of the natural commensal microflora and the microflora accumulated from the water by the bivalve during feeding immediately prior to harvest. As noted earlier, the microflora in the shellfish growing waters are highly variable, responding to temperature, salinity, nutrients, and pollution. Therefore, bivalves harvested from various locations or from the same location at different times may show considerable variation in microflora species composition and concentration.

SHELLSTOCK HARVESTING AND HANDLING

The method by which bivalve molluscan shellfish are harvested depends on the species and the depth of the overlying water. Commercial vessels employ various types of tongs, rakes, dredges, and mechanical harvesters to harvest shellfish. Some clams are dug by hand from mud flats exposed by low tides; and in areas of the Northwest, oysters are harvested by hand at low tide (McKee 1963).

Immediately after harvest, legal size bivalves are separated from undersized animals and dead-shell material, which are returned to the reefs. The legal size for harvest is based on individual state regulations.

Shellstock (bivalves in the shell) are given a brief wash with water from the harvest area before being readied for transport. The shellstock is transported dry, usually in burlap sacks or other open-mesh containers. However, if the harvest boat is docking at the processing plant, shellstock may be left in piles on the deck of the boat. Shellstock is generally not refrigerated on the harvest boat; but when the boat reaches the dock, the containers of shellstock are placed in refrigerated trucks for transport. The interval between harvest and receipt at a processing plant may vary from 1 to 5 days depending on the proximity of the harvest area to the processing plant. Oysters will remain alive in the shell for about 2 weeks if kept cool. However, long storage time decreases the microbiological quality of the bivalves.

Live shellstock harvested from approved shellfishing waters may be held in wet storage prior to sale or processing. Storage is accomplished in floating baskets held in approved shellfish growing waters or in tanks of clean seawater. Wet storage is usually used to de-sand those bivalve species that tend to accumulate sand in their mantles and gills, or to increase palatability by increasing salt content of bivalves harvested from low salinity waters. Care must be taken to prevent the bivalves from becoming contaminated during wet storage. Furthermore, wet storage should not be confused with depuration.

Harvesting, per se, does not change the microflora of bivalves, but it initiates a series of events that leads to changes. When disturbed by the harvest process, bivalves close their shells tightly, trapping water and its associated microflora. As the bivalve and its associated microflora respire, oxygen levels decrease and metabolic wastes accumulate in the shell liquid. The metabolic waste products from the bivalves serve as nutrients for growth of bacteria in the shell liquid. Growth of bacteria in shellstock is temperature mediated; therefore, rapid and complete cooling of the shellstock slows microbial growth and prolongs the storage life.

Hoff et al. (1967a) followed postharvest change in shellstock microflora in four Pacific Coast bivalve species: Pacific oyster, Olympia oyster, native littleneck clam *(Prototheca staminea),* and Manila clam *(Venerupis japonica).* At harvest, all species had SPCs between 10^3 and 10^4 cfu/g. After 15 days of storage at 50°F (10°C), all species had SPCs of <10^5 cfu/g, with native littleneck clams having the lowest SPC. At a storage temperature of 68°F (20°C), bacteria increased much more rapidly and exceeded the 10^5/g level after 3–6 days of storage, depending on bivalve species. Coliform and fecal coliform MPN counts remained relatively stable in all four species throughout the 15-day storage period. At higher storage temperatures of 68°F (20°C) and 81.5°F (27.5°C), multiplication of indicator bacteria was seen in some of the bivalve species studied.

At harvest, the SPC in Gulf Coast oyster meats ranges from 10^2 to 10^5 cfu/g with a mean of 10^3 cfu/g. Rarely do oyster meats exceed a SPC of 500,000/g at harvest (Cook and Ruple 1989; Presnell and Kelly 1961). Fecal coliform levels in oyster meats from approved harvesting areas are usually low, <50/100 g, and rarely exceed an MPN of 230/100 g at the time of harvest (Cook and Ruple 1989; Food and Drug Administration 1983; Paille et al. 1987).

Trace lot studies with Gulf Coast shellstock oysters have confirmed that the SPCs, fecal coliform counts, and *E. coli* counts increase in oysters while in sacks on the deck of the boat during transit to the dock (Table 2.3). These increases are more pronounced during the summer months. Further increases occur in these microbiological parameters in oysters during overland transport to processing plants when the oysters are not properly refrigerated during transport (Cook and Ruple 1989; Food and Drug Administration 1983).

Fecal coliforms other than *E. coli* found in shellstock oysters include *Klebsiella, Enterobacter,* and *Citrobacter* species. All these species are capable of multiplying in shellstock stored above 50°F (10°C), but *Klebsiella* sp. usually dominate. Storage of shellstock at 50°F (10°C) or below suppresses multiplication of fecal coliforms (Cook and Ruple 1989). Species of *Pseudomonas, Edwardsiella, Providencia,* and *Acinetobacter* possessing the characteristics of fecal coliforms, i.e. production of gas in EC medium (American Public Health Association 1970), when incubated at 122°F (44.5°C), have been isolated from oysters (Paille et al. 1987).

Unlike indicator bacteria, pathogenic enteric bacteria do not multiply in shellstock oysters. *Salmonella typhimurium* and *S. seftenberg* inoculated into

Table 2.3 Change in Levels of Bacteria in Shellstock Oysters During Commercial Transport from Harvest Area to Plant[a]

	Harvest	Dock	Plant
Fecal coliforms (MPN/100 g)	74	180	550
E. coli (MPN/100 g)	34	67	65
Standard plate count (cfu/g)	2,500	6,000	12,000

[a]Data represent geometric mean of 12 samples (Cook and Ruple, 1989).

oysters *(Crassostrea commercialis)* during feeding decreased by 100-fold during dry storage for 12 days at 68–77°F (20–25°C) (Son and Fleet 1980).

Levels of *Vibrionaceae* found in shellstock at harvest also change in response to storage temperature with time. In oysters held above 68°F (20°C), *V. parahaemolyticus* increased during the first few days of storage followed by a decline (Johnson et al. 1973; Son and Fleet 1980; Thompson et al. 1976). Along with *V. parahaemolyticus, A. hydrophilia* and lactose-positive vibrios showed increases during 7 days of storage at 68°F (20°C) (Hood et al. 1983). Normal commercial handling techniques do not prevent the multiplication of *V. parahaemolyticus, V. cholera,* and *V. vulnificus* or *A. hydrophilia* in shellstock oysters. However, chilling the shellstock to <50°F (10°C) prevents the growth of all these bacteria except *A. hydrophilia* (Cook and Ruple 1989).

Human enteric viruses that may be present in the shellfish at harvest do not multiply in the bivalve tissues. However, they may remain viable during the storage life of the shellstock (DiGirolamo et al. 1970b; Tierney et al. 1982).

BIVALVE PROCESSING

Raw Processing

Raw processing of bivalves is limited to removal of the meat from the shell (shucking), washing of the meats with potable water, and packing the meats in containers. Sealed containers are placed in boxes and covered with crushed ice for chilling and transport.

Shucking involves forcing a knife between the two valves of the mollusc shell and cutting the adductor muscle where it is attached to one of the valves. Care is taken not to cut into the meat. The knife is then used to cut the adductor muscle from the other valve thus freeing the meat.

Washing of the meats is accomplished by placing them on a stand-supported perforated tray called a skimmer and spraying with potable water. An alternate method of washing is "blowing." Meats are placed in a tank of water and agitated by the introduction of air from the bottom of the tank. Both washing methods remove mucus, pseudofeces, mud, and shell fragments from the meats. Meats are thoroughly drained before packing (Anon. 1986).

Shucking and washing of oyster meats under sanitary conditions reduces the SPC and the 77°F (25°C) aerobic plate count (APC) of the meats by about 1 log. However, the number of indicator bacteria, vibrios, and *A. hydrophilia* may not be significantly changed by processing (Hood et al. 1984; Thompson et al. 1976; Vanderzant et al. 1973).

Shellfish processors sometimes add chlorine to the water in the blowing tank in an attempt to reduce the bacterial load on bivalve meats. Chai et al. (1984) and Pace et al. (1988) demonstrated that blowing oysters in 150 ppm chlorine water for 20–30 minutes caused a reduction in the APC and altered the species composition of the oyster microflora. *Vibrio* sp. were apparently least affected by the chlorine treatment and increased from 23.0 to 63.7% of the population. This corresponded to the finding of Motes (1982), who demonstrated that blowing in chlorinated water did not significantly reduce the concentration of *V. cholerae* in oyster meats.

Maintenance of good sanitation in the processing plant prevents the bivalve meats from becoming contaminated during raw processing. Rapid chilling of the shucked meats to near 32°F (0°C) retards spoilage.

Microbiological Changes in Shucked Bivalve Meats

The storage temperature is an important factor in controlling bacterial growth in shucked bivalve meats. Storage of containers of meats in melting ice provides the longest shelf life.

Plate Counts. During the first few days of cold storage, bacterial populations undergo a change as indicated by a reduction in the SPC. Bacteria unable to tolerate the cold die, leaving a population of spoilage bacteria adapted for the storage temperature. Bacterial populations then increase in a logarithmic fashion. When the SPC exceeds 10^6/g, the meats are of substandard quality. Kelly (1964) found the time required for the SPC in Pacific and Eastern oysters to reach 10^6/g at 32°F (0°C), 37.4°F (3°C), and 50°F (10°C) to be approximately 16, 10, and 3 days respectively. Similar findings were reported for the Pacific oyster and the Olympia oyster (Hoff et al. 1967b).

Shucked soft-shell clam meats have a shorter shelf life than oysters. Clams take on a rancid, fishy odor after 5–7 days of storage at 33.8°F (1°C) and exceed a SPC of 10^6/g at 9–14 days (Cox 1965).

Indicator Bacteria. Coliforms increase in shucked bivalve meats during storage at all holding temperatures above freezing (Cox 1965; Hoff et al. 1967b; Hood et al. 1984; Kelly 1964). The fate of fecal coliforms including *E. coli* in shucked meats is temperature dependent and varies with the shellfish species. In meats held on ice or under refrigeration at 41°F (5°C) or less, multiplication of fecal coliforms is rare and only slight when it occurs (Cox 1965; Hoff et al. 1967b; Hood et al. 1984; Kelly 1964; Wilson and McCleskey 1951). At storage temperatures approaching 50°F (10°C), increases of fecal coliforms have been noted (Hoff et al. 1967b). Higher storage temperatures cause shellfish to spoil rapidly. While indicator bacteria may multiply in shucked bivalve meats, temperatures play a significant part in controlling that multiplication.

Yeasts.   Pink yeasts have been implicated in the discoloration and spoilage of raw processed oysters during storage (Hunter 1920). Yeasts have been documented to growth in oysters held at 35.6°F (2°C) (Hood 1983), but not in shucked soft-shell clam meats held at 33.8°F (1°C) (Cox 1965). McCormack (1950, 1956) isolated a pink yeast that could grow on frozen oysters, producing pink discoloration.

Pink discoloration of bivalve meats can arise from factors other than yeasts. Bivalves feeding on certain dinoflagelates may accumulate a red pigment in the tissues of their digestive gland. As these tissues are broken down by enzymatic and bacterial action, the pigment is leaked from the digestive gland. Leakage of the pigment is accelerated by cutting of the meat or by freezing (Sieling 1971). The discoloration does not affect the wholesomeness or flavor of the shellfish, but consumers often reject them. Heating to 131°F (55°C) for 25 minutes removes the abnormal color without noticeably affecting their taste or keeping quality (Burrell 1974).

Vibrionaceae.   The fate of *Vibrio parahaemolyticus* in shucked oyster meats has been studied using meats inoculated with laboratory strains via injection (Johnson and Liston 1973), surface contamination (Goatcher et al. 1974), and natural uptake (Johnson and Liston 1973). Meats naturally contaminated before harvest were also studied (Hood et al. 1984; Thompson et al. 1976). Storage temperatures in these studies ranged from 33.8°F (1°C) to 51.8°F (11°C). *V. parahaemolyticus* levels did not increase in any of the studies and typically showed a rapid decline. A reduction of the salt level in the oyster meats caused by washing in combination with cold temperature (Covert and Woodburn 1972) is probably responsible for the decrease in *V. parahaemolyticus*. However, *Pseudomonas* species that produce *V. parahaemolyticus* inhibiting substances have been isolated from oysters and may be partly responsible for the repression of *V. parahaemolyticus* (Goatcher and Westhoff 1975).

The survival of *Vibrio vulnificus* in sterile oyster homogenates held at 39.2°F (4°C) was studied by Oliver (1981). His results suggested that oyster tissues contained components that brought about a dramatic decrease in *V. vulnificus* not attributable to cold shock. When whole oysters were inoculated with *V. vulnificus* and held at 32.9°F (0.5°C), a reduction in numbers of *V. vulnificus* occurred, but it was no greater than the reduction caused by cold shock alone. Although only preliminary, this work suggests that cold storage of oysters will bring about a reduction in the level of *V. vulnificus*.

V. cholerae, unlike *V. parahaemolyticus* and *V. vulnificus,* is not affected by cold shock. Reily and Hackney (1985) demonstrated that laboratory-grown stains of *V. cholerae* persisted at high levels in oyster homogenates for 21 days at 44.6°F (7°C).

Unlike the other species of *Vibrionaceae* tested, *Aeromonas hydrophila* increased in oyster meats in cold storage. Significant increases were found in processed oysters after 7 days of storage at 39.2°F (4°C) (Hood et al. 1984).

Spoilage.   Bivalve molluscan shellfish differ in chemical composition from both teleost fish and crustacean shellfish in having a larger quantity of carbohydrate and a lower quantity of nitrogen in their flesh (Jay 1986). Bivalves store energy in their tissues in the form of a carbohydrate, glycogen. Levels of glycogen in oyster meats range from 0.47 to 6.8% depending on time within their spawning cycle (Sidwell et al. 1979). The chemical composition of food products affects the types of microorganisms that develop on and contribute to the spoilage of foods. Therefore, the types of bacteria associated with spoilage in bivalve meats are expected to differ from those in marine fish and crustaceans.

The early literature relating to deterioration of bivalve meats was reviewed by Fieger and Novak (1961). In the initial stage of oyster meat spoilage, the microflora are quite mixed. As spoilage progresses, acid is produced from the breakdown of the glycogen. Groups of microorganisms that are favored by low pH become dominant. These include lactobacilli, streptococci, and yeasts. Clams, like oysters, also undergo a fermentative type spoilage.

Lactobacillus were found by Shiflett et al. (1966) to be a major component of the microflora in Pacific oysters and accounted for 75% of the microflora after 2 days of storage at 44.6°F (7°C). *Pseudomonas, Achromobacter,* and *Flavobacterium* levels decreased during storage.

Clearly, there is a void in our knowledge of the microflora responsible for spoilage in bivalves. This is particularly true of clams and mussels.

HEAT PROCESSING

Heat Shock

The hand shucking of raw bivalves can be made easier by subjecting them to mild heat treatment just prior to shucking. The time and temperature of heat treatment depends on the bivalve species and size. This process does not kill the bivalve, but causes it to relax its adductor muscle so that the shells can be more easily opened. In addition, heat shock brings about a reduction in the number of bacteria in the shucked meat.

Heat shock is also used to facilitate meat removal from bivalves to be canned. In this case the heat treatment is more severe and may include steaming under pressure for several minutes. The juice lost by the bivalves during heat shock is collected and used in canning.

Canning

Canning is used as a long-term preservation method. Procedures for canning various bivalve species have been published (Jarvis 1943; Tanikawa and Doha 1965) and if recommended processing times and temperatures are followed, few microbiological problems are encountered.

Pasteurization

Pasteurization has been used on an experimental basis to prolong the shelf life of raw oysters without decreasing their texture and flavor. Chai et al. (1984) found that pasteurization at 161.6°F (72°C) for 8 minutes in a retortable pouch resulted in a product of superior microbiological quality that did not significantly differ from fresh oysters in odor, appearance, flavor, or texture. The aerobic and facultative anaerobic plate counts remained low for 3 months when the pasteurized oysters were stored at 32.9°F (0.5°C). However, results were not consistent among batches of pasteurized oysters, presumably because of a difference in the amounts and species of bacteria in individual oysters prior to processing.

Immediately following pasteurization, the majority of the bacteria in the oysters were *Bacillus* sp. During storage for up to 5 months, *Bacillus* sp. continued to dominate the aerobic microflora whereas *Clostridium, Corynebacterium, Listeria, Peptostreptococcus,* and *Staphylococcus* emerged in the facultative anaerobic microflora. Even though the microflora of the oysters was significantly modified by the heat treatment, the typical stages of spoilage were seen, including an increase in acidity, gas formation, and proteolysis (Pace et al. 1988).

Freezing

Freezing has been used effectively to retard the microbiological spoilage of a variety of fish and shellfish (Heen and Karsti 1965). If the products are properly glazed to prevent dehydration and kept under a constant temperature of –20°F (6.6°C), they will have a shelf life of many months.

Freezing is not intended to kill the bacteria in fishery products, but rather to prevent them from metabolizing, growing, and causing spoilage. However, the freezing process usually brings about an initial reduction in the number of bacteria followed by a gradual decline during storage. The survival of bacteria during freezing and subsequent storage depends on the species of bacteria.

In studies with Pacific oysters, DiGirolamo et al. (1970a) found that freezing and storage of the oysters for 24 hours at 0°F (–17.7°C) reduced salmonellae levels by 97%. Increasing the storage time to 1 week reduced levels to 99% of the initial. *E. coli* were found to be less sensitive to freezing than salmonellae.

Freezing significantly lowered the SPC of Eastern oysters, but after thawing their shelf life as measured by SPC was similar to that of oysters that had not been frozen (Hackney et al. 1988).

Irradiation

Low-dose ionizing radiation has been shown to retard the spoilage of Eastern oysters (Novak 1966), Pacific oysters (Shiflett et al. 1966), and soft clams

(Connors and Steinberg 1964). High-dose radiation sufficient to sterilize oysters produced "off" odors, "off" flavors, and destroyed nutrients (Liuzzo et al. 1966; Novak et al. 1966). Chapter 17 *(this volume)* provides a detailed discussion on the changes in microflora of bivalves brought about by irradiation.

INDICATORS OF SPOILAGE

"Spoilage refers to any change in the condition of food in which the latter becomes less palatable, or even toxic; these changes may be accompanied by alterations in taste, smell, appearance, or texture" (Singleton and Sainsbury 1978). Because toxic spoilage is rarely a problem with bivalves harvested from approved shellfishing areas, organoleptic evaluation appears to be the best measure of the quality of bivalve meats. However, as Liuzzo and Novak (1975) pointed out, organoleptic evaluations are subjective and vary considerably among individuals.

Microbiological Indicators

Freshly processed bivalve meats should have a SPC of <500,000 cfu/g (Food and Drug Administration 1989b), and when the SPC exceeds 1,000,000 cfu/g, meats are considered to be of substandard quality. Many of the bacteria that contribute to the spoilage of ice-stored bivalve meats are psychrophilic and do not grow at 95°F (35°C). The microbial load on ice-stored meats may be greater than that measured by the SPC; therefore, the SPC may not be measuring the bacterial population responsible for quality loss. Hunter and Linden (1923) suggested that it was "futile" to attempt to grade oysters as to decomposition by the use of bacteria counts. Rosen (1966) also found that the bacteria counts [77°F (25°C)] were unrelated to the organoleptic quality of soft clams.

Liuzzo et al. (1975) examined the correlation of organoleptic evaluation of ice-stored oyster meats with both psychrophilic [41°F (5°C)] and mesophilic [89.6°F (32°C)] plate counts. They observed that as the organoleptic scores decreased, the psychrophilic count increased, with the correlation between the two being significant. Unexpectedly, they noted a decrease in the mesophilic count during storage. This work emphasizes that the SPC which measures mesophilic bacteria is not an adequate tool with which to gauge the quality of oysters.

Chemical Indicators

pH. The spoilage process in bivalves, particularly oysters, is fermentative in nature, characterized by a gradual decrease in pH. Hunter and Linden (1923) found a relationship of the odor and appearance of Eastern oysters to the pH of their liquor. Oysters with liquor above pH 6.0 are classified as "good," and oysters with liquor of a pH below 5.0 are usually considered to be in an advanced stage of decomposition. Differences have been observed in the pH changes in oysters during different seasons of the year and in the pH of the oyster meat as opposed to the surrounding oyster liquor (Gardner and Watts 1956). However, Liuzzo and Novak (1975) confirmed that a significant correlation existed between a reduction in organoleptic quality and pH of both the oyster meats and oyster liquor.

Rosen (1966) demonstrated that organic acids increased in soft clams during cold storage. However, except in the case of extreme freshness and extreme spoilage, the organoleptic quality did not relate to the quantity of organic acid present.

Total Reducing Substances.   Studies by Liuzzo et al. (1975) demonstrated that the total reducing substance (TRS) content of oysters increases throughout ice storage of shucked meats. Significant correlations were observed between the TRS values and organoleptic scores, pH, and psychrophilic plate count. Although the TRS value appears to be useful in assessing oyster quality, no level has been set that could characterize an oyster as spoiled.

SUMMARY

The majority of the research on the microbiology of bivalve molluscan shellfish has centered around public health concerns related to pollution of shellfish growing waters with sewage and the resulting potential for disease transmission to humans. At present, investigations have shifted to the autochthonous estuarine pathogens. While these areas are important to address, there is still a lack of basic knowledge about the microflora that cause spoilage of bivalve meats and methods to critically evaluate spoilage.

REFERENCES

Abeyta, C., Kaysner, C., Wekell, M., Sullivan, J., and Stelma, G. 1986. Recovery of *Aeromonas hydrophila* from oysters implicated in an outbreak of foodborne illness. *J. Food Protect.* **49**:643–646.

American Public Health Association. 1970. *Recommended Procedures for the Examination of Sea Water and Shellfish,* fourth edition. American Public Health Association, New York.

Andrews, W., Diggs, C., Presnell, M., Miescier, J., Wilson, C., Goodwin, C., Adams, W., Furfari, S., and Musselman, J. 1975. Comparative validity of members of the total coliform and fecal coliform groups for indicating the presence of *Salmonella* in the Eastern oyster, *Crassostrea virginica. J. Milk Food Technol.* **38**:453–456.

Anonyomous. 1986. *Code of Federal Regulations* **21**:161.130.

Anonyomous. 1988. *Seafood Safety: Seriousness of Problems and Efforts to Protect Consumers.* GAO/RCED-88-135, United States General Accounting Office, Washington, D.C.

Bernard, F. R. 1970. Occurrence of the spirochaete genus *Cristispira* in western Canadian marine bivalves. *Veliger* **13**:33–36.

Buck, J. D., Bubucis, P. M., and Combs, T. J. 1977. Occurrence of human-associated yeasts in bivalve shellfish from Long Island Sound. *Appl. Environ. Microbiol.* **33**:370–378.

Burrell, V. G. 1974. Thermal bleaching of red algal pigment in shucked oysters. *Marine Fish. Rev.* **36**:19–22.

Chai, T., Pace, J., and Cossaboom, T. 1984. Extension of shelf-life of oysters by pasteurization in flexible pouches. *J. Food Sci.* **49**:331–333.

Colwell, R. R. and Liston, J. 1960. Microbiology of shellfish: bacteriological study of the natural flora of Pacific oysters *(Crassostrea gigas). Appl. Microbiol.* **8**: 104–109.

Connors, T. J., and Steinberg, M. A. 1964. Preservation of fresh, unfrozen fishery products by low-level radiation. II. Organoleptic studies on radiation pasteurized soft shell clam meats. *Food Technol.* **18**:113–116.

Cook, D. W. 1984. Fate of enteric bacteria in estuarine sediments and oyster feces. *J. Miss. Acad. Sci.* **29**:71–76.

Cook, D. W., and Ellender, R. D. 1986. Relaying to decrease the concentration of oyster-associated pathogens. *J. Food Protect.* **49**:196–202.

Cook, D. W., and Ruple, A. D. 1989. Indicator bacteria and *Vibrionaceae* multiplication in post-harvest shellstock oysters. *J. Food Protect.* **52**:343–349.

Covert, D., and Woodburn, M. 1972. Relationships of temperature and sodium chloride concentration to the survival of *Vibrio parahaemolyticus* in broth and fish homogenate. *Appl. Microbiol.* **23**:321–325.

Cox, J. R. 1965. Bacteriological studies on the shelf life of soft clams *(Mya arenaria). J. Milk Food Technol.* **28**:32–35.

DiGirolamo, R., Liston, J., and Matches, J. 1970a. The effects of freezing on the survival of *Salmonella* and *E. coli* in Pacific oysters. *J. Food Sci.* **35**:13–16.

DiGirolamo, R., Liston, J., and Matches, J. R. 1970b. Survival of virus in chilled, frozen, and processed oysters. *Appl. Microbiol.* **20**:58–63.

Dimitroff, V. T. 1926. Spirochaetes in Baltimore market oysters. *J. Bacteriol.* **12**:135–177.

Ellender, R., Cook, D., Sheladia, V., and Johnson, R. 1980a. Enterovirus and bacterial evaluation of Mississippi oysters. *Gulf Res. Rep.* **6**:371–376.

Ellender, R., Mapp, J., Middlebrooks, B., Cook, D., and Cake, E. 1980b. Natural enterovirus and fecal coliform contamination of Gulf Coast oysters. *J. Food Protect.* **43**:105–110.

Fieger, E., and Novak, A. 1961. Microbiology of Shellfish Deterioration. In: Georg Borgstrom (ed.). *Fish as Food, Vol.* 1. Academic Press, New York.

Fleet, G. H. 1978. Oyster depuration—a review. *Food Technol. Aust.* **30**:444–454.

Food and Drug Administration. 1983. *Bacteriological quality of approved area summer harvested Louisiana oysters during harvest and interstate shipment.* Food and Drug Administration, Shellfish Sanitation Branch, Northeast Technical Services Unit, North Kingston, R. I.

Food and Drug Administration. 1989a. *National Shellfish Sanitation Program Manual of Operations, Part I, Sanitation of Shellfish Growing Areas.* Public Health Service, Shellfish Sanitation Branch, Washington, D.C.

Food and Drug Administration. 1989b. *National Shellfish Sanitation Program Manual of*

Operations, Part II, Sanitation of the Harvesting, Processing and Distribution of Shellfish. Public Health Service, Shellfish Sanitation Branch, Washington, D.C.

Fraiser, M. B., and Koberger, J. A. 1984. Incidence of salmonellae in clams, oysters, crabs and mullet. *J. Food Protect.* **47**:343–345.

Fugate, K. J., Cliver, D. O., and Hatch, M. T. 1975. Enteroviruses and potential bacterial indicators in Gulf Coast oysters. *J. Milk Food Technol.* **38**:100–104.

Galtsoff, P. S. 1964. *The American Oyster.* Fishery Bulletin No. 64. U.S. Department of the Interior, U.S. Government Printing Office, Washington, D.C.

Gardner, E. A., and Watts, B. M. 1956. Correlation of pH and quality of shucked southern oysters. *Commer. Fish. Rev.* **18**:8–15.

Garland, C. D., Nash, G. V., and McMeekin, T. A. 1982. Absence of surface-associated microorganisms in adult oysters *(Crassostrea gigas). Appl. Environ. Microbiol.* **44**:1205–1211.

Goatcher, L. J., and Westhoff, D. C. 1975. Repression of *Vibrio parahaemolyticus* by *Pseudomonas* species isolated from processed oysters. *J. Food Sci.* **40**:533–536.

Goatcher, L. J., Engler, S. E., Wagner, D. C., and Westhoff, D. C. 1974. Effect of storage at 5°C on survival of *Vibrio parahaemolyticus* in processed Maryland oysters *(Crassostrea virginica). J. Milk Food Technol.* **37**:74–77.

Goyal, S. M., Gerba, C. P., and Melnick, J. L. 1979. Human enteroviruses in oysters and their overlying waters. *Appl. Environ. Microbiol.* **37**:572–581.

Greenberg, A. E., and Hunt, D. A. (eds.). 1985. *Laboratory Procedures for the Examination of Seawater and Shellfish,* fifth edition. American Public Health Association, Washington, D.C.

Hackney, C., Rippen, T., and Sanders, L. 1988. Quality of previously frozen oysters repacked for the fresh market. In: *Proceedings Twelfth Annual Conference of the Tropical and Subtropical Fisheries Technological Society of the Americas,* Florida Sea Grant College Program, SGR-92, Gainesville, Florida, pp. 421–428.

Heen, E., and Karsti, O. 1965. Fish and Shellfish Freezing. In: G. Borgstrom (ed.). *Fish as Food, vol. IV, Processing: Part 2.* Academic Press, New York.

Hoff, J. C., Beck, W. J., Ericksen, T. H., Vasconcelos, G. J., and Presnell, M. W. 1967a. Time–temperature effects on the bacteriological quality of stored shellfish. I. Bacteriological changes in live shellfish: Pacific oysters *(Crassostrea gigas),* Olympia oysters *(Ostrea lurida),* Native Littleneck clams *(Prototheca staminea),* and Manila clams *(Venerupis japonica). J. Food Sci.* **32**:121–124.

Hoff, J. C., Beck, W. J., Ericksen, T. H., Vasconcelos, G. J., and Presnell, M. W. 1967b. Time–temperature effects on the bacteriological quality of stored shellfish. II. Bacteriological changes in shucked Pacific oysters *(Crassostrea gigas)* and Olympia oysters *(Ostrea lurida). J. Food Sci.* **32**:125–129.

Hood, M. A. 1983. Effects of harvesting waters and storage conditions on yeast populations in shellfish. *J. Food Protect.* **46**:105–108.

Hood, M. A., Ness, G. E., Rodrick, G. E., and Blake, N. J. 1983. Effects of storage on microbial loads of two commercially important shellfish species, *Crassostrea virginica* and *Mercenaria campechiensis. Appl. Environ. Microbiol.* **45**: 1221–1228.

Hood, M. A., Baker, R. M., and Singelton, F. L. 1984. Effect of processing and storing oyster meats on concentrations of indicator bacteria, vibrios and *Aeromonas hydrophila. J. Food Protect.* **47**:598–601.

Hunter, A. C. 1920. A pink yeast causing spoilage in oysters. *U.S. Department of Agriculture Bulletin* No. 819, p. 24.

Hunter, A. C., and Harrison, C. W. 1928. Bacteriology and chemistry of oysters, with special reference to regulatory control of production, handling and shipment. *US Dept. Agric. Tech. Bull.* **64**:1–75.

Hunter, A. C., Linden, B. A. 1923. An investigation of oyster spoilage. *Am. Food J.* **18**:538–540.

Jarvis, N. D. 1943. Principles and methods in the canning of fishery products. *US Fish Wildlf. Serv. Res. Rep.* No. 7, 366 pp.

Jay, J. M. 1986. *Modern Food Microbiology,* third edition. Van Nostrand Reinhold, New York.

Johnson, H. C., and Liston, J. 1973. Sensitivity of *Vibrio parahaemolyticus* to cold in oysters, fish fillets and crabmeat. *J. Food Sci.* **38**:437–441.

Johnson, W. G., Salinger, A. C., and King, W. C. 1973. Survival of *Vibrio parahaemolyticus* in oyster shellstock at two different storage temperatures. *Appl. Microbiol.* **26**:122.

Kelly, C. B. 1964. Time–temperature effect on bacteriological quality of stored oysters. In: *Proceedings, Fifth National Shellfish Sanitation Workshop,* pp. 193–202.

Koburger, J. A., and Miller, M. L. 1984. *Plesiomonas shigelloides:* a new problem for the oyster industry? In: *Proceedings of the Ninth Annual Tropical and Subtropical Fisheries Conference of the Americas.* TAMU-SG-85-106, pp. 337–342.

Kueh, C., and Chan, K. 1985. Bacteria in bivalve shellfish with special reference to the oyster. *J. Appl. Bacteriol.* **59**:41–47.

Liuzzo, J. A., and Novak, A. F. 1975. Correlation of organoleptic evaluation with chemical and microbiological tests in oysters. *Food Prod. Dev.* **9**:78.

Liuzzo, J. A., Barone, W. B., and Novak, A. F. 1966. Stability of B-vitamins in Gulf oysters preserved by gamma radiation. *Fed. Proc.* **25**:722.

Liuzzo, J. A., Lagarde, S. C., Grodner, R. M., and Novak, A. F. 1975. A total reducing substance test for ascertaining oyster quality. *J. Food Sci.* **40**:125–128.

Lovelace, T. E., Tubiash, H., and Colwell, R. R. 1968. Quantitative and qualitative commensal bacterial flora of *Crassostrea virginica* in Chesapeake Bay. In: *Proceedings of the National Shellfisheries Association.* **58**:82–87.

McCormack, G. 1950. Pink yeast isolated from oysters grown at temperatures below freezing. *Commer. Fish. Rev.* **12**:28.

McCormack, G. 1956. Growth characteristics of the pink yeast that causes discoloration of oysters. *Commer. Fish. Rev.* **18**:21–22.

McKee, L. G. 1963. The oyster, clam, scallop and abalone fisheries. In: Stansby, M. E. (ed.). *Industrial Fishery Technology.* Van Nostrand Reinhold, New York, pp. 183–192.

Motes, M. L. 1982. Effect of chlorinated wash water on *Vibrio cholerae* in oyster meats. *J. Food Sci.* **47**:1028–1029.

Murchelano, R. A., and Brown, C. 1968. Bacteriological study of the natural flora of the Eastern oyster, *Crassostrea virginica. J. Invert. Pathol.* **11**:519–520.

Novak, A. F., Liuzzo, J. A., Grodner, R. M., and Lovell, R. T. 1966. Radiation pasteurization of Gulf Coast oysters. *Food Technol.* **20**:103–104.

Oliver, J. D. 1981. Lethal cold stress of *Vibrio vulnificus* in oysters. *Appl. Environ. Microbiol.* **41**:710–717.

Pace, J., Wu, C. Y., and Chai, T. 1988. Bacterial flora in pasteurized oysters after refrigerated storage. *J. Food Sci.* **53**:325–327.

Paille, D., Hackney, C., Reily, L., Cole, M., and Kilgen, M. 1987. Seasonal variation in the fecal coliform population of Louisiana oysters and its relationship to microbiological quality. *J. Food Protect.* **50**:545–549.

Peixotto, S. C. S., Finne, G., Hanna, M. O. and Vanderzant, C. 1979. Presence, growth and survival of *Yersinia enterocolitica* in oysters, shrimp and crab. *J. Food Protect.* **42**:974–981.

Presnell, M. W., and Kelly, C. B. 1961. *Bacteriological studies of commercial shellfish operations on the Gulf Coast.* U.S. Public Health Service Technical Report F-61-9.

Reily, L. A., and Hackney, C. R. 1985. Survival of *Vibrio cholerae* during cold storage in artificially contaminated seafoods. *J. Food Sci.* **50**:838–839.

Richards, G. P. 1988. Microbial purification of shellfish: a review of depuration and relaying. *J. Food Protect.* **51**:218–251.

Rippey, S. R., and Verber, T. L. 1986. *Shellfish Borne Disease Outbreaks.* Food and Drug Administration, Shellfish Sanitation Branch, Northeast Technical Services Unit, Davisville, Rhode Island.

Ronk, R. J. 1988. Emerging trends: *Vibrio vulnificus. J. Assoc. Food Drug Offic.* **52**:7–11.

Rosen, B. 1966. Accumulation of organic acids during cold storage of shucked soft clams, *(Mya arenaria)* (L.), in relation to quality. *Fish. Indust. Res.* **3**:5–11.

Rowse, A. J., and Fleet, G. H. 1982. Viability and release of *Salmonella charity* and *Escherichia coli* from oyster feces. *Appl. Environ. Microbiol.* **44**:544–548.

Shiflett, M. A., Lee, J. S. and Sinnhuber, R. O. 1966. Microbial flora of irradiated dungeness crabmeat and Pacific oysters. *Appl. Microbiol.* **14**:411–415.

Sidwell, V. D., Loomis, A. L., and Grodner, R. M. 1979. Geographic and monthly variation in composition of oysters, *Crassostrea virginica. Marine Fish. Rev.* **41**:13–17.

Sieling, F. W. 1971. Harmless coloration of oysters explained. *Commer. Fish. News* **4**:1–2.

Singleton, P., and Sainsbury, D. 1978. *Dictionary of Microbiology.* John Wiley & Sons, New York.

Sobsey, M., Hackney, C., Carrick, R., Ray, B., and Speck, M. 1980. Occurrence of enteric bacteria and viruses in oysters. *J. Food Protect.* **43**:111–113.

Son, N. T., and Fleet, G. H. 1980. Behavior of pathogenic bacteria in the oyster, *Crassostrea commercialis,* during depuration, re-laying, and storage. *Appl. Environ. Microbiol.* **40**:994–1002.

Tall, B. D., and Nauman, R. K. 1981. Scanning electron microscopy of *Cristispira* species in Chesapeake Bay oysters. *Appl. Environ. Microbiol.* **42**:336–343.

Tanikawa, E., and Doha, S. 1965. Heat processing of shellfish. In: G. Borgstrom (ed.). *Fish as Food, Vol. IV, Processing: Part 2.* Academic Press, New York.

Thompson, C. A., Vanderzant, C., and Ray, S. M. 1976. Effect of processing, distribution and storage on *Vibrio parahaemolyticus* and bacterial counts of oysters *(Crassostrea virginica). J. Food Sci.* **41**:123–127.

Tierney, J. T., Sullivan, R., Peeler, J. T., and Larkin, E. P. 1982. Persistence of polioviroses in shellstock and shucked oysters stored at refrigeration temperatures. *J. Food Protect.* **45**:1135–1137.

Vanderzant, C., Thompson, C. A., Jr., and Ray, S. M. 1973. Microbial flora and level of *Vibrio parahaemolyticus* of oysters *(Crassostrea virginica),* water and sediment from Galveston Bay. *J. Milk Food Technol.* **36**:447–452.

Vasconcelos, G. J., and Lee, J. S. 1972. Microbial flora of Pacific oysters *(Crassostrea gigas)* subjected to ultraviolet-irradiated seawater. *Appl. Microbiol.* **23**:11–16.

Vasconcelos, G. J., Jakubowski, W., and Ericksen, T. H. 1969. Bacteriological changes in shellfish maintained in an estuarine environment. In: *Proceedings of the National Shellfisheries Association* **59**:67–83.

Vaughn, J. M., Landry, E. F., Vicale, T. J., and Dahl, M. D. 1980. Isolation of naturally occurring enteroviruses from a variety of shellfish species residing in Long Island and New Jersey marine embayments. *J. Food Protect.* **43**:95–98.

Wait, D. A., Hackney, C. R., Carrick, R. J., Lovelack, G., and Sobsey, M. D. 1983. Enteric bacteria and viral pathogens and indicator bacteria in hard shell clams. *J. Food Protect.* **46:**493–496.

Wilson, T. E., and McCleskey, C. S. 1951. The effect of storage on indicator organisms in shucked oysters and a comparison of methods for the determination of enterococci. *Food Res.* **16:**377–382.

3

Microbiology of Crustacea Processing: Crabs

Ralph R. Cockey and Tuu-jyi Chai

The crab industry is one of the major components of the seafood industry in the United States. Among shellfish crab ranks second to shrimp in the dollar value of the harvested product. The dominant U.S. crab fisheries are blue crab, king crab, Dungeness crab, and tanner/snow crabs. Over the past 9–10 years the landings of blue and Dungeness crabs have remained rather constant while that of king crab has experienced a continual decline. At the same time the tanner and snow crab landings have increased, in part compensating for the losses in the king crab fishery.

The major products from king, Dungeness, and tanner/snow crabs are frozen sections, frozen claws, frozen meat, and canned meat; the main products from blue crabs are fresh and pasteurized meat. Blue crabs are harvested from estuarine and coastal waters, in contrast to the deeper ocean water from which the other four crabs are harvested. Therefore, blue crabs are more susceptible to environmental factors and their effects on the microbiological flora of the crabs.

Like other fresh seafood, crabmeat is highly perishable. It is contaminated by a wide variety of potential spoilage organisms during processing. The high value of the product and its susceptibility to spoilage have prompted investigations into the microbial flora and organisms responsible for spoilage. Crabmeat can also serve as a vector for food poisoning microorganisms of public health significance. Fresh and pasteurized crabmeat extracted from the body cavity constitute the major products. Claw meat and cocktail claw (claw fingers) are also produced but are of secondary economic value compared to the body meat. Most studies have centered on the fresh and pasteurized products which serve as the main focus of this review.

HARVEST AND UTILIZATION

Species of Economic Importance

The five major species of crabs of economic importance in the United States are blue crab *(Callinectes sapidus)*, king crab *(Paralithodes camtschatica)*, Dungeness crab *(Cancer magister)*, tanner crab *(Chionoecetes bairdi)*, and snow crab *(Chionoecetes opilio)*.

The annual catch for the five major crab fisheries is shown in Table 3.1 (NMFS 1980–1989). Blue crab landings have been gradually increasing over the last 10 years while that of Dungeness has remained constant. King crab has dropped from first in terms of pounds landed in 1979 to the last species in 1988. The snow/tanner fishery which started in the late 1960s has increased in importance, and in 1988 the total pounds landed was second only to blue crabs. A comparison of landings and economic value for crabs is shown in Table 3.2 (NMFS 1980–1989).

The blue crab is the smallest of the four, averaging about 5–6 inches carapace width in adult males. Blue crabs are found from Massachusetts Bay south down the Atlantic coast to Florida and along the Gulf of Mexico from Florida to Texas. The crabmeat processing industry is concentrated especially in the Chesapeake Bay area and in all the states bordering the Atlantic Ocean and the Gulf. The principal products are fresh and pasteurized crabmeat, whole steamed crabs, and live and frozen soft shelled crabs.

Dungeness crabs are much larger than blue crabs, weighing from 2–3 pounds each and measuring 8–9 inches across the carapace. The Dungeness is distributed in the coastal waters of northern California extending northward along the Pacific coast to central Alaska. It is harvested commercially in California, Oregon, Washington, and Alaska. The main products are fresh crabs, cooked whole crabs, and crab sections and crabmeat retailed as either fresh, frozen, or canned.

King crab, the giant among edible crabs, can weigh as much as 24 pounds and measure up to 5 feet from tip to tip of outstretched legs. It is found in the cold bottom waters of central and western Alaska ranging from southeastern Alaska to Prince William Sound, Kodiak Island, along the Aleutians, and into the Bering Sea. The major products are frozen sections containing four legs and one claw each and frozen and canned meat.

Tanner/snow crabs are similar in appearance to the king crab but are much smaller, averaging 2.5–3 pounds each. They are found along the West Coast in deep waters in roughly the same area as king crabs. There is also a fishery in the North Atlantic along the coast of Maine. The products are frozen cooked sections and frozen and canned meat.

Other crabs of lesser importance commercially are red crab *(Geryon quinquedens)*, Jonah crab *(Cancer borealis)*, rock crab *(Cancer irroratus)*, and stone crab *(Menippe mercenaria)*. The red crab is a deep sea crab most abundant on the continental slope off the north and middle Atlantic states. The Jonah and rock crabs are ocean crabs found along the coast from the north Atlantic to the south Atlantic states. The stone crab ranges from North Carolina southward along the coast and into the Gulf.

Table 3.1 Annual Landings of the Major Crab Species in the United States[a]

Crab Species	1979	1980	1981	1982	1983	1984	1985	1986	1987	1988
Blue (hard crabs)	152,830	163,206	194,114	195,476	191,754	201,556	190,524	184,491	197,826	218,663
Dungeness	38,690	38,278	35,576	32,868	28,763	24,595	28,282	22,408	29,460	47,440
King	154,589	185,624	88,054	38,492	25,581	17,204	15,363	25,909	29,065	20,973
Snow/tanner	131,395	121,674	107,474	68,767	61,077	48,765	85,742	110,000	113,812	146,326
Other	11,682	14,329	19,777	13,999	9,817	20,469	17,721	12,852	16,205	22,227
Total	489,184	523,111	445,995	349,602	316,992	312,953	337,632	355,660	386,368	455,629

[a]All amounts shown are in thousands of pounds.

Table 3.2 1988 Commercial Landings and
Economic Value for Crabs

Species	Pounds[a]	Dollars[a]
Blue crab	218,663	84,357
Dungeness	47,440	54,771
King	20,973	84,153
Snow/tanner	146,326	137,052
Other	22,227	23,227
Total crab	455,629	383,560

[a]Amounts are given in thousands.

Fishing Methods

Blue crabs are harvested primarily using baited crab pots and trot lines. During the winter months dormant crabs are fished in Virginia using bottom dredges, allowing plants to be in operation year round. Gulf Coast states harvest crabs mainly by potting year round because of the warmer climate in winter. The season in Maryland runs from April to about mid-November. Some plants in Maryland continue picking on a reduced scale through the winter months using dredged crabs imported from Virginia and potted crabs from the south.

Harvesting begins in the early morning. As they are brought on board, the crabs are inspected for size and separated by sex. In Maryland the legal minimum size for hard crabs is 5 inches and for peelers 3 inches. Most of the males are marketed whole and supply the steamed crabs for restaurants, seafood markets, and crab feasts. The bulk of the females are sold to processing plants for picking of crabmeat. Watermen deliver their catch on shore by late morning to mid afternoon. As they are received at the plants the crabs are placed in crates, cooked, cooled, and refrigerated for picking the next day. Crabs that cannot be cooked the same day as harvested are stored at 45–50°F (7.2–10°C) until they are cooked, usually the next morning.

King, Dungeness, and snow/tanner crab harvesting methods are similar. Baited pots are used, varying in size to accommodate the size of the species being harvested. Only male king crabs with a minimum 7-inch wide carapace can be legally harvested. Also only the male is legal for Dungeness crabs. All the crab types are placed in live wells with circulating seawater aboard ship. Snow crabs are easily injured and die quickly out of water, and postharvest handling procedures are critical to ensure a healthy live crab for processing (Dassow and Learson 1963).

Processing of Dungeness, King, Snow, and Tanner Crabs

At the processing plant Dungeness crabs are prepared for cooking by removing the carapace, gills, and viscera and breaking the crab in two (Dassow and Learson 1963). The sections or halves are cooked for 10–12 minutes in boiling fresh water, cooled briefly in cold water, and shaken to remove the meat. The meat is placed in brine to facilitate separation of shell fragments which sink to the bottom while the meat remains suspended. The meat then receives a fresh water spray to remove excess salt and is sealed in 5-pound cans, and stored at 32° to 40°F (0–4.4°C). Meat for freezing is packed in 1- to 5-pound cans and hermetically sealed. Frozen meat can be stored for 6 months at 0°F (–17.8°C) with no significant quality change.

King crabs to be processed are butchered and the sections cooked in boiling water for 20–22 minutes. Meat is removed from the warm sections by passing through rollers which break the shell and force the meat out while the empty shell proceeds through the roller mechanism. The meat is washed, inspected, and boiled a second time in dilute brine for about 4 minutes. After the second cook the meat is spray washed, inspected, and packaged. King crabmeat is much more amenable to freezing than the meat of other crab species. It can be stored for up to 1 year at 0°F (–17.8°C) with little loss of quality. Meat for freezing is prepared similar to the process described above except that after cooking, the sections are chilled prior to squeezing to minimize breakage of meat. Processing of snow/tanner crabs closely parallels that of king crab. Recovery of the meat from cooked sections is somewhat difficult, requiring both manual and mechanical roller techniques.

Processing of Blue Crabs

One of the most critical points in the processing of blue crabs is cooking. Cooking facilitates the removal of meat from the crab's body and claws. Cooking times, temperatures, and methods have all been shown to affect yield and subsequent refrigerated storage life of fresh crab meat. Boiling and cooking in steam under pressure are the two methods used throughout the industry for crabs to be picked for crabmeat. In Maryland, Virginia, and North Carolina, regulations state that crabs must be cooked by steam under pressure; however, cooking for 10 minutes after the retort reaches 250°F (121.1°C) is recommended based on reports by Littleford (1957a,b), Dunker et al. (1959), and Ulmer (1964). This process was found to give higher yields of meat than either 5 or 15 minutes at the same temperature. Ulmer found that both steam boiling for 15 minutes and retorting for 10 minutes produce meats with acceptably low bacterial levels initially. Steam boiling gave higher yields but the meat from retorted crabs had a longer shelf life. The higher yield from the boiled crabs was explained partially by a higher moisture content in the meat as compared to meat from retorted crabs. The higher moisture level may also explain the shorter shelf life for boiled crabs by favoring bacterial growth.

Miller et al. (1974) stated that steam pressure should be applied until the internal temperature of the centermost crab reaches 235°F (112.8°C). The minimum internal crab temperature recommended in Maryland is 240°F (115.5°C). (Maryland State Department of Health 1957). Ulmer (1964) reported an average internal crab temperature of 247°F (119.4°C) after retorting for 10 minutes at 15 psi steam pressure.

After cooking, crabs are cooled for 3–4 hours, usually in a well-ventilated screened area. As the cooked crabs approach ambient temperature they are placed in refrigerated storage for picking the next day. In some plants, especially along the Gulf Coast, as soon as the cooked crabs are cool the backs are removed and the bodies washed before refrigerated storage.

Whole crabs are placed on the picking table for meat removal. The pickers remove the carapace, clean away the gills and viscera, and finally remove the meat from the body cavities. The claws are saved for hand or machine picking at a later time. Practices vary considerably throughout the industry; however, most processors require that pickers deliver their meat to the packing room each hour or as soon as they pick 5 pounds, whichever occurs first. In the packing room the cans are weighed, lids applied, and the filled containers iced and refrigerated. Some plants inspect the meat for shell fragments before packing in ice and storing.

Some plants, particularly in Maryland, use the Quik-Pik machine (Fig. 3.1)

a   b

Figure 3.1 The Quik-Pik machine line. (a) Rotating cage for claw and leg removal. (b) Shaker machine.

for body meat removal from prepared cores in addition to handpicking. The machine picking process will be described in detail in a later section.

MICROBIOLOGICAL FLORA

Fresh Raw Crabs

The bacterial flora of freshly caught crabs reflect that of the environment from which they are harvested and show qualitative and quantitative differences depending on season, water quality, and geographic location (Tobin and McCleskey 1941).

The meat of newly caught healthy crabs has a very low bacterial count; in fact, it is almost sterile (Fieger and Novak 1961). Bacteria associated with fresh live crabs appear to be mainly on the body surfaces as relatively few were found in the intestinal tract (Goresline and Smart 1942). The hemolymph of 18% of 290 freshly caught blue crabs from Chincoteague Bay, Virginia, was found to be sterile, with 16% sterility in summer and 23% in winter crabs (Tubiash et al. 1975). Male crabs had a higher bacterial hemolymph burden than females. The mean hemolymph most probable number per milliliter was 2,756 for males, 1,300 for females, and 1,876 for both sexes. The higher bacterial levels for male hemolymph were explained, in part, by the higher incidence of injury and missing appendages in males than in females.

The microbiological flora found in raw crabs of different species are shown in Table 3.3 (Craig et al. 1968; Sizemore et al. 1975; Williams-Walls 1968). Sizemore et al. (1975) reported the predominant bacteria in blue crab hemolymph were *Vibrio parahaemolyticus* and related *Vibrio* sp. The highest levels were found from crabs examined during May, June, and July, the months shown by Kaneko and Colwell (1973) to be associated with high levels of *V. parahaemolyticus* in Chesapeake Bay. Sizemore et al. (1975) found the most dominant species to be *Vibrio, Pseudomonas,* and *Acinetobacter.* Other species found were *Aeromonas, Bacillus,* and *Flavobacterium. Clostridium botulinum* Type E was found in the gills and viscera of blue crabs by Lilly et al. (1971) and *C. botulinum* Type F in blue crabs by Williams-Walls (1968).

Faghri et al. (1984) studied crabs from cold waters and the relationship between the bacterial flora found on crabs taken from waters close to human habitation to those from more isolated areas away from human habitation. Greater numbers of bacterial species were found in Dungeness crabs from Kodiak Island and the Columbia River, and rock crabs from Maine whereas the least number of species were from the tanner crab from the Bering Sea. The greatest concentrations of bacteria for all crabs occurred in the gills, 1×10^3–1×10^7/g of tissue, compared to 1×10^1–4×10^2/g of muscle tissue. Whereas *Escherichia coli* was not found in crabs from the Kodiak Island area, *Klebsiella*

Table 3.3 Bacteria Found in Raw Crabs

Crabs and Source	Bacteria		
	Gill and Viscera	Muscle	Hemolymph
Blue crab, Chincoteague Bay, Virginia	*Clostridium botulinum* Type E *Clostridium botulinum* Type F		Acinetobacter spp. Aeromonas spp. Bacillus spp. Coliforms Flavobacterium spp. Pseudomonas spp. Vibrio spp.
Dungeness, Kodiak Island, Alaska	*Acinetobacter* spp. *Aeromonas hydroxyla* *Alcaligenes* spp. *Citrobacter freundii* *Enterobacter agglomerans* *Klebsiella pneumoniae* *Moraxella* spp. *Pasteurella* spp. *Pseudomonas maltophilia* *Pseudomonas fluorescens* *Sarcina* spp. *Staphylococcus* spp., coagulase negative *Streptococcus* group D *Yersinia enterolitica*	*Acinetobacter* spp. *Alcaligenes* spp. *Micrococcus* spp. *Moraxella* spp. *Pasteurella* spp. *Pseudomonas* spp. *Sarcina* spp. *Staphylococcus* spp., coagulase negative *Streptococcus* group D	

Dungeness, Columbia River	*Acinetobacter* spp. *Achromobacter* spp. *Aeromonas* spp. *Alcaligenes* spp. *Micrococcus* spp. *Moraxella* spp. *Pseudomonas* spp. *Staphylococcus* spp., coagulase negative *Vibrio* spp.	*Acinetobacter* spp. *Achromobacter* spp. *Alcaligenes* spp. *Micrococcus* spp. *Moraxella* spp. *Pseudomonas* spp. *Staphylococcus* spp., coagulase negative	*Acinetobacter* spp. *Vibrio* spp.
Rock, Maine	*Acinetobacter* spp. *Achromobacter* spp. *Aeromonas* spp. *Alcaligenes* spp. *Beneckea* spp. *Citrobacter* spp. *Klebsiella* spp. *Micrococcus* spp. *Moraxella* spp. *Pseudomonas* spp. *Pseudomonas fluorescens* *Staphylococcus* spp., coagulase negative *Vibrio* spp.	*Acinetobacter* spp. *Achromobacter* spp. *Micrococcus* spp. *Moraxella* spp. *Staphylococcus* spp., coagulase negative	*Acinetobacter* spp. *Achromobacter* spp. *Alcaligenes* spp. *Beneckea* spp. *Klebsiella* spp. *Micrococcus* spp. *Pseudomonas* spp. *Staphylococcus* spp., coagulase negative *Vibrio* spp.
Tanner, Bering Sea	*Acinetobacter calcoaceticus* *Alcaligenes* spp. *Micrococcus* spp. *Moraxella* spp. *Staphylococcus* spp., coagulase negative	*Micrococcus* spp. *Staphylococcus* spp., coagulase negative	

and other enteric bacteria were isolated from these crabs. Faghri et al. stated the general profile of *Klebsiella, Citrobacter,* and *Yersinia* species is suggestive of sewage contamination. *Vibrio* species, including *V. vulnificus* and *V. parahaemolyticus,* were found in crabs taken from the mouth of the Columbia River and the Maine coast but not in crabs from the Alaskan waters.

Spoilage of Crabmeat

After steam cooking crabs should be sterile. They become contaminated during cooling, storage, and subsequent processing and may contain appreciable numbers of bacteria by the time the product reaches the packing room. The normal shelf life may not be realized if the product is stored under less than optimum conditions. Bacteria found in fresh crabmeat are shown in Table 3.4 (Alford et al. 1942; Allen and Woodburn 1972; Kautter et al. 1974; Lee and Pfeifer 1975; Peixotto et al. 1979; Weagant et al. 1988). Lee and Pfeifer (1975) reported that during refrigerated storage of Dungeness crabmeat *Moraxella, Pseudomonas, Acinetobacter,* and *Flavobacterium/Cytophaga* species predominated. *Arthrobacter, Bacillus, Micrococcus, Staphylococcus,* and *Proteus* did not grow in the meat during storage. *Micrococcus, Staphylococcus,* and *Proteus* were introduced during processing while the remainder originated from the raw crabs.

Fellers (1927), in an investigation into spoilage of canned crabmeat, reported a type of nongaseous spoilage that was most prevalent. Spoilage was due to the growth of several species of "aerobic sporulation bacilli" such as *Bacillus cereus* and *B. mesentericus.* Spoilage was characterized by texture loss, turbid liquor, abnormal taste and odor, and slight discoloration. Harris (1932) reported that the *Proteus* group was most important in the spoilage of crabmeat with both

Table 3.4 Microorganisms Found in Fresh Crabmeat

Blue Crabmeat	*Dungeness Crabmeat*
Achromobacter sp.	*Acinetobacter* sp.
Acinetobacter sp.	*Arthrobacter* sp.
Bacillus sp.	*Bacillus* sp.
Clostridium botulinum Types E and F	*Flavobacterium—Cytophaga*
Listeria sp.	*Lactobacillus* sp.
Moraxella sp.	*Micrococcus* sp.
Proteus sp.	*Moraxella* sp.
Pseudomonas sp.	*Proteus* sp.
Sarcina sp.	*Pseudomonas* sp.
Staphylococcus sp.	*Staphylococcus* sp.
Streptococcus sp.	*Vibrio* sp.
Vibrio sp.	Yeasts
Yersinia enterocolitica	

pseudomonads and flavobacteria possibly contributing to the deterioration. *Streptococcus, Sarcina,* and *Achromobacter* were also reported by Harris as present in the crabmeat. Alford et al. (1942), in a study of the spoilage of fresh crabmeat, found the bacterial flora of the fresh product consisted of 6.40% cocci, 23.4% *Pseudomonas* and *Achromobacter* species, and 16.2% other bacilli. During storage at 37.4–41°F (3–5°C) there was a marked decrease in cocci accompanied by an increase in *Pseudomonas* and *Achromobacter* so that by 11–15 days the distribution was 96.3% *Pseudomonas* and *Achromobacter,* 3% cocci, and 0.7% other bacilli. Alford et al. concluded that, at refrigeration temperatures, *Pseudomonas* and *Achromobacter* were most active in crabmeat spoilage although *Proteus* may be the dominant genus in some spoiled crabmeat.

In the early investigations into crabmeat spoilage, identification of the responsible organisms was difficult due to the methodology, laboratory techniques, and general information relating to taxonomy used at that time. *Achromobacter* organisms listed in the early literature were later shown to be *Moraxella/Acinetobacter* (Chai 1970, 1981). Many organisms identified as *Proteus* and some *Alcaligenes* could in fact have been *Pseudomonas putrifaciens* or its related pseudomonads (Chai 1979; Chen and Chai 1982; Chai et al. 1968). In general, *Pseudomonas* and *Moraxella/Acinetobacter* are the major spoilage organisms in seafood.

Bacteria Involved in Foodborne Illness

Fresh, live crabs coming into the processing plants carry pathogenic bacteria indigenous to the marine environment. Type E *Clostridium botulinum* has been shown to be commonly found in the marine environment including areas from which blue crabs are fished (Cockey and Tatro 1974). *C. botulinum* type E was isolated from Dungeness crab (Craig et al. 1968), from the viscera and gills of blue crabs (Lilly et al. 1971), and also from processed meat of the blue crab (Kautter et al. 1974). Fortunately no cases of botulism have ever been attributed to consumption of crabmeat. Fresh meat generally spoils after 10–14 days of refrigerated storage [36–38°F (2.2–3.3°C)], well short of the time required for toxin production. The recommended pasteurization process of 185°F (85°C) for 3 minutes at the coldest point in the can of crabmeat is adequate to kill 10^6 spores/g of *C. botulinum* Type E that may have been present in the fresh meat. Type E spores can be introduced into the meat post-pasteurization during the cooling process if the can seams are defective; however subsequent refrigeration at 36°F (2.2°C) or below is required and protects the product from becoming toxic.

Six species of vibrios have been associated with foodborne disease (Hackney and Dicharry 1988). They are commonly found in the marine environment (Colwell and Kaper 1978; Colwell 1984) and are part of the natural flora of fresh raw seafoods including blue crabs (Krantz et al. 1969; Fishbein et al. 1970). An outbreak of cholera occurred in 1978 after consumption of cooked crabs recontaminated with *Vibrio cholerae* 01 El tor (Blake et al. 1980). Similarly an outbreak of food poisoning due to *V. parahaemolyticus* occurred in 1971. A number of people became ill following a crab feast. It was found that the steamed crabs had been contaminated with *V. parahaemolyticus* from live crabs during shipment (Dadisman et al. 1973). Another vibrio of great concern is *V. vulnificus*, the cause of a highly dangerous septicemia and an invasive form of wound infection with associated mortality rates as high as 40–60% in susceptible individuals (Hackney and Dicharry 1988). Vibrios are readily killed by cooking and should not present a problem if precautions are taken to prevent recontamination of the cooked product.

Due to the nature of crab processing and the involvement of hand picking most crabmeat contains some level of coagulase-positive *Staphylococcus aureus;* consequently, if the product is mishandled or temperature abused, growth with production of enterotoxin can occur (Cockey and Chai 1989). There are no statistics on the incidence of staphylococcal food poisoning related to crabmeat; however it should be assumed that the organism is present and thus the meat should be kept well refrigerated at all times.

There are a number of other pathogenic bacteria widespread in the environment that could contaminate crabmeat during processing. These include *Salmonella*, pathogenic *Escherichia coli*, *Yersinia*, *Listeria*, and *Campylobacter*. Peixotto et al. (1979) reported on the presence of *Yersinia enterocolitica* in 21% of the blue crab samples examined. Weagant et al. (1988), reporting on the incidence of *Listeria* in frozen seafoods, found 35 of 57 samples (including crabmeat) examined contained *Listeria* species and 15 of the 57 samples contained *L. monocytogenes*.

PROCESSING AND MICROBIOLOGICAL QUALITY CONTROL

Microbiological Standards

The microbiological standards for fresh and pasteurized crabmeat for the 10 producing states are shown in Table 3.5 (Cockey 1983). An aerobic plate count (APC) of 1×10^5/g for fresh crabmeat is generally considered the upper limit. Other standards pertain to either fecal coliforms (FC) or *E. coli* (EC) or both. Georgia is the only state with a standard for pathogenic *S. aureus*. A zero tolerance for *S. aureus* is unrealistic for a product that receives the high level of human contact that crabmeat does during the picking and inspection process. Over the last 3 years Maryland has been routinely checking fresh crabmeat for pathogenic *S. aureus* with an eye to setting a realistic standard that does not impose impossible constraints on industry while assuring protection of the consumer. Recently the National Advisory Committee on Microbiological

Table 3.5   Bacteriological Standards for Fresh and Pasteurized Crabmeat
in the Blue Crabmeat Producing States

State		APC/g^a	$TC/100 \text{ g}^b$	$FC/100 \text{ g}^c$	$EC/100 \text{ g}^d$	Other
Alabama	Fresh	1×10^5		93		
	Past.					
Florida	Fresh	$<1 \times 10^5$	$<1 \times 10^4$	<50		
	Past.	$<2.5 \times 10^5$				
Georgia	Fresh	1×10^5	1×10^4	0	0	*S. aureus* 0
	Past.					*Salmonella* 0
Louisiana	Fresh	1×10^5			36	
	Past.					
Maryland	Fresh	1×10^5			36	
	Past.	2.5×10^4			0	
Mississippi	Fresh	1×10^5			5×10^3	
	Past.					
North Carolina	Fresh	1×10^5			36	
	Past.	3×10^3		0	0	
South Carolina	Fresh	1×10^5		93	46	
	Past.	3×10^3		0	0	
Texas	Fresh				3.6 MPN/g	
	Past.				0	
Virginia	Fresh	1×10^5		50		
	Past.	$<3 \times 10^3$		<30		

^{a}APC, aerobic plate count per gram.
^{b}TC, total coliforms most probable number per 100 g.
^{c}FC, fecal coliform MPN/100 g.
^{d}EC, *Escherichia coli* MPN/100 g.

Criteria for Foods (1989) has been approaching the final approval of 1,000 pathogenic *S. aureus*/g of crabmeat as the maximum allowable level. For pasteurized crabmeat Maryland and Florida specify an APC/g of 2.5×10^4 whereas for the remaining states the APC is more realistic at 3×10^3/g or less and surely no pasteurized product should contain any fecal coliforms, *E. coli,* or pathogens.

Sanitation and Quality Control Considerations

Production of high-quality crabmeat begins with the processing plants itself. Its design should provide for an efficient product flow from the receiving area to the packing room and cold storage ensuring complete separation of live crabs and cooked products at all times. Walls, ceilings, and floors should be made of materials and constructed in a manner to facilitate good cleaning and sanitation and precautions taken to prevent entry of insects and rodents.

Of greatest importance in the maintenance of high-quality crabmeat is good plant sanitation and observance of sanitary practices throughout the process as shown by several surveys (Phillips and Peeler 1972; Ray et al. 1976; Wentz et al. 1985). In the Phillips and Peeler survey plants were divided into two groups, those with good plant sanitation and those with poor plant sanitation. The geometric mean values for the APC/g of fresh crabmeat ranged from 16,000 to 45,000 in plants with good sanitation compared to a range of 140,000–320,000 for meat from plants with poor sanitation. Wentz et al. (1983) reported an APC geometric mean of 20,000 for the finished product in plants maintaining high sanitary standards and Ray et al. (1976) reported an APC of 21,000/g of meat processed under carefully controlled sanitary conditions, but 150,000/g for meat obtained from other crabmeat plants. Our experience at the University of Maryland has similarly shown a direct relationship between good sanitation practices and low bacterial content of the product. Also, coupled with low sanitation, frequently there is a general laxity in picking room management, such as: meat is not delivered to the packing room for icing on a timely basis, and crabs on the picking table are not rotated properly so that some crabs may sit at room temperature several hours before picking. Phillips and Peeler (1972) reported similar findings in their survey of the blue crab industry. For over 10 years the University of Maryland, through the Seafood Science Program, has provided microbiological support to the seafood processing industry in an effort to improve the quality of their products. In addition to sanitation another factor affecting the bacterial level of crabmeat is the type and condition of crabs being processed. Paparella (1978) reported that sponge crabs (gravid females carrying an egg mass), taken from the picking tables and from overnight storage, contained bacterial levels as high as 1×10^6/g of whole crab. Meat picked from these crabs had a plate count of $(1-3) \times 10^5$/g. When retort cooking times were increased from 10 minutes to 15–16 minutes, the bacterial counts on the whole crab and the fresh meat dropped to acceptable values. Consistently we have found cooked sponge crabs to harbor greater numbers of bacteria than crabs without a sponge as shown in Table 3.6 (Pace et al. 1989). Processors are advised to increase their cooking time by several minutes when appreciable numbers of sponge crabs are received in the plant.

Similarly, cooked green crabs, sometimes called buckrams, carry higher levels of bacteria into the picking room. Buckrams are blue crabs that have recently molted and contain a higher level of water than fat crabs. They are wetter after cooking and the high moisture level encourages bacterial growth during overnight refrigerated storage. As with sponge crabs, many processors increase the cooking time for buckram crabs to produce a drier product that is more amenable to producing meat with lower bacteriological levels.

Table 3.6 Bacterial Count of Sponge Crab Compared to Crabs Without Sponge (APC/g)

| Year | Sponge Crabs | | Crabs with No Sponge | |
	Range	*Mean*	*Range*	*Mean*
1977	3.5×10^4–4×10^6	5.6×10^5	1.3×10^4–2.5×10^5	4.8×10^4
1978	2.7×10^4–6×10^5	3.4×10^5	3×10^3–1.7×10^5	2.8×10^4
1980	6.8×10^3–6.7×10^5	9.7×10^4	2.5×10^4–3×10^5	3.5×10^4
1981	1.6×10^4–2.3×10^5	7.3×10^4	7.1×10^3–4×10^5	5.8×10^4

The importance of observing proper retort cooking cannot be overstressed. Undercooking can lead to high survival of bacteria in the crabs and produce a wet cooked product. Both conditions favor high levels of bacteria in whole crabs, in the picking room and in the meat picked from them. The experience of one processor is shown in Table 3.7 (unpublished data, Cockey and Chai). In general, this is a well managed plant with excellent maintenance and sanitation programs. In an effort to increase yields the processor was using the short cooking time; however, when the cook was increased to 250°F (121.1°C) for 10 minutes the improvement was dramatic.

Pasteurization Process

The normal shelf life of fresh crabmeat is 7–10 days (Bernarde and Littleford, 1957; Webb et al. 1973); however, under optimum refrigerated storage meat with a low initial bacterial count may last 14 days. Early attempts to extend shelflife using antibiotics (Bernarde 1958; Bernarde and Littleford 1957) proved unsuccessful as did use of phosphoric acid to lower pH (Tobin and McClesky 1941). Puncochar and Pottinger (1938) attempted pasteurization in a boiling water bath but the meat was dry and discolored. Anzulovic and Reedy (1942) attempted pasteurization at temperatures below the boiling point of water. A

Table 3.7 Effect of Undercooking on Bacterial Quality of Crabmeat

| Date | Retort Temp (°F) and Cooking Time | Bacteria (APC/g) | |
		Whole Crab	*Meat*
8/10/87	240 (115.5°C) for 4 min	4.0×10^6/g	1.0×10^6
		2.8×10^6/g	1.6×10^6
		3.1×10^6/g	1.5×10^6
8/26/87	250 (121.1°C) for 10 min	8.3×10^3	4.5×10^3
			2.6×10^4
			1.7×10^4
			1.1×10^4

process of 160°F (71.1°C), meat temperature, for 10 minutes or 170°F (76.7°C) for 1 minute was successful in extending the refrigerated storage life to 6 weeks. Byrd (1951) patented the pasteurization process of holding crab meat for 1 minute at 171–210°F (77.2–98.9°C) depending on the desired shelf life (1–12 months). Tatro (1970) published a guideline for heating the meat in hermetically sealed cans in a water bath between 190 and 192°F (87.8–88.9°C) until the temperature at the coldest point reached 185°F (85°C) and holding for 1 minute.

Lerke and Farber (1971) reported growth and toxin production by *Clostridium botulinum* Type E spores in Dungeness crabmeat and shrimp. The inoculated meat was pasteurized for 5 minutes at 180°F (82.2°C) and held at 40°F (4.4°C) for 30–40 days before toxin developed. Cockey and Tatro (1974) concluded that a process of 185°F (85°C) for 1 minute in the center of a 1-pound can (401 × flat) is sufficient to reduce inoculated Type E spore levels of $1 \times 10^8/100$ g to <6/100 g and the pasteurized meat remained nontoxic for 6 months at 40°F (4.4°C). Duersch et al. (1981) recommended increasing the time at 185°F (85°C) to 3 minutes to provide for a 12 D cook based on the thermal death time studies of Type E *C. botulinum* reported by Lynt et al. (1977). Table 3.8 shows the thermal resistance characteristics of three strains of type E *C. botulinum*. Ward et al. (1982) recommended a pasteurization process based on an F 16/185 value of 31. This ensures that all crabmeat, regardless of size or shape of the container, receives a similar minimum process in terms of total lethality. Ward further emphasized the importance of the cooling part of the pasteurization cycle. Tatro (1970) recommended cooling the cans in an icewater bath until the meat temperature reached 100°F (37.8°C) in about 50 minutes and then placing the cans in refrigerated storage at 36°F (2.2°C). However, many hours of refrigeration are required to further reduce meat temperatures to safe levels below 38°F (3.3°C). Pasteurized meat is not sterile and during cooling time it is quite possible surviving microorganisms will start to grow and thus reduce the expected shelf

Table 3.8   Thermal Resistance of *Clostridium Botulinum* Type E Spores[a]

Temperature (°F)	Strain					
	Beluga		Alaska		G21-5	
	D	Z(15.0)[a]	D	Z(12.6)[a]	D	Z (15.2)[a]
165 (73.8°C)	12.97		10.39		6.8	
170 (76.6°C)	4.07		3.04		2.38	
175 (79.4°C)	1.65		1.35		1.10	
180 (82.2°C)	0.74		0.51		0.63	
185 (85°C)	0.29					

[a]Z value based on thermal death time in crabmeat.

life of the product. Ward recommended cooling the meat in an icewater bath to an internal meat temperature of 55°F (12.7°C) within 180 minutes of heat processing and then storing at 35°F (1.6°C) (Ward et al. 1982).

Pasteurized crabmeat is not sterile and each year some level of spoilage does occur. There are several conditions known to contribute to this. Should storage temperatures rise much above 38°F (3.3°C), even for short periods, viable bacteria may grow and bacterial spores may germinate and commence growth that leads to spoilage. Leakage from can defects allows entry of cooling water contaminated with psychrotrophic bacteria that may grow in the meat during refrigerated storage. Underprocessing due to operator negligence or equipment failures may produce a product with high levels of surviving bacteria. Finally, another contributing factor is poor meat quality going into the cans. If the meat contains high levels of bacteria originally, the chance of potential spoilage bacteria surviving pasteurization is greatly enhanced.

Pace et al. (1987, 1989) examined a number of cans of spoiled pasteurized crabmeat, both hand and machine picked. They reported several differences in the two spoiled products. The machine picked meat produced only CO_2 on spoilage whereas the hand picked meat produced CO_2 and sometimes H_2S as well. The pH of the hand picked spoiled meat was lower (6.69) than that of the machine picked spoiled product (7.38). The facultatively anaerobic counts in both spoiled products were similar (1.5×10^8/g, hand and 6.2×10^7, machine) whereas the aerobic count was much higher in hand picked meat (2.6×10^8/g) than machine picked (3.8×10^2/g). The different characteristics of spoilage of the two products may be due to the higher vacuum produced in cans of machine picked meat. Machine picked meat is between 90 and 100°F (32.2–37.8°C) when sealed compared to 60–70°F (15.6–21.1°C) for the hand picked product. The higher initial temperature results in increased vacuum on cooling, thus placing greater stress on can integrity enhancing the danger of leakage if seaming faults exist. Ninety percent of the cans exhibited seam abnormalities. Seam thickness was the value most often out of the acceptable range followed in frequency by the body hook. The authors concluded that leakage during the cooling cycle of the pasteurization process definitely played a role in the spoilage although other factors of initial meat quality, underprocessing, and temperature abuse could not be ruled out.

Machine Picking

Commercial machine picking of the body meat of blue crabs started on a trial basis in 1978. The Quik-Pik is the first machine of several types tested to successfully extract the body meat of blue crabs. At that time there were three machines operating in Maryland. Since then the number of machines has increased to 10. During the last 10 years much has been learned about the necessary management of the process needed to consistently produce a bacteriologically acceptable product. Whole cooked crabs to be machine picked are loaded into a round, rotating, slotted cage (Fig. 3.1). As the cage rotates the legs, fins and claws are broken off and drop through the slots. The rods are spaced so the legs and fins are separated from the claws which are saved for picking claw meat. The bodies then proceed through the debacking machine which removes the carapace, trims the ends of the cores, and, finally, the gills and viscera are removed by rotating brushes. The remaining cores are washed by jets of tap water and deposited onto a short conveyor belt which delivers them to handlers who place them open side down on special racks, 20 cores per rack. The loaded racks are then passed through a steam tunnel where they are heated for 1.5–2 minutes to a temperature range of 130–150°F (54.4–65.5°C). Heating is necessary to facilitate machine removal of the meat and to prevent shattering of the shell, because cold chitin is too brittle to withstand the stress during machine picking. As the racks come from the steam tunnel they are placed in the shaker (Quik-Pik Machine, Fig. 3.1) which vigorously vibrates throwing the meat down on a conveyor belt. As the belt moves to the packing table the meat is thoroughly inspected and any shell present is removed by hand. A doctor blade scrapes the meat from the belt onto the packing table where it is placed in containers and either packed in ice or further processed as pasteurized meat.

Machine picking requires constant attention to cleanliness and sanitation to produce a product that meets the bacterial guideline of 1×10^5 APC/g and 36 $E.$ $coli$ MPN/100 g. The two points in the machine picking process that have the greatest impact on bacterial quality are the shaker machine and the meat conveyor belt. It is essential that the machine be disassembled and thoroughly cleaned and sanitized at the end of the day and allowed to dry completely overnight. Liquid on the machine during operation creates an aerosol that can heavily contaminate the meat as it drops to the belt (Cockey 1980). The meat conveyor or inspection belt is continually washed with a tap water spray and sanitized in a chlorine bath (200 ppm or above) as it returns to the shaker after the meat has been removed. The bath solution should be changed at least hourly to ensure an adequate level of active chlorine for sanitation. A comparison of the bacterial levels of machine to hand picked meat is shown in Table 3.9 (Cockey and Chai 1989). The APC for machine picked meat was lower than for the hand product both years reported as were the levels of coagulase-positive $S.$ $aureus.$ The values reported are averages and the high $E.$ $coli$ values for the machine product were occasioned by a particularly severe contamination each year in one plant. In 1987 $E.$ $coli$ had become established in an area on the shaking machine and was overlooked during routine clean-up and sanitation. As the machine vibrated during shaking an aerosol developed, heavily inoculating the meat and conveyor belt below. Once the source was found and proper clean-up procedures instituted the meat $E.$ $coli$ levels dropped to <30 MPN/100 g. A similar situation was found in 1988. $E.$ $coli$ had become established in the tiny crevice between

Table 3.9 Comparison of Bacterial Quality of Machine Picked and Hand Picked Crabmeat[a]

Year	Hand Picked				Machine Picked			
	No. Samples	APC/g	EC/100	SA/100 g	No. Samples	APC/g	EC/100 g	SA/100 g
1987	146	1.4×10^5	33.4	10,000	139	5.8×10^4	153	3,402
1988	68	7.4×10^4	45	502	56	4.2×10^4	405	117

[a]All values shown are the arithmetic average. SA, Coagulase positive *Staphylococcus aureus*, MPN/100 g. Values for machine picked meat were determined just prior to pasteurization.

the doctor blade and the frame to which it was attached, thus inoculating the meat during removal to the packing table.

SUMMARY

Low bacterial level is of primary importance to protect the processor from losses due to spoilage and ensure the safety of the product for the consumer. During processing crabmeat is exposed to a wide variety of microorganisms from intrinsic and extrinsic sources, making control of bacterial quality extremely difficult. Many crab products such as fresh and pasteurized meat may be consumed without further cooking and the bacterial content of the product cannot be overemphasized. Rigorously applied sanitation programs, carefully managed processes that preclude cross contamination, and closely monitored refrigerated storage are essential. Generally the industry does a good job, providing safe wholesome products that are available throughout the year. This is particularly true when you consider that often the plants are old and were built with little consideration or knowledge of good sanitation and manufacturing practices that protect the bacterial quality of the product.

There is a need for improvement in inspection systems. Seafood inspection is addressed by Spencer Garrett in another chapter and will not be discussed at length here; however the inspection programs currently under development will ultimately benefit industry greatly, not only from improved product quality assurance but through consumer confidence as well.

REFERENCES

Alford, J. A., Tobin, L., and McClesky, C. S. 1942. Bacterial spoilage of iced fresh crabmeat. *Food Res.* **7**:353–359.

Allen, E., and Woodburn, M. 1972. Bacteriological quality and occurrence of *Vibrios* in Dungeness crabmeat in Oregon processing plants and markets. *J. Milk Food Technol.* **35**:540–543.

Anzulovic, J. V., and Reedy, R. J. 1942. Pasteurization of crabmeat. *Fish. Market News* **4**:3–6.

Bernarde, M. A. 1958. Comparison of tap and distilled water antibiotic dip solutions on storage life of fresh crabmeat. In: *Antibiotic Annual 1957–1958*. Medical Encyclopedia, Inc., New York, pp. 224–228.

Bernarde, M. A., and Littleford, R. A. 1957. Antibiotic treatment of crab and oyster meats. *Appl. Microbiol.* **5**:368–372.

Blake, P., Allegra, D. T., Snyder, J. D., Barrett, T. J., McFarland, L., Caraway, C. T., Feeley, J. C., Craig, J. P., Lee, J. V., Puhr, N. D., and Feldman, R. A. 1980. Cholora—a possible endemic focus in the United States. *N. Engl. J. Med.* **302**:305.

Byrd, G. C. 1951. Method of keeping the meat of shellfish in a fresh condition. Patent No. 2,546,428.

Chai, T. 1970. Studies on the bacterial flora of fish pen slime. MS Thesis, pp. 154. University of Massachusetts, Amherst, Massachusetts.

Chai, T. 1979. *Fishery Bacteriology*. Joint Commission Rural Reconstruction. Taipei, Taiwan.

Chai, T. 1981. Usefulness of electrophoretic pattern of cell envelope protein as a taxonomic tool for fish hold slime *Moraxella* species. *Appl. Environ. Microbiol.* **42**:351–356.

Chai, T., Chen, C., Rosen, A., and Levin, R. E. 1968. Detection and incidence of specific species of spoilage bacteria on fish. II. Relative incidence of *Pseudomonas putrefaciens* and fluorescent pseudomonads on haddock fillets. *Appl. Microbiol.* **16**:1738–1741.

Chen, H. C. and Chai, T. 1982. Microflora of drainage from ice in fishing vessel fish holds. *Appl. Environ. Microbiol.* **43**:1360–1365.

Cockey, R. R. 1980. Bacteriological assessment of machine picked meat of the blue crab. *J. Food Protect.* **43**:172:–174.

Cockey, R. R. 1983. Bacteriologic standards for fresh shellfish and crabmeat. University of Maryland Sea Grant Publication No. UM-SG-MAP-83-01.

Cockey, R. R. and Chai, T. 1989. Quik-Pik machine processing of crabmeat and quality control. In: *Proceedings of Interstate Seafood Seminars*. October, 1989. Ocean City, Maryland.

Cockey, R.R. and Tatro, M. C. 1974. Survival studies with spores of *Clostridium bolutinum* type E in pasteurized meat of the blue crab *Callinectes sapidus*. *Appl. Microbiol.* **27**:629–633.

Colwell, R. R. 1984. *Vibrios in the Environment*. John Wiley & Sons, New York.

Colwell, R. R. and Kaper, J. 1978. *Vibrio cholorae, Vibrio parahaemolyticus* and other vibrios: occurrence and distribution in Chesapeake Bay. *Science* **198**:394.

Craig, J. M., Hayes, S., and Pilcher, K. S. 1968. Incidence of *Clostridium botulinum* type E in salmon and other marine fish in the Pacific northwest. *Appl. Microbiol.* **16**:553–557.

Dadisman, T. A., Jr., Nelson, R., Molenda, J. R., and Garber, H. J. 1973. *Vibrio parahaemolyticus* gastroenteritis in Maryland. I. Clinical and epidemiological aspects. *Am. J. Epidemiol.* **96**:414–426.

Dassow, J. A., and Learson, R. J. 1963. The crab and lobster fisheries. In: M. E. Stansby (ed.) *Industrial Fisheries Technology*. Robert E. Krieger, New York, pp. 188–195.

Duersch, J. W., Paparella, M. W., and Cockey, R. R. 1981. Processing Recommendations for Pasteurizing Meat from the Blue Crab. Marine Products Laboratory, University of Maryland. UM-SG-MAP-81-02.

Dunker, C. F., Ulmer, D., Jr., and Wharton, G. W. 1959. Factors affecting yield of meat from the blue crab. In: *Proceedings of the Gulf and Caribbean Fisheries Institute*, November, 1959, pp. 40–46.

Faghri, M. A., Pennington, C. L., Cronholm, L. S., and Atlas, R. M. 1984. Bacteria associated with crabs from cold waters with emphasis on the occurrence of potential human pathogens. *Appl. Environ. Microbiol.* **47**:1054–1061.

Fellers, C. R. 1927. Non-gaseous spoilage in canned marine products. *U. Wash. Fish.* **1**:229–238.

Fieger, E. A., and Novak, A. F. 1961. Microbiology of shellfish deterioration. In: Borgstrom, G. (ed.) *Fish as Food, Vol. 11* Academic Press, New York, pp. 561–611.

Fishbein, F., Mehlman, I. J., and Pitcher, J. 1970. Isolation of *Vibrio parahaemolyticus* from the processed meat of Chesapeake Bay blue crabs. *Appl. Microbiol.* **20**:176–178.

Goresline, H. E., and Smart, H. F. 1942. Microbiological studies on the freezing and storage of soft shell crabs. *J. Bacteriol.* **43**:43–44.

Hackney, C. R. and Dicharry, A., 1988. Seafood borne bacterial pathogens of marine origin. *Food Technol.* **42**:104–109.

Harris, M. M. 1932. A bacteriological study of decomposing crabs and crabmeat. *Am. J. Hygiene* **15**:260–265.

Kaneko, T., and Colwell, R. R. 1973. Ecology and *Vibrio parahaemolyticus* in Chesapeake Bay. *J. Bacteriol.* **113**:24–32.

Kautter, D. A., Lilly, T., LeBlanc, A. J., and Lynt, R. K. 1974. Incidence of *Clostridium botulinum* in crabmeat from the blue crab. *Appl. Microbiol.* **28**:722.

Krantz, G., Colwell, R. R., and Lovelace, E. 1969. *Vibrio parahaemolyticus* from the blue crab *Callinectes sapidus. Chesapeake Bay Sci.* **164**:1286–1287.

Lee, J. S., and Pfeifer, D. K. 1975. Microbiological characteristics of Dungeness crab *(Cancer magister). Appl. Microbiol.* **30**:72–78.

Lerke, P., and Farber, L. 1971. Heat pasteurization of crab and shrimp from the Pacific coast of the United States: Public health aspects. *J. Food Sci.* **36**:277–279.

Lilly, T. Jr., Harmon, S. M., Kautter, D. A., Solomon, H. M., and Lynt, R. K. 1971. An improved medium for the detection of *Clostridium botulinum* type E. *J. Milk Food Technol.* **34**:492–497.

Littleford, R. A. 1957a. Retort cooking of blue crabs. Bulletin No. 1. University of Maryland Seafood Processing Laboratory, Crisfield, Maryland.

Littleford, R. A. 1957b. Studies on pasteurization of crabmeat. Bulletin No. 2. University of Maryland Seafood Processing Laboratory, Crisfield, Maryland.

Lynt, R. K., Solomon, H. M., Lilly, T., Jr., and Kautter, D. A. 1977. Thermal death time of *Clostridium botulinum* type E in meat of the blue crab. *J. Food Sci.* **42**:1022–1025.

Maryland State Department of Health. 1957. Regulations governing crabmeat. Maryland State Department of Health. Baltimore, Maryland.

Miller, T. M., Webb, N. B., and Thomas, F. B. 1974. Technical operations manual for the blue crab industry. University of North Carolina Sea Grant No. UNC-SG-74-12, Special Scientific Report No. 28.

National Advisory Committee on Microbiological Criteria for Foods. 1989. *Food Chem. News* July 10.

National Marine Fisheries Service 1980–1989. Fisheries of the United States 1979–1988. U.S. Department of Commerce, pp. 32–33.

Pace, J., Cockey, R., and Chai, T. 1987. Sources of spoilage of pasteurized crabmeat. In: *Proceedings of Interstate Seafood Seminars,* October 1987. Virginia Beach, Virginia.

Pace, J., Cockey, R. R., and Chai, T. 1989. Microbiological flora of machine and hand picked crabmeat. *J. Food Sci.* (submitted).

Paparella, M. W. 1978. Information Tips 78-4. University of Maryland, Marine Products Laboratory. Crisfield, Maryland.

Peixotto, S. S., Finne, G., Hanna, M. O., and Vandergant, C. 1979. Presence, growth and survival of *Yersinia enterocolitica* in oysters, shrimp and crab. *J. Food Protect.* **42**:974–981.

Phillips, F. A., and Peeler, J. T. 1972. Bacteriological survey of the blue crab industry. *Appl. Microbiol.* **24**:958–966.

Puncochar, J. F., and Pottinger, S. R. 1938. Commercial production of meat from the blue crab *(Callinectes sapidus).* U.S.F.W.S. Comm. Fish. T.L. 8.

Ray, B., Webb, N. B., and Speck, M. L. 1976. Microbiological evaluation of blue crab processing operations. *J. Food Sci.* **41**:398–402.

Sizemore, R. K., Colwell, R. R., Tubiash, H. S., and Lovelace, T. E. 1975. Bacterial flora of the hemolymph of the blue crab, *Callinectes sapidus:* numerical taxonomy. *Appl. Environ. Microbiol.* **29**:393–399.

Tatro, M. C. 1970. Guidelines for pasteurizing meat from the blue crab *(Callinectes sapidus)*. I. Water bath method. Contribution No. 419. Natural Resources Institute. University of Maryland, Crisfield, Maryland.

Tobin, L. C., and McCleskey, C. S. 1941. Bacteriological studies of fresh crabmeat. *Food Res.* **6:**157–167.

Tubiash, H. S., Sizemore, R. K., and Colwell, R. R. 1975. Bacterial flora of the hemolymph of the blue crab, *Callinectes sapidus:* most probable numbers. *Appl. Microbiol.* **29:**388–392.

Ulmer, D. H. B., Jr. 1964. Preparation of chilled meat from Atlantic blue crab. *Fish. Indust. Res.* **2:**21–45.

Ward, D. R., Flick, G. J., Hebard, C. E., and Townsend, P. E. 1982. Thermal processing pasteurization manual. Department of Food Science and Technology. VPI State University, Blacksburg, Virginia. VPI-SG-82-04.

Weagant, S. W., Sado, P. N., Colburn, K. G., Torkelson, J. D., Stanley, F. A., Krane, M. K., Shields, S. C., and Thayer, C. F. 1988. The incidence of *Listeria* species in frozen seafood products. *J. Food Protect.* **51:**655–657.

Webb, N. B., Stokes, S. J., Thomas, F. B., Moncol, N. B., and Hardy, E. R. 1973. Effect of sanitation procedures on bacterial levels in blue crab processing plants. *Proceedings of the Gulf and Caribbean Fisheries Institute,* May 1973, pp. 109–114.

Wentz, B. A., Duran, A. P., Swartzentruber, A., Schwab, A. H., and Read, R. B., Jr. 1983. Microbiological quality of fresh blue crab meat, clams and oysters. *J. Food Protect.* 46:978–981.

Wentz, B. A., Duran, A. P., Swartzentruber, A., Schwab, A. H., McClure, F. D., Archer, D., and Read, R. B., Jr. 1985. Microbiology of crabmeat during processing. *J. Food Protect.* **48:**44–49.

Williams-Walls, N. J. 1968. *Clostridium botulinum* type F: isolation from crabs. *Science* **162:**375–376.

4

Microbiology of Crustacean Processing: Shrimp, Crawfish, and Prawns

Russell J. Miget

Scientific interest in naturally occurring marine and freshwater microorganisms dates back to the decades following the discoveries of Louis Pasteur and his contemporaries in the 1850s. However, it was not until the turn of the century, when consumption of molluscan shellfish was associated with numerous outbreaks of cholera, that interest was generated in determining the relationship between the occurrence of enteropathogenic bacteria in seafood products and human health. When subsequent studies showed that the naturally occurring microflora in fish and shellfish from unpolluted waters were significantly different from those associated with warm-blooded animals, attention shifted once again to the public health problems associated with consumption of raw molluscan shellfish (Liston 1980).

However, because seafood products are highly perishable (losing much of their value even prior to spoilage as a result of off odors and flavors) there has been a continuing interest in the mechanisms of spoilage, and subsequent methods to increase shelf life. A review of work on seafood spoilage through the mid-1940s is presented by Reay and Shewan (1949), while more recent data are summarized by Shewan (1977) and Liston (1980).

The majority of literature on the microbiology of seafood products is concerned with marine finfish species, because these constitute the majority of commercially important products (landings of 80 million metric tons of finfish vs. 4 million metric tons for *all* crustaceans worldwide (Fisheries of the United States 1988). However, many similarities exist between the spoilage of finfish and spoilage of crustacean shellfish which is the topic of this chapter.

Organizationally, the topics to be covered will:

- Review investigations concerning the microbial flora on marine shrimp.
- Discuss how the microflora change both quantitatively and qualitatively as shrimp move through the harvest, processing, and distribution systems.
- Describe biochemical/microbial changes that lead to loss of freshness and ultimately spoilage.
- Identify microorganisms of public health significance associated with the spoilage of crustacean shellfish.
- Conclude with a section on retail handling procedures designed to minimize loss of freshness and prolong shelf life.

NATURALLY OCCURRING MICROFLORA

It has been concluded by numerous investigators that the microflora associated with live freshwater and marine crustaceans reflect the microbial population of the waters in which they are living (Cobb and Vanderzant 1971; Liston 1980; Shewan 1977). Table 4.1 summarizes the qualitative data describing the indigenous microflora of saltwater shrimp.

Differences in "normal" microflora for various shrimp studied are due to differences in the microflora of surrounding waters and sediment, species differences, and postharvest handling procedures, as well as plating media and incubation temperature.

In some instances classification changes have altered dominant species composition (e.g., several *Moraxella* and *Acinetobacter* species were formerly identified as *Achromobacter*). Similarly, Alvarez and Koburger (1979a) suggested that several flagellated cocci responsible for shrimp spoilage should be placed in the genus *Planococcus* rather than *Micrococcus*. Cobb et al. (1976) reported that when 0.5% NaCl was added to the plating media *Vibrio* species predominated, whereas in an earlier similar study (Vanderzant et al. 1970) using standard methods agar (without NaCl) cornyeforms dominated, along with *Pseudomonas*, *Moraxella*, and *Micrococcus*. They further suggested that

Table 4.1 Dominant Bacterial Genera in Fresh Marine Shrimp

Shrimp Genus/Location	Dominant Bacterial Genera	Reference
Penaeus/Gulf of Mexico	*Achromobacter, Bacillus, Micrococcus, Flavobacterium Pseudomonas*	Campbell and Williams (1952)
Penaeus/Gulf of Mexico	Coryneforms, *Pseudomonas Moraxella, Micrococcus*	Vanderzant et al. (1970)
Penaeus/Gulf of Mexico	*Vibrio, Pseudomonas, Moraxella, Acinetobacter*	Cobb et al. (1976)
Penaeus/Texas Ponds	*Aeromonas, Pseudomonas, Vibrio*	Vanderzant et al. (1973)
Sicyonia/Gulf of Mexico	*Flavobacterium, Cytophaga*	Koburger et al. (1975)
Pandalus/Pacific	*Arthrobacter, Micrococcus, Bacillus, Pseudomonas*	Lee and Kolbe (1982)
Pandalus/Pacific	*Moraxella, Acinetobacter, Flavobacterium, Cytophaga*	Lee and Kolbe (1977)
Metapenaeus/?	Coryneforms, *Miccrococcus*	Liston (1980)
Parapeneopsis/?	*Micrococcus, Achromobacter*	Liston (1980)
Pandalus/?	*Achromobacter, Pseudomonas,* Coryneforms	Liston (1980)
Pandalus/?	*Acinetobacter, Moraxella, Flavobacterium, Pseudomonas*	Harrison and Lee (1968) in Liston (1980)

changes in the nutritional properties (and/or salinity) on the surface of shrimp, or interactive microbial activities, are involved in the changes of the microbial flora of shrimp during iced storage. These activities are in addition to the emerging psychrotrophic population outcompeting mesophiles for available nutrients.

Most investigators (Cann 1977; Liston 1980; Shewan 1977) concluded that marine shrimp from warm waters, such as finfish, carry a microbial population composed primarily of Gram-positive bacteria, such as *Micrococcus,* coryneforms, and *Bacillus,* whereas cold-water species harbor predominately Gram-negative microbes, including *Moraxella, Acinetobacter, Pseudomonas, Flavobacterium,* and *Vibrio* (Cann 1974, 1977).

Quantitatively, warm-water marine shrimp often show total aerobic counts of 10^6/g when captured (Vanderzant et al. 1970) whereas cold-water Pandalid species are often several orders of magnitude lower, in the range of $10^2–10^3$/g (Zapatka and Bartolomeo 1973).

Studies concerning the microflora on prawns and crawfish have concentrated more on bacteria of public health significance than on indigenous populations that result in spoilage. This is no doubt due to the fact that these organisms, whether grown in ponds or captured in the wild, are more likely to harbor potentially harmful microorganisms due to their proximity to surface run off into their freshwater habitat. Thus, these populations will be described in the section dealing with microorganisms of public health significance.

YEASTS, MOLDS, FUNGI, AND VIRUSES

Liston's review of the literature (1980) concluded that there was limited information on the occurrence of these microorganisms on fish and shellfish. In a summary table he lists the following yeasts as most abundant on fish and shrimp: *Rhodotorula, Candida,* and *Torulopsis.* He also cites work (Koburger et al. 1975; Phaff et al. 1952) describing the isolation of molds of the genus *Pulluleria (Aureobasidium)* from fresh marine shrimp, as well as chitinoclastic fungi growing on crab shells. Although viral diseases of marine shrimp significantly affect aquaculture production, none of these have been implicated in spoilage or human health concerns.

MICROBIOLOGICAL CHANGES THROUGH THE DISTRIBUTION SYSTEM

Wild Caught Warm-Water Species

Shrimp landed offshore are generally deheaded at sea. One of the justifications for this practice is based on early reports by Green (1949) and Carroll et al. (1968) that indicated that removal of the heads reduced the bacterial load on the shrimp by 50–80%. However, heading also exposes shrimp tail muscle to bacteria from its own digestive system, shrimper's hands, and all contact surfaces on the boat including the deck, ice bins, and so on.

Although delayed heading results in a softening and discoloration of the cephalothorax, Koburger et al. (1973) and Alvarez and Koburger (1979b) found no significant difference (sensory and bacterial counts) in the edible portion of headed and whole shrimp stored on ice for up to 14 days.

In a study of whole and headed rock shrimp *(Sicyonia brevirostris)* Bieler et al. (1973) showed that shrimp stored with heads on actually had lower total bacterial counts, maintained higher solids contents, and had more favorable organoleptic acceptability than did headed shrimp. Therefore, it would seem that heading shrimp at sea is perhaps as much an economic practice as it is a quality maintenance procedure. However, as will be discussed later, these studies only reported on "spoilage," not "freshness" indicators.

Numerous investigators have shown that adequate washing at sea prior to icing of headed shrimp is quite important in maintaining quality. Because the tail muscle is exposed on headed shrimp, and because the process of heading by hand contaminates the exposed flesh, it is critical that the newly headed shrimp be thoroughly washed to remove as much of the bacterial load as possible prior to iced or frozen storage.

As early as 1949, Green (1949) reported the average bacterial loads on shrimp harvested at sea as follows: (1) fresh caught head-on, 42,000/g; (2) fresh caught tails, 10,000/g; (3) fresh caught tails washed, 7,400/g. A one log reduction in similar tests where the shrimp were thoroughly washed of adhering slime and debris.

Novak (1973) reported that shrimp harvested offshore Louisiana in the Gulf of Mexico averaged 4.2×10^4 bacteria/g whereas nearshore samples averaged 2.14×10^5/g and bay shrimp 1.2×10^6/g. Following heading by commercial fishermen on the boats, total counts were reduced by 75%. In this study, washing headed shrimp reduced the total aerobic count by another 25%. In a similar study Carroll et al. (1968) reported that thorough washing reduced the bacterial load up to 75%, whereas Green (1949) showed the slime adhering to unwashed shrimp to harbor 2.1×10^5 bacteria/g. Nickelson and Vanderzant (1976), reporting unpublished data from Nickelson and Graham, stated that proper washing of shrimp before iced or frozen storage was the single most important factor in affecting the bacterial numbers on shrimp arriving at the dock. Thorough washing is especially critical after shrimp are headed and the tail meat has been exposed to direct contamination from hands, deck surfaces, baskets, culling tools, etc.

Although the intestine sometimes comes out when the head is removed, more often than not the rich bacterial flora in the gut comes into direct contact with the exposed tail meat, usually from the header's hands or gloves.

Unlike the bacterial flora on the shell and gills, which are representative of the microflora of the environment, the bacteria found in the digestive system may differ considerably due to selection for species capable of growth at a lower pH and oxygen tension. Hood and Meyers (1973), for example, reported that the microbiota of the digestive tract in *Penaeus setiferus* was significantly different both quantitatively and qualitatively from that of surrounding water and sediment (Table 4.2).

Nickelson and Vanderzant (1976) summarized the changes in microbial properties on shrimp harvested in the Gulf of Mexico by citing data from Campbell and Williams (1952) as representative of work by various researchers. Figure 4.1 represents a typical bacterial population change on ice stored shrimp with time.

The initial decrease is due to the decline of mesophilic species at cold (melting ice) temperatures. After 4 days the population begins to increase due to outgrowth of psychrotrophic microorganisms. For examples, Campbell and Williams (1952) showed a shift of more than half of the original species made up of *Bacillus, Micrococcus,* and *Flavobacterium* to *Achromobacter* and *Pseudomonas* (total 98%) after 16 days of storage on ice. Cook (1970) showed no consistent changes in bacterial counts during the initial decrease on headless

Table 4.2 Characteristics of Bacteria Associated with the Digestive Tract of *P. setiferus* and Associated Water and Sediment Habitat

		P.setiferus $(2.9 \times 10^7/g)$	*Sediments* $(2.9 \times 10^6/g)$	*Waters* $(1.5 \times 10^5/ml)$
Characteristic Genera of Bacteria Present	Vibrio Pseudomonas Beneckea		Bacillus Pseudomonas Vibrio Aeromonas Others	Pseudomonas Flavobacterium Chromobacterium Micrococcus Aeromonas Alginomonas Vibrio Others
Percentage of Total Bacteria Showing Enzymatic Character- istics	Proteolytic Amylolytic Lipolytic Cellulolytic Chitinolytic	60 60 40 0 85–100	19 2 1 1 10	— — — 1 10
Generation Time at 22°C in Peptone Seawater Medium[a]	30 minutes	2 hours	—	
Minimum pH for Growth[1]		5.0	6.0	—

[a]Average for three selected species.

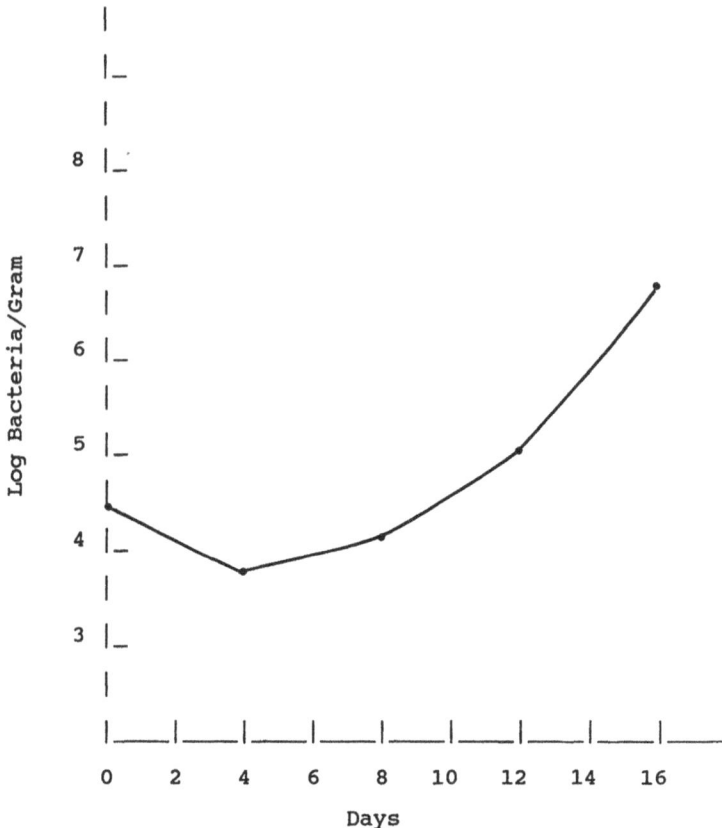

Figure 4.1   Bacterial population change on ice-stored shrimp.

brown shrimp *(Penaeus aztecus)*, but an increase of *Pseudomonas* species to 80–100% with increased time on ice.

Freezer Boat and Refrigerated Seawater System Shrimp

Recently, some shrimp boats in the Gulf and South Atlantic have installed freezers on board to both improve the quality of landed shrimp as well as allow for extended time at sea. Headed shrimp are first immersed in a brine solution to rapidly freeze them, then they are moved to a freezer hold for storage. Nickelson and Vanderzant (1976) report as much as an 85% reduction in bacterial load as a result of freezing on board the vessel, due in large part to the cold sensitivity of the mesophilic populations on warm-water shrimp.

Cold-water Pandalid shrimp are often stored whole in refrigerated seawater to maintain quality until offloaded at the processing plant. This practice eliminates the dependence on ice and reduces the labor coast to mix shrimp and ice. These small shrimp are then mechanically peeled at a shoreside processing plant. Thus, there is not the hand labor cost incentive to head at sea that is associated with the larger warm-water Penaeid species.

However, as indicated by Lee and Kolbe (1982), this system has the potential for producing a poor-quality shrimp for the following reasons:

1. Exposing the entire content of the shrimp hold to a common environment. Any "hot spot" or poorly refrigerated section of the hold will affect the entire catch.
2. New catches of shrimp added to the hold create temperature fluctuation.
3. An inadequately cleaned hold or plumbing associated with the recirculating seawater system can immediately contaminate the catch, as can inadequately cleaned shrimp.

The authors further reported that species identification of bacterial outgrowth in a "model refrigerated seawater system" installed on a fishing vessel showed a steady increase in *Pseudomonas* sp. population. During a 68-hour holding period the proportion of *Pseudomonas* sp. increased from 12 to 80% of the total. Thus, despite the fact that the total microbial population of the shrimp increased from 230/g to only 1,000/g after 68 hours, the potential for spoilage was much greater due to the shift in microbial species in the system. Exchanging water in a refrigerated seawater system is usually not practiced due to fluctuating temperatures or the costs of oversized refrigeration units to compensate for temperature fluctuations.

MICROORGANISMS ASSOCIATED WITH SEAFOOD SPOILAGE

Freshness

Although spoilage of shrimp, prawns, and crawfish is ultimately due to bacterial action, loss of freshness preceding spoilage primarily involves autolytic reactions controlled by naturally occurring enzymes in muscle tissue. Nucleotide catabolism in Penaeid shrimp has been correlated with loss of freshness by Cheuk et al. (1979), who compared adenosine deaminase and adenosine monophosphate deaminase activity with traditional spoilage parameters such as total volatile nitrogen, total plate count, and sensory evaluation. They concluded that activity of these enzymes gave a good indication of postharvest storage time and temperature. Other investigators (Fatima et al. 1981; Flick and Lovell 1972; Stone

1971) have reported similar correlations using other shrimp species. The sequence of nucleotide degradation has been determined to follow the reaction shown in Figure 4.2 (Flick and Lovell 1970). Whereas the first five reactions proceed relatively fast as a result of the actions of endogeneous enzymes in muscle tissue, the oxidation of hypoxanthine to xanthine and finally to uric acid is much slower and considered to be due primarily to bacterial enzymes (Kennish and Kramer 1987). Ehira and Uchiyama (1987) present an excellent review of the "K" value (based on a ratio of nucleotide catabolic products) which is widely used in Japan to estimate freshness of fish. They conclude that many studies on methods to estimate freshness of seafood have been incorrectly based on the concept that bacterial action causes freshness to deteriorate, whereas in fact freshness declines long before the onset of significant bacterial action. Least the concept be lost in semantics they suggest that seafood quality be defined using the terms biochemical freshness or enzymatic freshness versus bacterial freshness and/or spoilage.

Spoilage

With this distinction in mind the following describes the process of bacterial spoilage in crustacean shellfish. Once an animal dies, its body defenses cease to operate. In the case of finfish, it has been shown that bacteria present in the gills, gut, and skin begin to metabolize surrounding low-molecular-weight compounds, producing the volatile sulfur compounds associated with spoilage. Under chill storage conditions this is a surface phenomenon. However, when products are temperature abused, bacteria have been shown to significantly invade sterile muscle tissue through the lateral line pores as well as the gut, resulting in more rapid spoilage (Shewan and Murray 1979).

Spoilage of crustacean shellfish is also a surface phenomenon at chill temperatures. Crustacean shellfish in general exhibit a reasonably long chill storage shelf life. For example, Penaeid shrimp stored for 14 days on ice remained organoleptically acceptable (Novak 1973). Likewise, Pandalid species have been stored for 11 days (Lee and Kolbe 1982), crawfish for 7 days (Anderson and Grodner 1980), and *Macrobrachium* for 10 days (Angel et al. 1985).

Adenosine

(1)   (2)   (3)／   (4)＼ (5)   (6)   (7)

ATP → ADP → AMP →   Inosine → Hypoanthine → Xanthine → Uric Acid

(3) ＼   (4) ／
IMP

Figure 4.2 Sequence of nucleotide catabolism in penaeid shrimp.

The chilled shelf life of seafood products in general has been related to the following:

1. Initial bacterial flora, i.e., the number and types of spoilage organisms and subsequent contamination as a result of heading and further processing.
2. Certain enzymes that, in some species, continue to operate after death, thus effectively controlling the outgrowth of spoilage bacteria.
3. Low-molecular-weight muscle extractives in shrimp, crawfish, and prawns that are quantitatively and qualitatively different from those in more rapidly spoiling seafood products.
4. The integrity of the animal's "skin," i.e., how well it protects the underlying muscle from invasion by surface bacteria. Crustacean shellfish, with a chitinous exoskeleton, offer one extreme of protection from mechanical damage, whereas thin-skinned mackerel, for example, are easily damaged during harvesting and processing and thus spoil relatively quickly, i.e., within approximately 6 days (Shewan and Murray 1979).

As mentioned previously, numerous studies have indicated that crustacean shellfish from warm waters carry higher bacterial loads of mesophilic, Gram-positive bacteria, whereas cold-water species often have fewer total numbers, but the majority of the flora are Gram-negative including numerous psychrophiles. The dominant organisms associated with shrimp, prawn, and crawfish spoilage are psychrotrophic *Pseudomonas* and *Moraxella/Acinetobacter* species. Thus, it has been suggested (Liston 1980; Nickelson and Vanderzant 1976) that crustaceans harvested from cold waters spoil relatively more quickly than do their warm-water counterparts, because they are "pre-inoculated" with psychrotrophic Gram-negative spoilage bacteria. Such a flora would decrease the lag time before outgrowth and subsequent spoilage under refrigerated or iced storage.

Organoleptic evaluations of Penaeid shrimp (Cobb et al. 1976) and Pandalid shrimp (Matches 1982), both stored on ice, would tend to substantiate this suggestion. Warm-water Penaeids remained organoleptically acceptable for 14 days, whereas cold-water Pandalids spoiled after 11 days on ice. Bacteria isolated from each experiment showed a dramatic increase in the number of psychrotrophic Pseudomonads over the period of iced storage—constituting 92% of the microbial population of Penaeids and 67% of the Pandalids microflora at the termination of the experiments.

In the same study Matches (1982) showed that at higher "cold storage" temperatures [i.e., 42°F (5.6°C), 50°F (10°C), 62°F (16.7°C)] there was a shift away from Pseudomonads as primary spoilage organisms to *Moraxella* sp. At the highest temperature of 71.9°F (22.2°C) (organoleptically acceptable for only 2 days) *Proteus* species dominated the spoilage flora.

Ward and Baj (1988) likewise cited several studies (Disney 1976; Poulter et al. 1981; Summer et al. 1984) that indicate that fish harvested from tropical waters have a substantially increased iced storage shelf live vs. those from cold waters. However, the authors also cite a review of seafood storage by Lima dos Santos (1981) which suggests that experimental variables make comparisons of different studies difficult, and that the broad generalization of longer shefl life for tropical species should be reconsidered.

Cox and Lovell (1973) isolated and identified bacteria from commercially peeled crawfish tails, then determined which of them were actually responsible for crawfish spoilage. Of the 280 original isolates, 22.1% were found to be rapid spoilers, 16.4% were slow spoilers, and 61.5% were nonspoilers. The largest number of rapid spoilers belonged to the genus *Pseudomonas* followed by *Achromobacter* at less than one-half the number of isolates. Genera containing *no* rapid spoilers included *Aerobacter, Bacillus, Flavobacterium, Lactobacillus, Micrococcus, Sarcina,* and *Staphylococcus.* Slow spoilers, in addition to *Pseudomonas* and *Achromobacter,* included *Alcalignes, Flavobacterium, Aerobacter, Lactobacillus, Micrococcus,* and *Staphylococcus.*

In the same study the authors showed that while the fresh crawfish microflora was dominated by *Micrococcus, Staphylococcus,* and *Alcaligenes,* meat stored at 32°F (0°C) and 41°F (5°C) spoiled at 24 and 12 days respectively and were dominated by *Pseudomonas/Achromobacter* [32°F (0°C)] and *Achromobacter/Pseudomonas* [41°F (5°C)]. The data thus indicated that *Pseudomonas* spoilers may be more active than *Achromobacter* species at temperatures approaching 32°F (0°C).

In a study to determine the factors resulting in development of a mushy texture in *Macrobachium* stored on ice Premaratne et al. (1986) isolated an initial total plate count of 1.6×10^6/g which increased ato 8.6×10^7/g after 12 days. Proteolytic counts began to increase after 6 days comprising 22% of the total by day 8, 42% by day 10, and 87% by day 12. A definite off (spoilage) odor was detected after 6 days. Even though proteolytic bacteria eventually penetrated into the muscle, resulting in spoilage odors, it was concluded that these spoilage bacteria were not responsible for texture loss because mushiness occurred after only 2–3 days, whereas significant collagenolytic enzyme production did not occur until after 4–6 days of storage.

Angel et al. (1981), in an assessment of the shelf life of *Macrobrachium rosenbergii* stored in ice, reported that the prawns were judged acceptable by a sensory panel after 10 days of storage, but not after 14 days. Increases in total volatile base nitrogen, changes in proximate composition, and increases in aerobic psychrotrophic bacteria numbers run simultaneously correlated well with the taste panel evaluations.

Indole, a degradation product of tryptophan, has been correlated with the strong odor of decomposition indicative of shrimp spoilage. Smith et al. (1984) determined that production of indole in shrimp is a result of bacterial rather than endogenous enzyme action, and proceeds differently under chilled storage and temperature abuse conditions. Above 50°F (10°C) highly proteolytic, indole-positive organisms such as *Aeromonas* and *Proteus* attack muscle proteins, releasing typtophan which is subsequently converted to indole via bacterial metabolism. Much smaller amounts of indole are produced at chill storage temperatures by nonproteolytic microorganisms such as *Flavobacterium,* only after tryptophan is released via psychrotrophic proteolytic organisms such as *Pseudomonas.* Thus, high indole levels in shrimp are indicative of temperature abuse somewhere in the production/processing chain.

Bottino et al. (1979) showed that bacterial action during 18 days of iced storage did not significantly alter the fatty acid profile of brown shrimp *(Penaeus aztecus),* nor was it altered after 183 days of frozen storage. The relatively low concentration of fat in the tail meat of shrimp ($<1.0g/100$ g of raw meat vs. 2–5 g/100 g for most ocean finfish, to >10 g/100 g for certain mackerels), along with its resistance to oxidation under iced or frozen storage conditions no doubt contribute to the comparatively excellent storage qualities of shrimp.

In summary, the general scheme of spoilage of seafood products including shrimp, prawns, and crawfish is outlined by Liston (1980) and shown below.

1. Spoilage bacteria are naturally present on fish.
2. Amino acid and other non-protein nitrogen (NPN) substrate pools are present.
3. There is selective growth of organisms (mostly *Pseudomonas*) which actively oxidatively deaminate amino acids.
4. Repression of proteinase production is derepressed by selective use of amino acids by *Pseudomonas* bacteria.
5. Amino acid are recruited to substrate pool by bacterial hydrolysis of protein.
6. Ammonia and volatile fatty acid production sharply increases due to 5.
7. Specific "spoiler" types of bacterial produce sulfur-containing and other odorous compounds.

MICROORGANISMS OF PUBLIC HEALTH CONCERN

Illnesses or diseases resulting from consumption of shrimp, prawns, or crawfish are comparatively rare. Ninety percent of all reported illnesses resulting from consumption of seafood are attributable to molluscan shellfish and a few species of finfish, i.e., those associated with ciguatoxin and scombrotoxin (Garrett 1988).

Disease or illness resulting from consumption of shrimp, prawns, or crawfish can generally be attributed to product contamination from one or more of the following three sources.

• Naturally occurring pathogens, in particular *Clostridium botulinum* Type E, *V. cholerae, V. parahaemolyticus,* and *V. vulnificus.*

- Pathogens introduced into the aquatic environment as a result of improper disposal of human waste and/or land runoff. Organisms include *Clostridium perfringens, Staphylococcus, Erysipelothrix, Edwardsiella, Salmonella, Shigella, Franciscella,* and numerous *Vibrio* species (Liston 1980).
- Pathogens introduced via product handlers—from personnel on the harvest boat, through processing plant and food service handlers, and ultimately the consumer. Often postharvest handling introduces coliforms, fecal coliforms including *E. coli, Staphyloccus,* and *Salmonella.*

Pathogenic microorganisms have rarely been isolated from shrimp harvested in offshore waters. Green (1949) reported that although almost half the fresh caught shrimp stored in ice or trawlers contained coliforms, only rarely was *E. coli* confirmed. Most reports on the natural flora of wild caught shrimp verify the general absence of coliforms (Campbell and Williams 1952; Harrison and Lee 1968; Vanderzant et al. 1970). Certain studies noted that trawl caught shrimp sampled at the processing plant contained both coliforms and *E. coli,* perhaps indicating unsanitary conditions on board the boat (Surkiewicz et al. 1967). However, most cases of foodborne illness can be prevented (even if the product is initially contaminated with pathogens) by proper handling and cooking.

In most cases, a large number of food poisoning bacteria must be present to produce sufficient toxin to result in illness. Controlling initial populations, preventing their outgrowth by maintaining low storage temperatures, and proper cooking to destroy vegetative cells and toxins will prevent most foodborne illness.

However, special precautions must be taken with seafood products that are to be consumed raw, or products that are preprocessed (blanched, brined, cooked) and then handled prior to consumption without reheating or cooking.

Cooked shrimp, for example, are potentially harmful because they can become contaminated with pathogenic microorganisms after processing. Because cooking destroys most vegetative spoilage microorganisms, pathogens can proliferate and produce toxins before the microorganisms normally associated with raw products can produce characteristic spoilage odors. In addition, Troller and Frazier (1962a,b) showed that naturally occurring Pseudomonads responsible for spoilage actually inhibit pathogens by outcompeting for available nutrients. Liston (1980) summarized data from several studies related to changes in bacterial numbers and species during shrimp processing. In general the same events occurred; numbers declined dramatically following blanching/cooking of the raw or brined product, only to increase by one or more orders of magnitude during peeling, grading, and packing.

Grodner and Novak (1974) reported on the incidence of microorganisms of public health significance which were isolated from both crawfish and crawfish harvesting waters in South Louisiana. In summary they found that total coliforms, fecal *E. coli,* and fecal streptococci could be expected to be found in crawfish because these were commonly isolated in the harvest waters. However, coagulase-positive staphylococci, *Shigella* sp., *Salmonella* sp., and *Clostridium botulinum*, Type E, were not normally found in either crawfish or water samples.

However, when blanched peeled tail meat from commercial processing plants was tested, 24% of the product samples had total plate counts in excess of 1×10^6 colony-farming units (cfu)/g, 70% were positive for *E. coli*, 30% for coagulase-positive straphylococci, and 10% positive for fecal steptococci. The data thus showed the potential for human contamination of the product as a result of hand peeling.

In a more recent article on changes in bacterial populations throughout the crawfish processing line Anderson and Grodner (1980) assayed for standard plate counts, total coliforms, *E. coli*, and coagulase-positive *S. aureus*. Four commercial crawfish processing plants were surveyed once a month for 3 months. In summary, they found that:

1. Total counts were reduced by approximately 98% (2.83×10^7 cfu/g to 1.22×10^5 cfu/g) during the initial blanching prior to peeling.
2. The product was subsequently contaminated as a result of peeling yielding the following data: Total counts 4.18×10^5 cfu/g; total coliforms 4.72×10^2 cfu/g; *E. coli* 8.8×10^1 cfu/g (range $0.0–1.1 \times 10^3$ cfu/g); and *S. aureus* 2.12×10^3 cfu/g (range $0.0–1.1 \times 10^4$ with >79% samples having $>1.0 \times 10^2$ cfu/g) in packaged tailmeat. They concluded by emphasizing the need for improved sanitation in crawfish processing plants to ensure production of a wholesome and safe crawfish tailmeat.

An FAO report (Anon. 1988) containing abstracts of bacteriological surveys carried out on wild and pond reared shrimp and prawns in several developing countries indicate the need to maintain a quality assurance program during shrimp and prawn processing. Several of the abstracts discussed the contamination of raw and cooked products with *S. aureus* as a result of handing. At one plant in Sri Lanka >65% of the processing line staff carried *S. aureus* on their fingers. Several other reports caution against using animal manure as a source of fertilizer for aquaculture ponds because this practice introduces *Salmonella* which was later isolated from the harvested product.

Zuberi et al. (1983) reported on bacterial changes throughout the processing lines at two shrimp plants in Karachi, Pakistan. The study was an excellent example of how economics can influence bacteriological quality of a food product. One plant (A) processed Jaira shrimp *(Penaeus merguiensis)*, a relatively high value shrimp primarily exported, whereas the other plant (B) processed Kalri shrimp *(Metapenaeus monoceros)*, a smaller shrimp of considerably less value. Bacteriological data from each plant are summarized in Table 4.3.

A further description of the plants noted that in Plant B the processing area was small and equipment was poorly maintained. A foul odor prevailed through-

Table 4.3 Bacterial Data from Two Shrimp Plants in Karachi, Pakistan

Product	Bacterial sp.	Plant A	Plant B
Unwashed raw	APC/g	1.2×10^7	1.2×10^7
Washed raw	APC/g	6.7×10^6	9.3×10^6
Frozen raw	APC/g	9.4×10^6	8.3×10^7
Unwashed raw	Coliform/fecal coliform/g	110/85	94/24
Washed raw	Coliform/fecal coliform/g	88/60	153/124
Frozen raw	Coliform/fecal coliform/g	94/41	180/133

out the plant due to decomposing shrimp on the floor. Windows and doors were unscreened and open, allowing flies and other insects to enter. The shrimp were left at room temperature for several hours prior to processing. Chlorine was not added to tap water used for washing shrimp. Plant A, on the other hand, was larger, better maintained, used plenty of ice on the stored product, and frequently cleaned utensils and contact surfaces. Dominant bacterial species isolated from shrimp in both plants showed a *Pseudomonas/Micrococcus* dominance in the raw product which shifted toward *Pseudomonas* after washing and freezing.

Duran et al. (1983), reporting bacteriological survey results from 33 shrimp breading lines in the U.S., concluded that, in general, the plants produced a wholesome, safe product. A 10-fold decrease in the mean APC occurred during processing—a result of washing and the effect of adding batter and breading (which increased the weight/shrimp value). They cautioned, however, that the batter in some cases increased the coliform count and in one case was the source of *Salmonella* contamination.

Ridley and Slabyj (1978), in a survey of three different shrimp processing plants (brine-cooked shell on, hand-peeled raw, and machine-peeled cooked), concluded that each process reduced the viable bacterial count on the raw product resulting in a bacteriologically acceptable product. However, they cautioned that an increase in bacterial load after the catch and while in storage prior to processing significantly altered the quality of the final product.

EFFECT OF FREEZING ON MICROORGANISMS

Liston (1980) reported that the effect of freezing is to generally reduce viable counts, with the numbers continuing to fall during frozen storage. However, data on the effect of freezing on bacterial populations are often conflicting. Thus, it might be safest to regard the freezing of crustacean shellfish as maintaining the existing bacterial population. Ideal freezing procedures, i.e., rapid freeze and low nonfluctuating temperature, are also most protective of bacteria.

Quality changes during frozen storage of breaded shrimp were noted by Gates et al. (1985), who found that shrimp kept under retail frozen storage conditions [i.e., minimum temperature $\leq-4°F$ ($\leq-20°C$) with daily fluctuations of 53.6–64.4°F (12–18°C) range] showed sensory deterioration after 3–4 months of storage whereas replicate samples stored under warehouse conditions [i.e., below –4°F (–20°C) and daily fluctuations of 35.6–37.4°F (2–3°C)] maintained acceptable sensory quality for up to 13 months of storage. Initial and final microbial counts, however, were approximately the same for all storage treatments, indicating that quality changes under frozen storage were due to factors other than microbial activity.

Novak (1973) reported an initial bacterial reduction of 94% on shrimp blast frozen in 5-pound cartons. The population was further reduced to 99% after 11 months of frozen storage.

Swartzentruber et al. (1980) carried out a bacteriological survey of frozen shrimp and lobster tails pulled from retail shelves throughout the United States. The following summarize their findings: (1) Using the recommended acceptable limit of $10^6/g$ aerobic plate count at 86°F (30°C), 96% of 1,464 cooked shrimp samples, and 80% of 1,300 raw in-shell samples were in compliance, along with 46% of 1,315 lobster tail samples (at $10^5/g$ maximum APC). (2) Using suggested guidelines of 400/g (coliforms), 250/g *(E. coli),* and 100/g *(S. aureus),* 99+% of all four products were within acceptable levels except for raw in-shell shrimp, in which 97% of the samples met the coliform guideline.

In a similar survey of frozen raw and cooked shrimp collected in Singapore, Singh et al. (1987) reported the following based on the International Commission on Microbial Specification for Foods standards (ICMSF 1986): Shell-on prawns (SP) 98% APC acceptable; peeled and deveined prawns (P&DP) 93% APC acceptable; shell-on shrimp (SS) 99% APC acceptable; peeled & deveined shrimp (P&DS) 99% APC acceptabe; cooked than frozen shrimp (CS) 100% APC acceptable. The *E. coli* survey indicated only 2% (SP), 4% (P&DP), 3% (SS), 0% (P&DS), and 0% (CS) of the samples were of defective quality. Ninety-nine percent of the 657 shrimp samples tested were of good quality based on ICMSF standards for *S. aureus* for raw and cooked product. *Shigella* and *Vibrio cholerae* were not found in any samples. *Salmonella* was isolated in only three (0.5%) of the samples.

In a study to determine the time–temperature tolerance of frozen tropical shrimp, Riaz and Qadri (1987) reported that *Penaeus merguiensis,* when protected from dehydration and held at nonfluctuating temperatures, maintained a high quality life (HQL) of 310 days at –20.2°F (–29°C), 144 days at 15°F (–15°C), and 75 days at 20.6°F (–6.7°C). Results were based on sensory scores from a triangle taste test.

Madero (1977) compared the postharvest age of whole Pacific shrimp *(Pandalus jordani)* with quality deterioration during frozen storage and found that the loss of quality (texture, juiciness, flavor, overall desirability) during 12 months of frozen storage at −0.4°F (−18°C) was directly related to the period of cold storage prior to cooking the product (i.e., 1, 3, and 5 days). The quality deteriorationduring frozern storage was most closely correlated with the decomposition of trimethlamine oxide to dimethlamine and formaldehyde.

Hobbs (1976) summarized microbiological data on frozen cooked peeled prawns imported from five countries between 1968 and 1973. These data showed that *E. coli*, coagulase-positive staphylococci, *Salmonella* sp., *V. parahaemolyticus*, as well as noncholera Vibrios all survived freezing and frozen storage. Vanderzant and Nickelson (1972) reported a 2-log reduction of viable *V. parahaemolyticus* after 8 days of storage at −0.4°F (−18°C); however some cells remained viable. Likewise, Olivez (1981) and Boutin et al. (1985) observed a sixfold reduction in *V. vulnificus* after 40 days of frozen storage but not a complete kill of vegetative cells. Reily and Hackney (1985) showed *V. cholerae* survived freezing in a seafood homogenate for over 2 months. Thus, freezing of crustacean shellfish, particularly those that have been hand processed, cannot be relied on to destroy bacterial pathogens that can grow and produce toxins when temperature abused.

Fatima et al. (1988) evaluated partial freezing of Penaeid shrimp as a means of extending shelf life relative to products stored in ice. In a comprehensive study comparing iced and partially frozen [super chilled 26.6°F (−3°C)] products they concluded that the prime quality life of super chilled shrimp was 16 days compared with iced shrimp prime quality life of only 8 days.

RETAIL HANDLING PROCEDURES

Maintenance of quality at the retail level generally centers on maintaining proper holding temperatures. Ensuring proper holding temperatures at retail is important for several reasons. First, the line of seafoods handled by the retail sector will have a much greater range in "age" than other items in the meat mix. Specifically, some products may be 1–2 days "old" (the time interval between capture, death, and delivery) in the case of aquacultured products including shrimp, crawfish, and prawns, whereas others such as wild caught shrimp may arrive 15–20 days after harvest due to the distances that fisherman must travel in pursuit of targeted species. With proper handling, these products are safe, wholesome, and quite acceptable to consumers. However, as previously discussed, these products may arrive containing higher levels of psychrotrophic spoilage microorganisms, and therefore will require closer scrutiny and managerial oversight to ensure that proper temperatures are maintained between the time products are delivered until they are sold. Second, the spoilage organisms naturally occurring on seafoods tend to grow at refrigerated temperatures whereas spoilage organisms on "terrestrial" meats are more sensitive to cold temperatures, and in fact are shocked by refrigeration temperatures. Third, the retail sector is generally provided with the longest stewardship of the product, and customers expect wholesome seafoods whenever they shop.

There are three separate occurrences in the retail setting where proper temperatures must be ensured. These are: *(1) when the product is delivered, (2) while the product is being held as inventory away from the point of sale, and (3) while the product is being kept "on display."* Each component of temperature assurance at the retail level is critical, as the effects of holding temperatures are cumulative insofar as shelflife is concerned.

While there may be little doubt about what the proper holding temperature should be for fresh seafoods, there seems to be uncertainty about how best to achieve and maintain it in the display case. Control of product temperatures in the display case is a function of (1) the equipment used, (2) the procedures for stocking the case, and (3) the managerial oversight given between the time produced is placed in the case until it is sold.

But effective retail management also requires other programs such as merchandising the product line so that customers are attracted to the department and have positive perceptions of the seafoods being offered. Thus, management for maintenance of quality is only one managerial objective required of retailers. *In fact a tradeoff may exist between merchandising for impulse sales and merchandising for maintenance of quality.* Obviously, the rate at which inventory is turned should influence which merchandising option takes precedence. Regardless of which option is entertained, there is one law that holds true for refrigerated seafoods: *For every degree increase in product temperature in the lower range (i.e., temperatures increasing above 32°F (0°C) shelf life is reduced by 2 days* (Borgstrom 1955; Ronsavalli 1982). Thus, it is important to determine how different merchandising strategies impact on quality maintenance.

SIMULATED RETAIL DISPLAY PROCEDURES

Miget et al. (1987) carried out a study to determine optimal seafood product handling procedures by simulating a hypothetical retail environment (through the use of similar equipment and the adoption of certain practices that are customary in retail operations), and evaluating different procedures for stocking and close up as they relate to maintenance of product temperature. The findings outlined here have immediate utility to the retail food industry in (1) establishing procedures which maintain optimal holding temperatures and (2) more effectively using departmental talent.

ICE-ONLY SERVICE CASES

The use of ice-only equipment is appealing to some retailers who find themselves in situations where inventory is turning rapidly, or in those circumstances where allocation of square footage for a fully equipped department is beyond their current needs. In addition, certain retailers may opt for an ice-only merchandising approach to communicate a particular theme to clientele.

Ice-only cases have certain benefits. Clearly there is less surface area to clean and sanitize, and odor control may be easier because of this. Also, this type of display equipment is simple to operate, because ice is the only cooling medium. However, because this display option does not insulate product surfaces from warmer ambient temperatures in the store, this display option requires more continual managerial oversight to maintain optimal product temperatures.

Because of the product surface/ambient air interface, stocking procedures are required that take this warming feature into account. Results of this study, in which the internal temperatures of fish fillets stored in metal pans and displayed in an ice only case were continuously monitored, led to the following conclusion.

Placing products in pans surrounded with ice will not keep the contents of the container below the zone where psychrotrophic bacteria grow. However, with judicious use of top ice, this type of environment may be suitable for firms experiencing rapid case turnover; i.e., at least twice a day. Therefore, ice-only cases should not be used as an inventory storage facility, but rather a way to display small quantities of the product line.

REFRIGERATED DISPLAY CASE

The display case must simultaneously accomplish two different goals. First, it must present the seafood product line to customers in a pleasing, hygienic manner. Second, the display environment must ensure proper holding temperatures, particularly in situations where inventory is carried over several days.

Refrigeration equipment should be viewed as having two purposes: (1) to slow the melting rate of ice so that the seafood display does not have to be rebuilt during the course of a business day and (2) to maintain a colder airspace environment. While this colder airspace environment is beneficial, it should not be considered a substitute for the efficient chilling brought about by product–ice contact.

RETAIL DISPLAY SUMMARY

As related to practical display/storage methods for fresh seafood at the retail level, the study indicated the following:

1. Storage in or under melting ice provides maximum shelf life. Exposed meat items such as fillets should be wrapped in, or covered with, a plastic wrap first. Whole fish and crustacean shellfish as well as cans or jars of meats should be covered with ice.
2. None of the treatments evaluated in the study (lactobacilli inoculum, high pressure chlorinated water rinse, carbon dioxide atmosphere, or combinations thereof) could substitute for storage of product at 32°F (0°C) (i.e., melting ice); nor could any of the treatments prevent rapid spoilage under temperature abuse conditions likely to be encountered in retail establishments, especially those using ice-only display cases.

REFERENCES

Alvarez, R. J., and Koburger, J. A. 1979a. *Planococcus citreus:* its potential for shrimp spoilage. In: Nickelson R. (ed.) *Proceedings of the Fourth Annual Tropical and Subtropical Fisheries Technological Conference of the Americas,* Texas A&M University Sea Grant College Program, College Station, Texas, pp. 79–86.

Alvarez, R. J., and Koburger, J. A. 1979b. Effect of delayed heading on some quality attributes of *Penaeus* shrimp. *J. Food Protect.* **42:**407.

Anderson, D. C., and Grodner, R. M. 1980. Bacteriological survey of crawfish processing in Louisiana. In: Nickelson R. (ed.) *Proceedings of the Fifth Annual Tropical and Subtropical Fisheries Technological Conference of the Americas.* Texas A&M Sea Grant College Program, College Station, Texas, pp. 81–93.

Angel, S., Basker, D., Kanner, J., and Juven, B. J. 1981. Assessment of shelf life of freshwater prawns stored at 0°C. *J. Food Technol.* **16:**357.

Angel, S., Weinberg, Z. G., Juven, B. J., and Linder, P. 1985. Quality changes in the fresh water prawn, *Macrobrachium rosenbergii* during storage on ice. *J. Food Technol.* **20:**553.

Anon. 1988. Seventh session of the IPFC Working Party on Fish Technology and Marketing. Bangkok, Thailand, April 19–22, 1988. *Fish Technology News.* Melbourne, Australia. FAO Communication Services Unit.

Anon. 1989. *Fisheries of the United States 1988.* Current Fisheries Statistics No. 8800. U.S. Department of Commerce. U.S. Government Printing Office, Washington, D.C.

Bieler, A. C., Matthews, R. F., and Koburger, J. A. 1973. Rock shrimp quality as influenced by handling procedures. In: *Proceedings of the Gulf and Caribbean Fisheries Institute* **25:**56.

Borgstrom, G. (ed.) 1955. *Fish as Food, Vol.* IV. Academic Press, New York.

Bottino, N. R., Lilly, M. L., and Finne, G. 1979. Fatty acid stability of Gulf of Mexico brown shrimp *(Penaeus aztecus)* held on ice and in frozen storage. *J. Food Sci.* **44:**1778.

Boutin, B., Reyes, A., Peeler, J., and Twedt, R. 1985. Effect of temperature and suspending vehicle on survival of *Vibrio parahaemolyticus* and *Vibrio vulnificus. J. Food Protect.* **48:**875.

Campbell, L. O., and Williams, O. B. 1952. The bacteriology of Gulf Coast shrimp. *Food Technol.* **6:**125.

Cann, D. C. 1974. Bacteriological aspects of tropical shrimp. In: Kreuzer R. (ed.) *Fishery Products* Fishing News (Books), West Bayfleet, Surrey, England, pp. 338–344.

Cann, D. C. 1977. Bacteriology of shellfish with reference to international trade. In: *Handling, Processing and Marketing of Tropical Fish.* Tropical Products Institute, London, England, pp. 377–394.

Carroll, B. J., Reese, G. B., and Ward, B. Q. 1968. Microbiological study of iced shrimp: *Excerpts from the 1965 Iced Shrimp Symposium.* Circular No. 284. U.S. Department of Interior. U.S. Government Printing Office, Washington, D.C.

Chang, O., Lun Cheuk, W., Nickelson, R., Martin, R., and Finne, G. 1983. Indole in shrimp: effect of fresh storage temperature, freezing and boiling. *J. Food Sci.* **48:**813.

Cheuk, W. L., Finne, G., and Nickelson, R. 1979. Stability of adenosine deaminase during ice storage of pink and brown shrimp from the Gulf of Mexico. *J. Food Sci.* **44:**1625.

Cobb, B. F., and Vanderzant, C. 1971. Biochemical changes in shrimp inoculated with *Pseudomonas, Bacillus,* and a coryneform bacterium. *J. Milk Food Technol.* **34:**533.

Cobb, B. F., Vanderzant, C., Hanna, M. O., and Yeh, C.-P. S. 1976. Effect of ice storage on microbiological and chemical changes in shrimp and melting ice in a model system. *J. Food Sci.* **41:**29.

Cook, D. W. 1970. A study of the bacterial spoilage patterns in iced *Penaeus* shrimp. *Completion Report on Project 2-61-R* submitted to the U.S. Department of the Interior, Bureau of Commercial Fisheries.

Cox, N. A., and Lovell, R. T. 1973. Identification and characterization of the microflora and spoilage bacteria in freshwater crayfish, *Procambarus clarkii. J. Food Sci.* **38:**679.

Disney, J. G. 1976. The spoilage of fish in the tropics. In: Cobb B. F. and Stockton A. B. (eds.) *Proceedings of the First Annual Tropical and Subtropical Fisheries Technology Conference.* Texas A&M University Sea Grant Program, College Station, Texas, pp. 23–39.

Duran, A. P., Wentz, B. A., Lanier, J. M., McClure, F. D., Schwab, A. H., Scwartzentruber, A., Barnard, R. J., and Read, R. B., Jr. 1983. Microbiological quality of breaded shrimp during processing. *J. Food Protect.* **46:**974.

Ehira, S. and Uchiyama, H. 1987. Biochemical changes in relation to freshness. In: Kramer D. E. and Liston J. (eds.) *Seafood Quality Determination.* Elsevier, New York, pp. 185–204.

Fatima, R., Farooqui, B., and Quadri, R. B. 1981. Inosine monophosphate and hypoxanthine as indices of quality of shrimp *(Penaeus merguiensis). J. Food Sci.* **46:**1125.

Fatima, R., Kahn, M. A., Qadri, R. B. 1988. Shelf life of shrimp *(Penaeus merguiensis)* stored in ice (0°C) and partially frozen (–3°C). *J. Sci. Food Agric.* **42:**235.

Flick, G. J., and Lovell, R. T. 1970. Postmortem degradation of nucleotides and glycogen in Gulf shrimp. *Food Technol.* **30:**1743.

Flick, G. J., and Lovell, R. T. 1972. Postmortem biochemical changes in the muscle of Gulf Shrimp *(Penaeus aztecus). J. Food Sci.* **37:**609.

Garrett, E. S. 1988. Microbiological standards, guidelines, and specifications and inspection of seafood products. *Food Technol.* **42:**90.

Gates, K. W., EuDaly, J. G., Parket, A. H., and Pittman, L. A. 1985. Quality and nutritional changes in frozen breaded shrimp stored in wholesale and retail freezers. *J. Food Sci.* **50:**853.

Green, M. 1949. Bacteriology of shrimp. II. Quantitative studies on freshly caught and iced shrimp. *Food Res.* **14:**372.

Grodner, R. M., and Novak, A. F. 1974. microbiological guidelines for freshwater crayfish. *(Procambarus clarkii* Girard). In Auault J. W., Jr. (ed.) *Second International Symposium on Freshwater Crayfish.* Louisiana State University, Baton Rouge, pp. 161–170.

Harrison, J. M., and Lee, J. S. 1968. Microbial evaluation of Pacific shrimp processing. *Appl. Microbiol.* **18**:188.

Hobbs, B. C. 1976. Microbiological hazards of international trade. In: Skinner F. A. and Carr J. G. (eds.) *Microbiology in Agriculture, Fisheries and Food.* Academic Press, New York, pp. 161–180.

Hood, M. A., and Meyers, S. P. 1973. Microbial aspects of penaeid shrimp digestion. In: *Proceedings of the Gulf and Caribbean Fisheries Institute 26th Annual Session,* pp. 81–92.

ICMSF. 1986. *Microorganisms In Foods II.* University of Toronto Press, Toronto, Canada.

Kennish, J. M., and Kramer, D. E. 1987. A review of high pressure liquid chromatographic methods for measuring nucleotide degradation in fish muscle. In: Kramer D. E. and Liston J. *Seafood Quality Determination.* Elsevier, New York, pp. 209–219.

Koburger, J. A., Matthews R. F., and McCullough, W. E. 1973. Some observations on the heading of *Penaeus* shrimp. In: *Proceedings of the Gulf and Caribbean Fisheries Institute 26th Annual Session,* pp. 144–148.

Koburger, J. A., Norden, A. R., and Kempler, G. M. 1975. The microbial flora of rock shrimp—*Sicyonia brevirostris. J. Milk Food Technol.* **38**:147.

Kramer, D. E., and Liston, J. 1986. Determination of fish freshness using the "K" value and comments on some other biochemical changes in relation to freshness. In: Kramer D. E. and Liston J. (eds.) *Seafood Quality Determination.* Elsevier, New York, pp. 185–197.

Lee, J. S., and D. K. Pfeifer, D. K. 1977. Microbiological characteristics of Pacific shrimp *(Pandalus jordani). Appl. Environ. Microbiol.* **33**:853.

Lee, J. S., and Kolbe, E. 1982. Microbiological profile of Pacific shrimp, *Pandalus jordani,* stowed under refrigerated seawater spray. *Marine Fish. Rev.* **44**:12.

Lima dos Santos, C. A. M. 1981. The storage of tropical fish in ice A review. *Trop. Sci.* **23**:97.

Liston, J. 1980. Microbiology in fishery science. In: Connell J. J. and Staff of Torry Research Station (eds.) *Advances In Fish Science and Technology.* Fishing News Books Limited. Farnham, Surrey, England, pp. 138–157.

Madden, R. H., and Kinghan, S. 1987. Changes in the microbial quality of *Nephrops norvegicus* during processing for retail packs. *J. of Food Protect.* **50**:460.

Madero, C. F. 1977. Effect of initial quality on the frozen shelf life of Pacific shrimp *(Pandalus jordani).* M. S. Thesis Oregon State University, Corvallis, Oregon.

Matches, J. R. 1982. Effects of temperature on the decomposition of Pacific coast shrimp *(Pandalus jordani). J. Food Sci.* **47**:1044.

Miget, R., Wagner, T., and Haby, M. 1987. Determination of optimal display procedures for fresh and previously frozen seafoods using ice-only cases and seafood cases which require supplemental refrigeration. Final Report to *Gulf and South Atlantic Fisheries Development Foundation, Inc.* Contract 32-15-18160/1566. Tampa, Florida.

Nickelson, R., and Vanderzant, C. 1976. Bacteriology of shrimp. In: Cobb B. F. and Stochton A. B. (eds.) *Proceedings of the First American Tropical and Subtropical Fisheries Technology Conference.* Texas A&M Sea Grant College Program, College Station, Texas, pp. 254–271.

Novack, A. F. 1973. Microbiological considerations in the handling and processing of crustacean shellfish. In: Chicheter E. O. and Graham H. D. (eds.) *Microbial Safety of Fishery Products.* Academic Press, New York, pp. 59–73.

O'Bannon, B. K. (ed.) 1989. Fisheries of the United States 1988. Current Fisheries Statistics #8800. U.S.D.D., N.O.A.A., U.S. Govt. printing office.

Olivez, J. 1981. Lethal cold stress of *Vibrio vulnificus* in oysters. *Appl. Environ. Microbiol.* **40:**710.

Phaff, H. J., Mrak, E. M., and Williams, O. B. 1952. Yeasts isolated from shrimp. *Mycologia* **44:**431.

Poulter, R. G., Curren, C. A., and Disney, J. G. 1981. Chill storage of tropical and temperate water fish—differences and similarities. Advances in the refrigerated treatment of fish, especially underutilized species. *Bull. Int. Inst. Refrig.* **49:**111.

Premaratne, R. J., Nip, W. K., and Moy, J. H. 1986. Characterization of proteolytic and collagenolytic psychrotrophic bacterial of ice-stored freshwater prawn, *Macrobrachium rosenbergii. Marine Fish. Rev.* **48:**44.

Reay, G. A., and SHewan, J. M. 1949. The spoilage of fish and its preservation by chilling. *Adv. Food Res.* **11:**343.

Reily, L., and Hackney, C., 1985. Survival of *Vibrio cholerae* during cold storage in artifically contaminated seafoods. *J. Food Sci.* **50:**838.

Riaz, M., and Qadri, R. B. 1987. Time–temperature tolerance of frozen shrimps 1. High quality life and acceptability time of frozen glazed shrimps. *Trop. Sci.* **27:**167.

Ridley, S. C., and Slabyj, B. M. 1978. Microbiological evaluation of shrimp *(Pandalus borealis)* processing. *J. Food Protect.* **41:**40.

Ronsavalli, L. J. 1982. A recommended procedure for assuring the quality of fish fillets at point of consumption. *Marine Fish. Rev.* **44:**8.

Shewan, J. M. 1977. The bacteriology of fresh and spoiling fish and the biochemical changes induced by bacterial action. In: *Handling, Processing and Marketing of Tropical Fish*. Tropical Products Institute, London, England, pp. 51–66.

Shewan, J. M., and Murray, C. K. 1979. The microbial spoilage of fish with special reference to the role of psychrophiles. In: Russell A. D. and Fuller R. (eds.) *Cold Tolerant Microbes in Spoilage and the Environment*. Academic Press, New York, pp. 117–135.

Singh, D., Chan, M., Hoon Ng, H., and Ong Yong, M. 1987. Microbiological quality of frozen raw and cooked shrimps. *Food Microbiol.* **4:**221.

Smith, R., Nickelson, R., Martin, R., and Finne, G. 1984. Bacteriology of indole production in shrimp homogenates held at different temperatures. *J. Food Protect.* **47:**861.

Stone, F. E. 1971. Inosine monophosphate (IMP) and hypoxanthine formation in three species of shrimp held on ice. *J. Milk Food Technol.* **35:**354.

Sumner, J. L., Gorczyca, E., Cohen, D., and Brady, P. 1984. Do fish from tropical waters spoil less rapidly in ice than fish from temperature waters? *Food Technol. Aust.* **36:**328.

Swartzentruber, A., Schwab, A. H., Duran, A. P., Wentz, B. A., and Read, R. B., Jr. 1980. Microbiological quality of frozen shrimp and lobster tail in the retail market. *Appl. Environ. Microbiol.* **40:**765.

Surkiewicz, B. F., Hyndman, J. B., and Yancey. 1967. Bacteriological survey of the frozen prepared foods industry. II. Frozen breaded raw shrimp. *Appl. Microbiol.* **15:**1.

Troller, J. A., and Frazier, W. C. 1962a. Repression of *Staphylococcus aureus* by food bacteria. I. Effect of environmental factors on inhibition. *Appl. Microbiol.* **11:**11.

Troller, J. A., and Frazier, W. C. 1962b. Repression of *Staphylococcus aureus* by food bacteria. II. Cause of inhibition. *Appl. Microbiol.* **11:**163.

Vanderzant, C., and Nickelson, R. 1972. Survival of *Vibrio parahaemolyticus* in shrimp tissue under various environmental conditions. *Appl. Microbiol.* **23:**34.

Vanderzant, C., Mroz, E., and Nickelson, R. 1970. Microbial flora of Gulf of Mexico and pond shrimp. *J. Milk Food Technol.* **33:**346.

Vanderzant, C., Cobb, B. F., Thompson, C. A., and Parker, J. C. 1973. Microbial flora, chemical characteristics an shelf-life of four species of pon reared shrimp. *J. Milk Food Technol.* **36:**443.

Ward, D. R., and Baj, N. J. 1988. Factors affecting microbiological quality of seafoods. *Food Technol.* **42:**85.

Zapatka, F. A., and Bartolomeo, B. 1973. Microbiological evaluation of cold-water shrimp *(Pandalus borealis). Appl. Microbiol.* **25:**858.

Zuberi, R., Qadri, R. B., and Siddiqui, P. M. A. 1983. Influence of processing on bacteriological quality of frozen shrimp. *J. Food Protect.* **46:**5720.

5

Microbiology of Mince, Surimi, and Value-Added Seafoods

Steven C. Ingham

Before directly considering the microbiology of mince, surimi, and value-added seafoods, it is necessary to understand how these products are made. The processes used range from relatively simple to highly complex. Steps in the processes can have important, and sometimes conflicting, effects on the microbiology of the finished product. Therefore, this chapter describes the processes used to manufacture mince, surimi, and value-added seafoods, followed by an overview of the microbiology of these products.

MINCE

Minced fish flesh, commonly referred to as mince, is produced by mechanically deboning headed, gutted fish, or a portion thereof. In the mechanical deboning process, the fish is either pressed against a perforated drum by a moving belt or it is pressed against a perforated cylindrical head by an auger. In both systems, the pressure and perforation size are such that the fish flesh, fat, and any blood present are forced through the perforations for collection, but the skin, bones, and connective issue do not go through. More information about mechanical deboning processes and equipment can be found in Lanier and Thomas (1978).

The starting material for mince production can be fillets removed from fish, the frame remaining after filleting, or headed, gutted fish (see Fig. 5.1). Mince made from fillets (fillet mince) is lighter in color than that made from headed, gutted fish or frames. The desirability of this lighter color varies with the intended use of the mince. A light colored mince would be more desirable in fish sticks and portions than in recipes where mince is being used as a substitute for red meat.

Production of fillet mince is often not economically feasible. In order for a fish processor to make a profit on fillets or fillet mince, the costs of filleting must be passed on to the consumer. If the fillet is converted to mince, the costs of mechanical deboning must also be borne by the consumer. For many species, the market price for fillets far exceeds that of fillet mince and the demand for mince is insufficient to sell it profitably. Another drawback of fillet mince is that meat

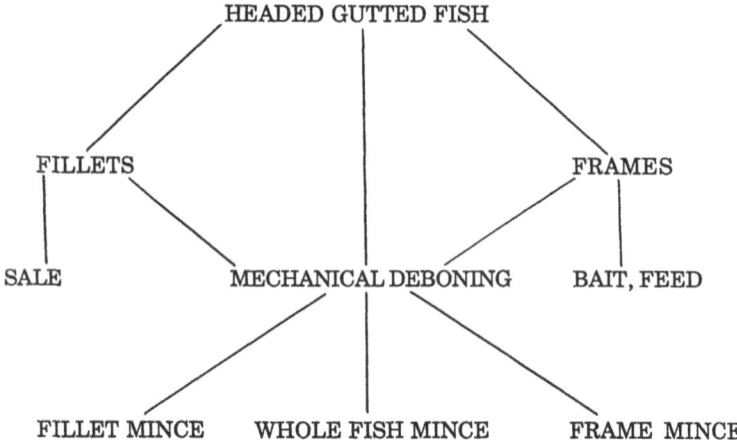

Figure 5.1   Role of mince production in fish processing.

remaining on the frame after filleting may not be utilized for human consumption but is instead used for bait or animal feed, or is discarded. Discarding frames and the high-quality protein on them can be viewed as an ethically questionable practice in light of continuing population pressures on the world food supply. Making mince from frames addresses this concern. Frame mince is dark in color because blood from the backbone and kidney is present. Frame mince is thus suitable for use as a red meat substitute. However, processors have often had little economic incentive to recover frame meat. It is conceivable, though, that increasing demand for mince, decreasing demand for frames as bait or feed, increasing disposal costs, or other economic factors could make frame mince production an important profit-making or cost reduction technique.

A third starting material for mince production is headed, gutted fish. When fillets of a particular species are not commercially valuable, use of the entire fish to make mince can be cost effective. This is especially true in cases where mince is used to make surimi.

Mince has potential application as an ingredient in many different foods (Martin 1976). At present, however, it is most widely used to prevent voids and enhance cohesiveness in frozen breaded fish sticks and portions. Legal standards for grade at present do not allow mince to be used in these foods (CFR 1988). It should be noted that these standards apply only to products under voluntary inspection by the United States Department of Commerce (USDC). Processors whose products are not inspected by USDC can legally use mince as an ingredient. Mince is also used commercially in fish cakes and seafood dishes as an extender. In addition, mince can be used as a substitute for meat in many traditional foods. Researchers at Cornell University have successfully formulated foods such as tacos, chili, spaghetti sauce, and "sloppy joes" which contain mince. These foods are heavily spiced and thus any distinctive fish flavor is masked. Supermarket demonstrations of these products have shown that consumers generally like recipes containing mince, provided they do not hold strong negative preconceived notions about eating fish (Regenstein 1986). At present, however, foods containing mince as a substitute for red meat are not commercially available.

SURIMI

Surimi is an intermediate food material that has been used in Japan for centuries to make several foods (Lee 1984). Surimi is made by mincing fish flesh, thoroughly washing the fish flesh, and then refining and dewatering it. Traditionally surimi was mixed with ingredients such as salt and spices, kneaded, and then steamed, fried, or broiled to make kamaboko, tempura, and chikuwa, respectively (Sonu 1986). The washing of minced fish flesh removes substantial amounts of water-soluble proteins, vitamins, and minerals as well as pigments and odoriferous compounds. The major components of surimi are thus the myofibrillar proteins actin and myosin. These proteins can readily form gels and can be manipulated by food processors to make foods that have a variety of textures and shapes. The excellent functionality of surimi has led to its use in making seafood analog products such as imitation crab, scallop, shrimp, and lobster (Pigott 1986). Before 1960, surimi was commonly not frozen, but was quickly made into heated foods, because denaturation occurred during frozen storage. This freeze-denaturation greatly reduced the surimi gel functionality. In 1960, Japanese researchers discovered that by adding antidenaturants such as sucrose, the freeze-denaturation problem could be practically eliminated (Sonu 1986). This discovery changed the surimi-based foods industry from one that was batch oriented and dependent on the uneven supply of fresh unfrozen surimi to one that had a more consistent supply of high quality frozen surimi. The steady supply of surimi allowed expansion of the surimi-based foods industry and made the price of surimi-based foods more consistent (Miyake et al. 1985).

Another important factor in the growth of the surimi industry both in Japan and worldwide has been the dramatic increase in consumption of surimi-based seafood analogs. Imports of simulated crab meat from Japan to the U.S. rose from approximately 4,000 metric tons in 1981 to well over 30,000 metric tons in 1985 (Sonu 1986; USDC 1987). In 1988 U.S. consumption of raw surimi and surimi-based seafood rose to 87,357 metric tons, with the U.S. producing 104,903 metric tons (Vondruska et al. 1989). Worldwide consumption of surimi-based foods is expected to exceed 1 million tons by 1990 (Holmes and Riley 1987).

The manufacture of surimi alters its composition from the original mince. Flesh from a typical white-flesh species of fish contains about 80% water and <1% carbohydrate by weight (Watts and Merrill 1963). Commercial surimi can contain between 75 and 85% water and between 5 and 10% carbohydrate, depending on its intended use (Holmes and Riley 1987; Lee 1986; Miyauchi et al. 1973; Sonu 1986). Surimi thus contains proportionally less protein than mince. Most of the lipid present in species typically used in surimi making is removed when wash water, lipid, and blood are decanted from the washing tanks (Miyake et al. 1985). Most surimi produced today contains very little sodium chloride and is referred to as "low-salt surimi." The small amount of sodium chloride present in most surimi comes from a very dilute sodium chloride solution used at the end of the washing step in order to aid in subsequent dewatering (Lee 1986; Pigott 1986). Small amounts of "salt-added surimi" are made commercially. Typical formulations for low-salt and salt-added surimis are shown in Table 5.1.

When surimi is converted into the several surimi-based foods its composition is again changed. Salt, sugar, starches, egg white, monosodium glutamate, and other flavorings are typical ingredients added to surimi in making traditional Japanese foods and modern seafood analogs. In addition, small amounts of crabmeat are occasionally added in the manufacture of imitation crab. Figure 5.2 highlights the steps done in making surimi and the products into which it can be made.

VALUE-ADDED SEAFOODS

Value-added seafoods include battered and breaded seafoods, smoked seafoods. dried fish, and precooked seafood entrees. Of these products, the two most economically important in the U.S. are battered breaded seafoods and precooked seafood entrees.

Battered breaded foods include fish sticks and portions, fillets, shrimp, scallops, and specialty products such as fish and chips, steaks, stuffed fillets and

Table 5.1 Additives Used to Make Low-Salt Surimi and Salt-Added Surimi

	Low-Salt Surimi	*Salt-Added Surimi*
NaCl	—	2.5%[a]
Sucrose	4.0%	5.0%
Sorbitol	4.0%	5.0%
Sodium tripolyphosphate	0.3%	—

[a]By weight.
Adapted from Sonu, 1986.

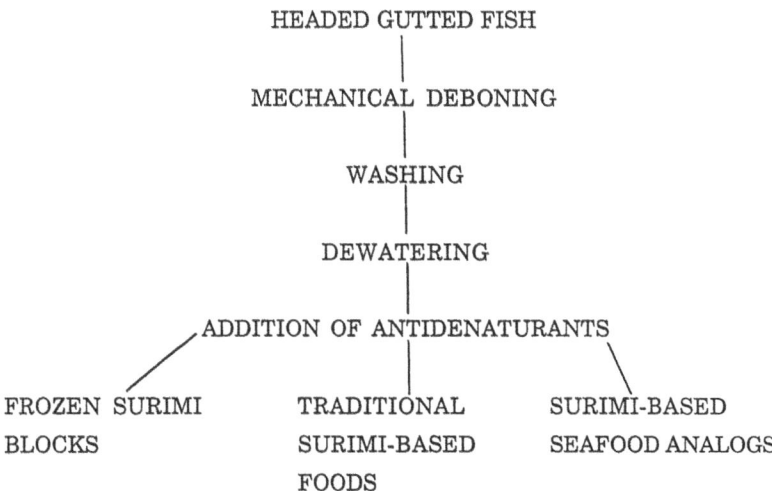

Figure 5.2 Production and uses of surimi.

shrimp, and crab sticks. The formulation of batters and breadings, as well as the processes associated with them, are very active areas in today's food industry.

Systems for the preparation of battered and breaded seafoods include single line and tandem line operations. Single line operations are used when the desired product weight gain attributable to batter and breading is <30%. The seafood is predusted with a flour, gluten, or dry batter mix and then is soaked in batter, coated with breading, and in some instances precooked. Batters used may be based on flour, starch, or gums. Batters may also be categorized according to whether they contain leavening agents. Breadings consist of various mixtures of flour, starch, and seasonings. The precooking step, typically frying, is done to set the batter and breading. Regardless of whether precooking is done, the food is frozen quickly and held frozen until it is fully cooked immediately prior to consumption (see Fig. 5.3). Tandem line operations follow the same sequence as the single line except batter and breading are applied twice before the optional precooking step. Pick-up (weight gain caused by batter and breading) in a tandem line exceeds 30%.

In the single and tandem line processes the last coating step uses a dry coating. In the batter fry or tempura line, the last coating step uses a wet coating. Batter-fried seafood is thoroughly predusted, coated with batter, and fried before being frozen (Johnson and Hutchison 1983).

Battered and breaded seafoods are very important in the fast food business and in institutional food service. With the increasing prevalence of microwave ovens in the U.S., research has been focused on developing battered and breaded products with coatings that remain attractive and crispy following microwave cooking.

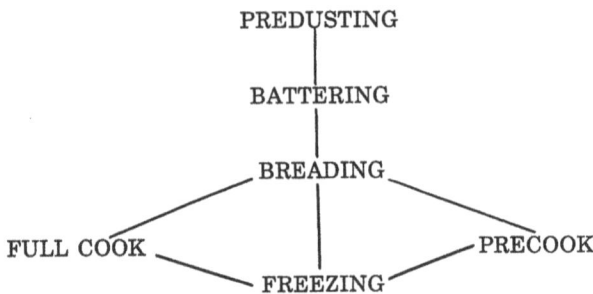

Figure 5.3 Steps for production of battered and breaded seafood.

Demographic trends such as increases in two-career families and single-parent households, in addition to the growing consumer demand for convenience foods, have made precooked entrees very important in today's food industry. Precooked seafood entrees include a wide range of items. The role of the particular seafood can range from minor, e.g., frozen linguini with clam sauce, to major, e.g., frozen oven-fried fillet dinner. In addition, specialty seafood items such as gumbo and jambalaya may be found in grocery freezer cases. The importance of convenient precooked frozen meals in the average American's diet is expected to increase in the future (Anonymous 1988a).

GENERAL MICROBIOLOGICAL ASPECTS OF MINCE, SURIMI, AND VALUE-ADDED PRODUCTS

Each step in the processing of a seafood can have an impact on its final microflora. In addition, the environment of the seafood before it is harvested can have profound quantitative and qualitative effects on the microflora of the raw material. The predominant genera of bacteria on fish at the time of catch are generally Gram-negative cocci such as *Moraxella* and *Acinetobacter* (Nickelson et al. 1980) and Gram-negative rods such as *Pseudomonas, Flavobacterium,* and *Vibrio* (Liston 1980). On occasion, Gram-positive cocci of the genus *Micrococcus* will predominate (Gillespie and Macrae 1975). During storage of fish in ambient atmosphere on ice or at refrigeration temperature, the genera *Pseudomonas, Moraxella,* and *Achromobacter* become dominant (Gillespie and Macrae 1975; Lerke et al. 1965; Shewan et al. 1960). Some *Pseudomonas* and *Achromobacter* species will produce large amounts of trimethylamine (TMA), an odorous compound associated with spoiled fish that has been much studied as an indicator of fish spoilage (Banwart 1979; Laycock and Regier 1971).

Exposure of seafood to sediments may lead to a high proportion of sporeforming species present in the microflora. Exposure to sediments may be the result of the seafood species living on the bottom or may result from harvesting techniques, such as bottom trawling, which stir bottom sediments up into the water column.

Other harvesting and handling techniques also contaminate seafoods with microbes. Rapid lifting of fish trawls up through the water column may result in "belly burst" and contamination of fish with fecal matter. In addition, fish and other seafoods are often left unprotected on the boat once harvested and thus may be contaminated with bird droppings. Furthermore, the boat, equipment, and workers are all sources of contamination.

The skin and slime layer of freshly caught marine fish may contain anywhere from 10^2 to 10^7 cells per square centimeter. Intestinal contents and gill tissue contain similar loads (Frazier and Westhoff 1988). Once fish flesh is exposed in the filleting operation, it can be inoculated by each of these sources in addition to being inoculated by workers and equipment. Cod *(Gadus morhua)* fillets have been reported to have 10^5–10^7 cells per gram (Spreekens 1969).

Exposure of seafood species to sewage-contaminated water may result in contamination of the seafoods with pathogenic bacteria and viruses of fecal origin. Flatfishes caught near effluent discharge points in the Baltic Sea were found to contain *Salmonella,* a genus that causes a high percentage of foodborne illnesses (Wuthe and Findel 1972). *Clostridium perfringens,* a sporeforming anaerobe capable of causing gastroenteritis, was found in all fish gut samples from fish caught near sewage outfalls in Puget Sound, WA. Gut samples from fish caught in areas of Puget Sound distant from sewage outfalls contained *C. perfringens* 33% of the time (Matches et al. 1974).

Fish may also contain spores of Type E *Clostridium botulinum,* one cold-tolerant type of this lethal toxin-producing bacterium. Type E *C. botulinum* is an anaerobic sporeforming bacterium found in marine and aquatic environments such as sediments and the intestinal tract of fish. It can grow and produce toxin at temperatures ranging from 37.9°F (3.3°C) to 113°F (45°C) (Banwart 1989). Many refrigerators do not maintain temperatures below 37.9°F (3.3°C), so growth of *C. botulinum* is a threat when fish is stored under anaerobic conditions. Outbreaks of botulism caused by Type E *C. botulinum* have involved smoked fish, vacuum-packaged fish, and canned fish (Anonymous 1974, 1976).

Another pathogenic bacterium that can grow at cold temperatures is *Aeromonas hydrophila.* This organism has been isolated from oysters and a variety of red meats (Abeyta et al. 1986; Okrend et al. 1987) and is also a common inhabitant of freshwater ponds, lakes, and rivers, where it can cause disease in fish and reptiles (Gordon et al. 1979; Rigney et al. 1978). *Aeromonas hydrophila* can grow on foods at temperatures at least as low as 37.4°F (3°C) (Eddy and Ketchell 1959). Since the early 1970s, *A. hydrophila* has been cited as a potential cause of foodborne gastroenteritis. The organism caused a 1983 outbreak of gastroenteritis in Louisiana in which raw oysters were the vehicle. In addition, a 472-case gastroenteritis outbreak in 1982 in Louisiana, which also involved oysters, may have been caused by *A. hydrophila* (Abeyta et al. 1986). Because *A. hydrophila* does not grow well at levels of NaCl above 1.5% (Ingham and Potter 1988a; Palumbo et al. 1985), it is most likely to be a contaminant of fish from fresh or estuarine waters.

Other pathogenic bacteria naturally present in the preharvest environments of seafoods include *Vibrio parahaemolyticus* and *V. vulnificus*. These species are found in warm waters and require soduim chloride at levels of 2–3% for optimal growth. Foodborne illness caused by *V. parahaemolyticus* and *V. vulnificus* is most likely to be caused by ingesting raw seafoods, or as a result of cross-contamination of cooked seafoods.

Evidence has also been presented to suggest that the pathogenic bacterium *Listeria monocytogenes* can frequently contaminate seafoods during processing (Weagant et al. 1988). No confirmed seafood-related outbreaks of listeriosis have yet been reported, although consumption of raw fish and shellfish was a possible cause of a New Zealand outbreak (Lennon et al. 1984). Instances of government regulatory actions concerning seafoods containing *L. monocytogenes* have occurred (Anonymous 1988c).

The effects of environment, harvesting, and preliminary processing on seafood microbiology are thoroughly covered elsewhere in this book. In the manufacture of mince, surimi, and value-added seafoods, a supply of high quality raw material is absolutely essential for making products of high quality.

Microbiology of Mince

Overall, the microbiology of mince is very similar to that of fish fillets. Nickelson et al. (1980) found that the flora of minces made from several species of warm-water fish was the same as for the whole fishes. The number of bacteria found in mince varies depending on the starting material. Cod frame mince was found to have a higher bacterial load than cod fillet mince, although the bacteriological quality of the mince reflected that of the starting material. Similar results were obtained with herring *(Clupea harengus)* mince. Scattered samples contained low counts of coliforms, fecal streptococci, and *Staphylococcus aureus* and populations of these organisms tended to increase during storage on ice (Cann and Taylor 1976).

Mechanical deboning of fish flesh ruptures muscle cells and liberates amino acids, organic acids, and other compounds that may be used in bacterial metabolism. Mince is thus an excellent bacterial growth medium. Another important microbiological aspect of mincing is that the process disperses bacteria already present on fillets, frames, or fish throughout the mince. The immediate effect of mincing on aerobic plate count (APC) appears to be variable. Cann and Taylor (1976) reported no significant bacteriological difference between cod fillets and fillet mince. However, Licciardello (1980) reported that mincing of red hake *(Urophycis chuss)* fillets increased APC. This increase more than offset a decrease in APC that occurred when the fillets were scaled and washed prior to mechanical deboning. Coliform populations paralleled APC during the red hake processing. These trends were presumed to be the result of workers handling fillets while loading the mechanical deboner (Licciardello 1980). In an earlier study, Raccach and Baker (1978) found that mechanical deboning of cod frames increased APC from an initial level of 1.5×10^5 to 1.7×10^6. The APC of this frame mince was similar to that of the frames prior to washing in preparation for deboning. Total coliform populations dropped 65% during this washing step, but after deboning had risen to 50% of the initial levels. Mechanical deboning of headed, gutted cod, pollock *(Pollachius virens)*, and whiting *(Merlangius merlangus)* also increased APC by approximately 1 log unit from initial levels of 10^4–10^5 (Raccach and Baker 1978). In another study, no seasonal variations were found in the bacteriological quality of cod mince. However, there was an increase in bacterial load during the processing day. Presumably, bacterial populations built up on the mechanical deboner throughout the day (DeClerck 1979). During deboning, friction between the fish and perforated head increases the temperature of the head and enhances bacterial growth. In addition, chunks of meat that remain on the machinery can contain high levels of bacteria and contaminate incoming products. Operating the mechanical deboner in a cold room and periodically cleaning and sanitizing it can mitigate these problems.

Bacterial growth progresses rapidly in mince during refrigerated storage. APC of cod, pollock, and whiting minces stored at 37.4°F (3°C) increased 10^3–10^4 above initial levels within 4 days (Raccach and Baker 1978). At the abusive temperature of 55.4°F (13°C), bacterial growth is even more rapid, with counts on pollock mince exceeding 10^7 cells/g within 24 hours (Ingham and Potter 1987)).

Frozen storage of mince can lead to undesirable texture when the mince is thawed. Textural problems are most severe for minces made from gadoid species such as cod and haddock *(Melanogrammus aeglefinus)*. However, decreasing the storage temperature of gadoid minces to –22°F (–30°C) can alleviate this problem (Regenstein 1986). In terms of bacterial growth, frozen storage causes a gradual decrease in bacterial load. The extent of this decrease is time dependent. Raccach and Baker (1978) found little change in mince APC after 2 weeks of storage at –13°F (–25°C), and only a slight decrease after 3 months at this temperature. A study done in the Soviet Union found that APC of Alaska pollock *(Theragra chalcogramma)* decreased about 10-fold during 3 months of frozen storage (Shkol'nikova 1970). Work done with cod fillets suggests that during longer frozen storage, APC can decrease by 10^2. This decrease can extend the iced shelflife of the thawed fillets (Licciardello and D'Entremont 1987).

Mince is traditionally shaped into blocks that are frozen. Estimates of the bacteriological quality of frozen mince blocks vary. One study found that 40% of mince samples contained over 10^6 cells/g (Blackwood 1973). Licciardello and Hill (1978) found that frozen mince blocks generally had APC of $<10^6$ cells/g.

The qualitative aspects of mince microbiology involve the ability of pathogenic species to grow on uncooked and cooked mince. Under certain conditions, competition from the natural microflora of mince reduces the likelihood of significant growth of *S. aureus* and *Salmonella sp.*

It is well known that *S. aureus* does not compete well with other bacteria and thus will not grow on foods unless the foods have been cooked or contain enough salt to inhibit other species of bacteria (Banwart 1989; Bryan 1976). Lynch and Potter (1982) found that *S. aureus* added to cod mince was quickly outgrown by the native microflora even when 0.5% potassium sorbate was added. This phenomenon was seen at 44.6°F (7°C) and 59°F (15°C).

At temperatures at or slightly above refrigeration, *Salmonella* is outgrown by psychrotrophic bacteria. In fact, few pathogenic species grow at good refrigeration temperatures, and thus refrigeration has historically been the major weapon in preventing bacterial foodborne illness associated with fish. However, recent studies have shown that *A. hydrophila* and *L. monocytogenes* can grow at refrigeration temperatures and thus could be hazardous in both raw and cooked mince.

Some pathogenic bacteria will only grow on mince if certain conditions exist. *Clostridium botulinum* is anaerobic, although it can grow on fish in the presence of subatmospheric levels of oxygen during storage at abusive temperatures (Post et al. 1985). Thus, botulism resulting from ingestion of mince is unlikely unless the mince has been stored under a vacuum or modified atmosphere at abusive temperatures. *Vibrio parahaemolyticus* and *V. vulnificus* will grow only in foods containing at least 0.3–1% NaCl (Beuchat 1974; Ingham and Potter 1988a; Nelson and Potter 1976; Nickelson and Vanderzant 1971). However, these species are often naturally part of the fish flesh microflora and processes such as addition of salt to mince, or rinsing fillets in brine prior to mincing could permit these vibrios to grow if the storage temperature was warmer than typical refrigeration.

The cooking of mince drastically reduces its bacterial load and thus reduces competition for contaminating pathogenic bacteria. Pathogens can be introduced to cooked mince through cross-contamination or improper handling. The growth of pathogenic bacteria on cooked seafoods and other muscle foods has become an increasingly important area of research as the popularity of precooked refrigerated entrees and meals increases. *Aeromonas hydrophila* has been found to grow well on cooked pollock mince over a wide range of temperatures. From inoculum levels of 3,000–7,000 cells/g, *A. hydrophila* populations on cooked Atlantic pollock mince exceeded 10^6 cells/g in less than 5 days at 41°F (5°C), 36 hours at 55.4°F (13°C), and 9 hours at 77°F (25°C) (Ingham and Potter 1988a). In addition, *A. hydrophila* was found to compete well with the psychrotrophic spoilage bacterium, *Pseudomonas fragi*, on cooked pollock mince stored at 41°F (5°C) and 55.4°F (13°C) (Ingham and Potter 1988b).

Cooked pollock mince was a less than ideal medium for growth of *V. parahaemolyticus*. Populations declined during storage at 41°F (5°C) as expected for this cold-sensitive organism. During 77°F (25°C) storage, *V. parahaemolyticus* populations increased 10^2–10^3 cells/g but only after 27 hours. This growth was unexpected because of the low NaCl level (0.27%) in the mince (Ingham and Potter 1988a). However, Nelson and Potter (1976) reported that two strains of *V. parahaemolyticus* grew in laboratory media containing 0.5% NaCl and were viable in media with 0.3% NaCl. Beuchat (1974) also found that *V. parahaemolyticus* grew in trypticase soy broth containing 0.5% NaCl.

Staphylococcus aureus grew well on cooked pollock mince stored at 77°F (25°C). As expected, *S. aureus* did not grow on the cooked mince during 41°F (5°C) storage (Ingham and Potter 1988a).

Mince Preservation Techniques

Mince is a supportive medium for growth of both spoilage and pathogenic bacteria. Thus, researchers have tried to develop techniques to improve the microbiological safety and keeping quality of mince. A dehydrated minced fish product was developed that had 5% moisture and did not require refrigeration. Acceptable rehydration, binding, and sensory properties for this product were obtained by adding modified tapioca starch, texturized soy fiber, and salt (Bello and Pigott 1979). Researchers found that reducing the water activity of mince to 0.7 by adding salt and sorbic acid resulted in acceptable shelflife even at abusive temperatures (Varga et al. 1979). To prevent deterioration of salt minced fish, packaging impermeable to light, air, and water was found to be essential (Bligh 1977). Another type of salt mince product was made by heating mince, pressing it to remove moisture, adding 15% NaCl, and hot-packing into cans that were flushed with nitrogen and sealed. This product was tested as an ingredient in fish cakes. Salt-tolerant and sporeforming bacteria were found to survive the heat treatment used (Avery et al. 1981). Salted minced fish cakes have been made from the freshwater African species, *Haplochromis* spp. These cakes were considered to be a viable alternative to traditional processing of small bony fish (Dhatemwa et al. 1985).

The addition of 0.5% potassium sorbate to uncooked cod mince was found to decrease APC slightly during storage at 44.6°F (7°C) and 59°F (15°C). The preservative effect of potassium sorbate was much greater for cooked cod mince. Sorbate also slowed the growth of *S. aureus* added to cooked cod mince and delayed the production of thermostable nuclease by this pathogen (Lynch and Potter 1982).

Two promising techniques for extending mince shelf life are irradiation and modified atmosphere storage. Both of these subjects are covered in detail elsewhere in this book.

Microbiology of Surimi and Surimi-Based Foods

The initial step in the transformation of mince to surimi is a thorough washing. This step can have several microbiological effects. During washing, paddles stir the mince and distribute bacteria. The wash water removes significant numbers of bacterial cells from the mince (Licciardello and Hill 1978) just as washing of fillets reduces APC. Traditionally, the temperature of the wash water is no less than 50°F (10°C). Use of colder water reduces the speed of the dewatering step that follows washing. In addition, more water can be removed during dewatering if water warmer than 50°F (10°C) is used (Sonu 1986). Thus, psychrotrophic bacteria present in the mince may be capable of growth during the wash step.

Another step in surimi-making that has microbiological ramifications is the addition of carbohydrate antidenaturant compounds. Typically a mix of sucrose and sorbitol is used. Use of this blend decreases problems of excessive sweetness and browning that result when only sucrose is used (Sonu 1986). These compounds may occasionally be contaminated with bacterial spores, and constitute an abundant carbon source in addition to the proteins and amino acids already present. It is well known that many bacteria grow best in carbohydrate-rich media, and it has been found that the APC of Atlantic pollock surimi rises slightly faster than that of Atlantic pollock mince during storage at 41°F (5°C) and 55.4°F (13°C) (Ingham and Potter 1987). Acidic compounds are frequently produced during bacterial metabolism of carbohydrates, and as a result, the pH of Atlantic pollock surimi dropped relative to that of Atlantic pollock mince. This decrease in pH occurred when bacterial populations had risen to near 10^7 colony forming units (cfu) g and thus a drop in pH could be used as an index of surimi spoilage (Ingham and Potter 1987). Most surimi is packed into blocks and then frozen, so spoilage of surimi should not be a problem unless it is held too long before freezing or improperly thawed.

The native microbial load of surimi varies depending on the length of time the fish is held before processing. Japanese surimi made onboard ship had a mean APC of 10^4 cfu/g whereas Japanese surimi produced onshore had a mean APC of 10^6 cfu/g. Surimi produced onshore in Alaska, closer than Japan to the fishing area, had APC values between these extremes with 37% of the samples having APC of 10^5 cfu/g and 34% of the samples having APC of 10^6 cfu/g (Elliot 1987). Samples of U.S.-made surimi tested prior to freezing had between 4.3 and 9,300 "Most Probable Number" (MPN)/g coliforms and two of the 35 samples had at least 4 MPN/g fecal coliforms. About 90% of the bacteria isolated from the samples of surimi were Gram-negative and the predominant genera were *Pseudomonas, Acinetobacter,* and *Moraxella* (Elliot 1986). Similarly, Japanese researchers reported that total viable cell counts for blocks of frozen surimi were 10^5–10^6 cfu/g. One prolific strain of *Enterobacter cloacae* was found to cause browning of surimi (Fujita et al. 1974).

When blocks of surimi are thawed in preparation for further processing, psychrotrophic bacteria begin to grow in the thawed portions. Thawed surimi is mixed with salt and several other ingredients and processed into a surimi-based food. Many ingredients used in making surimi-based foods can contain high levels of aerobic sporeforming bacteria and their spores. Starch, spices, and dried egg white are particularly prone to contamination by *Bacillus* species (Fung 1983). Of particular concern is the pathogenic species *B. cereus*. This organism has been found in several surimi and meat-product additives (Shinagawa et al. 1988).

During processing surimi and surimi-based foods are prone to contamination by airborne microorganisms. In one study, Japanese researchers found that agar plates 9 cm in diameter exposed to the air in manufacturing plants for 30 minutes had between 0.26 and 2.09 *S. aureus* colonies per plate and between 0.34 and 5.70 coliform colonies per plate (Fujita et al. 1979). These results show the importance of minimizing exposure of cooked surimi-based foods to the processing plant environment.

A variety of cooking methods are used in making surimi-based foods. To make kamaboko, surimi is mixed with salt and other ingredients and then thoroughly kneaded. During steaming, a homogeneous protein gel forms. This gel gives kamaboko a very rubbery texture. Steam cooking is also done in the manufacture of hampen, a spongy product that contains entrapped air. Frying is the cooking method used in making tempura and satsuma-age, two other traditional Japanese foods. An example of a traditional broiled surimi-based food is chikuwa, which is traditionally shaped like a hollow bamboo stem. Just as in the making of kamaboko, a rubbery protein gel forms and this gel is set by the heat processing (Sonu 1986).

In making seafood analogs, a two-step cooking procedure is used. During the first step, the mix of surimi and other ingredients is heated during extrusion in order to set the gel and allow further molding, fiberizing, or composite-molding. In the second cooking step, the seafood analog is heated by steam or hot [176–194°F (80–90°C)] water for 20 to 30 minutes (Lee 1984, 1986). Japanese law requires all surimi-based foods to be cooked during manufacture to an internal temperature of at least 167°F (75°C) (Suzuki 1981). Typical heat-processing treatments do not sterilize surimi-based foods. *Bacillus* spores have been found to survive in kamaboko after 90 minutes at 203°F (95°C), germinate during storage, and cause spoilage (Motegi and Matsubara 1970). Genera of bacteria isolated from kamaboko include *Pseudomonas, Flavobacterium, Corynebacterium, Lactobacillus, Bacillus,* and *Micrococcus* (Sasayama 1973).

The spoilage of kamaboko during storage has been thoroughly studied. One characteristic type of kamaboko spoilage is browning caused by bacterial growth. Several species of bacteria are known to cause browning, including *Achromobacter brunificans, Serratia marcescens,* and *Pseudomonas* sp. (Mori et al. 1974). The *Pseudomonas* sp. caused rapid browning of kamaboko containing either glucose or sucrose. This species produced a compound, tentatively identified as 2,5-diketogluconic acid or a related compound, that participated in the browning reaction (Nabetani et al. 1974). Another type of spoilage reported to affect kamaboko was softening and slime production resulting from growth of *B. licheniformis*. The slime was a levan made from sucrose in the surimi (Mori et al. 1973b). The source of the *B. licheniformis* was found to be potato starch used in the preparation of the kamaboko (Mori et al. 1973a). *Leuconostoc* has also been found to produce dextran slime on kamaboko (Uchiyama and Amano 1959).

The spoilage of surimi-based imitation crab has also been studied. Two freshly processed imitation crab products, crableg and flaked crab, initially had APC of 10^2–10^3 cfu/g. The flaked crab had the higher APC of the two products because it was handled more during processing. Bacterial growth on these foods during 59°F (15°C) and 50°F (10°C) storage was very rapid, with counts exceeding 10^9 CFU/g within 2 and 4 weeks, respectively (Yoon et al. 1988). As storage progressed, *Bacillus* became the dominant genus at 59°F (15°C) and 50°F (10°C), and *Pseudomonas* became the dominant genus at 41°F (5°C) and 32°F (0°C). The growth of these bacteria caused increases in lactic acid, acetic acid, and butanediol and a drop in pH during storage. The production of volatile bases by bacteria during storage was much lower than typically observed in studies of fish flesh (Yoon et al. 1988).

Seafood analogs are often manually incorporated into sandwiches, salads, and dips, which receive no cooking before being consumed. Under these conditions, the analogs are subject to postcooking contamination with pathogens. A major concern involves contamination with the pathogen *Listeria monocytogenes*. This organism has caused outbreaks of listeriosis involving cole-slaw (Schlech et al. 1983), pasteurized milk (Fleming et al. 1985), and soft cheese (Janes et al. 1985). *Listeria monocytogenes* survives well during refrigeration and on some refrigerated foods it can grow to large numbers. It is a potential environmental contaminant of many foods, including surimi-based foods. A study of seven surimi-based food samples divided into 66 subsamples found that 20 of the subsamples contained *L. monocytogenes* (Weagant et al. 1988). Surimi-based seafood analogs are usually frozen for distribution and then thawed and refrigerated prior to use. Growth of *L. monocytogenes* on these foods during refrigerated storage is being studied.

Along with *L. monocytogenes*, *A. hydrophila* is a pathogen capable of growth on foods during refrigeration. *Aeromonas hydrophila* was found to grow well on Atlantic pollock low-salt surimi (0.07% NaCl) at 41°F (5°C), 55.4°F (13°C), and 77°F (25°C). However, the level of salt in Atlantic pollock salt-added surimi (2.44% NaCl) had a marked inhibitory effect on growth of *A. hydrophila* at all three temperatures (Ingham and Potter 1988a).

Staphylococcus aureus is a toxin-producing Gram-positive coccus whose presence in food usually indicates cross-contamination or mishandling. This organism grows well in protein-rich foods and is tolerant of high levels of salt. However, it only grows well when there is little competition from other bacteria.

Inoculated *S.aureus* did not grow on cooked Atlantic pollock surimi stored at good refrigeration temperatures but it did grow well at abusive temperatures. At 77°F (25°C), *S. aureus* actually grew better on Atlantic pollock low-salt surimi and salt-added surimi than on Atlantic pollock mince. *Staphylococcus aureus* prefers to metabolize carbohydrates rather than proteins and amino acids, and the difference in *S. aureus* growth for mince (0% added carbohydrate) and surimis (7 and 11% added carbohydrate) reflects the different carbohydrate contents of these products (Ingham and Potter 1988a).

Vibrio parahaemolyticus is a potential cross-contaminant of surimi-based foods. Because it does not grow at refrigeration temperatures, it presents a risk only when a contaminated food is stored at abusive temperatures. In addition, it will only grow on foods containing more than approximately 0.5% salt. Cooked Atlantic pollock salt-added surimi supported growth of *V. parahaemolyticus* during 77°F (25°C) storage, but no growth occurred on cooked low-salt surimi (Ingham and Potter 1988a).

Preliminary results indicate that growth of non-proteolytic types of *C. botulinum* on surimi-based crab analog is inhibited by the inclusion of at least 2.4% (water phase) NaCl and by the heat processing done in making the crab analogs. These steps do not appear to inhibit growth of proteolytic *C. botulinum* on crab analogs (Eklund 1987). *Clostridium botulinum* toxin production was found to occur in inoculated kamaboko stored at 86°F (30°C), regardless of the redox potential of the kamaboko (Sasajima et al. 1978). Proper refrigeration [at or below 37.9°F (3.3°C)] remains the major method for preventing growth of *C. botulinum* on surimi-based foods (Eklund 1987).

Several preservation techniques for surimi-based foods have been studied. Japanese researchers have thoroughly studied the use of irradiation for preservation of these foods, and their work is described elsewhere. Modified atmosphere packaging may find use with surimi-based foods, and its use is also described elsewhere. An additive mix of refined alcoholic rice fermentation products and amino acids was tested as a preservative for smoked kamaboko and was found to inhibit growth of coliforms (Ishida and Watanabe 1981). Dipping of kamaboko in lysozyme solution has also been found to increase shelf life (Akashi and Oono 1972).

Microbiology of Value-Added Seafoods

Freezing is the primary preservation method used for preserving battered and breaded seafood products. In many cases the product is also cooked before it is frozen. The microbiological effects of freezing battered and breaded seafoods are similar to those already discussed for mince, surimi, and surimi-based foods. A slow decrease in vegetative bacterial population occurs during frozen storage (Vanderzant et al. 1973). Bacterial spores survive freezing in proportionally higher numbers than vegetative cells. The severity of heat processing done prior to freezing affects the relative numbers of spores and vegetative cells in the products at the beginning of freezing. Some heat processes will heat shock spores and cause them to germinate before being frozen.

The microflora of battered and breaded seafoods can arbitrarily be divided into the organisms naturally associated with the seafood, organisms associated with the batter and breading ingredients, and organisms associated with processing workers and equipment. The microbiology of batter and breading has been thoroughly reviewed by Fung (1983). The major ingredients in batter and breading are flour, starch, spices, milk, eggs, and water.

Several genera of bacteria may be found in flour and starch. Probably the most important genus is *Bacillus*. *Bacillus* spores can survive in flour and starch during storage and may also survive cooking treatments used for battered and breaded seafoods.

Spices can be heavily contaminated with bacterial cells and spores of bacteria and molds. Even though some spices contain compounds with antimicrobial activity, the addition of spices to a cooked food can contaminate that food and possibly lead to foodborne illness. Spores of the foodborne pathogens *C. perfringens* and *B. cereus* have been frequently found on spices (Powers et al. 1975, 1976). The heating of foods containing spices will decrease the microbial load of the spices. Spices can be pasteurized using irradiation or ethylene oxide. Pasteurization by either method markedly reduces the microbial load and risk of foodborne illness associated with spices.

Liquid milk is an excellent medium for bacterial growth, but in its dried form it does not support microbial growth. Once dried milk is reconstituted, microbial growth can occur rapidly. Thus, dry milk should not be moistened until immediately before it is to be used in the battering operation. In addition, batches of batter containing milk should not be left sitting for extended periods at temperatures conducive to microbial growth.

Eggs, once they are broken, are also an excellent medium for microbial growth. Liquid eggs, which are pasteurized to kill *Salmonella* cells, should be stored frozen and be used as soon as possible after being thawed under refrigerated conditions. Dried eggs do not support growth, but, like dried milk, should be used promptly once they are rehydrated. As in all food manufacture, water used for making batter should meet applicable microbiological standards.

The sanitary condition of workers and equipment has a very important influence on the microbiological quality of battered and breaded seafoods. It is imperative that equipment be cleaned and sanitized often, and that workers practice good hygiene. In addition, extended periods during which products are held at microbial growth temperatures must be avoided.

The microbiological quality of frozen breaded raw shrimp has been thoroughly studied. The major sources of microbes in processing this product were found to be the shrimp and the batter. Breading was found to have very low microbial populations. The temperature at which batter was held during the battering of the shrimp had a major effect on its microbial load. Unchilled batter that was held for extended times was found to frequently have bacterial populations above 10^6 cfu/g, whereas batter that was chilled and periodically replaced had bacterial loads 10–1,000 times lower. The APC of the frozen raw breaded shrimp was found to be consistently higher on products processed in unsanitary conditions (Surkiewicz et al. 1967). In two studies, a high percentage of frozen raw breaded shrimp samples were found to contain more than 1 million cells per gram. Surkiewicz et al. (1967) reported that 85% of samples exceeded this level, and Nickerson and Pollak (1972) found that 48% of samples contained at least 10^6 cells/g. Researchers found that there is no consistent relationship between APC and populations of enterococci, coliforms, and *Escherichia coli* (Nickerson and Pollak 1972; Vanderzant et al. 1973). Numbers of enterococci and coliforms reportedly found on frozen breaded raw shrimp varied considerably, probably because of differences in raw materials and workmanship. Another study found that *Salmonella* and *V. parahaemolyticus* were absent from frozen raw breaded shrimp and that the predominant genera of bacteria included *Bacillus,* coryneforms, and *Micrococcus* (Vanderzant et al. 1973).

Bacterial counts on precooked battered and breaded seafoods are considerably lower than on frozen breaded raw shrimp. In a very comprehensive study, Baer et al. (1976) found that average bacterial counts ranged from 450 cfu/g to 15,000 cfu/g on seven precooked battered and breaded seafoods (fish sticks, fish cake, crab cake, scallops, clams, haddock, and fish and chips). Of these products, crab cakes were most likely to have high coliform counts (only 39.2% of samples had <3 MPN/g). For precooked battered and breaded clams, 92.8% of samples had <3 MPN/g coliforms.

The application of sound food handling and processing techniques is necessary to ensure the microbiological safety and keeping quality of battered and breaded seafoods. These techniques are also vitally important for precooked seafood entrees. Sanitary handling, thorough heat processing, proper use of high-quality ingredients, and avoidance of postcooking cross contamination must be the norm in preparation of these entrees. The current trend in favor of precooked refrigerated (nonfrozen) foods has placed an added emphasis on these practices in the production of safe value-added seafood products.

REFERENCES

Abeyta, C. Jr., Kaysner, C. A., Wekell, M. M., Sullivan, J. J., and Stelma, G. N. 1986. Recovery of *Aeromonas hydrophila* from oysters implicated in an outbreak of foodborne illness. *J. Food Protect.* **49**:643–646.

Akashi, A., and Oono, A. 1972. The preservative effect of egg white lysozyme on unpackaged kamaboko. *J. Agric. Chem. Soc. Jpn.* **46**:177–183.

Anonymous. 1974. *Botulism in the United States, 1899–1973. Handbook for Epidemiologists, Clinicians, and Laboratory Workers*. Centers for Disease Control, Atlanta, Georgia.

Anonymous. 1976. Botulism—Alaska. *US Morbid. Mortal. Wk. Rep.* **25**:399–400.

Anonymous. 1988a. Food consumption. *Natl. Food Rev.* 2–15.

Anonymous. 1988b. Certain dried fish deemed adulterated. *Food Technol.* **43**:36.

Anonymous. 1988c. Frozen cooked shrimp recalled due to *Listeria* contamination. *Food Chem. News* 8/15/88:37.

Avery, K. W. J., Lamprecht, E., and Riley, F. R. 1981. *Salted Minced Pressed Fish Packed Under Nitrogen*. Annual Report, Fishing Industry Research Institute, Cape Town, **35**:17–18.

Baer, E. F., Duran, A. P., Leninger, H. V., Read, R. B., Jr., Schwab, A. H., and Swartzenruber, A. 1976. Microbiological quality of frozen breaded fish and shellfish products. *Appl. Environ. Microbiol.* **31**:337–341.

Banwart, G. J. 1989. *Basic Food Microbiology*. AVI Publishing, Westport, Connecticut.

Bello, R. A., and Pigott, G. M. 1979. A new approach to utilizing minced fish flesh in dried products. *J. Food Sci.* **44**:355–358.

Beuchat, L. R. 1974. Combined effects of water activity, solute, and temperature on the growth of *Vibrio parahaemolyticus*. *Appl. Microbiol.* **27**:1075–1080.

Blackwood, C. M. 1973. Utilization of mechanically separated fish flesh—Canadian experience. In: R. Kreuzer (ed.) Fishery Products, Fishing News Books, Ltd., London England, pp. 325-329.

Bligh, E. G. 1977. A note on salt minced fish. Paper read at Meeting of Tropical Products Institute, 5–9 July, 1976 in London, England.

Bryan, F. L. 1976. *Staphylococcus aureus*. In: Defiguerido M. P. and Splittstoesser D. F. (eds.) *Food Microbiology: Public Health and Spoilage Aspects*. AVI Publishing, Westport, Connecticut, pp. 12–128.

Cann, D. C., and Taylor, L. Y. 1976. The bacteriology of minced fish prepared and stored under experimental conditions. In: *The Production and Utilization of Mechanically Recovered Fish Flesh*, Torry Research Station, Aberdeen, Scotland, pp. 39–45.

Code of Federal Regulations. 1988. 21 CFR 151-175.

DeClerck, D. 1979. Quality assessment of losses occurring during portioning of cod *(Gadus morhua)*. *Rev. Agric.* **32**:1257–1265.

Dhatemwa, C. M., Hanson, S. W., and Knowles, M. J. 1985. Approaches to the effective utilization of *Haplochromis* spp. from Lake Victoria. II. Production and Utilization of dried, salted minced fish cakes. *J. Food Technol.* **20**:1–8.

Eddy, B. P., and Ketchell, A. G. 1959. Cold-tolerant fermentative Gram-negative organisms from meat and other sources. *J. Appl. Bacteriol.* **22**:57–63.

Eklund, M. 1987. Microbiology of surimi-based analogs. Quarterly Report of Northwest and Alaska Fisheries Center, National Marine Fisheries Service. Jan–March, 1987, p. 29.

Elliot, E. L. 1987. Microbiological quality of Alaska pollock surimi. In: Kramer D. E. and Liston J. (eds) *Seafood Quality Determination, Proceedings of an International Symposium held 10–14 November, 1986*. Elsevier Science Publishers, Amsterdam, Netherlands.

Fleming, D. W., Cochi, S. L., MacDonald, K. L., Brondum, J., Hayes, P. S., Plikaytis, B.D., Holmes, M. B., Audurier, A., Broome, C. V., and Reingold, A. L. 1985. Pasteurized milk as a vehicle of infection in an outbreak of listeriosis. *N. Engl. J. Med.* **312**:404–407.

Frazier, W. C., and Westhoff, D. C. 1988. *Food Microbiology*, fourth edition. McGraw-Hill, New York.

Fujita, Y., Miyamoto, M., and Matsuda, T. 1974. The brown discoloration in fish jelly products. VI. A new browning bacteria isolated from frozen surimi. *Bull. Jpn. Soc. Sci. Fish.* **40:**825–833.

Fujita, Y., Miyazaki, W., and Kanayama, T. 1979. Microbial control at 'kamaboko' (and fish sausage) processing plants. I. Airborne microorganisms at the processing plants. *Bull. Jpn. Soc. Sci. Fish.* **45:**891–899.

Fung, D. Y. C. 1983. The microbiology of batter and breading. In: Suderman D. R. and Cunningham F. E. (eds) *Batter and Breading.* AVI Publishing, Westport, Connecticut, pp. 106–119.

Gillespie, N. C., and Macrae, I. C. 1975. The bacterial flora of some Queensland fish and its ability to cause spoilage. *J. Appl. Bacteriol.* **32:**91–100.

Gordon, R. W., Hazen, T. C., Esch, G. W., and Fliermans, C. B. 1979. Isolation of *Aeromonas hydrophila* from the American alligator, *Alligator mississippiensis. J. Wldlf. Dis.* **15:**239–243.

Holmes, C., and Riley, C. 1987. Surimi plant. In: Bartholomai A. (ed) *Food Factories: Processes, Equipment, Costs.* VCH Publishers, New York, pp. 207–214.

Ingham, S. C., and Potter, N. N. 1987. Microbial growth in surimi and mince made from Atlantic pollock. *J. Food Protect.* **50:**312–315.

Ingham, S. C., and Potter, N. N. 1988a. Survival and Growth of *Aeromonas hydrophila, Vibrio parahaemolyticus,* and *Staphylococcus aureus* on cooked mince and surimis made from Atlantic pollock. *J. Food Protect.* **51:**634–638.

Ingham, S. C., and Potter, N. N. 1988b. Growth of *Aeromonas hydrophila* and *Pseudomonas fragi* on mince and surimis made from Atlantic pollock and stored under air or modified atmosphere. *J. Food Protect.* **51:**966–970.

Ishida, K., and Watanabe, K. 1981. Preservation of foods by means of a high-alcohol seasoning (miro). *N. Food Indust.* **23:**82–85.

Janes, S. M., Fannin, S. L., Agee, B. A., Parker, E., Vogt, J., Run, G., Williams, J., Lieb, L., Pendergast, T., Werner, S. B., and Chin, J. 1985. Listeriosis outbreak associated with Mexican-style cheese—California. *US Morbid. Mortal. Wk. Rep.* **34:**357–359.

Johnson, R. T., and Hutchison, J. 1983. Batter and breading processing equipment. In: Suderman D. R. and Cunningham F. E. (eds) *Batter and Breading.* AVI Publishing, Westport, Connecticut, pp. 124–125.

Lanier, T. C., and Thomas, F. B. 1978. Minced fish: its production and use. North Carolina State University Sea Grant College Program Publications, Raleigh, North Carolina.

Laycock, R. A., and Regier, L. W. 1971. Trimethylamine-producing bacteria on haddock *(Melanogrammus aeglefinus)* fillets during refrigerated storage. *J. Fish. Res. Board Can.* **28:**305–309.

Lee, C. M. 1984. Surimi process technology. *Food Technol.* **38:**69–80.

Lee, C. M. 1986. Surimi manufacturing and fabrication of surimi-based products. *Food Technol.* **40:**115–124.

Lennon, D., Lewis, B., Mantell, C., Becroft, D., Dove, B., Farmer, K., Tonkin, S., Yeates, N., Stamp, R., and Mickleson, K. 1984. Epidemic perinatal listeriosis. *Pediatr. Infect. Dis.* **3:**30–34.

Lerke, P., Adams, R., and Farber, L. 1965. Bacteriology of spoilage of fish muscle. III. Characterization of spoilers. *Appl. Microbiol.* **13:**625–630.

Licciardello, J. J. 1980. Microbiological aspects of minced fish. Paper read at third National Technical Seminar on Mechanical Recovery and Utilization of Fish Flesh, 1–3 December, 1980, at Raleigh, North Carolina.

Licciardello, J. J., and D'Entremont, D. L. 1987. Bacterial growth rate in iced fresh or frozen-thawed Atlantic cod, *Gadus morhua. Marine Fish. Rev.* **49:**43–45.

Licciardello, J. J., and Hill, W. S. 1978. Microbiological quality of commercial frozen minced fish blocks. *J. Food Protect.* 41:948–952.

Liston, J. 1980. Microbiology of seafoods. In: Connell J. J. (ed.) *Advances in Fish Science and Technology.* Surrey, England: Fishing News (Books), Surrey, England, pp. 138–157.

Lynch, D. J., and Potter, N. N. 1982. Effects of potassium sorbate on normal flora and on *Staphylococcus aureus* added to minced cod. *J. Food Protect.* 45:824–828.

Martin, R. E. 1976. Mechanically-deboned fish flesh. *Food Technol.* 30:64–70.

Matches, J. R., Liston, J., and Curran, D. 1974. *Clostridium perfringens* in the environment. *Appl. Microbiol.* 28:655–660.

Miyake, Y., Hirasawa, Y., and Miyanabe, M. 1985. Technology of surimi manufacturing. *Infofish* 5/85:29–32.

Miyauchi, D., Kudo, G., and Patashnic, M. 1973. Surimi—a semiprocessed wet fish protein. *Marine Fish. Rev.* 35:7–9.

Mori, K., Sawada, H., Nabetani, O., Maruo, S., and Hirano, T. 1973a. Studies on the spoilage of fish jelly products. I. Softening spoilage of film packaged kamaboko due to *Bacillus licheniformis. Bull. Jpn. Soc. Sci. Fish.* 39:1063–1069.

Mori, K., Nabetani, O., Maruo, S., and Hirano, T. 1973b. Studies on the spoilage of fish jelly products. II. The inner slimy substance in softened spoilage parts of kamaboko, produced by *Bacillus licheniformis. Bull. Jpn. Soc. Sci. Fish.* 39: 1071–1076.

Mori, K., Nabetani, O., and Hirano, T. 1974. Studies on the spoilage of fish jelly products. III. Browning of kamaboko by *Pseudomonas* sp. *Bull. Jpn. Soc. Sci. Fish.* 40:959–962.

Motegi, S., and Matsubara, M. 1970. Bacteria isolated from the film packaged fish-paste product kamaboko. *J. Food Hyg. Soc. Jpn.* 11:49–51.

Nabetani, O., Hirano, T., and Mori, K. 1974. Studies on the spoilage of fish jelly products. IV. Browning of kamaboko by *Pseudomonas* sp. *Bull. Jpn. Soc. Sci. Fish.* 40:963–967.

Nelson, K. J., and Potter, N. N. 1976. Growth of *Vibrio parahaemolyticus* at low salt levels and in nonmarine foods. *J. Food Sci.* 41:1413–1417.

Nickelson, R., and Vanderzant, C. 1971. *Vibrio parahaemolyticus:* a review. *J. Milk Food Technol.* 34:447–452.

Nickelson, R., Finne, G., Hanna, M. O., and Vanderzant, C. 1980. Minced fish flesh from nontraditional Gulf of Mexico finfish species: bacteriology. *J. Food Sci.* 45:1321–1326.

Nickerson, J. T. R., and Pollak, G. A. 1972. Bacteriological examination of raw breaded frozen shrimp. *J. Milk Food Technol.* 35:167–169.

Okrend, A. J. G., Rose, B. E., and Bennett, B. 1987. Incidence and toxigenicity of *Aeromonas* species in retail poultry, beef and pork. *J. Food Protect.* 50: 509–513.

Pace, P. J., and Krumbiegel, E. R. 1973. *Clostridium botulinum* and smoked fish production. *J. Milk Food Technol.* 36:42–49.

Palumbo, S. A., Morgan, D. R., and Buchanan, R. L. 1985. Influence of temperature, NaCl, and pH on the growth of *Aeromonas hydrophila. J. Food Sci.* 50: 1417–1421.

Pigott, G. M. 1986. Surimi: the "high tech" raw materials from minced fish flesh. *Food Rev. Int.* 2:213–246.

Post, L. S., Lee, D. A., Solberg, M., Furgang, D., Specchio, J., and Graham, C. 1985. Development of botulinal toxin and sensory deterioration during storage of vacuum and modified atmosphere packaged fish fillets. *J. Food Sci.* **50**:990–996.

Powers, E. M., Lawyer, R., and Masuoka, Y. 1975. Microbiology of processed spices. *J. Milk Food Technol.* **38**:683–687.

Powers, E. M., Thomas, G. L., and Brown, T. 1976. Incidence and levels of *Bacillus cereus* in processed spices. *J. Milk Food Technol.* **39**:668–670.

Raccach, M., and Baker, R. C. 1978. Microbial properties of mechanically deboned fish flesh. *J. Food Sci.* **43**:1675–1677.

Regenstein, J. M. 1986. The potential for minced fish. *Food Technol.* **40**:101–106.

Rigney, M. M., Zilinsky, J. W., and Rouf, M. A. 1978. Pathogenicity of *Aeromonas hydrophila* in red leg disease in frogs. *Curr. Microbiol.* **1**:175–179.

Sasajima, M., Shiba, M., Matsushita, A., Arat, K., Yokoseki, M., and Takamizawa, M. 1978. The effect of packaging style on the production of *Clostridium botulinum* Type A toxin in kamaboko. *Bull. Tokai Reg. Fish. Res. Lab.* **95**:85–89.

Sasayama, S. 1973. Irradiation preservation of fish meat jelly products. II. Classification of spoilage bacteria in irradiated kamaboko. *Bull. Tokai Reg. Fish. Res. Lab.* **75**: 39–46.

Schlech, W. F., Lavigne, P. M., Bortolussi, R. A., Allen, A. C., Haldane, E. V., Wort, A. J., Hightower, A. W., Johnson, S. E., King, S. H., Nicholls, E. S., Broome, C. V. 1983. Epidemic listeriosis—evidence for transmission by food. *N. Eng. J. Med.* **308**:203–206.

Shewan, J. M., Hobbs, G., and Hodgkiss, W. 1960. The *Pseudomonas* and *Achromobacter* groups of bacteria in the spoilage of marine white fish. *J. Appl. Bacteriol.* **23**:463–468.

Shinagawa, K., Konuma, H., Tokumaru, M., Takemasa, N., Hashigiwa, M., Shigehisa, T., and Lopes, C. A. M. 1988. Enumeration of aerobic sporeformers and *Bacillus cereus* in meat product additives. *J. Food Protect.* **51**:648–650.

Shkol'nikova, S. S. 1970. Microbiological control of minced fish products. *Rybnoe Knozyaistvo* **46**:66–69.

Sonu, S. 1986. Surimi. National Oceanic and Atmospheric Administration Technical Memorandum NMFS–SWR–013.

Spreekens, K. J. A. van. 1969. Changes in the quality of ice-stored cod and prepacked fresh cod fillets. *Conserva* **17**:327–333.

Surkiewicz, P. F., Hyndman, J. B., and Yancey, M. V. 1967. Bacteriological survey of the frozen prepared foods industry 11. Frozen breaded raw shrimp. *Appl. Microbiol.* **15**:1–9.

Suzuki, T. 1981. *Fish and Krill Protein: Processing Technology*. Applied Science Publishers, London, England.

Uchiyama, H., and Amano, K. 1959. The mechanism of the slime formation on sugared kamaboko. Identification of dextran with enzyme preparation from *Penicillium funiculosum*. *Bull. Jpn. Soc. Sci. Fish.* **24**:840.

United States Department of Commerce. 1987. European and American surimi developments. *Marine Fish. Rev.* **49**:69–72.

Vanderzant, C., Matthys, A. W., and Cobb, B. F. 1973. Microbiological, chemical and organoleptic characteristics of frozen breaded raw shrimp. *J. Milk Food Technol.* **36**:253–261.

Varga, S., Sims, G. G., Michalik, P., and Regier, L. W. 1979. Growth and control of halophilic microorganisms in salt minced fish. *J. Food Sci.* **44**:47–50.

Vondruska, J., Kinoshita, R., and Mileazzo, M. 1989. Situation and outlook for surimi and surimi seafood. National Marine Fisheries Service, United States Department of Commerce, Washington, D.C.

Watts, B. K., and Merrill, A. L. 1963. *Composition of Foods: Raw, Processed, Prepared.* United States Department of Agriculture, Washington, D.C.

Weagant, S. D., Sado, P. N., Colburn, K. G., Torkelson, J. D., Tanley, F. A., Krane, M. H., Shields, S. C., and Thayer, C. F. 1988. The incidence of *Listeria* species in frozen seafood products. *J. Food Protect.* **51:**655–657.

Wuthe, H. H., and Findel, G. 1972. Salmonellae in flat fishes from coastal waters. *Arch. Lebens. Hyg.* **23:**110–111.

Yoon, I. H., Matches, J. R., and Rasco, B. 1988. Microbiological and chemical changes of surimi-based imitation crab during storage. *J. Food Sci.* **53:**1343–1346.

6

U.S. Seafood Inspection and HACCP

E. Spencer Garrett and
Martha Hudak-Roos

Currently there is increasing consumer concern and public perception that seafood is unsafe and requires more federal inspection effort. Such perceptions stem from highly vocal consumer organizations such as "Public Voice" and others calling for congressional remedy (Haas 1989, a,b,c; Public Voice 1986, 1987, 1989; Uva 1989). Seafood inspection is therefore one of the most discussed topics in regulatory food control systems. However, contrary to popular belief, there are federal/state regulatory seafood surveillance programs carried out by federal and state agencies.

FOOD AND DRUG ADMINISTRATION (FDA)

The FDA had one of the first programs to deal specifically with seafoods at the federal level, which began in 1926 and was augmented with the passage of the Public Health Services Act of 1946. This special program was a voluntary federal/state/industry initiative aimed toward protecting consumers from salmonellosis (mainly typhoid) in the consumption of raw molluscan shellfish. The cooperative program became known as the National Shellfish Sanitation Program (NSSP) and during its zenith in the late 1950s and early 1960s gained wide acceptance among federal, state, and industry participants, evidenced by the deployment of over 1,000 personnel utilizing 50 laboratories, 500 boats, and 8 planes taking part in some aspect of growing water classifications, shoreline sanitary surveys, patrol of harvest areas, plant inspections, and product evaluations (Garrett 1986). This program, initially operated by the U.S. Public Health Service (USPHS), was transferred to the FDA in 1968. The tripartite basis for management of the NSSP has been modified substantially by the formation of the Interstate Shellfish Sanitation Conference (ISSC) in 1982. The ISSC is modelled after the Interstate Milk Shippers Conference. The purpose of the ISSC is to foster and improve shellfish sanitation by providing to participating states a more direct and organized forum and opportunity to exert more direct leadership.

Another cooperative seafood surveillance program operated by FDA, states, and industry participants is the Canned Salmon Control Plan. This specialized program was initiated in the mid-1930s and has been modified over the intervening years. The genesis of the program was to initiate a practical and effective regulatory system for canned salmon, given the extreme difficulty of short harvesting and processing seasons and the geographical remoteness of processing facilities scattered over wide areas in the state of Alaska.

In addition to the two aforementioned voluntary cooperative specialized seafood surveillance programs, FDA also operates a mandatory general compliance program to ensure that domestically produced and imported seafoods are in compliance with the Food, Drug, and Cosmetic Act of 1938 as amended. That act represents the most recent legislative authority for mandatory surveillance of all fishery products and provides the basis for FDA to operate a general compliance program for seafoods.

U.S. DEPARTMENT OF COMMERCE (USDC)

In 1946, the Congress enacted the Agricultural Marketing Act to create a voluntary inspection program for fishery products. That act was focused toward developing voluntary U.S. Grade Standards for Fishery Products and inplant process and product inspections. These standards allow for the use of U.S. inspectional and grading marks on fishery packages as one way to promote products as safe, wholesome, and of high quality. In addition to being a voluntary program, this activity is also a fee for service program in that participants must pay for the federal inspection and grading services. The program was transferred to the U.S. Department of Interior's Bureau of Commercial Fisheries with enactment of the Fish and Wildlife Act of 1956, and subsequently to the U.S. Department of Commerce's National Marine Fisheries Service (NMFS) with the execution of Executive Reorganization Plan Number Four of 1970. The USDC program initially provided for continuous inspection of processing plants and establishments, lot inspection, and grading. Over the last decade, however, the program has been modified to encourage participating plants to develop and implement their own quality control and assurance systems so that NMFS could move away from the historical practice of "continuous" inspection.

U.S. DEPARTMENT OF AGRICULTURE (USDA)

While USDA [through their Food Safety and Inspection Service (FSIS)] has extensive public health inspection systems that combine animal health and human health considerations, they do not inspect seafoods. They do have an agreement with NMFS that authorizes cross-licensed and NMFS-trained FSIS inspectors to certify fishery plants and products under the NMFS program. Also, although not directly related to seafood inspection per se, USDA is the agency within the federal government to develop "Federal Purchasing Specifications" or "Commercial Item Descriptions" for all foods, including seafoods, which the federal government purchases as a large institutional buyer. Often these buying documents are adopted by individual seafood processors as company or "brand" processing specifications.

U.S ENVIRONMENTAL PROTECTION AGENCY (EPA)

EPA has the responsibility to establish pesticide residue tolerances in foods, including seafoods. In this matter it works closely with FDA. In addition, the agency has broad responsibility for restoring and maintaining water quality to provide for the protection and propagation of fish, shellfish, and wildlife. In this effort the agency's primary focus has been on the reduction of pollutants in fish and shellfish growing waters, and conducting research to better employ "micro-biological indicators" in classifying harvest waters.

DEPARTMENT OF DEFENSE (DOD)

The DOD represents one of the largest institutional purchasers of food in the United States. The commodities purchased are used in either institutional food service programs such as troop feeding in military establishments, or for retail sale in military commissaries. Prior to 1977 for fishery products, DOD maintained a large cadre of inspectors to examine plant sanitation and products destined for troop feeding. For reasons of economy, responsibility for this activity relative to seafoods for troop feeding was transferred to NMFS in 1977. Nonetheless, DOD still maintains limited inspectional capability for seafood commodities destined for retail sale in military commissaries.

STATE SEAFOOD INSPECTION ACTIVITIES

Nearly every state has codified all or part of the Food, Drug, and Cosmetic Act of 1938 as amended, and associated regulations. Therefore, every state also engages in some form of seafood inspection. Generally, such inspection programs are modeled after the compliance type approach used by FDA. However, at the state level the multijurisdictional authorities (agencies) extant at the federal level also occur.

INSPECTION APPROACHES

Philosophical and regulatory agency approaches to providing consumer protection in the consumption of animal proteins differ significantly. Fields (1977) pointed out that, in the case of land based animal products, there is a logical federal inspection system approach that links animal health and human health considerations. He further pointed out that such a program is administrated by a single federal agency that allows for uniform state participation. Garrett (1986), on the other hand, indicated that for seafoods no such logical approach is being used, which is resulting in a myriad of differing standards and compliance schemes.

A further complicating factor in terms of seafood consumer protection schemes is that such great reliance is placed on general compliance programs resulting in infrequent on-site inspections. As Hudak-Roos and Garrett (1989) pointed out, although general compliance programs are often accused of providing less protection for consumers than "continuous" inspection, such is not always the case. However, she indicated further that general compliance programs have three major weaknesses, i.e., they are antithetical to what U.S. consumers expect of their food protection systems; they encourage the multiplicity of regulations and regulatory agencies, requiring overlapping resources at the federal and state levels; and they are not based on current technological food protection rationale, namely the Hazard Analysis Critical Control Point (HACCP) concept.

WHAT IS HACCP?

HACCP is an acronym for Hazard Analysis Critical Control Point theory. It is a logical, simple, but highly specialized system of food control designed in a systematic fashion for preventing public health and other problems from occurring (FSIS 1989). The technique applies to production through consumption and is unique from a regulatory perspective in that it is a nontraditional type of noncontinuous inspection representing innovative state-of-the-art-technology.

HACCP has been evolving in the food industry since 1971 and is still an evolving concept. It was deliberated initially by the First National Food Protection Conference (APHA 1972). It received increased regulatory acceptance in 1973 and 1974 due to the threat of botulinum in canned mushrooms. In the U.S. (on three different occasions from 1985 through 1987) it has been recommended by different subcommittees of the National Academy of Sciences (NAS 1985a,b,

1987) to be employed in foods as the inspectional technique of choice. Since 1987, the concept has been further rapidly evolving and is being addressed by a number of organizations including the Campden Food Preservation Research Association (CFPRA), the Food Standards Program of the Codex Alimentarius Commission (CAC), the International Commission for Microbiological Specifications for Foods (ICMSF), the National Oceanic and Atmospheric Administration's National Marine Fisheries Service (NOAA/NMFS), the National Advisory Committee on Microbiological Criteria for Foods (NACMCF), the National Fisheries Institute (NFI), and the National Food Processors Association (NFPA) (CAC 1989; CFPRA 1987; ICMSF 1988; NACMCF 1989; NFI 1989; NFPA 1990; NMFS 1988).

The theory behind the HACCP concept is that, if properly implemented, governmental inspection frequencies should be much less in facilities or on products employing HACCP as opposed to those operating solely under Good Manufacturing Practices (GMPs) or when examining food products for which there is an unknown production and control history (Hudak-Roos and Garrett 1989). Further, proper deployment of a HACCP system overcomes many of the weaknesses inherent in traditional inspection schemes, by a focus of resources that should offer more appropriate cost/benefit ratios (ICMSF 1988).

One of the benefits of a properly operating in-plant HACCP system is that the system separates the essential from the superfluous aspects of microbiological control (NAS 1985a) and negates the need for a great deal of microbiological testing during production with such procedures being generally restricted to verification (Bauman 1989; NACMCF 1989).

HOW IS HACCP APPLIED?

The HACCP approach is a two-step system requiring extensive technological knowledge in the production, processing, and end-use of the specific food products to be covered. The first step of the process is to conduct a comprehensive hazard analysis of the food relative to its intended end-use, including considerations of raw materials, ingredients, role of manipulative processes to control hazards, consumer populations at risk, and epidemiological evidence relative to the potential safety considerations of the food (Hudak-Roos and Garrett 1988). This hazard analysis, sometimes referred to as a "Structured Hazard Analysis," can be conducted in accordance with existing formal procedures (Baird-Parker 1987; Corlett 1989).

The second step of the HACCP process is the determination of each step of a processing (or other application consideration i.e., distribution, food services, etc.) operation; the hazard(s) associated with each step; definition of the preventive measures that can be achieved at each processing step to minimize the hazard(s) to acceptable levels; identification of the critical control points where the hazard(s) can be controlled; determination of monitoring procedures, either by observation and/or physical measurement, which can be relied on to demonstrate control of hazards; and initiation of necessary verification procedures (including records to be shared with a regulatory agency having jurisdiction) to ensure effectiveness of the controls (Corlett 1989; Garrett 1988; Hudak-Roos and Garrett 1989; ICMSF 1988; NAS 1985a). For example, Figure 6.1 indicates one possible generic operational flow diagram of a cooked shrimp process (other variations would relate to different product forms or sequential steps of processing, or the elimination of specific operational steps). Should the "Receiving" step under the indicated cooked shrimp processing scenario be considered one of the critical control points, then Table 6.1 indicates a scope of necessary considerations that should be considered under a HACCP concept, provided the hazards are defined in terms of biological, chemical, and physical parameters (NMFS 1989).

The HACCP concept requires a thorough technical understanding of the individual commodities to which it is to be applied. Due to individual in-plant processing scenarios, when HACCP is to be applied at the processing level each plant will have to have an individual HACCP program.

The National Academy of Sciences has indicated that although the HACCP concept has worked well for low acid canned foods, it has not been successfully transferred to other food commodities for several reasons (NAS 1985a). Included among those reasons was that in a regulatory HACCP development process there must be an understanding and mutually agreed on differentiation of roles between a regulatory agency and the industry to be regulated. The NAS pointed out, for example, that to implement successfully HACCP programs in food commodities the regulatory agency's role should be restricted to:

- Identifying the basic elements of the program on a commodity by commodity basis;
- Determining the adequacy of industry developed HACCP plans;
- Requiring only minimal record keeping requirements to determine compliance;
- Requiring HACCP programs to be mandatory beyond just the processing level;
- Ensuring regulatory personnel are properly trained in HACCP.

The industry responsibilities include:

- Assuming the lead role in developing actual details of the HACCP program;
- Assuring that management and other personnel are properly trained in HACCP concepts;
- Assuring regulatory access to required records for verification of the HACCP system;
- Committing to work cooperatively with the regulatory agency on HACCP modifications.

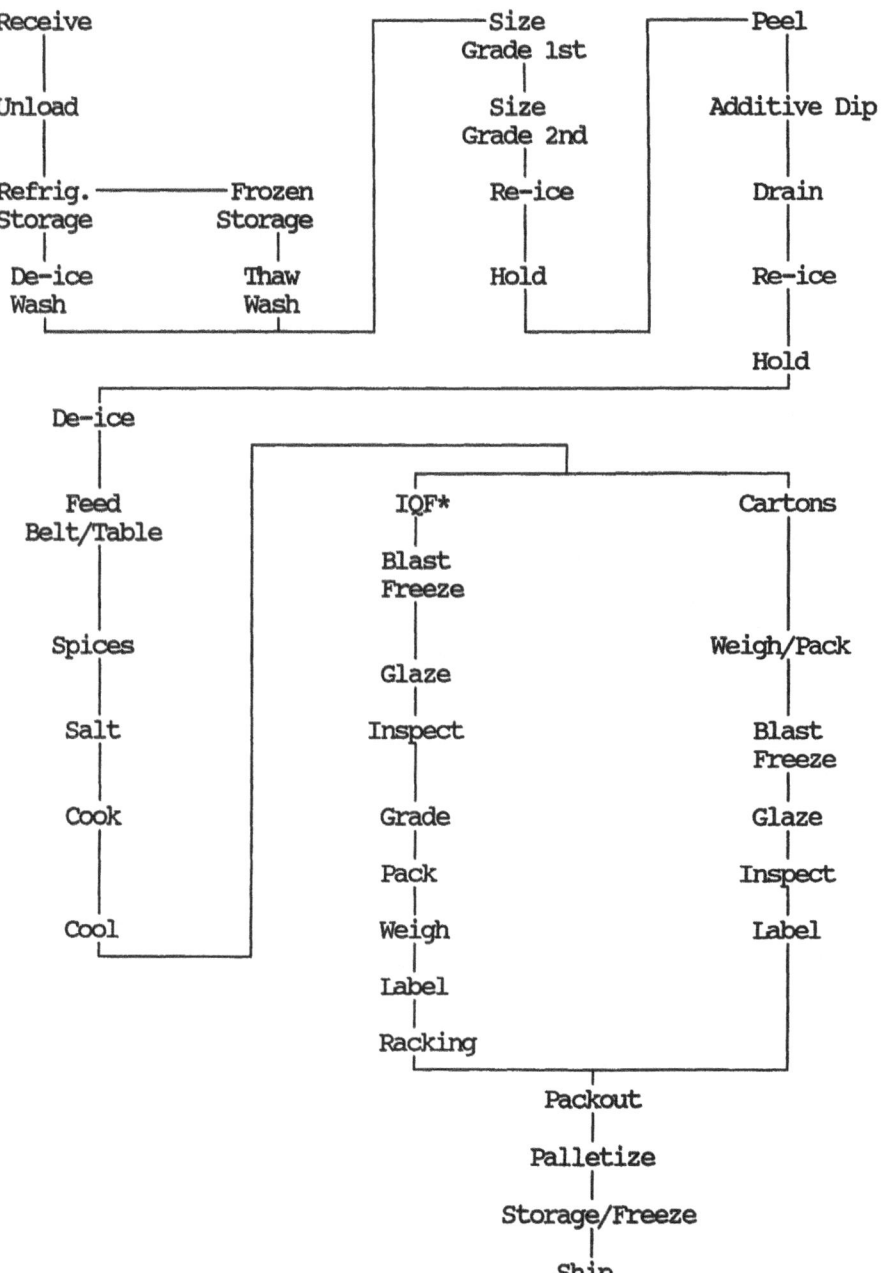

Figure 6.1 A cooked shrimp process flow variable.

Table 6.1 Plant Receiving Steps—HACCP Considerations

Hazard	Control Points	Importance	Preventive Measures	Monitoring	Records
Thermal abuse	Incoming raw materials	CCP	Purchasing specifications	Temperature checks and visual observation	Raw material control report
Decomposition	Receiving room		Control time/temperature abuse	Testing for specification compliance	
Microbial pathogens	Wooden boxes		Control product movement	Sensory and visual examinations for decomposition	
Additives abuse	Unloading area		Adequate physical separation of raw from cooked shrimp		
a. Bisulfite					
b. Sodium Hydroxide					
c. Borates					
d. Phosphates					
e. Chlorine					
Contaminants—filth/extraneous materials					
Integrity of package					
Copacking problems (short weight, dehydration, etc.)					
Adequacy of cooked raw material					
Mixing raw with cooked products					

The NAS further indicated that ". . . There is no fundamental reason why the broad application of HACCP throughout the Food industry should not occur . . ." (NAS 1985a).

Regardless of the evolutionary nature of the HACCP concept as applied to in-plant and regulatory food control systems, it is generally agreed that the concept must be premised upon the following seven basic principles (NACMCF 1989):

1. Assess hazards associated with growing, harvesting, raw materials and in-gredients, processing, manufacturing, distributing, marketing, preparation, and consumption of the food.
2. Determine the Critical Control Points (CCP) required to control the identified hazards.
3. Establish the critical limits that must be met at each identified CCP.
4. Establish procedures to monitor CCPs.
5. Establish corrective action procedures to be taken when there is a deviation identified by monitoring a CCP.
6. Establish effective record keeping systems that document the HACCP plan.
7. Establish procedures for verification that the HACCP system is working correctly.

HACCP BEYOND MICROBIOLOGY

Varying approaches to the development and implementation of HACCP systems are being initiated. However, because HACCP is evolving beyond the scope of low acid canned foods, there currently appear to be points of confusion and controversy concerning definitions, which is not dissimilar to that experienced by the inclusion of microbiological criteria into Codex Alimentarius Codes of Practice (Garrett 1988).

Traditionally, the evolution of the HACCP concept has been developed by microbiologists. While the early proponents of the concept are to be applauded for their conceptual contributions, they limited their focus to controlling microor-ganisms or unacceptable by-products of microbiological etiology. There is one school of opinion that such a narrow focus is hindering the evolution of the HACCP concept potential into full incorporation in food protection programs (Garrett 1989).

Bryan (1988), for example, in his explanation of what the HACCP system is and is not, defines a hazard as either unacceptable contamination (which receives little further definition), and/or microbiological unacceptability. His definition of a critical control point (CCP) has a microbiological focus. On the other hand, Bryan diffuses the role of HACCP in inspection. However, it could be viewed that what Bryan describes as verification can be the regulatory agency inspection role. His conclusion, then, that ". . . Food protection programs must evolve from the inspection mentality that is so frequently associated with them . . ." can be based upon HACCP and become institutionalized.

The question remains: Should the HACCP concept be focused only on microbiological issues in foods, or should the concept be expanded successfully to take into account the full range of consumer hazards in food consumption regarding food safety, plant/food hygiene, and economic fraud issues? This view, at least in part, seems supported by Corlett (1989) when he indicated that ". . . Generally any hazard may be addressed using HACCP, including chemical and physical hazards." However, after making the statement Corlett still focuses only on microbiological hazards in his risk assessment and assignment of hazard categories protocol.

In terms of CCP definitions there are several more inconsistencies. For example, in the Low Acid Canned Food (LACF) Regulations (CFR 1987a), which are often referred to as exemplary of HACCP regulations, a critical control point is not defined. Rather, those regulations refer to critical factors primarily relating to scheduled processes and the attainment of commercial sterility. Parallel regulations dealing with FDA's umbrella GMP define a critical control point in extremely broad fashion. There it is defined as points in a food process where there is a high probability that loss of control may allow or contribute to a hazard to health or in the decomposition of the final food (CFR 1987b). One difficulty with the latter definition is that any processing step may materially contribute to either entity.

On the other hand, the ICMSF defines critical control points in a bifurcated fashion in that a CCP1 will ensure absolute control of a hazard generally through a physical or chemical process, whereas a CCP2 will minimize but not ensure control of any specific hazard (ICMSF 1988). Again, this is an example of a process control type of definition incorporating a lethal process focusing primarily on microbiological considerations.

In the USA, the National Marine Fisheries Service (which was mandated by the Congress to design an improved seafood surveillance system for seafoods utilizing the HACCP concept) has taken a departure from traditional CCP definitions and has added plant/food hygiene and economic risk considerations by defining a critical control point as a ". . . specific operational step of a food manufacturing process, the failure of which may automatically result in an unacceptable consumer health or economic risk . . ." (NMFS 1989b).

Further confusion in terms of HACCP understanding relate to harmonizing the various definitions currently being used in the area of sanitation versus process CCPs, so that sanitation considerations as well as those of process control can be fully integrated into the HACCP concept. For example, as previously indicated, the LACF GMP (in terms of critical areas) has a process control orientation. Because that regulatory program has been successfully implemented since 1974, its orientation has been allowed to permeate the HACCP concept to the exclusion of the cross-cutting sanitation CCPs. The ICMSF CCP definition whereby its CCP1 must ensure control of a hazard whereas its CCP2 cannot ensure control but only minimize the occurrence of the hazard probably represents a transitional evolution to broaden the CCP definition to incorporate sanitation or food hygiene issues. Nonetheless, because HACCP is a total systems approach, it must, in the final analysis, evolve further into addressing all hazards whether they be of microbiological, chemical, physical, or even of economic concern. Critical control points can, in fact, control all of the consumer hazards.

Regardless of the aforementioned divergent view points, HACCP is a workable system, and when properly applied can serve as the major vehicle to focus scarce resources and provide increased consumer protection in the consumption of foods.

HACCP APPLICATIONS TO MARINE FOOD PRODUCTS

Other than for low acid canned foods, there have been few historical attempts to apply HACCP concepts to marine food products. Some early concepts for applying HACCP to the seafood industry were developed in the mid 1970s (Peterson and Gunnerson 1974). Also during the mid 1970s, an attempt was made to classify fishery products by their microbial risk beyond the traditional approach, so that more detailed risk evaluations could be factored into focusing regulatory microbiological monitoring efforts (Garret, et al. 1977).

For seafoods, Lee (1977) was probably the first to publish a detailed explanation on how the HACCP technique could be employed. In conducting a hazard analysis, Lee arrayed fishery products in order of decreasing consumer risk and concluded that heat processed foods usually comsumed with no additional cooking prior to consumption were of the highest priority as indicated in Table 6.2.

He next formulated operational process flow charts for various fishery products by listing each step of a processing scenario for specific seafood commodities. Figure 6.2 indicates his flow chart for the smoked fish industry. Lee pointed out that, although generically there are 13 steps involved in the commercial process of smoking fish, only three of those are critical in relation to microbial pathogens. Those were:

1. Brining to ensure proper salt penetration of the product;
2. Smoking to ensure proper thermal penetration of the product;
3. Storing to ensure proper refrigeration temperatures to prevent microbiological growth and possible toxin elaboration.

Table 6.2 Lee's Seafood Analysis

	Hazard Analysis	
Category	*Description*	*Example*
1	Heat processed foods, usually no additional cooking	
2	Non-heat-processed raw foods often eaten with no additional cooking	Shucked molluscan shellfish eaten raw
3	Formulated foods usually eaten after cooking	Fish sticks, breaded shrimp
4	Non-heat-processed raw foods usually eaten after cooking	Fresh or frozen fish fillets and cooked molluscan shellfish
5	Raw seafoods usually eaten after cooking	Live crustacean and molluscan shellfish

Despite these early efforts, HACCP has yet to be widely transferred to the seafood industry as either a method for in-plant control or regulatory compliance. To date, the major stumbling blocks on transferring the HACCP concept have centered on the lack of regulatory agency and industry understanding of the concept, and joint active participation in its development.

Nevertheless, it is anticipated that the transfer of the HACCP concept to the seafood industry will be greatly accelerated during the 1990s. In large part this is

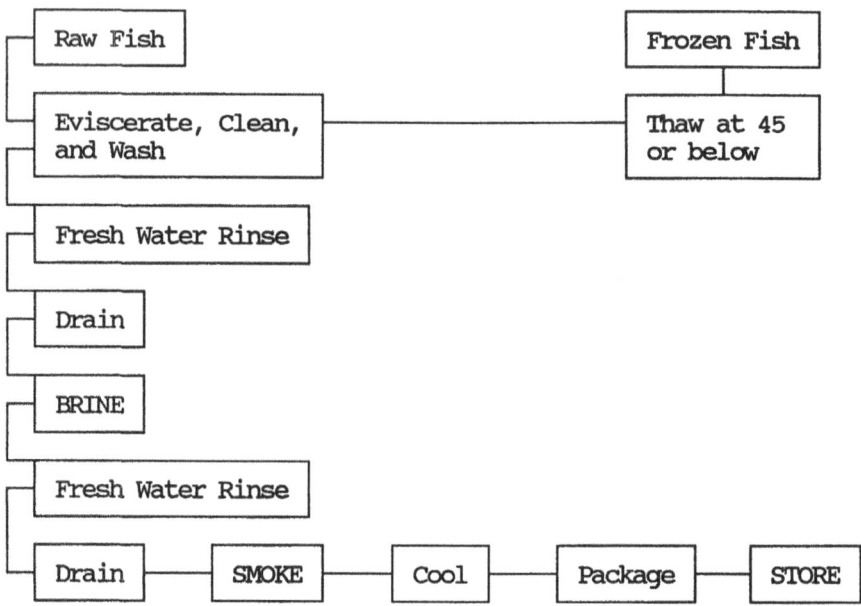

Figure 6.2 Smoked fish processing steps.

due to Congressional interest spurred on by industry trade associations (NFI 1989) and consumer organization requests to enact some form of mandatory federal seafood inspection based upon the HACCP concept.

Congress has directed NOAA/NMFS to develop a program of mandatory seafood surveillance based on HACCP. NMFS (1988) has initiated a study entitled the Model Seafood Surveillance Project (MSSP) to accomplish this task. To overcome the regulatory pitfall of scant industry involvement in developing HACCP regulatory mechanisms, the MSSP has conducted a series of joint industry/federal/ state HACCP application workshops for seafoods on a commodity by commodity basis as well as for regional harvesting vessels. The final results of this study are anticipated to be reported to the U.S. Congress at the close of 1990 (NMFS 1989b), though an interim progress report was published in early 1990 (NOAA, 1990).

To date, numerous workshops have been held and regulatory HACCP models have been developed for raw shrimp, breaded shrimp, cooked shrimp, fresh and frozen fish, breaded and specialty items, blue crab, and imports (NMFS 1989b,c,d, 1990a,b). Each of those HACCP models details a hazard analysis for the commodity as well as for each operational step of the process. In addition, the preventive measures that can be taken at each operational step to reduce the hazards to acceptable levels are defined. Further, the models identify each CCP of the process and list the necessary types of monitoring procedures at the CCPs as well as records necessary to share with regulatory agency to prove the HACCP system is working. The NMFS regulatory HACCP models also detail the necessary industry commitments, regulatory controls, research focus, and consumer education requirements necessary to make a regulatory HACCP scheme work for seafoods (NMFS 1989a).

HACCP SAMPLING PROCEDURES

Sampling is the process of examining randomly selected sample units from a lot (or process) to infer product compliance with some stated criteria. Although a myriad of differing sampling plans can be developed for this purpose, three of the most important questions in sampling are:

1. Is the sampling plan and associated decision criteria feasible and cost effective?
2. What is a lot in relationship to a sample unit and how are the sample units selected?
3. How well will the plan identify lots of varying quality?

Inherent in the application of sampling plans is the risk of making an incorrect decision on the actual lot status (i.e., rejection of a conforming lot or acceptance of a nonconforming lot). Decision makers must be aware that an inadequate understanding of these risks may lead to a false sense of security. For any given sampling plan, it is possible to describe mathematically the probabilities of making incorrect or undesirable decisions when confronted with lots of varying quality. This description is usually referred to as the sampling plan's "performance characteristics" and can be plotted as an operating characteristic (OC) curve and/or presented in tabular form. There are numerous references in the literature that elaborate on performance characteristics, but for the lay person the authors recommend those published by the NAS (1985a) and the ICMSF (1986).

For the simplest type of sampling plan, a 2-class attribute plan (i.e., each sample unit selected is judged to be conforming or nonconforming or the number of defects per sample unit is counted), a few selected values on the OC curve can be used to describe the performance of that plan. Conventionally, these so-called reference values are the AQL (Acceptable Quality Level, i.e., the actual percent nonconforming sample units in the entire lot for which the sampling plan will indicate lot acceptance 95% of the time), the IQL (Indifference Quality Level, i.e., the actual percent nonconforming sample units in the entire lot for which the sampling plan will indicate lot acceptance 50% of the time), and the RQL (Rejectable Quality Level, i.e., the actual percent nonconforming sample units in the entire lot for which the sampling plan will indicate lot acceptance 10% of the time). In the following 2-class sampling plan example, these reference values are given and further explained.

Suppose a 10,000-pound lot containing 2,000 5-pound cartons of a certain product is to be evaluated for one or more criteria based on the 2-class plan $n = 6$ and $c = 1$, where n equals the number of cartons drawn from the lot, and c equals the acceptance number (the number of nonconforming sample units that will be allowed before rejection of the lot). This sampling plan has the following reference performance characteristics: AQL = 6.3%; IQL = 26%; RQL =51%. *Interpretation:* For a stated criterion:

AQL: If the lot (2,000 5-pound cartons) actually has 6.3% (126) nonconforming cartons, this sampling plan will indicate lot acceptance 95% of the time (or lot rejection 5% of the time). *Note:* Implicit in the selection of a sampling plan is that the plan's AQL represents for the producer, consumer, seller, buyer, etc., an agreed on percent nonconforming for which the sampling plan should accept the lot most of the time (usually taken as 95% of the time).

IQL: If the lot (2,000 5-pound cartons) actually has 26% (520) nonconforming cartons, then this sampling plan will indicate lot acceptance 50% of the time (or lot rejection 50% of the time).

RQL: If the lot (2,000 5-pound cartons) actually has 51% (1020) nonconforming cartons, then this sampling plan will indicate lot acceptance 10% of the time (or lot rejection 90% of the time.) *Note:* Implicit in the selection of a sampling plan is that the plan's RQL represents for the producer, consumer, seller, buyer, etc., an agreed on percent nonconforming for which the lot should be rejected most of the time (usually taken as 90% of the time).

It has been the authors' experience that many individuals unfamiliar with the intricacies of sampling assume that to increase the sensitivity of a sampling plan, more samples must be taken. Such is not necessarily the case, for one could change the decision criteria. For example, rather than using the sampling plan $n = 6$ and $c = 1$, suppose an alternative plan of $n = 6$ and $c = 0$ were considered [i.e., accept the lot if none of the six sample units (cartons) are nonconforming; otherwise, reject the lot]. Without changing sample size ($n = 6$), what affect would changing the acceptance number c from 1 to 0 have on the reference performance characteristics? This information is given in Table 6.3.

This alternative plan ($n = 6$, $c = 0$) is certainly more favorable to the consumer or buyer in that less nonconforming product will be accepted; however, the relatively small AQL may be too restrictive on the producer or seller. For $n = 6$, an increase in sampling plan sensitivity by changing the decision criteria (c from 1 to 0) is currently employed by the FDA in the microbiological examination of fresh and frozen crabmeat for adulteration with filth as evidenced by the presence of *Escherichia coli* (FDA 1982). In this regulatory scheme, the critical limit for *E. coli* is 3.6/g (Imvic confirmed). It has a two-part application depending on whether inspectional information is sufficient to indicate the most probable source of the *E. coli*. When this is not the case, as could be expected for imported product, then the sampling plan $n = 6$, $c = 1$ is used. On the other hand, where regulatory inspectional evidence can indicate the most probable source of microorganisms, the sampling plan $n = 6$, $c = 0$ is used.

Of concern to many contemporary microbiologists with this current FDA regulatory scheme is not the sensitivity of the sampling plans, but rather of the critical limit of *E. coli* being targeted at 3.6/g. When determining *E. coli*, the most probable number (MPN) technique is used. This statistical estimation procedure provides a population estimate of the *E. coli* actually present in the product. Unfortunately, this estimate is subject to a wide range of variation. For any given analysis, the estimated MPN value may differ appreciably from the actual value (Speck 1984). Furthermore, the value of 3.6/g *E. coli* is at the lowest end of the sensitivity measurement for the microbiological methods

Table 6.3 Effect on AQL, IQL, and RQL of Changing c from 1 to 0 for $n = 6$

Sample Plan	AQL (%)	IQL (%)	RQL (%)
$n = 6, c = 1$	6.3	26	51
$n = 6, c = 0$	0.85	11	32

employed. Some microbiologists (Archer 1989) favor a modest increase in the critical limit of 3.6/g to compensate for the statistical variation inherent in the MPN method.

Not all sampling plans are of the 2-class type. Frequently in practice, 3 (or more) class plans are used. This is particularly the case when food lots are examined microbiologically for compliance with a microbiological standard, tolerance or guideline. In a 3-class (or more) plan, the concept of lowercase ("little") m and uppercase ("big") M is employed. The value m usually represents the numeric limit for a target microorganism in a product manufactured under GMPs. Values less than m imply the process is in control, whereas values greater than m but not M indicate that the process may be going out of control. M represents the numeric limit (of the same target microorganism) that is considered unacceptable due either to a microbial hazard or a level of contamination resulting from inappropriate hygiene practice or storage. The entire lot is rejected when any sample unit exceeds M (ICMSF 1986).

For example, consider the following 3-class plan for *Staphylococcus aureus;* five sample units are analyzed. For the lot to be in compliance, none may exceed 1,000/g and no more than one of the five may exceed 500/g. This 3-class plan is conventionally represented as follows: $n = 5, c = 1, m = 500/g, M = 1,000/g$.

For 3 or more class sampling plans, a simple description of their reference performance characteristics using one set of values for the AQL, IQL, and RQL is not feasible. This representation would have to be reported for selected values between m and M at fixed levels greater than M. This description soon becomes complicated and of questionable practical use. Generally, the most useful representation of performance characteristics for 3 or more class plans is to employ two or more way tables.

For example, the performance characteristics for the 3-class plan given above, i.e., $n = 5, c = 1, m = 500/g, M = 1000/g$, can be presented as follows:

Let

$P = $ the actual percent of all possible sample units in the lot exceeding m but not $M;$ and

$Q = $ the actual percent of all possible sample units in the lot exceeding M.

Then the probabilities (%) of lot acceptance, for selected values of P and Q, are given in Table 6.4.

The value of a 3-class sampling plan is that it allows for an agreed on variance to a stated standard in order to provide for production inconsistencies. Further, when no sample unit exceeds M and the number of sample units exceeding m is less than or equal to c but greater than zero (i.e., lot is accepted but transitional concern), regulatory agencies may initiate some action directed toward the manufacturer/distributor/retailer/institution/chain. This action may

Table 6.4 Probabilities (%) of Lot Acceptance for the 3-class Plan $n = 5$, $c = 1$, $m = 500/g$, $M = 1000/g$ for Selected Values of P and Q

	Q			
P	0	10	20	30
0	100	59	33	17
10	92	53	29	14
20	74	41[a]	21	9
30	53	27	12	5
40	34	16	6	2

[a]*Interpretation:* Assume a lot consists of 40,000 pounds of product. When this lot contains $P = 20\%$ (8,000 pounds) exceeding 500/g but not exceeding 1,000/g and $Q = 10\%$ (4,000 pounds) exceeding 1,000/g, this 3-class plan will indicate lot acceptance 41% of the time. In like manner, the other percentages in the table can be interpreted.

range from simple notification of findings to increasing inspectional frequency, resampling, or a warning letter.

APPENDIX: GLOSSARY OF TERMS

Commodity: An article of trade or goods. Within the context of the MSSP—seafood products.

Control Point (NACMCF): Any point in a specific food system where loss of control does not lead to an unacceptable health risk.

Control Point (NMFS): Specific operational steps of a food manufacturing process whereby biological, chemical, physical, and/or economic factors may be controlled.

Critical Control Point (NACMCF): Any point or procedure in a specific food system where loss of control may result in an unacceptable health risk.

Critical Control Point (NMFS): Specific operational steps of a food manufacturing process, the failure of which may automatically result in an unacceptable consumer health or economic risk.

Critical Control Point (FDA-CFR, Title 21, Part 110): A point in a food process where there is a high probability that improper control may cause, allow, or contribute to a hazard or to filth in the final food or decomposition in the final food.

The majority of definitions of terms are taken from the NMFS Plan of Operations from the Model Seafood Surveillance Project. Where definitions differ from those developed by the National Advisory Committee on Microbiological Criteria for Foods, or other organizations, the differences are noted.

Critical Limit (NACMCF): One or more prescribed tolerances that must be met to ensure that a critical control point effectively controls a microbiological health hazard.

HACCP Plan (NACMCF): The written document that delineates the formal procedures to be followed in accordance with the HACCP general principles.

HACCP (NMFS): Hazard Analysis Critical Control Point. A specific non-traditional inspectional approach to control biological, chemical, physical, and/or economic hazards in foods. It is a two-part process done on a commodity-by-commodity basis. The first deals with defining the consumer hazards relative to the intended end use of the specific food commodity. The second part deals with flow charting each operational step of a food manufacturing process, defining the hazard associated with each step and assessing its relative importance, determining the preventive measures that can be used to reduce the hazard to acceptable levels, identifying the critical control points of the manufacturing process, determining the monitoring procedures either by observation or measurement to determine that the hazard is being controlled, and delineating the specific manufacturing records necessary for monitoring to ensure hazards are being controlled.

Food Hygiene: Characteristics of a product or process relating to quality, wholesomeness, or sanitation.

Food Quality: The inherent properties of a food that affects its merchantability.

HACCP System (NACMCF): The result of the implementation of the HACCP principles.

Hazard (NACMCF): Any biological, chemical, or physical property that may cause an unacceptable consumer health risk.

Hazard (NMFS): A chance for, or the risk of, an unacceptable biological, chemical, physical, and/or economic property in a food product that may cause consumer illness or distress.

Hazard Analysis: The identification of biological, chemical, physical, and/or economic chances or risks relative to a food product or manufacturing process that takes into consideration the intended end use of the food product.

Monitoring (NACMCF): A planned sequence of observations or measurements of critical limits designed to produce an accurate record and intended to ensure that the critical limit maintains product safety.

Monitoring (NMFS): The act of observation or measurement to record the control of hazards.

PLANOP: Plan of Operations—A planning document detailing specific operating protocols. A PLANOP focuses and delineates the objectives regarding a given goal by describing the problem and the steps necessary for its investigation, assigning responsibility for completion, and establishing methods and time schedules for the systematic completion of the protocol. The PLANOP also provides criteria by which the success or failure of the protocol can be evaluated.

Quality Control: The establishment and maintenance of optimum product and process attributes through systematic and coordinated efforts.

Quality Assurance: The sum of all those activities in which one engages to ensure that information and data generated is correct and reliable.

Risk: An estimate of the likely occurrence of a hazard or danger.

Risk Category: One of six categories prioritizing risk based on food hazards.

Sample: Any number of sample units to be used for examination.

Sample Unit: A container and/or its entire contents, a portion of the contents of a container or other unit of commodity, or a composite mixture of a product to be used for examination.

Sensitive Ingredient: Any ingredient historically associated with a known micro-biological hazard.

Significant Risk: Posing moderate likelihood of causing an unacceptable health risk.

Wholesomeness: The sound condition of a food product. The item is clean, free from adulteration, and otherwise suitable for use as human food.

Workshop: Scheduled planning meetings necessary for the industry identification of operational steps, hazards, preventative measures, critical control points monitoring procedures, and records relevant to a specific seafood commodity.

REFERENCES

APHA. 1972. *Proceedings of the 1971 National Conference on Food Protection.* Food and Drug Administration. Washington, D.C.

Archer, D. 1989. Director of Division of Microbiology, Center for Food Safety and Applied Nutrition, Food and Drug Administration. Washington, D.C.

Baird-Parker, A. C. 1987. The application of preventative quality assurance. In elimination of pathogenic organisms from meat and poultry. In: F.J.M. Smulders, (ed.) *Proceedings of a Symposium on Contamination, Prevention and Decontamination in the Meat Industry*, Zeist, The Netherlands, 2–4 June 1986. Elsevier Biomedical, Amsterdam, pp. 149–161.

Bauman, H. 1989. HACCP Implementation Considerations. Unpublished proceedings of FSIS Management Seminar. U.S. Department of Agriculture, FSIS, Washington D.C.

Bryan, F. L. 1988. Hazard analysis critical control point: what the system is and what it is not. *J. Environ. Hlth.* **50**:400.

CAC. 1989. Codex Alimentarious Commission. Alinorm 89/13. Report of the twenty-third session of the Codex Committee on Food Hygiene, Via delle Terme di Caralacalla, Rome.

CFPRA. 1987. Guidelines to the Establishment of Hazard Analysis Critical Control Point (HACCP): Technical Manual No. 19. Campden Food Preservation Research Association, Gloucestershire.

CFR. 1987a. Code of Federal Regulations, Title 21, Part 113. U.S. Government Printing Office, Washington, D.C.

CFR. 1987b. Code of Federal Regulations, Title 21, Part 110. U.S. Government Printing Office, Washington, D.C.

Corlett, D. 1989. *The Hazard Analysis and Critical Control Point System.* Food Beverage Technology International USA, pp. 220–222.

FDA. 1982. *Fish and Seafood Compliance Policy Guides: Crabmeat—Fresh and Frozen—Adulteration with Filth Involving the Presence of the Organism* Escherichia coli, *CPG 7108.02.* Food and Drug Administration. Washington, D.C.

Fields, F. R. 1977. Public Health Constraints on Food Production from the Sea. Presented at Fifth Food-Drugs from the Sea Conference, University of Oklahoma. Norman, Oklahoma, September 1977.

FSIS. 1989. The Hazard Analysis and Critical Control Point (HACCP) System and the Food Safety and Inspection Service concept paper. U.S. Department of Agriculture, Washington, D.C.

Garrett, E. S. 1986. Role of diseases in marine fisheries management. In: *Transactions of 50th North American Wildlife and Natural Resources, NOAA Conference, Tech. Memo. NMFS, F/NWR-16.* National Marine Fisheries Service, Washington D.C.

Garrett, E. S. 1988. Microbiological standards, guidelines, and specifications and inspection of seafood products. *Food Technol.* **42:**90–93, 103.

Garrett, E. S., and Hudak-Roos, M. 1989. Seafood surveillance and certification based upon the HACCP system. Unpublished Proceedings of Institute of Food Technologists Annual Meeting, June 25–29, 1989. Chicago, Illinois.

Garrett, E. S., Haines, G., Brooker, J., Hamilton, R., Billy, T., and Stocks, P. 1977. A new method to classify fishery products by microbial risk. In: *Proceedings of the 50th Anniversary of the Torrey Research Laboratory,* July. Aberdeen.

Haas, E. 1989a. Unpublished statement before NAS Subcommittee on Evaluation of Fishery Products. January 31, National Academy of Sciences, Washington, D.C.

Haas, E. 1989b. Testimony before Subcommittee on Oversight and Investigations, unpublished, U.S. House of Representatives, May 31, Washington, D.C.

Haas, E. 1989c. Testimony before Subcommittee on Fishery, Wildlife and Conservation, U.S. House of Representatives, unpublished, June 7, Washington, D.C.

Hudak-Roos, M., and Garrett, E. S. 1988. Model seafood surveillance project: an update. In: *Proceedings of the 13th Annual Conference of the Tropical and Subtropical Fisheries Technological Society of the Americans. SGR-94: 6-13.* Gulf Shores, Alabama.

Hudak-Roos, M., and Garrett, E. S. 1989. New approaches to seafood inspection. In: B. A. Twigg (Am. ed.) and Alan Turner (European ed.) *Food Beverage Technology International USA.* Sterling Publications Limited, Garfield House, London, pp. 223–224.

ICMSF. 1986. *Microorganisms in Foods 2, Sampling for Microbiological Analysis: Principles and Specific Applications,* second edition. International Committee on Microbiological Specifications for Food. University of Toronto Press, Toronto.

ICMSF. 1988. *Microorganisms in Foods 4, Application of the Hazard Analysis Critical Control Point (HACCP) System to Ensure Microbiological Safety and Quality.* International Committee on Microbiological Specifications for Food. Blackwell Scientific Publications, Oxford.

Lee, J. 1977. Hazard analysis and critical control point applications to the seafood industry. December 1977, ORESU-H-77. Oregon State University, Sea Grant College Program.

NACMCF. 1989. Hazard Analysis and Critical Control Point System. Adopted Report. U.S. Department of Agriculture, Washington, D.C.

NAS. 1985a. *An Evaluation of the Role of Microbiological Criteria for Foods and Food Ingredients.* National Academy of Sciences, National Academy Press, Washington, D.C.

NAS. 1985b. *Meat and Poultry Inspection, The Scientific Basis of the Nation's Program.* National Academy Press, National Academy of Sciences, Washington, D.C.

NAS. 1987. *Poultry Inspection, The Basis for a Risk-Assessment Approach.* National Academy of Sciences, National Academy Press, Washington, D.C.

NFI. 1989. Testimony by Mr. Lee Weddig, Executive Vice President, National Fisheries Institute before House Subcommittee on Fisheries and Wildlife Conservation and the Environment on Consumer Seafood Safety Act of 1989, H.R. 2511, June 7, unpublished, Washington, D.C.

NFPA. 1990. *A Technical Manual for Implementing Hazard Analysis Critical Control Systems.* National Food Processors Association, Washington, D.C. (in press).

NMFS. 1988. Plan of operations—Model Seafood Surveillance Project. National Marine Fisheries Service, Office of Trade and Industry Services, Washington, D.C.

NMFS. 1989a. Plan of operations—Model Seafood Surveillance Project. National Marine Fisheries Service, Office of Trade and Industry Services, Washington, D.C.

NMFS. 1989b. HACCP model cooked shrimp. Report of the Model Seafood Surveillance Project. National Oceanic and Atmospheric Administration, National Marine Fisheries Service, Office of Trade and Industry Services, September, Washington, D.C.

NMFS. 1989c. HACCP model for raw shrimp. Report of the Model Seafood Surveillance Project, National Oceanic and Atmospheric Administration, National Marine Fisheries Service, Office of Trade and Industry Services, Washington, D.C.

NMFS. 1989d. HACCP model for breaded shrimp. Report of the Model Seafood Surveillance Project, National Oceanic and Atmospheric Administration, National Marine Fisheries Service, Office of Trade and Industry Services, Washington, D.C.

NMFS. 1990a. HACCP model for fresh and frozen fish. Report of the Model Seafood Surveillance Project, National Oceanic and Atmospheric Administration, National Marine Fisheries Service, Office of Trade and Industry Services, Washington, D.C. (in press).

NMFS. 1990b. HACCP model for breaded and specialty fishery products. Report of the Model Seafood Surveillance Project, National Oceanic and Atmospheric Administration, National Marine Fisheries Service, Office of Trade and Industry Services, Washington, D.C. (in press).

NOAA. 1990. Meeting the challenge—new directions for seafood inspection. Report to Congress on the Model Seafood Inspection Project. National Oceanic and Atmospheric Administration, National Marine Fisheries Service. Pascagoula, Miss.

Peterson, A. C., and Gunnerson, R. 1974. Microbiological critical control points in frozen foods. *Food Technol.* **28:**37–44.

Public Voice. 1986. The great American fish scandal: health risks unchecked. A report by Public Voice for Food and Health Policy, Washington, D.C.

Public Voice. 1987. Hazardous fish: the raw facts. A report by Public Voice for Food and Health Policy, Washington, D.C.

Public Voice. 1989. Contaminated catch: holes in the shellfish safety net. A report by Public Voice for Food and Health Policy, Washington, D.C.

Speck, M. 1984. *Compendium of Methods for the Microbiological Examination of Foods,* second edition. American Public Health Association, Washington, D.C.

Uva, M. 1989. Unpublished statement before NAS Subcommittee on Evaluation of Fishery Products, Jan 31, National Academy of Sciences, Washington, D.C.

Part 2

7

Indicators and Alternate Indicators of Growing Water Quality

Howard Kator and
Martha W. Rhodes

The coliform group of indicator bacteria has traditionally been the basis for the microbiological growing area standard for shellfish waters since the early 1900s. Since that time researchers have identified deficiencies in its use as an indicator of fecal contamination or potential health risk in aquatic systems (e.g., Berg 1978; Cabelli 1978b; Dutka 1973). Responses to these criticisms are evident in a measured progression to adopt the most fecal-specific coliform organisms as approved indicators. Thus, the fecal coliform has supplanted the total coliform group, and eventually, *Escherichia coli* may replace the fecal coliform. However, recent studies challenge the fundamental assumption that *E. coli* is a valid predictor of enteric disease in marine waters (Cabelli et al. 1983). Other workers have identified processes whose effects on indicator densities seriously question the validity of the coliform group as the basis for the growing water standard.

As a result, interest in alternate indicators of fecal contamination has intensified, although information concerning the survival characteristics and recovery methods for alternate indicators is generally restricted to freshwater environments. In Section 901 A entitled "General Discussion" of Standard Methods (APHA 1985), the statement is made with regard to pollution in ". . . estuaries and other bodies of saline water . . ." that "In the following sections, application of specific techniques to saline water are not discussed because the methods used for fresh waters also can be used satisfactorily with saline waters" (APHA 1985, 828). Contrary to Standard Methods (APHA 1985) there is no a priori basis to conclude that enumeration methods for freshwaters will be effective in shellfish growing waters. In fact, much of the literature suggests this assumption is unwarranted.

The purpose of this chapter is to provide an overview of approved and alternate indicators now used or considered for use in shellfish growing waters. We present our discussion from an ecological perspective, because we believe that understanding the interactions and responses of indicators in saline environments is essential to their selection and development of enumeration methods.

HISTORICAL PERSPECTIVES

Use of microbiological water standards to minimize shellfish-transmitted enteric disease has been a rather successful public health strategy, eliminating major outbreaks of gastroenteritis caused by salmonellae. The indicator concept, first elaborated for drinking water where contamination was assessed as an operational response using a multiple-tube enumeration procedure, was later applied to the waters of Raritan Bay in the early 1900s following incidents of typhoid associated with consumption of raw clams (Kehr et al. 1941). A target density of total coliforms was derived based on the dilution of a large point source of domestic sewage with a sufficient volume of water to yield a theoretical final ratio of indicator to bacterial pathogen. Dilution was anticipated to lower the pathogen density to a value yielding an unknown but significantly reduced potential public health risk. Use of a water-based rather than shellfish-based standard was supported by coliform data suggesting that bivalves (i.e., hard clams) grown in water at or below that standard indicator level would not bioconcentrate pathogens to densities exceeding a presumed minimum infective dosage. Adoption of a surrogate (indicator) for the pathogen *Salmonella typhi* was expedient for reasons that included the absence of an accurate and selective method for the recovery of this organism, and the inability to predict its occurrence in sewage effluent. Kehr et al. (1941) recognized the standard as an arbitrary index that does not index a predetermined level of risk, requiring verification through epidemiological investigation. "It is believed, therefore, that the most favorable method of reducing the danger of infection from the ingestion of raw hard clams is through the adoption of an *arbitrary* standard that would reduce to a satisfactory degree the estimated coliform content of the annual production of Raritan Bay hard clams" (Kehr et al. 1941, 94) (italics our own).

EMERGING ENVIRONMENTAL ISSUES
THAT QUESTION BASIC ASSUMPTIONS

A number of basic assumptions are implicit in the indicator concept. It is assumed that the index or standard is applicable only to a diluted waste effluent, that there exists a constant density ratio of indicator to the pathogen of concern (i.e., *S. typhi*) in the effluent, and that the ratio is conserved in the environment, at least within the immediate vicinity of a discharge (Kehr et al. 1941). However, the authors recognized that if pollution sources departed from these conditions, sole reliance on the bacteriological standard to reflect health risk was unwarranted. Analysis of contributing sources, the interdiction of judgment, and skepticism concerning the ability of a numerical standard value to offer unequivocal health protection were needed to address these circumstances.

Initially, dilution was considered the major physical factor affecting the densities of indicator bacteria, i.e., indicator bacteria were treated as "quasi" conservative elements. We now recognize that the approved coliform indicator for shellfish growing areas does not behave as a conservative element and can exhibit changes in density that are functions of (1) physiological responses to properties of the environment that result in loss of viability, change in culturability, persistence, or aftergrowth, and (2) interactions between the indicator and the indigenous microbiota. These are important processes affecting indicator densities in estuaries where entrainment and flushing characteristics can prolong contact time between indicator organisms and the environment.

Other deficiencies of coliform indicators include evidence for the extraenteral origins of some species that comprise the fecal coliform group (Lopez-Torres et al. 1987), the capacity of organisms defined as fecal coliforms to persist or even grow in aquatic environments under favorable temperature and nutrient regimens (Hazen 1988; Knittel et al. 1977; Rhodes and Kator 1988; Verstraete and Voets 1976), the inability of the indicator to differentiate vertebrate source, the effect of environmental exposure on recoverability (Rhodes et al. 1983; Xu et al. 1982), and the recognition that allochthonous bacterial indicators are prey to microbial grazers (Anderson et al. 1983; McCambridge and McMeekin 1979, 1980; Rhodes and Kator 1988).

If these criticisms promote uncertainty about the validity of the fecal coliform indicator, they must also diminish confidence in its use as a numeric standard applicable to all shellfish growing waters. Arguably, other bacterial indicators may be found that offer advantages in terms of source specificity, determination of source age, or recovery methodology. However, considering the importance of the environment on bacterial indicator fate and recovery, and the inherent regional and temporal variations that characterize dynamic estuarine environments, it is difficult to find credible the notion that a standard based on a bacterial indicator can be "universally applied" to all growing areas (Dutka 1979).

SOURCES AND COMPOSITION OF INDICATOR ORGANISMS IN SHELLFISH GROWING WATERS

Microorganisms used as indicators of fecal contamination are necessarily (but not exclusively) inhabitants of the alimentary tracts of warm-blooded animals. The bacterial composition of mammalian feces, which varies with host species, diet, and geographic location (Feachem et al. 1983), is dominated by obligate anaerobic bacteria belonging to major taxa that include *Bacteroides, Bifidobacterium, Clostridium,* and *Eubacterium.* Facultative anaerobes include the lactobacilli, *Enterobacteriacae,* and the fecal streptococci. Concentrations of these microorganisms range from 10^5–10^{11} organisms/g in the 150 g/day feces produced daily by humans in industrial societies (Feachem et al. 1983). Considering that obligate anaerobes are numerically dominant in human feces (Holdeman et al. 1976; Moore and Holdeman 1974), the use of aerobic recovery methods and facultative anaerobes as fecal indicators appears somewhat paradoxical. There is little doubt this reflects the latter's ease of recovery from aquatic environments compared to the rigorous cultivation procedures needed for obligate anaerobes and the presumption of poor survival.

Domestic and feral animals are important potential sources of fecal contamination in urban and rural areas (Feachem et al. 1983; Geldreich 1972). The microbial composition of domestic animal feces has been examined to catalog dominant genera (Barnes 1986), to compare ratios of common indicators to those in humans (Geldreich and Kenner 1969), and to identify indicator microorganisms unique to animals (e.g., Cooper and Ramadan 1955; Geldreich and Kenner 1969; Wheater et al. 1979). Similar data for feral warm-blooded animals living adjacent to or in shellfish growing waters (e.g., harbor seals, Calambokidis et al. 1989) are rare.

Anthropogenic sources of fecal organisms to shellfish growing areas include discharges of treated municipal sewage and releases of partially treated or raw sewage that occur owing to mechanical failure, combined sewer systems, failing septic systems, and vessel discharges. Bypassing, the release of untreated sewage from a treatment plant during periods of high rainfall, is not uncommon. Levels of bacterial indicators in effluents from a properly functioning sewage treatment plant using disinfection will be considerably reduced compared with those in the influent (Miescier and Cabelli 1982). However, disinfection is not a stoichiometric process and its effectiveness varies with waste composition, volume, target organism, and flow. Other sources include agricultural and storm water runoff, which can transport fecal wastes from humans and domestic and feral animals into shellfish growing areas. Resuspension and transport of contaminated sediment by storms or flooding may also be a source to growing waters.

Much of the literature concerning indicators in shellfish growing waters developed around the concept that fecal pollution is primarily derived from large point sources of sewage or river discharges. Polluted rivers, which integrate multiple point sources, are typical of conditions in highly urbanized northeastern coastal areas. Diffuse or nonpoint pollution is now recognized as a major source of fecal contamination to all types of receiving waters (Faust 1976; Geldreich et al. 1968; Gilliland and Baxter-Potter 1987). Indeed, in the mid-Atlantic and Southeast regions, significant proportions of the total shellfish growing acreage are closed to direct harvesting because of runoff ·from rural, agricultural, and wildlife sources (Leonard et al. 1989). Improperly functioning septic systems are implicated as important sources of fecal contamination. Issues of concern are the degree of treatment afforded by such systems, which is minimal and does not inactivate viruses, and the potential migration of effluent to estuarine waters because of inadequate drainfields, improper soil characteristics, or leaching through subsurface water affected by tidal fluctuations. Resolution of these issues awaits the application of rigorous and imaginative experimental approaches.

The public health significance of coliform bacterial levels in growing areas contaminated by diffuse sources remains unclear. The paradigm of a diluted sewage effluent does not apply, because sources and ratios of indicators to pathogens are variable. Fecal coliform densities that exceed the growing area standard are not uncommon in Chesapeake Bay subestuaries lacking identifiable point sources of human fecal pollution. Although direct evidence is lacking, it is believed estuaries can promote indicator survival (Erkenbrecher 1981). Contributing factors include inorganic and organic nutrient loading, high suspended solids, elevated temperatures, the presence of fine grained organic rich sediments, and poor tidal flushing. Indeed, the need to identify and validate indicator systems to assess the sanitary quality of growing areas impacted by nonpoint sources may be the greatest challenge to sanitarians and shellfish microbiologists since the adoption of the coliform growing area standard. An evaluation of indicators in nonpoint source impacted areas must consider the effects of microbial food webs on allochthonous bacteria, the possible extraenteral origins of indicators, and contamination derived from multiple sources including wild and domestic animals. The paucity of data to evaluate these concerns has engendered criticism of the current indicator and its validity as a public health standard.

INFLUENCE OF THE ESTUARINE–MARINE ENVIRONMENT ON THE FATE OF ALLOCHTHONOUS INDICATORS OF FECAL CONTAMINATION

Sanitary engineers and microbiologists have expended considerable effort to identify environmental factors and processes that affect densities of enteric bacteria and their recovery. Much of the early literature on coliform survival suggested that "dieoff" (or "decay") was the only functional response of coliforms exposed to marine or estuarine environments. This was primarily attributed to the "bactericidal" property of seawater (Ketchum et al. 1952). Immediate decreases in concentrations of coliform organisms discharged from a sewage treatment plant can result from turbulent transport and mixing, processes that presumably occur over intervals of minutes or hours. At some distance from the discharge, the effluent becomes dynamically passive, and light, temperature, salinity, bacterivory, antagonism, inhibition, sedimentation, and autecological responses become important factors affecting indicator fate.

The term "dieoff" is inappropriate to describe changes in indicator densities over time. The apparent reduction of recoverable counts from estuarine waters is not exclusively the result of "true" cell death, i.e., cells that are nonviable, but is also a function of:

1. Enumeration methodology
2. Physiological adaptation to an adverse environment
3. Complex interactions of physical, biological, and chemical processes

The net effect of these processes is a function of their degree of interaction, relative dominance, and the unique characteristics of each environment.

Effects of various environmental parameters on indicator survival have been studied by many investigators but it is often difficult to unequivocally assess their "true" significance. Specific concerns arise with experimental designs that employ in vitro techniques, laboratory-adapted strains, artificial or filtered menstrua, and exposure devices incapable of preventing external contamination and containment of test strains (Anderson et al. 1983; Roper and Marshall 1979). Given the probability that the microbiota can be a major factor affecting bacterial numbers, it seems unwise to generalize based on experimental designs that exclude this component. This is not to discredit the value of in vitro studies, which are useful to establish basic principles, but ecological hypotheses should be tested under conditions that duplicate open or natural systems as closely as possible. Another concern focuses on methods used for preparation of test cells. A majority of survival studies involved procedures for preparation of test cells known to result in sublethal stress or be physiologically debilitating. Culture age, growth conditions, and laboratory manipulations are important factors contributing to sublethal injury. Use of cells pregrown in rich media and harvested during exponential growth, cold-shocking, and harvesting by repeated centrifugation in unfavorable solutions are known to compromise physiological indices, increase sublethal stress, reduce adenylate levels, reduce enzyme activities, produce changes in membrane integrity leading to leakage of cell constituents, and render cells sensitive to free radicals or heavy metals (Anderson et al. 1979; Granai and Sjogren 1981; Postgate 1967; Rhodes et al. 1983; Strange 1976). Failure to recognize these factors as sources of experimental bias has led to incorrect inferences of causality. Similar reservations apply to using laboratory-adapted strains because these cells may exhibit survival characteristics that are "atypical" compared with "fresh" fecal isolates (Anderson et al. 1979).

In view of the importance accorded test cell preparation, a relevant hypothesis that should be tested is the effect of indicator origin (exposure prehistory) on its fate. Are there differences in the survival of indicator cells prepared under laboratory conditions and cells derived from "natural" sources such as sewage treatment plant (STP) effluent, or agricultural or stormwater runoff? The STP provides a nutrient rich environment, at times maintained at higher temperatures than receiving waters, and exposes cells to varying degrees of toxic compounds and disinfectant-mediated injury. Could it be that indicator bacteria from diffuse natural sources are physiologically adapted to adverse nutrient conditions, intermittent lack of water, unfavorable temperatures, etc.? Such adaptation could significantly alter indicator persistence and recoverability compared with laboratory-grown cells.

Finally, it is difficult to see how investigations that treat the effects of environmental factors as if they were independent operators can lead to developing basic principles of cell survival. Treatments that isolate single variables will only provide information concerning the *potential* role of that variable. In the environment survival will be affected by the interaction of that factor with other perhaps more dominant processes. Laboratory studies must be augmented by admittedly more complex and variable in situ exposure studies that integrate physiochemical and biological factors. Conversely, in situ experiments must be carefully interpreted because of the potential for undetected processes that may affect test cell densities and remain undetected. Examples include breaching of exposure devices by autochthonous organisms owing to design or structural damage (Anderson et al. 1983; Roper and Marshall 1979) and physical penetration through the membrane "pores" by bacterivorous nannoflagellates (Cynar et al. 1985) or other procaryotes (Li and Dickie 1985). Chambers may exhibit highly variable "bottle" effects because of nutrients contributed by fouling communities on chamber surfaces, blooms caused by exclusion of autochthonous organisms or predators, and attenuation of incident light. Finally, estuaries are unique dynamic systems, characterized by temporal and spatial heterogeneity. Conclusions based on indicator fate studies performed in Puget Sound may not be applicable to Chesapeake Bay growing waters.

The following sections summarize important aspects of the effects of physical, chemical, and biological factors on indicator survival, primarily under estuarine and marine conditions.

Physical and Chemical Factors

Temperature. At first glance the literature dealing with the role of temperature on in situ survival of *E. coli* in estuarine waters appears equivocal. Investigators have reported both positive (Anderson et al. 1983; Rhodes and Kator 1988) and negative (Faust et al. 1975; Lessard and Sieburth 1983; Vasconcelos and Swartz 1976) correlations of temperature and survival. This confusion can be resolved if it is understood that temperature has both indirect and direct effects on indicator fate. It can have a direct effect on bacterial activity, so that under appropriate conditions multiplication may occur at warmer seasonal temperatures. *E. coli* cells prepared under conditions to minimize stress initially exhibit multiplication (large negative values of k) in membrane filtered water (0.2 μm) at elevated temperatures in Chesapeake Bay (Table 7.1). Maximum negative values of k correspond to increases in viable counts of approximately 1.0–1.5 log units. Although it may be assumed that low environmental temperatures [<10°C (<50°F)] would reduce coliform metabolic activity based on Q_{10} values, thereby favoring persistence, low environmental temperatures do not appear to have this effect. The indirect effects of low environmental temperature will be addressed in the context of sublethal stress but exposure of bacterial cells to low temperatures is known to compromise cell envelope integrity and physiological indices (Strange 1976). Indeed, positive values of k (Table 7.1) were obtained at temperatures below 10°C (50°F) in filtered water.

An indirect and important relationship between temperature and indicator recovery and enumeration is the effect of temperature on the development of sublethal stress. Sublethal stress reflects the inability of a microorganism to be cultured in a medium at a specified temperature because of prior injury, impairment, or damage that resulted from exposure to unfavorable environmental conditions. Sublethally stressed cells are particularly sensitive to recovery methodologies that use selective temperatures, lack resuscitative protocols, or use inhibitory substances to enhance media specificity and selectivity (Hackney et al. 1979). We have quantified the development of sublethal stress in *E. coli* as a function of temperature and salinity (Anderson et al. 1979; Rhodes et al. 1983). Sublethal stress and mortality are inversely related to temperature. At temperatures below 10°C (50°F) transiently acute or progressive development of sublethal stress is detectable using a variety of techniques. These include differential counts on selective vs. nonselective recovery media and an assay technique (Anderson et al. 1979) based on the observation that sublethally stressed cells

Table 7.1 Survival of *E. coli* exposed in situ in Filtered and Nonfiltered Estuarine Water During Two Exposure Intervals

| | | Mortality Rate Coefficient k/day[a] Exposure Interval (days) | | | |
| | | 0–3 | | 3–15 | |
Month	*Mean Temperature*	*Filtered*	*Nonfiltered*	*Filtered*	*Nonfiltered*
February	6.7	−0.15	0.03	0.68	0.92
March	8.8	0.36	0.23	0.58	0.72
April	17.8	−0.84	−0.30	0.01	0.58
May	25.1	−0.65	−0.48	0.03	0.80
July	28.2	−0.86	−0.56	0.07	0.57
November	10.3	−0.17	−0.08	0.09	0.51

[a] $-k = \ln(C_2/C_1)/(T_2-T_1)$ where C_2 and C_1 are the bacterial densities at final (T_2) and initial (T_1) times of exposure. (From Rhodes, M. W., and Kator, H. *Appl. Environ. Microbiol.* **54**, 1988; with permission.)

require a longer period of time to produce an electrochemically induced potential difference compared with nonstressed cells [electrochemical detection time (EDT)] (Figure 7.1). In practical terms, sublethally stressed cells may not be detected using enumeration methods that incorporate selective procedures. The effect of exposure to estuarine water on enumeration efficiency using experimental and approved recovery methods is shown in Figure 7.2. These data show that stress can be progressive over time. The decreased stress (increased recovery efficiency) occurring at favorable environmental temperatures between 6 and 12 days suggests these populations are survivors adapted to seawater. Note the poor enumeration efficiency for the direct M-FC procedure, an approved method (APHA 1985). Although resuscitation procedures in nonselective media have been employed to diminish the impact of selective enumeration, such measures may not be completely effective.

Sublethal stress must be considered if conventional selective enumeration procedures are used for recovery of indicator cells. Results from survival studies with *E. coli,* where viable counts were not "corrected" for the effects of sublethal stress, tend to overestimate mortality (underestimate cell densities), with the error being most significant at temperatures below 10°C (50°F). Consequently, the responses of cells to physical parameters or treatments where sublethal injury was unrecognized must be interpreted with caution because the observed mortality may have been incorrectly attributed to another variable and not the enumeration process. These concerns reinforce the notion that sublethal stress is an important factor affecting indicator choice when viable recovery methods are used and observed cell densities have regulatory significance.

Temperature-induced sublethal stress also renders cells sensitive to stressors. Mackey and Derrick (1986) showed that cold-shock sensitizes *E. coli* to very low concentrations of hydrogen peroxide and as a consequence reduces recovery of viable cells on an organic-rich medium. Postgate (1967) cautioned that chilling of samples and cultures promotes the death of stressed cells. The significance of these observations apparently remains unheeded because Standard Methods (APHA 1985) still suggests icing of microbiological samples if they cannot be processed within 1 hour. A systematic examination of sample storage parameters to optimize recovery under a variety of seasonal temperature regimens should be part of the indicator evaluation process.

Indirect effects of temperature on coliform fate arise from the effects of seasonal temperature on the densities, composition, and activities of the indigenous microbiota (Anderson et al. 1983; Rhodes and Kator 1988; Verstraete and Voets 1976). The role of the microbiota on indicator persistence and survival has been until recently a topic of considerable speculation. Thus, although the activities of antagonistic substances, parasitic and lytic microorganisms, and protozoans were recognized as potentially important factors controlling indicator abundance (as reviewed by Mitchell and Chamberlin 1975), it is only recently that microbial ecologists have confirmed bacterivory as an important carbon and

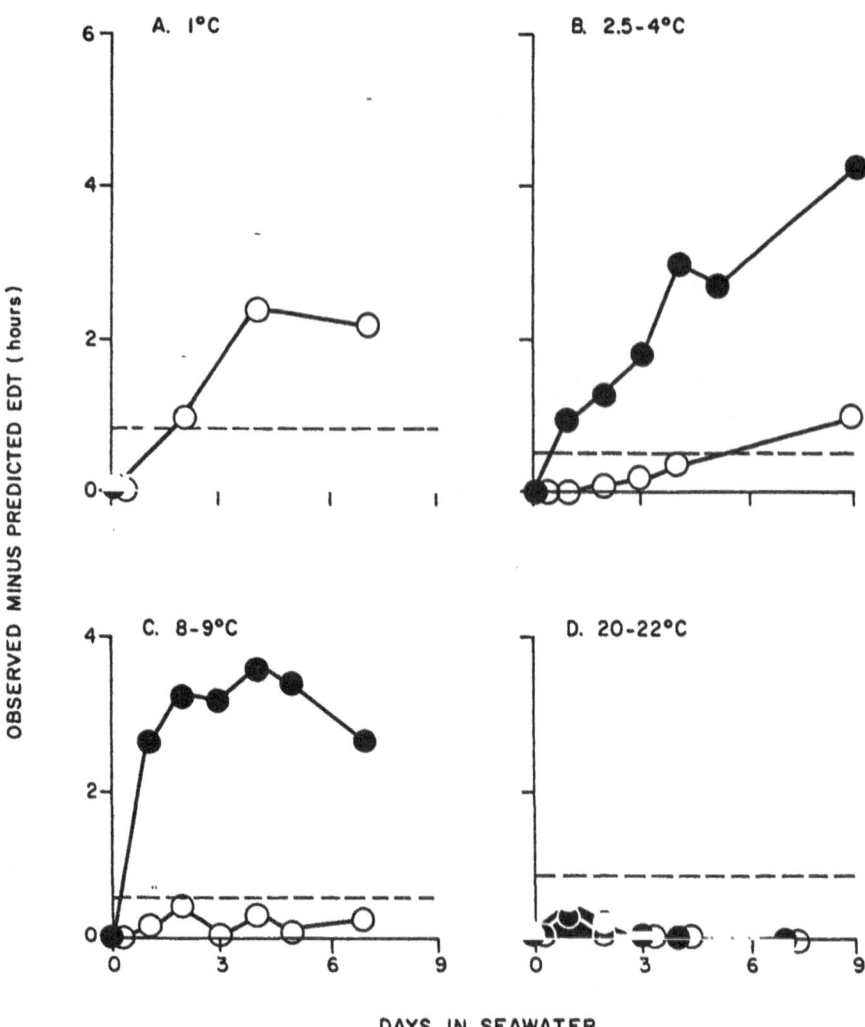

Figure 7.1 Incidence of stress in *E. coli* as measured by the electrochemical detection method for cells pregrown in M-9 minimal medium and exposed to seawater at various temperatures. Stress was defined as the difference between electrochemical detection time (EDT) for cells exposed in test media and predicted EDT based on standard curves. Predicted EDT values were derived from a linear regression line based on a range of viable counts and nonstressed cells. Upper 95% confidence limit for predicted EDT is shown as (- - -). Mean salinity for each experiment was 24 psu. Test media were TSB (tryptic soy broth) (○) and EC (●) incubated at 44.5°C (112.1°F). (From Rhodes, M. W., Anderson, I. C., and Kator, H. *Appl. Environ. Microbiol.* **45**:1872, 1983; with permission.)

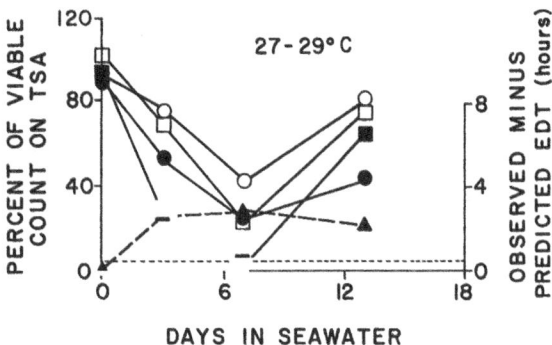

DAYS IN SEAWATER

Figure 7.2   Incidence of stress in *E. coli* as measured by enumeration efficiency using experimental and approved recovery procedures and the electrochemical detection technique. *E. coli* was pregrown in TSB and exposed at 27–29°C (80.6–84.2°F) in 20 psu salinity seawater. Enumeration efficiency presented as percent viable count on TSA. EDT (▲) determined in EC at 44.5°C (112.1°F), upper 95% confidence limit for predicted EDT indicated as (- - -). Enumeration methods were: (○), VRBA (violet red bile agar) overlay; (●), VRBA direct; □, M-FC overlay; ■, M-FC direct. (From Rhodes, M. W., Anderson, I. C., and Kator, H. *Appl. Environ. Microbiol.* **45:**1873, 1983; with permission.)

energy pathway (e.g., Wright and Coffin 1984). The importance of bacteria in coastal waters as food for heterotrophic microflagellates has been demonstrated (Anderson and Fenchel 1985; Fenchel 1982). Studies using diffusion chambers, perhaps somewhat artifactual in terms of community development and enhanced encounters between predator and prey, reveal effects of the microbiota on *E. coli* survival (Rhodes and Kator 1988). This is demonstrated (Figure 7.3, Table 7.1) by comparing survival of *E. coli* test cells in filtered (0.2 μm) and nonfiltered estuarine water contained in diffusion chambers deployed in situ. Initially, multiplication occurs in both filtered (predator-free) and nonfiltered water. After 3 days, mortality in the nonfiltered treatment is significantly greater. Generally, times of maximum decline in nonfiltered water coincide with maximum densities of heterotrophic flagellates or other predators. Although predation and the in situ growth of enteric bacteria are direct functions of temperature, the net combined effect of increased temperature is indicator removal. Failure to recognize both direct and indirect effects of temperature can lead to the erroneous conclusion that elevated temperature, per se, does not favor coliform survival.

Salinity.   Efforts to determine the effect of salinity on indicator survival constitute a small and somewhat inconclusive literature (Carlucci and Pramer 1960; Faust et al. 1975; Orlob 1956; Vasconcelos and Swartz 1976). Results range from no effect of salinity on *E. coli* viability to reduced survival with increasing salinity. This literature remains problematic for many of the same

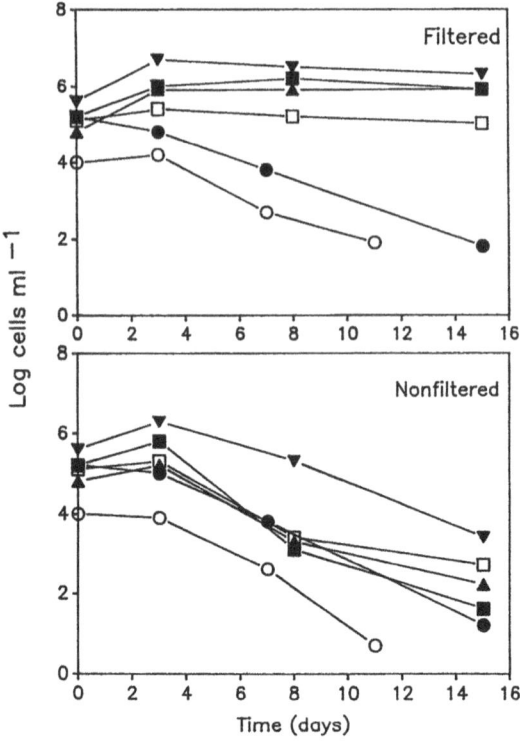

Figure 7.3 Survival of *E. coli* in diffusion chambers containing filtered and nonfiltered estuarine water and exposed in situ during various months. (○, February; ●, March; △, April; ▲, May; □, July; ■, November. (From Rhodes, M. W., and Kator, H. *Appl. Environ, Microbiol.* **54:**2904, 1988; with permission.)

reasons noted for temperature studies. The combination of all-or-none viable counting methods with factors known to be stressors, e.g., selective recovery methods, artificial seawater with possible trace toxicants, and harsh inocula preparation methods, have contributed to the apparent disappearance of test cells. These problems are avoided using techniques that measure graded responses and use cell preparation techniques that minimize injury (Anderson et al. 1979). Graded responses included EDT in selective and nonselective media, beta-galactosidase specific activity, and growth rate. Increased salinity is accompanied by decreased recovery of viable cells (Table 7.2), an increase in sublethal stress expressed as larger EDT values (Figure 7.4) and as the difference in cells recovered on selective versus nonselective media, and as reduced beta-galactosidase specific activity (Anderson et al. 1979). Cells exposed to 30 psu seawater for 2–9 days exhibit an apparent mortality about nine times greater in EC medium than in TSB.

Table 7.2   Percent Survival of *E. coli* in Seawater Adjusted
to Selected Salinities

Salinity (psu)	Exposure Time (days)	%Survival[a]
10	2	100.6
	5	87.6
	8	53.5
15	2	27.9
	5	11.7
	8	7.1
25	2	8.6
	5	5.1
	8	4.3
30	2	1.7
	5	0.7
	8	2.0

[a]Based on zero time densities of *E. coli* in seawater (ca. 10^7 cells per ml). Each datum is the mean of two experiments. (From Anderson et al. *Appl. Environ. Microbiol.* **38,** 1979; with permission.)

The idea that exposure of *E. coli* to saline water yields cells with altered physiological properties, thereby requiring modified enumeration methods, has been suggested. Dawe and Penrose (1978) observed that salinity-induced debility in coliforms is reduced by incorporation of seawater into the recovery medium. Gauthier et al. (1987) increased recovery of *E. coli* cells adapted to seawater by including sodium chloride in an enumeration medium, although the degree of response was strain and time dependent.

E. coli cell envelope composition and functional characteristics are different in cells grown on an estuarine water-based medium compared with those grown in a standard medium (Chai 1983). The effect of high osmolarity in *E. coli* is to repress synthesis of OmpF porin protein, a protective mechanism to reduce cell permeability to naturally occurring detergents, i.e., bile salts (Nikaido and Vaara 1987). This adaptive process retains membrane permeability, albeit reduced, to nutrients with molecular weights of 100–200. An indirect consequence of reduced membrane permeability under hyperosmotic conditions could be to alter significantly or reduce the uptake of molecules used in direct cell viability assays such as the tetrazolium salts. In summary, the studies mentioned support an emerging hypothesis that survival of *E. coli* in seawater is an active process involving physiological adaptation to a hyperosmotic environment.

Strange (1976) in his monograph on stress noted that osmotic shock causes loss of cell viability, reduces active transport of solutes, and releases a variety of metabolites and enzymes. Roth et al. (1988) demonstrated that a large proportion of *E. coli* cells exposed to hyperosmotic conditions are not recoverable on a nonselective medium. However, if the cells are osmotically upshocked in the presence of betaine (*N,N,N*-trimethylglycine, a naturally occurring and

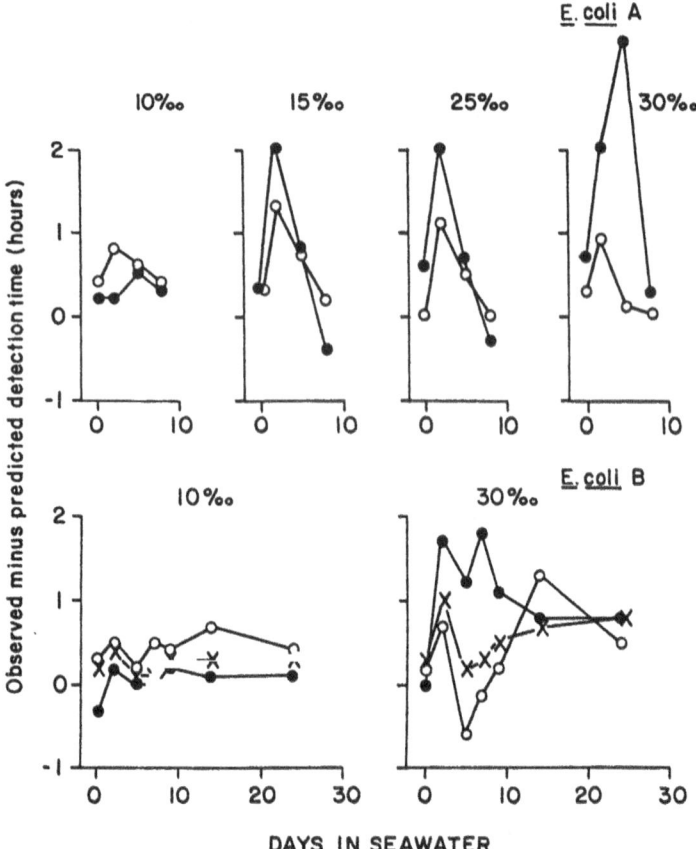

Figure 7.4 Incidence of stress in two strains of *E. coli* during exposure to seawater adjusted to various salinities. Stress was measured as EDT using TSB (o), EC (●), and A-1 media (X) at 44.5°C (112.1°F) (From Anderson, I. C., Rhodes, M. W., and Kator, H. *Appl. Environ. Microbiol.* **38**:1149, 1979; with permission.)

ubiquitous nitrogen-containing compound produced in microorganisms, animals, and plants), the cells remain culturable. Betaine accelerates uptake of glucose and reduces intracellular ATP that accumulates in osmotically upshocked cells and facilitates protein synthesis. Roth et al. (1988) developed a resuscitation method using a medium incorporating betaine, chloramphenicol (to prevent changes in cell density during resuscitation), ammonium and glucose as nitrogen and carbon sources. At this writing we are unaware if this method has been evaluated with estuarine samples.

Adsorption and Sedimentation.   The effect of estuarine particulates on the persistence of allochthonous bacteria continues to be an interesting area of research. Rubentschik et al. (1936) suggested that adsorption of *E. coli* to particulates with subsequent deposition in sediments is an important aspect of self-purification in salt lakes. Weiss (1951) investigated *E. coli* adsorption on river and estuarine silts and concluded that adsorption enhances bacterial sedimentation rate and is a function of particle type and size. Although flocculation of silts increase in seawater, the adsorptive capacity of silts toward bacteria decreases with increased salinity and the bacteria desorb. Milne et al. (1986) concluded that removal of fecal coliforms from estuarine water by deposition is directly related to the concentration of naturally occurring suspended solids. A similar relationship was not observed in higher salinity seawater and attributed to differences in the depositional behavior of suspended solids in the two systems. Roper and Marshall (1974, 1979) observed that *E. coli* adsorb to sediments at high electrolyte concentrations and desorb below a critical concentration. They hypothesized that in low electrolyte (freshwater) systems electrostatic forces allow bacteria to exist as stable colloidal dispersions. As salinity is increased, flocculation and sedimentation of bacteria and particles occur. This straightforward model can be complicated by organics present on the microorganism's "surface." Thus, adsorbed bacteria can adhere very strongly to sediment particles owing to production of organic exopolymers (Marshall, 1985). Viruses also adsorb to estuarine sediments. LaBelle and Gerba (1979) observed >99% adsorption of various enteric viruses to sediment at salinities of 1–35 psu. Alterations in salinity and pH produce small but variable effects on adsorption and desorption of most virus types examined. Because enteric viruses do not readily desorb they concluded that viral transport would be dependent on particle resuspension and transport.

Based on the above observations it is not surprising that densities of fecal indicator bacteria (Erkenbrecher 1981; Shiaris et al. 1987) and enteric viruses (LaBelle et al. 1980; Rao et al. 1984) are elevated in estuarine sediments compared to overlying waters. Sediments from an urban shellfish growing area in the Chesapeake Bay contained fecal coliforms at densities two orders of magnitude larger than in the water column (Erkenbrecher 1981). Variable but generally smaller values of the ratio of fecal coliform densities in sediment to water were observed in a small subestuary subject to nonpoint pollution (Table 7.3). Ratio values appeared to increase from the headwaters (Station 23) to the mouth (Station 2), perhaps reflecting closer coupling between the two phases in

Table 7.3 Ratio of Fecal Coliform Densities in Sediment (/100 g Dry Sed) to Water (per 100 ml) Simultaneously Sampled in a Small Subestuary of the York River, Virginia

Station	Mean Salinity	Date						Mean
		4-19-88	5-17-88	6-06-88	7-18-88	8-01-88	9-13-88	
23	14.0	5.2	1.0	3.9	2.1	ID	7.7	5.4
10	18.0	7.6	20.0	12.0	4.3	95.7	0.3	15.5
9	16.0	37.6	0.5	3.1	2.4	34.2	3.9	13.6
7	18.0	4.8	1.1	18.3	0.3	48.9	52.3	28.8
2	18.0	8.9	≥22.8	60.0	≥4.0	12.9	2.6	19.7

the more polluted, shallow headwaters, and the opposing effect of tidal flushing and very low fecal coliform densities at the mouth. Although fecal coliforms are infrequently isolated from York River, Virginia, sediments (a tributary to the lower Chesapeake Bay) in areas meeting the approved water quality standard, salmonellae are also recovered in these samples on occasions. LaBelle et al. (1980) found no correlation between densities of bacterial indicators and viruses in estuarine waters. There was a positive correlation between the numbers of viruses and fecal coliforms in sediments.

In vitro and in situ experiments (Gerba and McLeod 1976; Perez-Rosas and Hazen 1988; Roper and Marshall 1974, 1979) have shown the positive effect of marine or estuarine sediments on *E. coli* persistence. Similarly, adsorption to estuarine sediments enhances survival of enteroviruses (Smith et al. 1978). Protective effects are attributed primarily to accumulation of nutrients on particle surfaces and reduced predation and antibiosis. As previously mentioned, Roth et al. (1988) reported the organic compound betaine minimizes osmotic stress and restores the culturability of *E. coli* exposed to seawater. Le Rudulier and Bouillard (1983) have shown that betaine concentrations as low as 1 m*M* eliminate osmotic stress in *K. pneumoniae, S. typhimurium,* and *E. coli* over a range of high NaCl concentrations (0.65–1 *M*). There was no evidence that betaine is utilized as a growth substrate by *K. pneumoniae.* Betaine is a ubiquitous and relatively abundant compound in benthic organisms and is an important substrate in fermentation pathways coupled to methanogensis in marine sediments (Heijthuijsen and Hansen 1989; King 1984). The apparent enrichment and persistence of *E. coli* (and other enteric bacteria) in sediments may be augmented by the osmotolerance afforded by betaine or other osmolytes (Ghoul et al. 1990; Munro et al. 1989). The effects of anaerobic sediments, which are sources of inorganic nutrients and small-molecular-weight fermentation products, on indicator persistence is obviously a research issue requiring study.

The accumulation and enhanced survival of sewage microorganisms in sediments has been employed as a justification for questioning the validity of a water quality criterion as the basis for classification of shellfish growing waters (Shiaris et al. 1987). Sediments provide an integrated recent history of fecal pollution compared with overlying waters that reflect transient pollution conditions. Sediments serve as a reservoir for enteric pathogens that could be transported after resuspension by rainfall, currents, dredging, turbulence associated with recreational activities, etc. (Erkenbrecher 1981; LaBelle et al. 1980). Metcalf et al. (1973) noted that fecal coliform concentrations and the likelihood of isolating salmonellae in the water column are functions of tidal stage, suggesting an association between tidal currents and resuspension and transport of particulate material. Particle-associated pathogen transport is emphasized by greater viral accumulation by shellfish exposed to resuspended sediment compared to those exposed to undisturbed sediment (Landry et al. 1983).

Toxic Compounds. Although the lethality of disinfectants toward indicator bacteria is well described, the effects of environmental toxicants in shellfish waters and sediments on these bacteria is a rather impoverished area of research. This is surprising considering that estuarine areas can be polluted by high levels of toxic heavy metals, xenobiotics, and other materials. Runoff from agricultural and urban areas contains a variety of pesticides of varied toxicity and persistence characteristics. Effluents from kraft pulp mill plants may elicit a broad range of toxic effects (Sodergren 1989). Oil shale process waters are lethal to coliforms and *S. faecalis* under in vitro exposure conditions (Adams and Farrier 1982). Jones and Cobet (1975) noted the toxicity of naturally occurring heavy metals in Caribbean seawater to enteric bacteria. Widespread use of organotin compounds as antifouling paints has been recognized as contributing to high tributyltin concentrations in estuarine waters and sediments frequently exposed to vessels. Pettibone amd Cooney (1986) concluded that organotin compounds are not acutely toxic toward *E. coli* and *E. faecalis* isolates at naturally occurring concentrations but act as stressors. Chai (1983) noted that *E. coli* grown in a medium containing estuarine water manifests changes (adaptations) in cell envelope composition, changes that afforded decreased sensitivity to bacteriophage and colicins but also rendered cells more sensitive to heavy metals and detergents. In vitro exposure of *E. coli* to estuarine water containing various toxic chemicals alters cell envelope protein composition and detectability of plasmids, and affects carbohydrate and amino acid metabolism (Palmer et al. 1984). Creosote contamination of estuarine sediments is detrimental to microbial communities as reflected in reduced secondary production and biomass (Koepfler and Kator 1986). Although definitive information is lacking, sediments may be sources of chemical or biological stressors toward indicator bacteria although it is likely the degree of effect will be site specific.

Organic Compounds and Nutrients. In a review of factors affecting the survival of enteric microorganisms in marine and estuarine environments. Mitchell and Chamberlin (1975, p. 241) concluded a section on "Nutrient deficiencies" as follows: "These results coupled with laboratory findings tend to suggest a significant role for nutrients in determining survival of enteric bacteria in seawater." Until recently the literature did not refute this conclusion, but did little to amplify it or provide data supporting this hypothesis. Fortunately, this area has been the focus of a number of research groups such as Lopez-Torres et al. (1987), who describe a positive statistical association between nutrient concentrations and persistence of *E. coli*. *E. coli* starved in seawater manifest a variety of physiological and structural responses including gradual loss of beta-galactosidase activity, increased activity of other enzymes, and altered sensitivities to antibiotics, phages, and heavy metals (Munro et al. 1987). Fiksdal et al. (1989) concluded the increased activity of *E. coli* in seawater toward 4-methylumbelliferyl heptanoate during starvation reflects adaptation to low nutrient conditions. Studies are still needed to evaluate the direct effects of inorganic nutrients and organic carbon on enteric survival, especially in nonpoint impacted shellfish growing areas.

Light.   Direct lethal effects of light on enteric bacteria in seawater have been demonstrated under in situ (Bellair et al. 1977; Fujioka et al. 1981; Gameson and Gould 1975; Gameson and Saxon 1967) and in vitro experimental treatments (Fujioka et al. 1981; Fujioka and Siwak 1987; Gameson and Gould 1975; Kapuscinski and Mitchell 1981, 1983; McCambridge and McMeekin 1981). The overall significance of these studies to the mortality of indicator organisms in shellfish growing waters remains somewhat equivocal. This is because major estuaries where shellfish are produced in the eastern United States and Gulf of Mexico are highly turbid and may be dominated by complex microbial food webs. The importance of these factors resides in the attenuating capacity of suspended and dissolved material for lethal wavelengths of light and the light-stimulating effect on components of the microbiota that can lead to enhanced bacterial mortality. Interactive effects of light and the autochthonous microbiota have been demonstrated in fresh (Barcina et al. 1986) and estuarine waters under in vitro conditions (McCambridge and McMeekin 1981). A series of in situ experiments were performed in Chesapeake Bay using diffusion chambers specially modified to maximize light penetration and employing light and dark, filtered and nonfiltered treatments (Rhodes and Kator, 1990). Compared with cells suspended at 1.0 cm below the water surface, mortality from sunlight was essentially insignificant at 25.0 cm except during periods (fall, winter) of minimal light attenuation. Cells at 1.0 cm also exhibited significant sublethal stress. During the warm seasons significantly greater mortality occurred in the presence of light and the microbiota than with either alone. Enhanced mortality of the combined treatment can be attributed to stimulation of predation by sunlight, light-dependent release of antagonistic substances, or formation of photochemically induced toxicants. Therefore, in turbid shellfish growing waters sunlight-induced injury and mortality result from direct and indirect effects whose relative influence will be a function of local conditions.

Biological Factors

Biological factors that affect indicator fate have been discussed or mentioned in other sections. A complete survey of this literature is beyond the scope of this chapter. Biological interactions between indicator organisms and components of the microbiota are complex and definitive rate measurements of processes are lacking. Processes effecting removal through predation, parasitism or lytic activity (e.g., Anderson et al. 1983; Berk et al. 1976; Enzinger and Cooper 1976; Mitchell et al. 1967; Rhodes and Kator 1988; Roper and Marshall 1977, 1978) or antibiosis (Aubert et al. 1975; Sieburth and Pratt 1962), and competition for nutrients (Jannasch 1968) have been described. A need remains to determine the relative importance of biological interactions compared to other processes that affect indicator survival. Because biological processes in estuaries are highly variable in temporal and spatial dimensions, studies to assess these effects must be performed with sufficient replication and seasonal coverage to ensure the collection of representative data. Technical impediments to in situ experimentation that require resolution are uncontrolled contamination of exposure chambers by autochthonous microorganisms, inadequate sample volumes, and the need to unequivocally differentiate test cells from autochthonous cells. Approaches combining rapid direct viable counting methods with those permitting unequivocal identification of test organisms (e.g., fluorescent antibody, gene probes, etc., Roszak and Colwell 1987) may provide solutions.

Effect of Environmental Exposure on Indicator Recovery

The role of sublethal stress on the apparent "dieoff" and enumeration of indicator bacteria has been adequately covered in previous sections. Another important concept aspect concerning the fate and culturability of enteric microorganisms following environmental exposure was described by Xu et al. (1982). These workers noted a differential between numbers of viable cells recovered from seawater using a direct measure of viability and the viable count obtained using nonselective traditional culture methods. It was demonstrated that a significant proportion of cells starved in seawater rapidly enter a "nonrecoverable" stage, i.e., do not grow in standard media, but remain viable as shown by a direct viable count assay (Kogure et al. 1979). The phenomenon of nonrecoverability has been observed by other workers (Lopez-Torres et al. 1987; Martinez et al. 1989; Munro et al. 1987; Roth et al. 1988). Culturable cell counts drop significantly within 2–4 days of exposure. At these times culturable counts are 20% or less of total direct viable counts. Although significant numbers of culturable cells may remain (depending on the initial cell density), these responses suggest the "true" numbers of indicator organisms could be significantly underestimated because of the low culturable densities found in approved waters. A detailed discussion of this phenomenon can be found in Grimes et al. (1986) and Roszak and Colwell (1987). Recognition of this physiological adaptation has done much to stimulate rethinking the fate of allochthonous bacteria in marine and estuarine environments as well as the efficacy of traditional enumeration methods. Entry of a significant proportion of an enteric indicator population in seawater into a nonrecoverable stage would seriously question the validity of bacterial indicators.

We have not observed this phenomenon with either *E. coli* or *S. enteritidis* after comparatively long exposure periods, under both in vitro and in situ conditions. The effect of prolonged exposure on *E. coli* densities in 0.22-μm

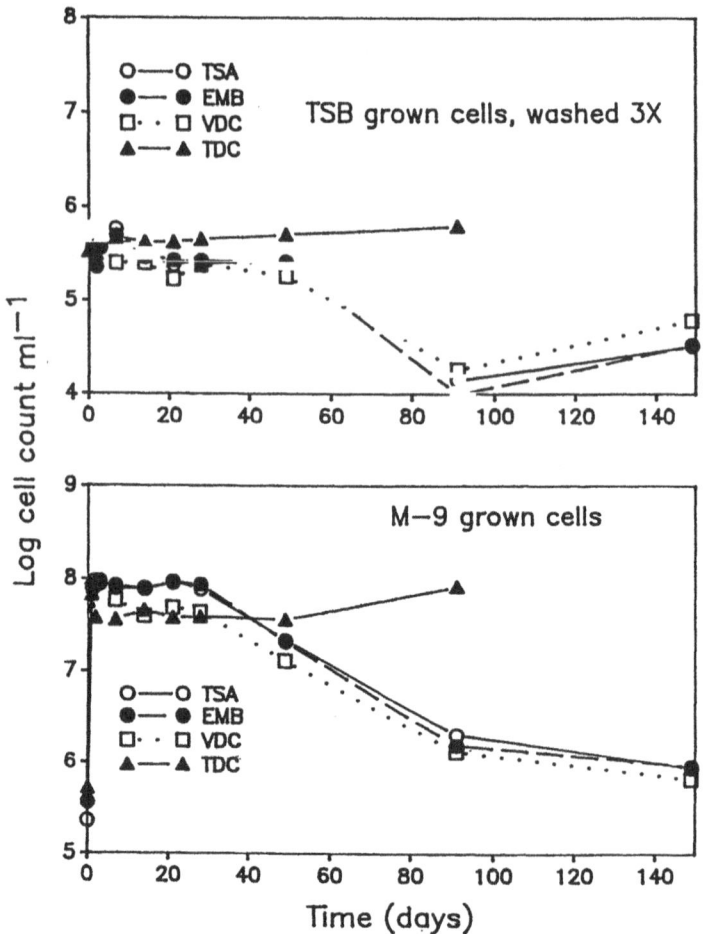

Figure 7.5 Survival of *E. coli* in filtered (0.22 μm) estuarine water (18 psu) incubated in vitro at 20–25°C (68–77°F). Isolates were pregrown in either M-9 and diluted to an O.D. of 0.1 with sterile estuarine water or prepared in TSB after Xu et al. (1982). Culturable cells were enumerated on TSA (○) or EMB (●) by spread plating and incubated at 25°C (77°F). Direct counts consisted of total direct count (▲) (Hobbie et al. 1977) and viable direct count (□) (Kogure et al. 1979).

filter-sterilized estuarine water (18 psu) at ca. 23°C (73.4°F) are shown in Figure 7.5. Enumeration was by acridine orange direct count (AODC) (Hobbie et al. 1977), viable direct count (Kogure et al. 1979). Culturable cells were recovered on tryptic soy agar (TSA) and eosin methylene blue agar (EMB, Difco) by spread plating. The effects of pretreatment on survival were compared using two different methods for preparation of test inocula. In the first treatment cells were grown in a rich medium, tryptic soy broth, and washed three times by centrifugation in 0.85% saline (pH 7.2). A second inoculum was prepared by growth and dilution into minimal medium M-9 as previously described by Anderson et al. (1979). The influence of inoculum preparation was first reflected in the initial proliferation of cells grown in M-9. It is unlikely this proliferation was the result of reductive division because about seven divisions would have been required and the average cell size after day 1 remained about 0.8 × 1.5 μm. *E. coli* recoveries on EMB and TSA at 25°C (77°F) were essentially independent of inoculum preparation. Our results also showed recovery and enumeration of *E. coli* was initially independent of enumeration method. Only after 50 days of exposure were total direct counts and viable culturable counts significantly different. There was no (M-9 grown) or little (TSB grown) difference between viable and culturable counts after 5 months (149 days) exposure. Similarly, *E. coli* and six different *Salmonella* servovars, exposed in diffusion chambers containing 0.2 μm filtered site water and deployed in the York River, showed extended persistence based on viable plate counts (Figure 7.6). Flint (1987) reported that *E. coli* survived in sterile river water for 260 days in the absence of predators and that direct and cultural counts were similar over the first 40 days of exposure. Fiskdal et al. (1989) observed both rapid and slow declines in culturable counts of *E. coli* in filtered seawater. Rapid decline was described as about a 1 log decrease in cell count over a 23-day period. Extended culturability (ca. 5 months) of *E. coli* in seawater was also reported by Munro et al. (1989). Munro et al. suggested that genetic differences in osmoregulatory capabilities, acting in response to test cell pretreatment and exposure conditions, are responsible for differences in survival and culturability.

We are uncertain about the causes of variation concerning entry of indicator cells into a nonculturable state. Some evidence suggests these may be related to properties of test strains, inocula preparation methods, or to water quality properties such as the concentration or type of organic matter. Because of the importance of this phenomenon to indicator selection and culturable enumeration methods, an investigation of factors affecting the entry of cells into a nonculturable state is desirable. This should focus on issues such as:

1. The effect of experimental protocols on the initiation of nonculturability, e.g., strain, laboratory adaptation, and inoculum preparation.
2. The effects of temperature, osmotic stress, light, organics, and other environmental factors on the transformation of cells to nonculturability.
3. Whether culturable and non-culturable cells are affected the same way by biological (e.g., predation) and physical removal mechanisms (e.g., sedimentation).
4. Determination of the susceptibility of different indicators to the phenomenon of nonculturability.

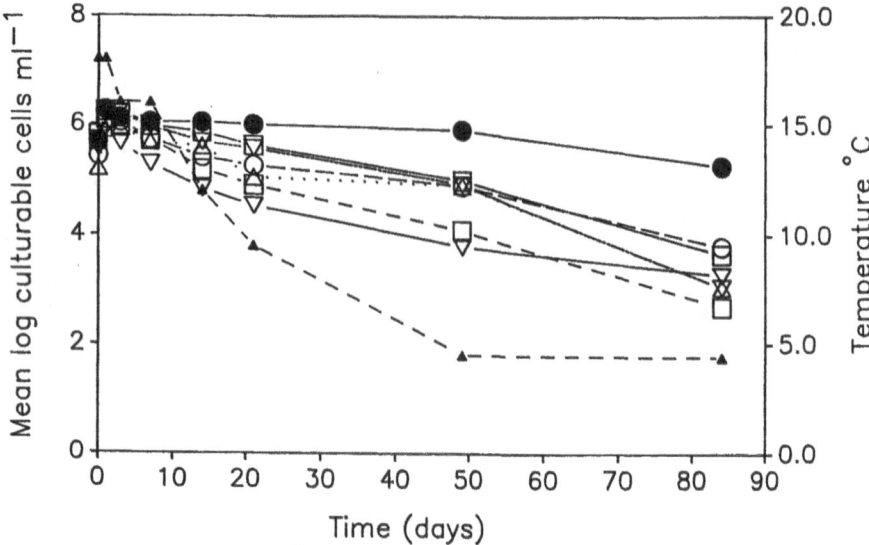

Figure 7.6 Survival of *E. coli* (●) and *Salmonella* serovars (open symbols; *S. enteriditis, montevideo, tennnessee, hadar, typhimurium, infantis*) in separate diffusion chambers containing filtered estuarine water and deployed at 1 m. Three chambers were used for each isolate and counts are means. Inocula were prepared by passing cells twice through M-9 minimal medium and diluting stationary phase cells to an O.D. = 0.1. *E. coli* and *Salmonella* culturable counts were initially determined on TSA by spread plating and incubation at 35°C (95°F). At 49 days, background interference was reduced by using a VRBA overlay on TSA (for *E. coli*), followed by resuscitation for 3 hours at 35°C (95°F) and incubation at 45°C (113°F). *Salmonella* serovars were enumerated at 49 days by overlayering BGSA on TSA, resuscitated as above and incubated at 42.5°C (108.5°F). Presumptive salmonellae were confirmed biochemically. Ambient water temperatures are shown as (▲).

INDICATOR SYSTEMS

Considerations Affecting Usefulness

Many authors (Berg 1978; Bonde 1977; Cabelli 1978a; Wheater et al. 1980) have listed those characteristics that an ideal indicator should possess. Most have agreed that no single indicator is "ideal" or can be universally applied. Indicators possess unique characteristics whose applicability depends on the question being asked and the specific environment tested. An indicator extremely sensitive to chlorination (i.e., a poor indicator of disinfection) would be inappropriate for sewage effluents but could be perfectly adequate in a nonpoint source polluted estuary. Very persistent indicators such as *Clostridium perfringens* spores or coprostanol may not be appropriate in a nonpoint source area but can be useful as indices of sewage plume dispersion or transport of particulate-bound material.

One paradigm of an indicator in shellfish waters would be to provide a criterion to index health risk (Cabelli et al. 1983). However, epidemiological studies necessary to establish such predictive relationships are lacking and the standards now used can reflect the presence of fecal contamination and disease-causing "potential." As noted, this usage is based on a quantitative relationship between coliform bacteria and an enteric pathogen in diluted sewage from a large population. However, in nonpoint source impacted estuarine and marine receiving waters, the constancy of this association may be lost because of variations inherent in small multiple sources, which are by nature intermittent, stormwater runoff, the interaction of indicators and pathogens in the environment, and fecal pollution from nonhuman sources. Application of indicators and standards to such growing areas requires that we understand the sources and responses of indicators to these systems.

Indicators may also serve to provide information regarding source or age of pollution in nonpoint impacted areas. An indicator used as an investigatory tool may provide information relevant to fecal source, i.e., human versus animal or animal type and the age of pollution or its "freshness." Indicators found inappropriate as indices of health risk may be useful for this purpose.

Traditional Indicators

Total Coliform Group. Although the total coliform group can be found in most current compilations of approved microbiological methods, the overwhelming thrust of a large literature is rather clear in authenticating the extrafecal origins and growth potentials of members of this group, its poor correlation with pathogens and health effects, and the superiority of the fecal coliform as an indicator of feces. Based on this information we conclude that although the group remains an accepted indicator and standard for drinking water, its continued application to shellfish growing waters is not justified.

Fecal Coliform Group. The fecal coliform group is the most frequently used indicator to classify shellfish growing waters and has largely replaced the total coliform group as a more specific indicator of fecal pollution. However, it possesses several characteristics that have engendered criticism of the total coliform group. The fecal coliform group is operationally defined (APHA 1985) and includes genera that are not restricted to fecal habitats and grow in saline receiving waters. Standard procedures for enumerating fecal coliforms do not differentiate component genera. Because the numeric value of the total coliform standard was not based on a quantitative assessment of health risk, the same must apply to the fecal coliform standard, the latter derived by correlative analysis of paired total and fecal coliform data from growing areas (Hunt and Springer 1974). The absolute value chosen reflects a methodological bias dictated by the constraints of the most probable number (MPN) distribution table. Compared with the enterococci, fecal coliforms (and *E. coli)* were considered an inferior indicator of sewage pollution because of large reductions in density that occur during sewage treatment, its greater sensitivity to chlorination, and its higher rate of dieoff (Miescier and Cabelli 1982). Fecal coliforms are capable of extended persistence and aftergrowth in estuarine waters (e.g., Figure 7.6). Shipments of shellfish harvested from approved waters in the Gulf of Mexico during periods of seasonally warm temperatures have been rejected because fecal coliform levels in the meats exceeded the market guideline as a result of multiplication after harvesting (FDA 1984). Concerns have been expressed about klebsiellae and estuarine "fecal coliform-mimicking" bacteria that proliferate in approved growing areas in Gulf of Mexico and other areas during periods of seasonally warm temperatures (Miescier et al. 1985).

Escherichia coli.   This organism is the dominant fecal coliform in human and animal feces and is generally considered an indisputable indicator of fecal contamination from warm-blooded animals. Its thermotolerance facilitates separating it from most other members of the coliform group using selective and differential methods. Replacement of the fecal coliform indicator with *E. coli* has been proposed because of thermotolerant-fecal coliforms whose presence in aquatic environments is not necessarily related to fecal contamination (Dufour, 1977). It is reasonable to assume that the development of rapid and direct enumeration methodologies would be simplified if the target indicator was a single species. For these and other reasons *E. coli* has been suggested as a more specific indicator of fecal contamination in estuarine waters.

However, we must ask if the information available is sufficiently compelling to support this conclusion. As noted, *E. coli* can enter a nonculturable but viable physiological state, physiologically adapt to saline conditions, and is susceptible to sublethal stress. Sublethal stress can significantly reduce its recovery when temperatures fall below 10°C (50°F), at high salinities, in the presence of toxic chemicals, or due to light-induced damaged.

E. coli does not fulfill the requirements of an ideal indicator for other reasons despite the assumption of a fecal origin. Besides its persistence and aftergrowth capabilities, its numbers in water are impacted significantly by the natural microbiota. In contrast, we would not anticipate that enteric viral pathogens or viral indicators would enter the microbial food web and be removed through predator–prey or other microbially mediated interactions. This hypothesis implies a fundamental difference between viral and bacterial indicator removal mechanisms. This selective removal process may contribute to enteroviral isolations in the absence of indicator bacteria. Therefore, although *E. coli* may be a useful indicator for enteric bacterial pathogens downstream from definitive point sources of fecal pollution, its value as a quantitative standard in shellfish growing waters impacted by diffuse sources appears equivocal. This conclusion must be considered in view of anticipated approaches to "fine-tune" the specificity, selectivity, and accuracy of bacterial enumeration methods.

Fecal Streptococci. The term "fecal streptococci" is used to describe a group of taxonomically diverse streptococci that are Gram-positive, catalase negative, nonspore-forming, facultative anaerobes associated with the gastrointestinal tracts of humans and animals. Functionally, the fecal streptococci are variously divided into groups and subgroups on the basis of serology and physiological characteristics. Thus, the serologically defined group-D streptococci of the fecal streptococci (Clausen et al. 1977) are divided into an "enterococcal" subgroup (*S. faecalis, S. faecium,* and *S. durans*) and a "nonenterococcal" subgroup (*S. bovis* and *S. equinus*). The latter term is curiously inappropriate and has unnecessarily contributed to confusion regarding the meaning of the term "enterococci" (Deibel and Hartman 1984).

Recent studies have shown that although the "nonenterococcal" species of the genus *Streptococcus* share the group-D antigen and a fecal habitat, these similarities do not reflect underlying taxonomic relationships. Analysis of oligonucleotides from 16S-rRNA of representative streptococci (Ludwig et al. 1985), and nucleic acid hybridization studies, have led to the creation of a new genus, *Enterococcus* (Schleifer and Kilpper-Balz 1984), which includes members of the enterococcal group as well as other species not associated with humans. The genus *Enterococcus* now includes nine valid species and proposals for including three newly discovered species have been made (Facklam and Collins 1989). *S. bovis* and *S. equinus* are taxonomically distinct and not considered valid members of this genus owing to differences in physiology and metabolic characteristics (Bergey's Manual of Systematic Bacteriology 1986). With the creation of three distinct genera from *Streptococcus,* i.e., *Enterococcus, Streptococcus,* and *Lactococcus,* the trivial terminology used to describe these organisms becomes more confusing and would benefit from revision.

Although far from being numerically dominant in human feces, constituting 0.1% or less of the gut microbiota (based on cell densities) (Moore and Holdeman 1974), fecal streptococci have long been considered an important indicator of fecal pollution. This is attributable to its recognized association with feces, observed differences in ratios of coliforms to fecal streptococci in human and animal feces (Geldreich and Kenner 1969), and its inability to grow in seawater (Slanetz and Bartley 1965). The complex nutritional requirements of this group are generally considered to preclude extraenteral growth (Mundt 1982).

The ratio of fecal coliforms to fecal streptococci has been accorded considerable importance in the literature and much discussion concerned its validity as an index of human versus animal fecal contamination (Feachem 1975; Geldreich and Kenner 1969). Its utility was predicated on observations that the relative proportion of fecal coliforms to fecal streptococci differs among mammalian species, human feces generally containing proportionately more coliforms than animal feces (Geldreich and Kenner 1969). However, data has been presented that critically questions assumed values of the ratio in feces and the validity of the ratio in freshwater environments (McFeters et al. 1974; Wheater et al. 1979). Values of the ratio in feces from a variety of animals and man vary considerably and certain animal species exhibit mean values that are similar to human sources (Kjellander 1960; Wheater et al. 1979). Furthermore, the ratio does not remain stable in receiving waters (because of differential survival) and is difficult to interpret in systems with multiple sources of fecal pollution.

Few studies have focused on the survival characteristics of fecal streptococci in marine and estuarine waters and these have yielded contradictory results (Lessard and Sieburth 1983; Slanetz and Bartley 1965; Vasconcelos and Swartz 1976). Consequently, it is difficult to reach a conclusion regarding the utility of the ratio in these waters. Differences in persistence will likely yield similar variations in the ratio such as found in freshwaters. It is difficult to appreciate how the recommended criteria for use of the fecal coliform to fecal streptococcus ratio (APHA 1985; Geldreich and Kenner 1969), no doubt derived with defined point sources in mind, can be met in growing areas polluted by multiple, diffuse sources of fecal contamination. Finally, as discussed below, enumeration of fecal streptococci in estuarine waters is method sensitive and results in analytical uncertainty that must be minimized if the ratio is to be a meaningful parameter. Reducing analytical variability is also important because fecal streptococci generally occur at lower densities than fecal coliforms in growing waters.

Although certain genera (e.g., *E. faecalis*, *E. faecium*) of fecal streptococci are considered specific to feces from human or other warm-blooded animals, phenotypically similar strains and biotypes can be isolated from other environmental sources (Beaudoin and Litsky 1981; Clausen et al. 1977; Mundt 1982). Strains and biotypes of *E. faecalis* and of *E. faecium* occur in a variety of plant materials, reptiles and insects (Geldreich and Kenner 1969; Geldreich et al. 1964; Mundt 1963; Mundt et al. 1958). Moreover, Mundt et al. (1962) reported growth of *E. faecalis* biotypes on plants and vegetables. In comparison, the survival of *S. bovis* and *S. equinus* in extraenteral environments and aquatic habitats is very limited (Geldreich and Kenner 1969; Slanetz and Bartley 1965; Wheater et al. 1979). Use of the fecal streptococci as an indicator in shellfish waters will probably depend on development of enumeration methods that distinguish between fecal and nonfecal strains.

Numerous methods for enumeration of fecal streptococci are available and include MPN, direct pour plate, and membrane filtration techniques, some methods incorporating resuscitative steps to minimize sublethal stress (Brezenski 1973; Yoshpe-Purer 1989). Comparisons of the various media and methods can be found in references such as Brezenski (1973), Switzer and Evans (1974),

Clausen et al. (1977), Pagel and Hardy (1980), Volterra et al. (1985), and Yoshpe-Purer (1989). Considerable differences between methods exist in terms of recovery efficiency, specificity, and selectivity, and each method requires some degree of confirmatory testing. Use of the fecal streptococci as an indicator in marine and estuarine shellfish waters will require improvements in the selectivity and specificity of media and an understanding of factors affecting its survival. Method validation should be performed in shellfish growing areas to identify typical fecal streptococcal species recovered and autochthonous bacteria that produce false positives. Although KF is considered the medium of choice for saline waters (APHA 1985), many investigators (including ourselves) and recently Yoshpe-Purer (1989) have concluded that KF medium is not sufficiently selective for use in marine waters. We note with approbation and mutual frustration that a number of our recommendations were promulgated quite a few years ago by Kjellander (1960) and Hartman et al. (1966).

Use of fecal streptococci as an indicator will probably depend on adoption of a common operational definition of fecal streptococci (as with fecal coliforms) that is based on a formalized and unique method for recovery of specific enterococcal or streptococcal species associated with feces. We view this as a challenging and formidable task.

Enterococci. The term "enterococci," with its connotation of fecal origin, is "understood" by some workers to exclude those fecal streptococcal species (*S. equinus* and *S. bovis*) associated almost exclusively with animal gastrointestinal tracts. This usage is misleading because the term implies streptococci sharing a fecal habitat. Deibel and Hartman (1984) consider the terms group-D streptococci and enterococci the same. Creation of the genus *Enterococcus* spp. (whose constitutents are "enterococci"?), which contains several nonhuman species (Facklam and Collins 1989) but excludes *S. equinus* and *S. bovis,* does little to resolve this semantic confusion.

Functionally, the enterococci are defined on the basis of selective recovery using one of several methods promulgated for this purpose (e.g., D'Aoust and Litsky 1975; Isenberg et al. 1970; Levin et al. 1975; Slanetz and Bartley 1965). The range of species selected varies with each method, some being more selective for the classic "enterococcal" *(E. faecalis, E. faecium)* group organisms than others. In reviewing this literature the reader must be wary of the operational nature of the results and current taxonomic relationships.

The most recent use of enterococci in marine waters is by Cabelli et al. (1983). mE medium (as modified by Levin et al. 1975) was used in polluted marine and estuarine waters to derive a statistical relationship between indicator density and the incidence of swimming-associated gastrointestinal disease. Cabelli et al. (1983) concluded enterococcal densities were better predictors of disease risk than fecal coliforms or *E. coli*. Similar studies to evaluate the enterococci and other indicators as predictors of enteric disease associated with consumption of raw shellfish are now being performed (A. Dufour, USEPA, personal communication).

Validation of the enterococci as an indicator of public health risk in shellfish growing areas will require improved recovery methods. The range of selectivity and specificity characteristics of methods for enumeration of the enterococci has been mentioned. Current recovery methods require confirmation of presumptive isolates (Ericksen and Dufour, 1986), increasing the time and cost of analysis. Although the mE-based method has been applied to marine and estuarine waters (Cabelli et al. 1983), its utility in nonpoint source impacted shellfish growing areas is undetermined. Similarly, enterococcal presence in stormwater runoff (Geldreich et al. 1968; Pagel and Hardy 1980) and differences concerning its persistence in estuarine and marine waters (Lessard and Sieburth 1983; Slanetz and Bartley 1965; Vasconcelos and Swartz 1976) demonstrate a need for distribution and survival studies. Research requirements include: (1) improving methods for recovery of stressed cells, (2) improving specificity for various *Enterococci* spp. and *Streptococci* spp., (3) reducing the incidence of false-positives, which can vary with temperature, season, and geographical region, and (4) minimizing background growth. The relative occurrence and densities of nonfecal biotypes of *E. fecaelis* and *faecium* in nonpoint source and stormwater runoff impacted areas should be determined. Rapid methods for confirmation of selected enterococcal species, based on serological or biochemical characteristics, could significantly improve its candidacy as an indicator in shellfish growing waters. Aside from a few reports describing the use of fluorescent antibody-based methods for species identification or enumeration (e.g., Abshire and Guthrie 1971; Pugsley and Evison 1975), direct methods for selective enumeration of this group from environmental samples remain to be developed.

Alternate Indicators

Obligate Anaerobes.   Although obligate anaerobes are numerically dominant constituents of the human fecal microbiota (Moore and Holdeman 1974), attempts to use these organisms as indicators of fecal pollution are rare. This is attributable to the belief that anaerobes do not persist in oxic waters, a lack of selective enumeration media, and the additional effort required for their cultivation and identification.

Bifidobacterium.   *Bifidobacteria* spp. are obligately anaerobic, Gram-positive, nonmotile bacteria that possess a characteristic pleomorphic and branching cell morphology. They are potentially important candidate indicators of fecal pollution as the habitat of many species appears restricted to the intestinal tracts of adult humans and infants and major groups of warm-blooded animals (Bergey's Manual of Systematic Bacteriology 1986; Cabelli 1978a; Levin 1977). In humans, bifidobacteria are a major component of the intestinal microbiota occurring at densities greater than 10^{10} cells/g dry feces and some species may comprise more than 6% of the culturable microbiota (Holdeman et al. 1976). A unique metabolic feature of this genus is that glucose is metabolized exclusively and characteristically through the fructose-6-phosphate shunt (bifid shunt) (Bezkorovainy 1989). The first reaction step is mediated by the enzyme fructose-6-phosphate phosphoenolketolase (F6PPK). This enzyme can be detected using a colorimetric assay and is a reliable phenotypic characteristic of the genus (Bergey's Manual of Systematic Bacteriology 1986; Scardovi 1981). Bifidobacteria can also utilize ammonium as a sole source of nitrogen. The sensitivity of these organisms to oxygen varies with species and strains, the presence of CO_2 allowing certain species to tolerate relatively high O_2 partial pressures.

A potentially valuable property of this group concerns its ability to differentiate human from animal fecal pollution based on "human" and "animal" bifiobacterial strains (Levin 1977). Levin and Resnick (1981) could not isolate bifidobacteria dominant in human feces (i.e., *B. longum, B. adolescentis*) from a variety of domestic and wild animals except swine. Mara and Oragui (1983) confirmed the low occurrence of these organisms in animal feces and developed a medium based on sorbitol fermentation, which selects for bifidobacteria derived from human sources. The ability of human-specific strains to ferment sorbitol can vary and not all sorbitol-fermenting bifidobacteria are derived from human feces (Bergey's Manual of Systematic Bacteriology 1986). Nevertheless, the possibility of differentiating human from animal fecal pollution should provide an impetus for the evaluation of this anaerobe.

Use of bifidobacteria as an indicator of fecal contamination in shellfish growing waters has been considered (Cabelli 1978a), but recent studies have focused primarily on its occurrence and recovery from freshwater environments (Carrillo et al. 1985; Mara and Oragui 1983; Munoa and Pares 1988). Information on the distribution of these bacteria, albeit limited to natural and sewage contaminated freshwaters, suggests its use as an indicator of very recent fecal pollution is valid as the organisms appear incapable of entraenteral growth.

Although obligate anaerobes, Gyllenberg et al. (1960) reported bifidobacteria survive as well as *E. coli* in freshwater. In contrast, Levin and Resnick (1981) observed that *B. longum* and *adolescentis,* exposed in vitro to fresh and marine water samples, are less persistent than *E. coli. B. adolescentis* populations decline considerably after 48 hours of in situ exposure in tropical freshwaters (Carrillo et al. 1985). Under these same conditions *E. coli* densities remain constant or increase. Results of an in vitro experiment to compare the survival of *B. adolescentis* (Fig. 7.7) with *E. coli* (Fig. 7.8) in fresh and estuarine water suggested temperature plays an important role in bifidobacterial survival. *B. adolescentis* cells were incubated in flasks containing membrane filtered (0.2 μm) water at two water temperatures characteristic of winter [6°C (42.8°F)] and late spring and summer [25°C (77°F)]. *B. adolescentis* persistence

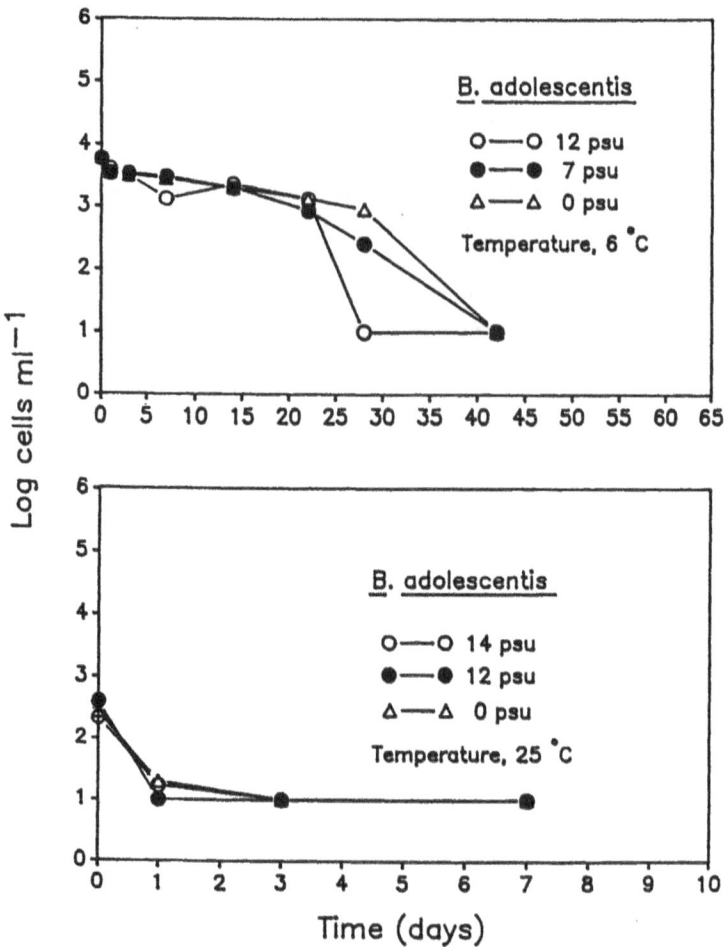

Figure 7.7 Survival of *Bifidobacterium adolescentis* in water collected at sites of contrasting salinities in a nonpoint source impacted shellfish growing area. *B. adolescentis* (ATCC 15703) was grown overnight at 35°C (95°F) in sorbitol broth modified after HBSA (Mara and Oragui 1983) and lacking antimicrobials or dye. The culture was diluted in double membrane filtered (0.45 and 0.22 μm) freshly collected site water and replicate volumes incubated at either 6°C (42.8°F) or 25°C (77°F). At intervals duplicate samples were removed and enumerated by spread plating onto RCA (reinforced clostridial agar) and incubated anaerobically at 35°C (95°F) for 72 hours.

at 6°C (42.8°F) was moderately better than *E. coli*, but at the higher temperature was significantly worse. Survival was generally similar for all values of salinity. In contrast, Levin and Resnick (1981) observed considerably less persistence of bifidobacteria from sewage exposed in vitro to membrane-filtered seawater (32 psu). Survival was also independent of temperature, with approximately 15% of

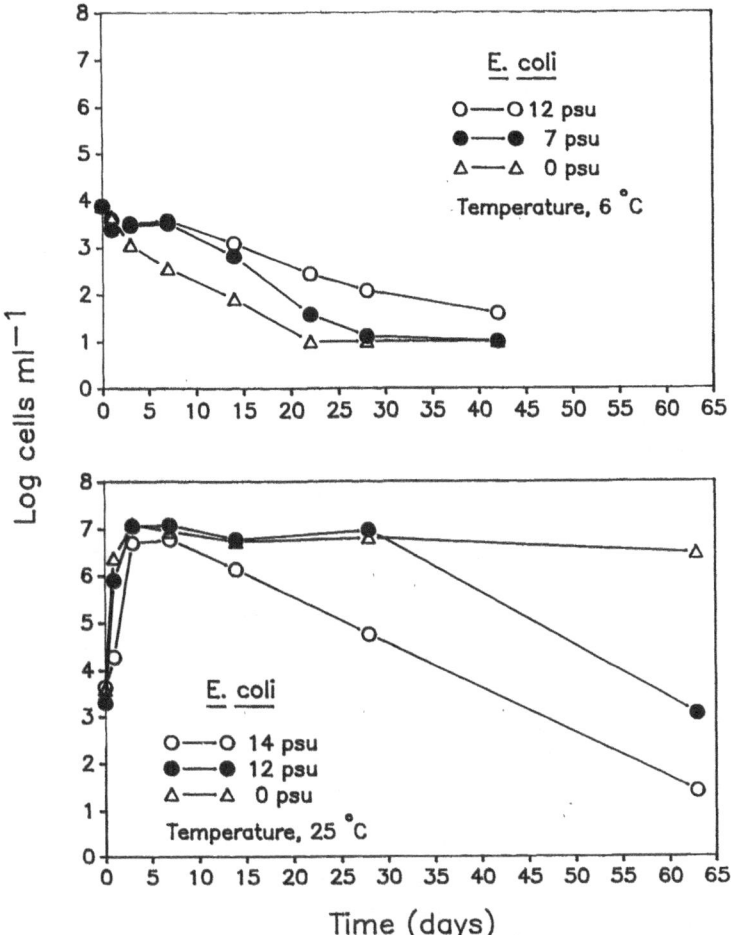

Figure 7.8 Survival of *Escherichia coli* in water collected at sites of contrasting salinities in a nonpoint source impacted shellfish growing area. *E. coli* was grown overnight at 35°C in trypticase soy broth. Cultures [environmental isolate at 6°C (42.8°F), human fecal isolate at 25°C (77°F)] were diluted in double membrane filtered (0.45 and 0.22 μm) freshly collected site water and replicate volumes incubated at either 6°C (42.8°F) or 25°C (77°F). At intervals duplicate samples were removed and enumerated by spread plating on TSA and incubated at 35°C (95°F) for 72 hours.

the initial density remaining after about 6 hours. Individual species and strains varied in survival capability.

The generally low bifidobacterial densities found in receiving waters may be attributed in part to poor recovery owing to oxygen toxicity and sublethal stress, factors that are exacerbated by selective media and harsh recovery methods. Support for this hypothesis is illustrated by the effect of enumeration methods on *B. adolescentis* recovery (Table 7.4). Munoa and Pares (1988) demonstrated the inability of *Bifidobacterium* spp. cells to produce colonies on a selective medium was caused by sublethal injury following exposure to seawater. To minimize the effect of sublethal stress on enumeration, a two-layer recovery procedure was designed that incorporated plating and incubation of the sample on a resuscitative medium (reinforced clostridal agar). This was followed with an overlay of BIM-25, a selective medium developed to improve the poor selectivity of YN-6 medium with environmental samples (Levin and Resnick 1981). Unfortunately, *B. adolescentis* did not grow as well on BIM-25 as other *Bifidobacterium* spp. [Munoa and Pares (1988)]. Improving recovery methods for this group of organisms, especially human-specific species, should be an objective of future research efforts. Resusitative techniques should be combined with membrane filtration because of the need for sample concentration. Despite the recognition that the bifidobacteria are sensitive to oxygen, use of prereduced media or an oxygen-free environment for recovery manipulations has not been prescribed. Although Levin and Resnick (1981) suggest oxygen toxicity is not a concern with recovery of bifidobacteria, sorbitol-fermenting strains are catalase negative, and our data (Table 7.4) and those of Munoa and Pares (1988) suggest factors that exacerbate sublethal stress should be minimized. Differences in the tolerance of bifidobacterial species or strains to oxygen (Bergey's Manual of Systematic Bacteriology 1986) may target the choice of an indicator bifidobacterial species. The role of particle association in protecting *Bifidobacterium* spp. and other anaerobes from oxygen toxicity in the environment is undetermined. In situ exposure studies are needed to measure the persistence of selected bifidobacterial species in natural waters and sediment.

Table 7.4 Percent Recovery of *B. Adolescentis*[a] Following Immediate Dilution in Various Diluents and Enumeration on Nonselective RCA (Reinforced Clostridial Agar) and Selective (HBSA) Human Bifid Sorbitol Agar Media Using Spread Plating and Membrane Filtration

| | *Percent Recovery*[b] | | | |
| | *Spread Plating* | | *Membrane Filtration* | |
Diluent[c]	RCA	HBSA	RCA	HBSA
Phosphate buffer	100	24	26	1
Ringer	93	15	26	1
Gelatin	101	26	39	8

[a]Prepared under conditions to minimize stress; grown and passed twice in PYG broth (peptone yeast extract glucose broth) at 35°C.
[b]Percent recovery compared to that obtained using phosphate buffer as the diluent and spread plating on RCA. Each value is the mean of four experiments.
[c]Phosphate buffer (APHA 1985); Ringer (Collins and Lyne 1984); Gelatin (Dowell and Hawkins 1968)

If using indicators to identify sources of fecal pollution to nonpoint impacted estuarine growing areas, the unequivocal fecal origin and limited survival properties of bifidobacteria are potentially valuable because their detection implies immediate fresh sources. The appearance of bifidobacteria in feeder streams may be used to locate sources of fecal pollution and to potentially differentiate human from nonhuman sources. Although discrimination of vertebrate source will require studies to verify the restricted host range of the sorbitol-fermenting bifidobacteria, the ability to differentiate human from animal pollution could aid management strategies directed toward reduction of all pollutant sources. Finally various bacteriophages, including those lytic to the anaerobe *Bacteroides fragilis,* are being considered as candidate indicators of fecal pollution. If bifidobacteria-specific bacteriophages occur in feces or sewage, these could be used as a human-specific viral indicator of fecal pollution.

Bacteroides. *Bacteroides* spp. are nonsporulating, obligately anaerobic Gram-negative bacteria which are among the predominant genera in the human colon (Holdeman et al. 1976; Salyers 1984). Like the bifidobacteria, *Bacteroides* is an important candidate indicator of fecal pollution as its habitat is restricted to the feces of humans and warm-blooded animals. As a group, *Bacteroides* may be more tolerant of oxygen than the bifidobacteria; some *Bacteroides* spp. produce catalase and *B. fragilis* and *B. distasonis* sythesize superoxide dismutase (Salyers 1984).

Until recently, the term *"Bacteroides fragilis* group" (BFG) was used to describe dominant bacteroides found in the human colon that included *B. fragilis* and its subspecies. The subspecies were subsequently accorded species rank on the basis of DNA homology studies (Salyers 1984) and dominant species now include *B. vulgatus, B. fragilis, B. thetaiotaomicron,* and *B. distasonis.* The term BFG still appears in current literature, and although taxonomically incorrect, has utility as a simple phrase referring to the dominant human colonic bacteroides species. *B. vulgatus, B. thetaiotaomicron,* and *B. distasonis* are numerically dominant (10^{10} cell/g dry weight feces) and *B. vulgatus* can constitute $>10\%$ (as cell count) of the culturable fecal microbiota (Holdeman et al. 1976). *B. fragilis,* which is more aerotolerant than *B. vulgatus,* occurs at densities similar to *E. coli.* The observation that *Bacteroides* species other than those found in humans dominate in ruminants (Macy 1981) supports the use of BFG species as a human-specific indicator. Macy (1981) provides a summary of host-related species composition and methods for recovery and cultivation of nonpathogenic *Bacteroides* spp.

Although some *Bacteroides* spp. are medically significant and can be isolated from a variety of clinical specimens, there is little information describing the occurrence of nonpathogenic bacteroides in natural waters. Post et al. (1967) first suggested use of the BFG as an indicator of fecal pollution. Allsop and Stickler (1985) evaluated BFG as a fecal indicator based on its presence in sewage, various freshwater and marine environments. Organisms belonging to this group were readily isolated downstream from waters impacted by sewage. Because *Bacteroides* spp. disappeared faster from these waters than *E. coli*, Allsop and Stickler (1985) concluded the ratio of BFG to *E. coli* counts reflects "aging" of a fecal pollutant and also proximity to the source. The presence of BFG organisms in natural waters was attributed to human fecal pollution because analysis of feces from a variety of domestic animals and feral birds revealed that BFG species occur at much lower densities (i.e., 10^5–10^{10}-fold lower) in common domestic farm animals (cattle, horses, pigs, chickens, and sheep)(Allsop and Stickler 1985). Higher densities occur in domestic pets and seagulls but these were still lower than levels in human feces. These observations, admittedly limited, support the potential use of BFG as a human-specific indicator in nonpoint source impacted shellfish growing areas.

General comments concerning *Bifidobacteria* spp. recovery methods also apply to this group. Perhaps more work on identification methods has occurred with *Bacteroides* spp. owing to its clinical importance as a pathogen and possible involvement in colon cancer. Allsop and Stickler (1984) developed an improved medium for recovery of BFG organisms from natural waters. Penicillin is incorporated into the medium to inhibit clostridia and membrane filtration (0.22 μm) is used for sample concentration. Application of this medium to samples of sewage and fresh and marine waters revealed interference by obligate and facultative anaerobes. Increasing the concentration of gentamicin improves the selectivity of the medium but decreases recovery efficiency. A resuscitation period is incorporated to minimize the effects of sublethal injury. Serious consideration of BFG as an indicator requires a detailed assessment of this recovery method, including its selectivity, validation of confirmatory steps required, and data describing the occurrence and persistence of BFG in estuarine waters.

Chromosomal DNA probes have been developed to speciate *Bacteroides* spp. as an alternative to conventional phenotypic methods. Roberts et al. (1987) and Morotomi et al. (1988) used whole cell dot blot assays to correctly differentiate *Bacteroides* by species. Kuritza and Salyers (1985) used a cloned DNA probe to enumerate *B. vulgatus* in feces and concentrations measured were similar to viable counts. Although the probe was highly specific for a segment of the *B. vulgatus* genome, its use for enumeration of cells in environmental samples seems unlikely because of its low sensitivity (ca. 10^7–10^8 cells required). A sample concentration procedure that would eliminate problems associated with particulates in estuarine water, e.g., clogged filters, and interfering overgrowth by the autochthonous microbiota (Amy and Hiatt 1989) will be needed. However, these probes should be valuable in supplanting or augmenting the specialized and labor-intensive biochemical characterization of presumptive isolates. Use of *Bacteroides fragilis* bacteriophage as an alternate indicator is discussed in the section on viral indicators.

Clostridium perfringens. *C. perfringens* is an anaerobic, Gram-positive, spore-forming rod whose presence in receiving waters has been linked to contamination by feces and wastewaters (Bisson and Cabelli 1980; Cabelli 1977a). Bisson and Cabelli (1980) developed a membrane filtration method for *C. perfringens* recovery incorporating a highly selective medium (mCP). The composition of this medium was later modified (Armon and Payment 1988) to reduce the concentration of a costly component without compromising selectivity. Bisson and Cabelli (1980) concluded that *C. perfringens* had value as an indicator of chlorination efficiency, unchlorinated sewage, fecal contamination, and as a "conservation tracer" delineating the areal impact and transport of wastewater effluents. Moreover, detection of vegetative cells in the environment reflects fresh and untreated fecal matter because the persistence of vegetative cells is short (Bisson and Cabelli 1980). Fujioka and Shizumura (1985) confirmed the utility of *C. perfringens* to detect wastewater discharge into freshwater streams. However, there is doubt *C. perfringens* would be a useful indicator in nonpoint impacted growing areas because (1) it is so persistent that it may be difficult to index to current pollution conditions, (2) it is widely distributed in soils and sediments, and (3) it is carried into growing areas from extraenteral sources by stormwater runoff and transport of suspended sediment (Matches and Liston 1974; Smith 1975).

Aerobes

Rhodococcus coprophilus. *Rhodococcus coprophilus* is an aerobic nocardioform actinomycete proposed as an indicator of domestic farm animal fecal pollution (Mara and Oragui 1981; Oragui and Mara 1983; Rowbotham and Cross 1977b). As a fecal indicator this organism is unique because it is associated with the feces of grazing farm animals but is not considered an active component of the rumen microbiota (Rowbotham and Cross 1977b). Rowbotham and Cross (1977b) noted that numbers (colony forming units/g) of *R. coprophilus* in pasture grass and in dung collected from cattle fed this pasture grass were similar, implying no multiplication occurs during passage through the animals. In contrast, significant increases in densities of *R. coprophilus* that occur during incubation of fresh dung at a moderate temperature [ca. 20°C (68°F)] with water confirms the coprophilic habitat of this organism. Consequently, this organism is commonly found in pasture grass and soils grazed by cattle and other domestic farm animals. Studies have shown that *R. coprophilus* is absent from human feces but consistently found in feces of cattle, sheep, pigs, horses, donkeys, farm-raised poultry and sporadically in dog and seagull feces (Mara and Oragui 1981).

    R. coprophilus can survive both dessication and temperatures of 2–4°C (35.6–39.2°F) in dung up to 6 weeks (Rowbotham and Cross 1977b). Goodfellow and Williams (1983) concluded that the coccal stage of *R. coprophilus*, carried by runoff into either fresh or salt water habitats, does not grow but remains viable. *R. coprophilus* persists in vitro for 17 weeks in nonfiltered freshwater incubated at 5°C (41°F), 20°C (68°F), and 30°C (86°F), whereas *E. coli* and fecal streptococci disappeared within 5 weeks (Oragui and Mara, 1983). These investigators suggested these and other properties of the organism could be used (in conjunction with other bacterial indicators) to determine the temporal characteristics, location, and source of fecal contamination to freshwater systems. Information concerning the distribution and persistence of this organism in estuarine waters is insufficient to evaluate its utility in shellfish growing waters. In an experiment using a recovery medium (MM3) devised by Mara and Oragui (1981), we observed extended persistence (ca. 30 days) of *R. coprophilus* in 0.2-μm filtered estuarine water incubated at 6°C (42.8°F) and 25°C (77°F) (Fig. 7.9) compared with *B. adolescentis* (Fig. 7.7) or *S. bovis* (Fig. 7.10), particularly at the higher temperature. Although *R. coprophilus* (Fig. 7.9) appeared to multiply more slowly than *E. coli*, (Fig. 7.8) and manifested less initial aftergrowth at 25°C (77°F), its persistence in water over the range of salinities and temperatures tested appeared generally better than *E. coli*. Whether the observed density increases [particularly at 6°C (42.8°F)] were due to growth or perhaps an artifact of the organism's fragmentable nocardioform morphology (see methodological considerations below) must be addressed in future studies. Similarly, the persistence of this organism in marine and estuarine sediments has not been examined, although we and other investigators (Attwell and Colwell 1984; Goodfellow and Haynes 1984; Goodfellow and Williams 1983) have isolated it and other actinomycetes from marine and estuarine sediments. We have isolated *R. coprophilus* in a small subestuary of the York River, Virginia, more frequently and at higher densities in water and sediment adjacent to a livestock farm compared to locations where this activity was absent. Our observations support its proposed use to identify domestic farm animals as a source of fecal pollution. Studies of *R. corprophilus* persistence in estuarine water and sediment, related particularly to temperature, are needed to assess its possible value as an index of current pollution conditions.

Disadvantages now associated with *R. coprophilus* are primarily limitations inherent in the current enumeration procedure (Mara and Oragui 1981). Only small volumes (0.2 ml) of sample can be spread plated onto MM3 agar, currently the medium of choice. Direct spread plating limits detection to 1 cfu/ml (e.g., five replicate plates containing 0.2 ml each). A recovery procedure is needed for concentrating *R. coprophilus* from water and using MM3 or other suitable medium for expressing characteristic colony morphology. The combination of membrane filtration (Pisano et al. 1986) with the novel selective method of Hirsch and Christensen (1983) could be evaluated for enumerating *R. coprophilus*. Hirsh and Christensen (1983) described a "selective" method for recovery of actinomycetes that eliminates bacterial contamination. This is based on the

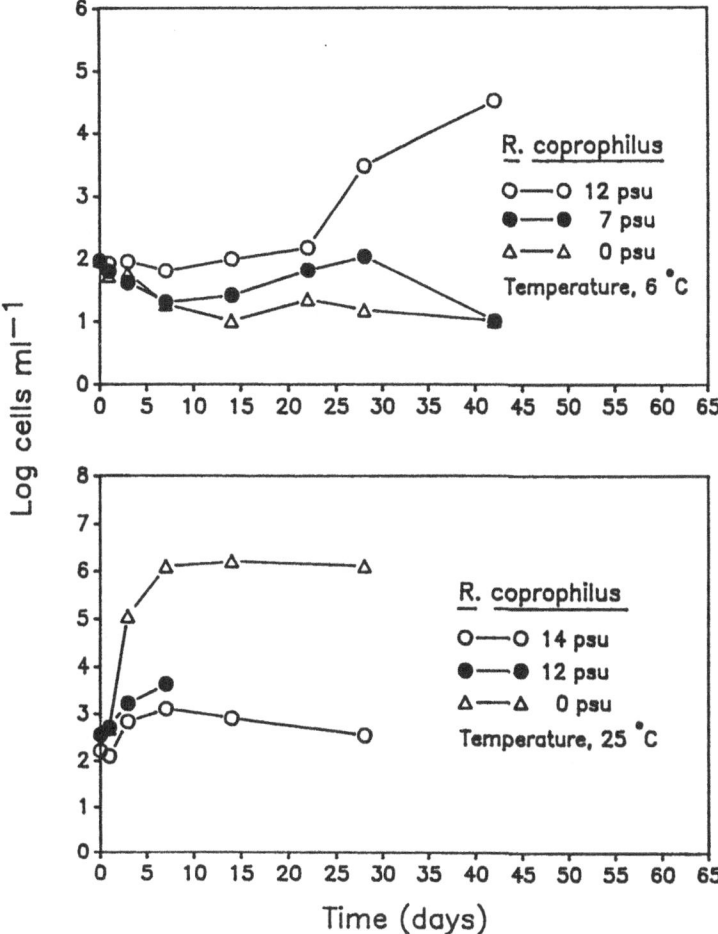

Figure 7.9 Survival of *Rhodococcus coprophilus* in water collected at sites of contrasting salinities in a nonpoint source impacted shellfish growing area. *R. coprophilus* (ATCC 29080) was grown in Bennett's broth at 30°C (86°F) for 3–5 days. The culture was diluted in double membrane filtered (0.45 and 0.22 μm) freshly collected site water and replicate volumes incubated at either 6°C (42.8°F) or 25°C (77°F). At intervals duplicate samples were removed and enumerated by spread plating onto Bennett's agar and incubated at 30°C (86°F) for 7 days.

ability of actinomycetes to penetrate the "pores" of a membrane filter on a growth medium, whereas bacteria remain on the filter surface. After an incubation period the filter is discarded and only actinomycetes that penetrate the filter form colonies. Another limitation noted by Mara and Oragui (1981) is the long incubation period (17–18 days) required for development of characteristic stellate colonies having both substrate hyphae and bright orange central papillae

(Rowbotham and Cross 1977a). The prolonged incubation is a consequence of (1) the low organic content of the medium, (2) the presence of inhibitors that are used to suppress bacterial growth, (3) the photochromogenic nature of pigment production, and (4) the time required to visibly develop the characteristic colony morphology. In our hands, MM3 recovered *R. corophilus* from freshwater and estuarine samples but not without some equivocal results and background interference. Time for appearance of typical *R. coprophilus* aesteroidal colonies varied for pure cultures and isolates recovered from environmental samples. The presence of numerous nonstellate pigmented and stellate nonpigmented colonies on MM3 required confirmatory testing. Biochemical confirmation of nocardioforms to the generic level is very time-consuming and identification on the basis of colonial morphology alone is neither reliable nor valid (Bergey's Manual of Systematic Bacteriology 1986; Goodfellow and Williams 1983). Rapid and direct methods for species verification (and enumeration) are needed. An indirect method to detect or identify *R. coprophilus* using actinophage appears doubtful because of the broad cross-reactivities of actinophage to both *Nocardia* and *Rhodococcus* (Prauser 1984).

Precise enumeration of *R. coprophilus* is complicated by the inherent morphogenetic nature of nocardioforms, and particularly for *R. coprophilus*, which forms extensively branched hyphae and coccoid elements that are connected in chains of varying length. Thus, a given colony-forming unit in an aqueous or sediment sample can originate from a variety of cell configurations. Rowbotham and Cross (1977a) observed that a typical colony enumerated from dung is derived from single or multiple coccoid elements, the latter being the most prevalent form. The complicating effect of *R. coprophilus* morphology on enumeration is neither a desirable indicator characteristic nor conducive to analytical precision.

Streptococcus bovis. *S. bovis* may be the dominant fecal streptococcus in warm-blooded animals (Kjellander 1960; Wheater et al. 1979). Use of *S. bovis* as a specific indicator of animal fecal pollution was first suggested by Cooper and Ramadan (1955). Since their report few investigators have evaluated its use as an indicator of animal fecal pollution in freshwater and none that we know of for marine and estuarine waters. Wheater et al. (1979) claim the proportion of *S. bovis* to total fecal streptococci is largest in ruminants but note the important role of diet and geographic location on this ratio. However, the association of *S. bovis* with ruminants is not unique as it has been be isolated from feces of dogs, cats, various birds, horses, and pigs (Clausen et al. 1977; Kenner et al. 1960). Geldreich and Kenner (1969), Clausen et al. (1977), and Wheater et al. (1979) were unable to isolate *S. bovis* from human feces. Oragui and Mara (1981) drew similar conclusions but did isolate *S. bovis* from sewage effluent. This contradiction was resolved when they located an abattoir discharging into the sewage treatment facility. However, other workers have isolated *S. bovis* from the feces of healthy humans (Kjellander 1960; Switzer and Evans 1974). Kjellander (1960) recovered *S. bovis* from about 15% of humans examined; Dalton et al. (1986) observed a fecal carriage rate in healthy humans of 10–16%. Some workers suggest these contrasting results are due in part to dietary variations and regional effects. Another explanation is that investigators used methods with different recovery efficiencies, selectivity, and specificity characteristics. Based on the literature the hypothesis that *S. bovis* is a unique indicator of animal fecal pollution remains somewhat equivocal.

Greater concurrence exists concerning the survival of *S. bovis* in natural waters compared with other enteric aerobic cocci. *S. bovis* mortality in freshwater is much greater than that of *E. fecalis* or *E. coli* (Clausen 1977; Geldreich and Kenner, 1969; McFeters et al. 1974; Wheater et al. 1979). Results of an in vitro exposure experiment using an isolate of *S. bovis,* exposed to filtered feeder stream or estuarine water and incubated at 6°C (42.8°F) or 25°C (77°F) (Fig. 7.10), confirmed its inability to persist, especially at the higher temperature. As stated earlier, poor survival may be viewed as a positive attribute of an indicator because detection of *S. bovis* implies fecal pollution of very recent origin. Conversely, its poor survival violates the requirement that an ideal indicator be at least as persistent as enteric pathogens.

Although *S. bovis* can be isolated with other *Enterococcus* spp. and *Streptococcus* spp. on some of the media used for recovery of these groups (e.g., Switzer and Evans 1974), only one medium is specifically designed for its isolation (Oragui and Mara 1981). Littel and Hartman (1983) described a selective medium for the fecal streptococci that differentiates *S. bovis* from other streptococci and enterococci by hydrolysis of a fluorogenic substrate and a colorimetric indicator of starch hydrolysis, amylose azure. The ability of *S. bovis* to ferment starch is considered an important phenotypic characteristic separating it from *Enterococcus* spp. and other fecal streptococci. Membrane-bovis agar (m-BA) of Oragui and Mara (1981) is a selective medium for *S. bovis* whose specificity is based on the ability of *S. bovis* under anaerobic conditions to utilize NH_4^+ as the sole source of nitrogen and the absence of a requirement for exogenous vitamins. A resuscitation step is incorporated to minimize the effect of temperature stress. m-BA is more specific and more efficient for recovery of *S. bovis* from freshwater and sewage than KF. Subsequently, Oragui and Mora (1984) described a modified m-BA medium (called mm-BA) containing less sodium azide because of reports noting its inhibitory effect on *S. bovis* strains from different geographical regions. Using m-BA Oragui and Mara (1981) reported that more than 65% of the fecal streptococci in selected animal feces are *S. bovis*. Moreover, a proportion (10%) of typical isolates later confirming as *S. bovis* fail to ferment starch. Isolates lacking this characteristic could lead to underestimation of *S. bovis* densities using the medium of Littel and Hartman (1983). A significant proportion of isolates we confirmed as *S. bovis* that were recovered on mm-BA from freshwater feeder streams and shellfish growing waters in Chesapeake Bay did not hydrolyze starch (Table 7.5).

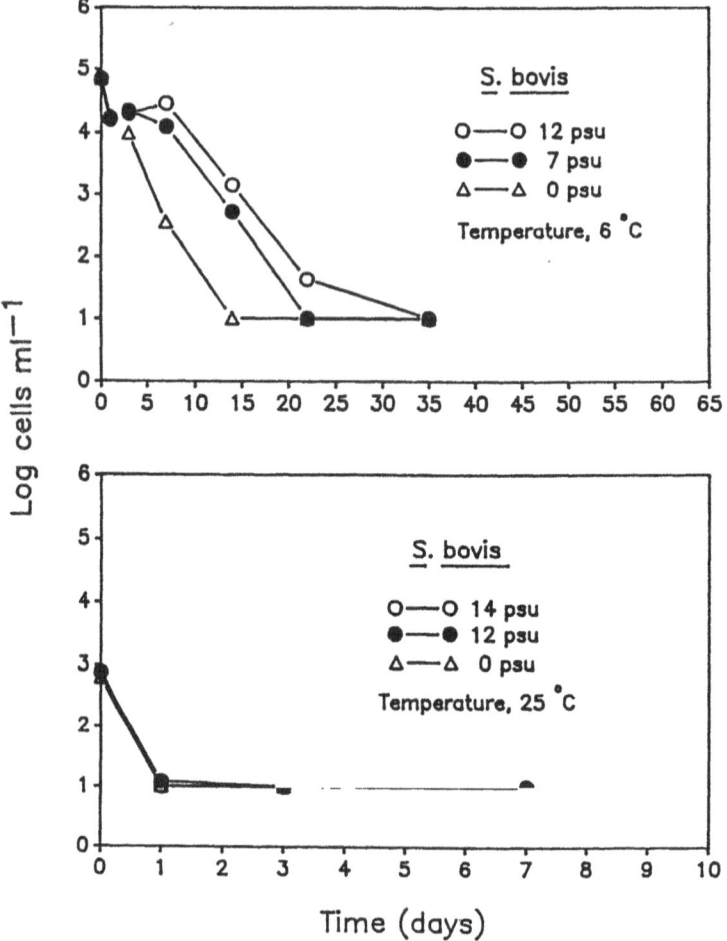

Figure 7.10 Survival of *Streptococcus bovis* in water collected at sites of contrasting salinities in a nonpoint source impacted shellfish growing area. *S. bovis* (ATCC 9809) was grown overnight in BHI broth at 35°C (95.0°F). The culture was diluted in double membrane filtered (0.45 and 0.22 μm) freshly collected site water and replicate volumes incubated at either 6°C (42.8°F) or 25°C (77.0°F). At intervals duplicate samples were removed and enumerated by spread plating onto BHI agar and incubated at 35°C (95°F) for 72 hours.

Our experience using mm-BA on samples from freshwater feeder streams and estuarine shellfish growing areas revealed departures from selectivity and specificity characteristics reported by Oragui and Mara (1981, 1984). Thus, in the analysis of approximately 400 mm-BA isolates, a significant proportion (65.1%) of "typical yellow colonies," i.e., presumptive *S. bovis*, were confirmed

Table 7.5 Proportion of Confirmed *S. bovis* Isolates from Feeder Stream and Estuarine Samples That Hydrolyzed Starch

	Number	*% of Isolates*
Strong +	45	30
Weak +	58	39
Negative	46	31
Total	149	100

as *E. faecium, S. salivarius,* and non-group-D streptococci. This proportion is considerably higher than the 1.3% false-positive rate observed by Oragui and Mara (1981). Conversely, in terms of false-negatives 8.9% of the atypical, nonyellow colonies were confirmed as *S. bovis. E. faecium* was recovered on mm-BA, as subsequently noted by Oragui and Mara (1984). Although *S. bovis* has potential as a source-specific and time-sensitive indicator, and the mm-BA method has the advantage of sample concentration, use in shellfish waters requires an evaluation of the medium's recovery efficiency, specificity, and selectivity over seasonal regimens and in different geographic regions. Also needed is development and validation of rapid screening methods for confirmation of presumptive *S. bovis* colonies.

Bacteriophages.   Occurrences of enteric viral disease attributed to shellfish consumption (Richards 1985) have stimulated research for an appropriate viral indicator. Inadequacies of the current bacterial standard have been emphasized by the lack of parity in survival characteristics of fecal coliforms and enteric viruses (Feachem et al. 1983) and the detection of enterovirus in shellfish from approved growing waters (Ellender et al. 1980; Goyal et al. 1979; Vaughn et al. 1979). Richards (1985) has advocated the direct use of enterovirus as indicators, in particular poliovirus that is easily cultivated and prevalent in sewage due to universal vaccination. Although cost, method, and time constraints preclude routine detection of enteric virus, such analyses could be used to assess potential health risk in growing areas lacking identifiable sources of human fecal pollution and closed to direct harvesting because of elevated indicator densities.

Compared with the costs and limits of direct enteroviral detection, bacteriophage indicators are appealing due to their resistance to disinfection (Keswick et al. 1985) and physical conditions that eliminate bacterial indicators, their ease of detection, low analytical cost, and short assay periods. Justification for use of coliphage as an indicator of human enteric virus (Kott 1981; Scarpino 1975) has been reviewed by Gerba (1987), who identified a number of deficiencies in this concept. Gerba (1987) expressed concern over the paucity of basic data describing ratios of specific coliphages to viral pathogens and the occurrence, persistence, and seasonal stability of coliphage in shellfish growing waters. The following studies illustrate the basis of his concern. Vaughn and Metcalf (1975) recovered enterovirus from shellfish growing waters free of coliphage. Seeley and Primrose (1980) described a subpopulation of coliphages apparently capable of replication at temperatures found in freshwater environments. In situ replication of somatic coliphages in estuarine water has been reported (Vaughn and Metcalf 1975). It is evident that studies of coliphage ecology in marine and estuarine waters are necessary to evaluate its utility as an indicator.

Use of male-specific RNA coliphages as indicators of sewage pollution has been discussed (Furuse 1987; Havelaar and Hogeboom 1984). FRNA coliphages share some properties with human enteroviruses such as type of nucleic acid, structure, and size although these similarities do not necessarily translate to similar functional properties in saline environments. FRNA phage may also be source specific. Furuse (1987) classified FRNA coliphages into four major serological groups having characteristic host distribution patterns. Group I is isolated exclusively from the feces of mammals other than man, and group III from human feces. The validity of this classification scheme should be examined and, if corroborated, could provide a method to assess the extent of human or animal pollution in nonpoint impacted estuarine waters.

Two coliphage assay systems have been recently developed that are selective for F-specific phage. Debartolomeis (1988) developed and applied a F-specific sensitive host strain to estuarine and marine waters. The assay uses a male *E. coli* host mutant (called Famp), which is also apparently resistant to lysis by DNA somatic coliphages. RNAse has to be incorporated in a parallel assay to distinguish between FRNA and FDNA filamentous phages. A second F-specific assay was developed by Havelaar and Hogeboom (1984) through addition of an *E. coli* plasmid coding for sex pilus production to a F⁻ *Salmonella typhimurium* host strain (WG45). *S. typhimurium* was chosen as the host specifically to avoid interference by somatic coliphage which are abundant in sewage. Theoretically, plaques observed on this host would be attributed primarily to FRNA phage because somatic salmonellae and FDNA filamentous phages occur at much lower densities in sewage. This assay has been extensively applied to studies of feces and various wastewaters (Havelaar and Nieuwstad 1985; Havelaar et al. 1984, 1986) but not to marine or estuarine waters.

Recently we compared densities of fecal coliforms and phages lytic against F⁺ *S. typhimurium* WG49 from samples collected along a salinity gradient in a subestuary subject to nonpoint pollution. Phages were infrequently recovered from 100-ml samples of feeder stream or estuarine water with elevated fecal coliform densities. Higher levels of phage were detected in estuarine sediments compared with feeder stream sediments. Although fecal coliform densities in freshwater feeder stream sediments were one to four orders of magnitude larger than phage densities, densities of phage and fecal coliforms in estuarine sediments were similar. This pattern of phage distribution was inconsistent with the fecal coliform data which identified the feeder streams as unequivocal sources of fecal pollution. Plaques were randomly picked for confirmation as male-specific phage. Of the total number of plaques produced on *S. typhimurium* WG49, 99% (293 of 294) were produced by RNAse-resistant phage. Phages purified from these plaques were also lytic against the female parent *S. typhimurium* WG45, were not lytic against a male strain of *E. coli*, and were lytic against environmental *Salmonella* spp. isolates. The susceptibility of the *S. typhimurium* WG49 host strain to MS2 was retested and confirmed. As a result, not only is the sanitary significance of our data uncertain, but the application of this particular assay system to nonpoint source impacted growing areas may be inappropriate. The identity and origin of these phages is now under investigation.

Our experience confirms the critical importance of the bacterial host strain to the coliphage assay. Host strains vary as to the accessibility, specificity and location of phage receptor sites (Havelaar and Hogeboom 1983). For example, Vaughn and Metcalf (1975) found the abundance of coliphages in shellfish growing waters varied over a 3-year period as a function of the *E. coli* host employed. Evaluation of coliphage assay systems should be conducted and verified using field samples collected to obtain seasonal coverage.

If FRNA phages are to be used as fecal indicators more must be known about their ecology and fate. A possible limitation concerns their comparative occurrence in feces and sewage; FRNA phages are an infrequent component of human (and some animal) feces (Furuse 1987; Havelaar 1987; Havelaar et al. 1986) and occur at densities significantly larger in sewage. Their abundance in sewage treatment plants may be a consequence of phage multiplication at environmental temperatures on bacterial hosts that formed pili at temperatures above 30°C (86°F) (Havelaar 1987; Havelaar and Pot-Hogeboom, 1988; Havelaar et al. 1986). Therefore, the extent to which FRNA phages can infect and lyse enteric bacteria in wastewater facilities, septic systems and growing area environments is an issue of concern because such multiplication would alter the ratios of indicator to bacterial and viral pathogens. Thus, although FRNA phages may serve as an indicator of wastewater and sewage contamination, its use as an indicator of fecal contamination or a predictor of health risk in shellfish growing waters requires careful analysis. Unless it can be demonstrated that FRNA phages occur at reasonably high densities in septic leachate, FRNA phages may be a poor indicator of human fecal contamination in nonpoint impacted shellfish growing areas. On the other hand, it may have potential use as an indicator of animal contamination because it is significantly more abundant in the feces of certain domestic farm animals (Havelaar and Pot-Hogeboom 1988; Havelaar et al. 1986).

Quantitative assessment of FRNA phages in estuarine waters will probably necessitate concentration. Using a single-agar-layer method (Grabow and Coubrough 1986), 100 ml is the largest practical sample volume that can be directly processed. Methods for phage concentration have relied primarily on filtration techniques developed for enteric viruses. Positively charged filters have been successfully used to concentrate coliphage from potable and freshwater systems (Logan et al. 1980; Singh and Gerba 1983). Although positively charged filters were considered effective for concentrating enterovirus from estuarine waters (Kilgen and Cole 1983), Havelaar (1986) reported they were ineffective for recovery of coliphage from artificial seawater. Concentration methods for enteric viruses from sea and estuarine waters have relied primarily on the use of electronegative microporous filters (Sobsey 1987). Debartolomeis (1988) evaluated a method for recovery and concentration of FRNA bacteriophages modified after the procedure of Purdy et al. (1984, 1985) for recovery of *Bacillus* bacteriophage from water samples. The procedure involves adsorption of phages to host cells added to a water sample, concentration of infected cells by centrifugation or membrane filtration, and plaque assay of the concentrate by agar overlay. Debartolomeis (1988) recovered 104 and 88% of naturally occurring F-specific phages in seawater and river water, repectively, compared with direct sample plating using the soft agar overlay technique. Application of these methods to enumeration of FRNA phage requires further evaluation before being considered for routine use in shellfish growing waters. Similarly, methods for the recovery of coliphages from a variety of representative estuarine and marine sediments must be developed and verified. In a study of this type, Armon and Cabelli (1988) compared the effectiveness of various eluants to release f2 phage experimentally adsorbed to purified clay minerals and in natural sediment.

In contrast to coliphage, bacteriophages active against the anaerobe *Bacteroides fragilis* demonstrate a high degree of host strain specificity and appear to lack activity against other species of *Bacteroides* spp. (Booth et al. 1979; Cooper et al. 1984; Keller and Traub 1974; Kory and Booth 1986; Tartera and Jofre 1987; Tartera el al. 1989). As a potential indicator of fecal contamination, *B. fragilis* phages were detected exclusively in human feces and sewage and appear to reflect the dominance of the *Bacteroides fragilis* host in human feces (Booth et al. 1979; Cooper et al. 1984; Salyers 1984; Tartera and Jofre 1987; Tartera et al. 1989). Phages lytic to the most efficient host strain examined, *B. fragilis* HSP 40, were recovered only from environmental areas subjected to sewage and were never detected in nonpolluted areas or those occupied exclusively by wild animals (Jofre et al. 1986; Tartera and Jofre 1987). These observations, coupled with the apparent inability of *B. fragilis* phage to multiply in freshwater, seawater, or sediment habitats (Jofre et al. 1986; Tartera et al. 1989), suggest this phage is a promising subject for verification as an indicator of human fecal pollution. The low isolation frequency of HSP 40 phages in humans, i.e., 10%, will require a concentration method and may limit its effectiveness as an indicator in shellfish growing areas impacted by nonpoint pollution.

The occurrence of phages active against alternate bacterial indicators such as *B. adolescentis* or *B. breve, S. bovis,* and *R. coprophilus* appears to have received scant attention. Phage assay systems could provide an alternative detection method to viable enumeration procedures which are unsatisfactory for reasons previously discussed. Enumeration of phages active against human specific sorbitol-fermenting bifidobacteria could provide an assay system similar to that used for *B. fragilis* (Jofre et al. 1986). As noted, cross-reactivities between *Nocardia* spp. and *Rhodococcus* spp. phages may preclude development of a *Rhodococcus*-assay system.

Although validation of a viral indicator will depend on establishing a statistically significant relationship between a given phage and target enteric pathogen or risk of enteric disease, such information will at first be very difficult and expensive to obtain. Therefore, candidate viral phage indicators must be carefully evaluated before these studies are begun to understand their ecology and to confirm their target specificities with samples collected from regionally characteristic growing areas over a variety of seasonal conditions.

Chemical Indicators of Fecal Pollution

Coprostanol. Coprostanol (5β(H)-cholestan-3β-ol) is formed by bacterial reduction of cholesterol and is one of the principal sterols in the feces of man and other mammals (Murtaugh and Bunch 1967; Walker et al. 1982). It can be the dominant sterol in domestic wastes. Unlike microbial indicators subject to chemical, biological, and physical stressors that influence their numbers and detection, coprostanol offers advantages as a abiotic marker of sewage pollution. The areal distribution of coprostanol has been used to delineate impacts of point sources such as sewage outfalls or dumping areas (Boehm 1983; Brown and Wade 1984; Holm and Windsor 1986; Kanazawa and Teshima 1978; Pierce and Brown 1984) but has not been applied to the study of nonpoint pollution. Attempts to correlate densities of bacterial indicators to coprostanol concentrations in estuaries subject to known inputs have yielded conflicting results (Churchland and Kan 1982; Goodfellow et al. 1977; Yde et al. 1982).

Concentrations of coprostanol in estuarine (Holm and Windsor 1986) and marine (Kanazawa and Teshima 1978) waters decrease with increasing distance from sources. Brown and Wade (1984) demonstrated that coprostanol in sewage effluent is primarily associated with particulate matter, which may be deposited immediately proximate to an outfall or transported out of the area depending on the dynamics affecting the distribution of fine-grained sediments. Increases of corprostanol concentration with depth in estuarine waters is attributed to settling of sewage-associated particulates and resuspension (Wade et al. 1983). The distribution of coprostanol concentrations in waters adjacent to the entrances of major estuaries results from small- and large-scale physical processes and the buoyancy characteristics of particulate matter. The hydrophobic nature of coprostanol causes it to be associated with particles whose transport subsequently affects its spatial distribution.

Serious consideration of coprostanol as an indicator in nonpoint impacted shellfish growing areas is hampered by a lack of data describing background levels in such environments and possible origins from nonfecal sources. Laboratory studies using estuarine water and sediment amended with radiolabeled cholesterol demonstrate its conversion to coprostanol by microorganisms (Teshima and Kanazawa 1978). Nishimura and Koyama (1977) suggested that stanols in recent sediments are derived from phytoplankton and sterol conversion. Under anaerobic conditions 5 β-isomers of stanols are produced from autochthonous sedimentary organic matter (Nishimura 1982). Pocklington et al. (1987) concluded that corpostanol is an equivocal indicator of fecal pollution, identifying the marine biota as a probable source because patterns of coprostanol occurrence in particulate matter are conjunctive with those of various chemical indicators of primary production, including natural phytosterols.

Qualitative and quantitative aspects of corprostanol degradation in estuarine systems remain poorly understood. Coprostanol appears to persist, particularly under anaerobic conditions, in both lacustrine (Nishimura and Koyama 1977) and marine (Bartlett 1987) sediments. Coprostanol attributed to marine mammals (Venkatesan et al. 1986) dating to 3500 B.C. has been found in Antarctic sediments remote from anthropogenic inputs. Recovery of 5β-stanols from late Pleistocene sediments has also been reported (Nishimura 1982).

Observations that demonstrate the longevity of coprostanol as a geochemical marker raise questions concerning its applicability as a quantitiative indicator of health risk or as a marker for growing areas affected by highly variable pollution sources such as storm or agricultural runoff. Eganhouse et al. (1988) noted concentrations of coprostanol and linear alkyl benzenes (the anionic surfactants found in household laundry detergents) were rather variable in samples of sewage sludge. This was attributed to differences in concentrations of coprostanol in human feces and the differential effects of biosynthetic and degradative processes that can take place in feces and during waste processing. Another argument against the use of coprostanol is that although samples can be preserved to delay analysis, the analysis remains technically detailed, lengthy, and costly. Eganhouse et al. (1988) sounded a cautionary note concerning prior and future quantitative surveys of coprostanol; unless its coeluting epimer, epicoprostanol, is clearly resolved, reported concentrations have been and will be overestimated. Overall, the preponderance of information suggests use of coprostanol as an indicator of fecal pollution in shellfish harvesting waters should be viewed as low priority, especially because there are no data supporting its use in nonpoint source polluted areas.

Acanthamoeba.   Free-living protozoans belonging to *Acanthamoeba,* a genus ubiquitous in soils and freshwater environments, have been detected in sewage contaminated sediments from freshwater, estuarine and ocean dump sites and outfalls (Daggett et al. 1982; O'Malley et al. 1982; Sawyer et al. 1977; Sawyer 1980; Sawyer et al. 1987). Generally, the occurrence of *Acanthamoeba* spp. has been associated with elevated levels of fecal bacteria although sediments positive for *Acanthamoeba* spp., enteroviruses, coprostanol, polychlorinated biphenyls, and heavy metals were sometimes negative for bacterial indicators (Sawyer et al. 1987). Sawyer et al. (1987) proposed *Acanthamoeba* as a monitoring indicator based on persistance of its cysts (for periods up to 2.5 years), the pathogenicity of some strains, and its association with sewage. These amoebae are widely distributed in nature. The significance of their association with fecally polluted marine samples requires further study to determine if in situ *Acanthamoeba* densities reflect utilization of bacteria growing in dumpsite sewage sludge or the direct input of populations carried in the sludge (Daggett et al. 1982; Sawyer et al. 1982). Perhaps, the most likely answer will be a combination of both processes. Quantitative data regarding *Acanthamoeba* densities in human feces, sewage, agricultural runoff, and nonpoint impacted areas would be needed to assess its applicability as an indicator in shellfish waters. The complexity and cost of this undertaking, drawbacks associated with indicator organisms that form resistant cysts, the lengthy incubation periods, and degree of expertise necessary to distinguish *Acanthamoeba* spp. from other genera of free-living amoebae are concerns that do not support its use as a broadly based indicator of fecal pollution.

Indicators and New Methods: Concerns

A review of new methods is beyond the scope or intent of this chapter. New methods must stand the test of application to environmental samples from representative shellfish-producing regions over a range of seasonal conditions. However, the use of new methods raises certain issues whose consequences must be kept in perspective, remembering that the goal of growing area classification is to protect the health of the shellfish-consuming public.

Some investigators (Berg 1978; Richards 1985) have proposed direct detection of microbial pathogens as an alternative to indicators, citing documented examples of instances where current indicators failed to predict pathogen presence or pathogens were detected in the absence of indicators. In our view, arguments to retain surrogate microbiological indicators (Cabelli 1977b; James 1979) remain valid because of the unpredictable occurrence of pathogens, variations in pathogen virulence, morbidity rates, and minimum infective dosages within a target population. Arguments based on the lack of accepted methods for direct and accurate routine detection of pathogens are less persuasive. Recent advances in biotechnology suggest the feasibility of direct pathogen detection will soon be reality. Experimental gene probes for hepatitis A virus (HAV) have been used to detect HAV in polluted estuarine waters (Jiang et al. 1987; Metcalf and Jiang 1988) and a gene probe for *Salmonella* spp. (GENE-TRAK Systems, 31 New York Avenue, Framingham, MA 01701) is commercially available.

Emergent methods may revolutionize some aspects of microbiology that traditionally have been associated with frustrating drudgery and repetitive procedures, i.e., enumeration and phenotypic characterization of isolates. After more than half a century of using essentially unchanged cultivation methods for recovery of the coliform indicator group, new methods for rapid detection are becoming available. These may be based on traditional cultivation methods, direct enumeration procedures, target phenotypic characteristics such as constituitive enzymes or antigenic factors, or use recombinant DNA technology for high specificity. We welcome these new methods, anticipating their promise of ultimate specificity and rapidity, but feel compelled to offer the following cautionary remarks.

First, the best detection systems will not eliminate the need to understand the ecology of an indicator; its survival characteristics and the relationship between environmental exposure and the assay. For example, allochthonous bacteria exposed to marine and estuarine environments lose their ability to grow on selective and conventional bacteriological media (Rhodes et al. 1983; Roszak and Colwell 1987) and important diagnostic characteristics may be altered or lost. False-negatives or misidentification can be a consequence of enumerative or identification schemes based on expression of a sensitive phenotypic characteristic. *E. coli* loses its ability to ferment lactose (Kasweck and Fliermans 1978) and β-galactosidase activity is reduced in cells exposed to seawater (Anderson et al. 1979; Munro et al. 1987). Loss or alteration of plasmids encoding for antibiotic and heavy metal resistance occurs in various *Enterobacteriaceae* during long-term starvation (Caldwell et al. 1989; Chai 1983) and streptococcal (and enterococcal) isolates from animal and environmental sources do not consistently yield valid reactions in minaturized biochemical testing systems (Molitoris et al. 1985).

Finally, it is conceivable that advances in sensitivity and accuracy of analytical methods could lead to requests for growing area closures based on improved detection of pathogens. Such action could produce an untenable situation for the shellfish industry and regulatory bodies. Detection of pathogens in approved shellfish growing waters using new (or old) methods should not be construed as prima-facie evidence of health risk. Cabelli's (1978b, 1979) arguments that standards must be based on functional relationships, i.e., correlating the results of prospective epidemiologic investigations with indicator densities, must be heeded.

CONCLUDING REMARKS

The purpose of this chapter has been to focus on the important role of indicator ecology on the selection, verification, and use of indicators in shellfish growing waters. This encompasses the influence of geographical region, aspects of sample storage and transport, enumeration, and the interpretation of quantitative results. The dynamic and variable nature of estuarine environments suggest it is unlikely that a single universal indicator will be found that can be applied to assess pollutant impact or risk in all shellfish growing waters. It is likely that a variety of indicators will be required, each with a specific niche or purpose. Given sufficient resources and comprehensive regional coverage, validation of a number of alternate indicators presented in this article should be attainable.

In the context of nonpoint polluted growing areas, we stated there are potential uses for alternate indicators besides indexing or predicting potential health risk. Nonpoint areas require indicators in addition to fecal coliforms or *E. coli* because traditional indicators cannot be used to distinguish sources of fecal contamination, i.e., animal versus human, or reliably provide information about the relative proximity and "age" of fecal pollution under all conditions. Use of multiple indicators will provide more information about a system, allow for intercomparison of different indicator results, and can be used to evaluate the success of management strategies designed to reduce various types of fecal pollution. We note in closing that comprehensive studies to examine the effects of temperature, salinity, the microbiota, and other important environmental variables on the survival, recovery, and enumeration of promising candidate indicators is a requisite for their use in shellfish growing waters.

REFERENCES

Abshire, R., and Guthrie, R. K. 1971. The use of fluorescent antibody techniques for detection of *Streptococcus faecalis* as an indicator of fecal pollution of water. *Water Res.* **5**:1089–1097.

Adams, J. C. and Farrier, D. S. 1982. The effect of some oil shale process waters upon the viability of indicator bacteria. *J. Environ. Qual.* **11**:171–174.

Allsop, K., and Stickler, D. J. 1984. The enumeration of *Bacteroides fragilis* group organisms from sewage and natural waters. *J. Appl. Bacteriol.* **56**:1524.

Allsop, K., and Stickler, D. J. 1985. An assessment of *Bacteroides fragilis* group organisms as indicators of human faecal pollution. *J. Appl. Bacteriol.* **58**:95–99.

American Public Health Association. 1985. *Standard Methods for the Examination of Water and Wastewater,* 16th edition. American Public Health Association, Washington, D.C.

Amy, P. S. and Hiatt, H. D. 1989. Survival and detection of bacteria in an aquatic environment. *Appl. Environ. Microbiol.* **55**:788–793.

Andersen, P., and Fenchel, T. 1985. Bacterivory by microheterotrophic flagellates in seawater samples. *Limnol. Oceanogr.* **30**:198–202.

Anderson, I. C., Rhodes, M., and Kator, H. 1979. Sublethal stress in *Escherichia coli:* a function of salinity. *Appl. Environ. Microbiol.* **38**:1147–1152.

Anderson, I. C., Rhodes, M., and Kator, H. 1983. Seasonal variation in survival of *Escherichia coli* exposed in situ in membrane diffusion chambers containing filtered and nonfiltered estuarine water. *Appl. Environ. Microbiol.* **45**:1877–1883.

Armon, R., and Cabelli, V. J. 1988. Phage f2 desorption from clay in estuarine water using nonionic detergents, beef extract, and chaotropic agents. *Can. J. Microbiol.* **34:**1022–1024.

Armon, R., and Payment, P. 1988. A modified m-CP medium for enumerating *Clostridium perfringens* from water samples. *Can. J. Microbiol.* **34:**78–79.

Attwell, R. W., and Colwell, R. R. 1984. Thermoactinomycetes as terrestrial indicators for estuarine and marine waters. In: Ortiz-Ortiz, L., Bojalil, L. F., and Yakoleff, V. (eds.) *Biological, Biochemical and Biomedical Aspects of Actinomycetes.* Academic Press, New York, pp. 441–452.

Aubert, M., Pesando, D., and Gauthier, M. G. 1975. Effects of antibiosis in a marine environment. In: A. L. H. Gameson (ed.) *Discharge of Sewage from Sea Outfalls.* Pergamon Press, New York, pp. 191–197.

Barcina, I., Arana, I., Iriberri, J., and Egea, L. 1986. Influence of light and natural microbiota of the Butron river on *E. coli* survival. *Antonie van Leeuwenhoek J. Microbiol.* **52:**555–566.

Barnes, E. L. 1986. Anaerobic bacteria of the normal intestinal microflora of animals. In: Barnes, E. M., and Mead, G. C. (eds.) *Anaerobic Bacteria in Habitats Other than Man.* Blackwell Scientific Publications, Oxford, England, pp. 425–238.

Bartlett, P. 1987. Degradation of coprostanol in an experimental system. *Mar. Pollut. Bull.* **18:**27–29.

Beaudoin, E. C., and Litsky, W. 1981. Fecal streptococci. In: Dutka, B. J. (ed.) *Membrane Filtration: Applications, Techniques, and Problems.* Marcel Dekker, New York, pp. 77–118.

Bellair, J. T., Parr-Smith, G. A., and Wallis, I. G. 1977. Significance of diurnal variations in fecal coliform die-off rates in the design of ocean outfalls. *J. Water Pollut. Control Fed.* **49:**2022–2030.

Berg, G. 1978. Chapter 1. The indicator system. In: G. Berg (ed.) *Indicators of Viruses in Water and Food.* Ann Arbor Science Publishers, Ann Arbor, Michigan, pp. 1–13.

Bergey's Manual of Systematic Bacteriology, Vol. 2. 1986. Sneath, H. A., Mair, P. H. A., Sharpe, M. E., and Holt, J. G. (eds.) Williams & Wilkins, Baltimore.

Berk, S. G., Colwell, R. R., and Small, N.E.B. 1976. A study of feeding responses to bacterial prey by estuarine ciliates. *Trans. Am. Microscop. Soc.* **95:**514–520.

Bezkorovainy, A. 1989. Chapter 4. Nutrition and metabolism of bifidobacteria. In: Bezkorovainy, A. and Miller-Catchpole, R. (eds.) *Biochemistry and Physiology of Bifidobacteria.* CRC Press, Boca Raton, Florida, pp. 93–130.

Bisson, J. W., and Cabelli, V. J. 1980. *Clostridium perfringens* as a water pollution indicator. *J. Water Pollut. Control Fed.* **52:**241–248.

Boehm, P. D. 1983. Coupling of organic pollutants between the estuary and continental shelf and the sediments and water column in the New York Bight Region. *Can. J. Fish. Aquat. Sci.* **40:**262–276.

Bonde, G. J. 1977. Bacterial indication of water pollution. In: Droop, M. R., and Jannasch, H. W. (eds.) *Advances in Aquatic Microbiology, Vol. 1* Academic Press, London, pp. 273–364.

Booth, S. J., Van Tassell, R. L., Johnson, J. L., and Wilkins, T. D. 1979. *Bacteriophages of Bacteroides. Rev. Infect. Dis.* **1:**325–336.

Brezenski, F. T. 1973. Fecal streptococci. In: *Proceedings of the First Microbiology Seminar on Standardization of Methods.* San Francisco, California. U.S. Environmental Protection Agency, Washington, D.C., pp. 47–68.

Brown, R. C., and Wade, T. L. 1984. Sedimentary coprostanol and hydrocarbon distribution adjacent to a sewage outfall. *Water Res.* **18:**621–632.

Cabelli, V. J. 1977a. *Clostridium perfringens* as a water quality indicator. In: Hoadley, A. W. and Dutka, B. J. (eds.) *Bacterial Indicators/Health Hazards Associated with*

Water. Special Technical Publication 635, American Society for Testing and Materials, Philadelphia, pp. 65–79.

Cabelli, V. J. 1977b. Indicators of recreational water quality. In: Hoadley, A. W., and Dutka, B. J. (eds.) *Bacterial Indicators/Health Hazards Associated with Water*. Special Technical Publication 635, American Society for Testing and Materials, Philadelphia, pp. 222–238.

Cabelli, V. J. 1978a. Obligate anaerobic bacterial indicators. In: Berg, G. (ed.) *Indicators of Viruses in Water and Food*. Ann Arbor Science Publishers, Ann Arbor, Michigan, pp. 171–200.

Cabelli, V. J. 1978b. New standards for enteric bacteria. In: Mitchell, R. (ed.) *Water Pollution Microbiology, Vol. 2*. John Wiley & Sons, New York, pp. 233–271.

Cabelli, V. J. 1979. Evaluation of recreational water quality, the EPA approach. In: James, A., and Evison, L. (eds.) *Biological Indicators of Water Quality*. John Wiley & Sons, Chichester, pp. 14-1–14-23.

Cabelli, V. J., Dufour, A. P., McCabe, L. J., and Levin, M. A. 1983. A marine recreational water quality criterion consistent with indicator concepts and risk analysis. *J. Water Pollut. Control Fed.* **55**:1306–1314.

Calambokidis, J., McLaughlin, B. D., and Steiger, G. H. 1989. Bacterial contamination related to harbor seals in Puget Sound, Washington. A final report to Jefferson County and Washington Department of Ecology, Cascadia Research, Olympia, Washington.

Caldwell, B. A., Ye, C., Griffiths, R. P., Moyer, C. L., and Morita, R. Y. 1989. Plasmid expression and maintenance during long-term starvation-survival of bacteria in well water. *Appl. Environ. Microbiol.* **55**:1860–1864.

Carlucci, A. F., and Pramer, D. 1960. An evaluation of factors affecting the survival of *Escherichia coli* in seawater. II. Salinity, pH, and nutrients. *Appl. Microbiol.* **8**:247–250.

Carrillo, M., Estrada, E., and Hazen, T. C. 1985. Survival and enumeration of the fecal indicators *Bifidobacterium adolescentis* and *Escherichia coli* in a tropical rain forest watershed. *Appl. Environ. Microbiol.* **50**:468–476.

Chai, T.-J. 1983. Characteristics of *Escherichia coli* grown in bay water as compared with rich medium. *Appl. Environ. Microbiol.* **45**:1316–1323.

Churchland, L. M., and Kan. G. 1982. Variation in fecal pollution indicators through tidal cycles in the Fraser River estuary. *Can. J. Microbiol.* **28**:239–247.

Clausen, E. M., Green, B. L., and Litsky, W. 1977. Fecal streptococci: indicators of pollution. In: Hoadley, A. W., and Dutka: B. J. (eds.) *Bacterial Indicators/Health Hazards Associated with Water*. Special Technical Publication 635, American Society for Testing and Materials, Philadelphia, pp. 247–264.

Collins, C. H., and Lyne, P. M. 1984. *Microbiological methods*. Fifth edition. Butterworths. London.

Cooper, K. E., and Ramadan, F. M. 1955. Studies in the differentiation between human and animal pollution by means of faecal streptococci. *J. Gen. Microbiol.* **12**:180–190.

Cooper, S. W., Szymczak, E. G., Jacobus, N. V., and F. P. Tally. 1984. Differentiation of *Bacteroides ovatus* and *Bacteroides thetaiotaomicron* by means of bacteriophage. *J. Clin. Microbiol.* **20**:1122–1125.

Cynar, F. J., Estep, K. W., and Sieburth, J. McN. 1985. The detection and characterization of bacteria-sized protists in "protist-free" filtrates and their potential impact on experimental marine ecology. *Microb. Ecol.* **11**:281–288.

Daggett, P.-M., Sawyer, T. K., and Nerad, T. A. 1982. Distribution and possible interrelationships of pathogenic and nonpathogenic *Acanthamoeba* from aquatic environments. *Microb. Ecol.* **8**:371–386.

Dalton, H. P., Archer, G. L., Slifkin, M., Harris, R. C., Welshimer, H. J., Sosnowski,

K. M., Nottebart, H. C., Jr., Duma, R. J., Clark, R. B., Warren, N. G., Kerkering, T. M., Swenson, P. D., Ager, A. L., Jr., and May, R. G., Jr. 1986. Blood specimens. In: Dalton, H. P., and Nottebart, H. C., Jr. (eds.) *Interpretive Medical Microbiology*. Churchill Livingstone, New York, pp. 28–29.

D'Aoust, R. A., and Litsky, W. 1975. Pfizer selective enterococcus agar overlay method for the enumeration of fecal streptococci by membrane filtration. *Appl. Environ. Microbiol.* **29**:584–589.

Dawe, L. L., and Penrose, W. R. 1978. "Bactericidal" property of seawater: death or debilitation? *Appl. Environ. Microbiol.* **35**:829–833.

Debartolomeis, J. 1988. Enumeration of F male-specific bacteriophages from sewage and fecally polluted waters. Ph.D. Dissertation, University of Rhode Island, Kingston, Rhode Island.

Deibel, R. H., and Hartman, P. A. 1984. The Enterococci. In: Speck, M. L. (ed.) *Compendium of Methods for the Microbiological Examination of Foods*. American Public Health Association, Washington, D.C., pp. 405–410.

Dowell, V. R., Jr., and Hawkins, T. M. 1968. *Laboratory Methods in Anaerobic Bacteriology*. U.S. Department of Health, Education, and Welfare, National Communicable Disease Center, Atlanta, Georgia.

Dufour, A. P. 1977. *Escherichia coli:* the fecal coliform. In: Hoadley, A. W., and Dutka, B. J. (eds.) *Bacterial Indicators/Health Hazards Associated with Water*. Special Technical Publication 635, American Society for Testing and Materials, Philadelphia, pp. 48–58.

Dutka, B. J. 1973. Coliforms are an inadequate index of water quality. *J. Environ. Hlth.* **36**:39–46.

Dutka, B. J. 1979. Microbiological indicators, problems and potential of new microbial indicators of water quality. In: James, A., and Evison, L. (eds.) *Biological indicators of water quality*. John Wiley & Sons, Chichester, pp. 18-1–18-24.

Eganhouse, R. P., Olaguer, D. P., Gould, B. R., and Phinney, C. S. 1988. Use of molecular markers for the detection of municipal sewage sludge at sea. *Marine Environ. Res.* **25**:1–22.

Ellender, R. D., Mapp, J. B., Middlebrooks, B. L., Cook D. W., and Cake. E. W. 1980. Natural enterovirus and fecal coliform contamination of Gulf Coast oysters. *J. Food Protect.* **43**:105–110.

Enzinger, R. M., and Cooper, R. C. 1976. Role of bacteria and protozoa in the removal of *Escherichia coli* from estuarine waters. *Appl. Environ. Microbiol.* **31**:758–763.

Ericksen, T. H., and Dufour, A. P. 1986. Methods to identify waterborne pathogens and indicator organisms. In: Craun, G. F. (ed.) *Waterborne Diseases in the United States*. CRC Press, Boca Raton, Florida, pp. 195–214.

Erkenbrecher, C. W., Jr. 1981. Sediment bacterial indicators in an urban shellfishing subestuary of the lower Chesapeake Bay. *Appl. Environ. Microbiol.* **42**:484–492.

Facklam, R. R., and Collins, M. D. 1989. Identification of *Enterococcus* species isolated from human infections by a conventional test scheme. *J. Clin. Microbiol.* **27**:731–734.

Faust, M. A. 1976. Coliform bacteria from diffuse sources as a factor in estuarine pollution. *Water Res.* **10**:619–62.

Faust, M. A., Aotaky, A. E., and Hargadon, M. L. 1975. Effect of physical parameters on the in situ survival of *Escherichia coli* MC-6 in an estuarine environment. *Appl. Microbiol.* **30**:800–806.

Feachem, R. 1975. An improved role for faecal coliform to faecal streptococci ratios in the differentiation between human and non-human pollution sources. *Water Res.* **9**:689–690.

Feachem, R. G., Bradley, D. J., Garelick, H., and Mara, D. D. 1983. Sanitation and disease. Health aspects of excreta and wastewater management. *World Bank Studies in Water Supply and Sanitation.* John Wiley & Sons, New York.

Fenchel, T. 1982. Ecology of heterotrophic microflagellates. II. Bioenergetics and growth. *Marine Ecol. Prog. Ser.* **8:**225–231.

Fiksdal, L., Pommepuy, M., Derrien, A., and Cormier, M. 1989. Production of 4-methylumbelliferyl heptanoate hydrolase by *Escherichia coli* exposed to seawater. *Appl. Environ. Microbiol.* **55:**2424–2427.

Flint, K. P. 1987. The long term survival of *Escherichia coli* in river water. *J. Appl. Bacteriol.* **63:**261–270.

Food and Drug Administration. 1984. Shellfish sanitation interpretation 34: Interpretation of bacteriological market standards for shellfish. Shellfish Sanitation Branch, Washington D.C., 12 pp.

Fujioka, R. S. and Shizumura, L. K. 1985. *Clostridium perfringens:* a reliable indicator of stream water quality. *J. Water Pollut. Control Fed.* **57:**986–992.

Fujioka, R. S., and Siwak, E. B. 1987. Bactericidal properties of long ultraviolet and visible wavelengths of sunlight. *J. Am. Water Works Assoc.* **79:**56.

Fujioka, R. S., Hashimoto, H. H., Siwak, E. B., and Young, R. H. F. 1981. Effect of sunlight on survival of indicator bacteria in seawater. *Appl. Environ. Microbiol.* **41:**690–696.

Furuse, K. 1987. Distribution of coliphages in the environment: general considerations. In: Goyal, S. M., Gerba, C. P., and Bitton, G. (eds.) *Phage Ecology.* John Wiley & Sons, New York, pp. 87–124.

Gameson, A. L. H., and Gould, D. J. 1975. Effects of solar radiation on the mortality of some terrestrial bacteria in sea water. In: Gameson, A. L. H. (ed.) *Discharge of Sewage from Sea Outfalls.* Pergamon Press, Oxford, pp. 209–219.

Gameson, A. L. H., and Saxon, J. R. 1967. Field studies on effect of daylight on mortality of coliform bacteria. *Water Res.* **1:**279–295.

Gauthier, M. J., Munro, P. M., and Mohajer, S. 1987. Influence of salts and sodium chloride on the recovery of *Escherichia coli* from seawater. *Curr. Microbiol.* **15:**5–10.

Geldreich, E. E. 1972. Water-borne pathogens. In: Mitchell, R. (ed.) *Water Pollution Microbiology.* Wiley-Interscience, New York, pp. 207–241.

Geldreich, E. E., and B. A. Kenner. 1969. Concepts of faecal streptococci in stream pollution. *J. Water Pollut. Control Fed.* **41:**R336–R352.

Geldreich, E. E., Kenner, B. A. and Kabler, P. W. 1964. Occurrence of coliforms, fecal coliforms, and streptococci on vegetation and insects. *Appl. Microbiol.* **12:**63–69.

Geldreich, E. E., Best, L. C., Kenner, B. A., and van Donsel, D. J. 1968. The bacteriological aspects of stormwater pollution. *J. Water Pollut. Control Fed.* **40:**1861–1872.

Gerba, C. P. 1987. Phage as indicators of fecal pollution. In: Goyal, S. M., Gerba, C. P., and Bitton, G. (eds.) *Phage Ecology.* John Wiley & Sons, New York, pp. 197–209.

Gerba, C. P., and McLeod, J. S. 1976. Effect of sediments on the survival of *Escherichia coli* in marine waters. *Appl. Environ. Microbiol.* **32:**114–120.

Ghoul, M., Bernard, T., and Cormier, M. 1990. Evidence that *Escherichia coli* accumulates glycine betaine from marine sediments. *Appl. Environ. Microbiol.* **56:** 551–554.

Gilliland, M. W., and Baxter-Potter, W. 1987. A geographic information system to predict non-point source pollution potential. *Water Resources Bull.* **23:**281–291.

Goodfellow, M., and Haynes, J. A. 1984. Actinomycetes in marine sediments In: Ortiz-Ortiz, L., Bojalil, L. F., and Yakoleff, V. (eds.) *Biological, Biochemical and Biomedical Aspects of Actinomycetes.* Academic Press, New York, pp. 453–472.

Goodfellow, M., and Williams, S. T. 1983. Ecology of actinomycetes. *Ann. Rev. Microbiol.* **37:**189–216.

Goodfellow, R. M., Cardoso, J., Eglinton, G., Dawson, J. P., and Best, G. A. 1977. A faecal sterol survey in the Clyde Estuary. *Marine Pollut. Bull.* **8:**272–275.

Goyal, S. M., Gerba, C. P., and Melnick, J. L. 1979. Human enteroviruses in oysters and their overlying waters. *Appl. Environ. Microbiol.* **37:**572–581.

Grabow, W. O. K., and Coubrough, P. 1986. Practical direct plaque assay for coliphages in 100-ml samples of drinking water. *Appl. Environ. Microbiol.* **52:**430–433.

Granai, C., III, and Sjogren, R. E. 1981. In situ and laboratory studies of bacterial survival using a microporous membrane sandwich. *Appl. Environ. Microbiol.* **41:**190–195.

Grimes, D. J., Atwell, R. W., Brayton, P. R., Palmer, L. M., Rollins, D. M., Roszak, D. B., Singleton, F. L., Tamplin, M. L., and Colwell, R. R. 1986. The fate of enteric pathogenic bacteria in estuarine and marine environments. *Microbiol. Sci.* **3:**324–329.

Gyllenberg, H., Niemela, S., and Sormunen, T. 1960. Survival of bifid bacteria in water as compared with that of coliform bacteria and enterococci. *Appl. Microbiol.* **8:**20.

Hackney, C. R., Ray, B., and Speck, M. L. 1979. Repair detection procedure for enumeration of fecal coliforms and enterococci from seafoods and marine environments. *Appl. Environ. Microbiol.* **37:**947–953.

Hartman, P. A., Reinbold, G. W., and Saraswat, D. S. 1966. Media and methods for isolation and enumeration of the enterococci. In: Umbreit, W. W. (ed.) *Advances in Applied Microbiology, Vol. 8.* Academic Press, New York, pp. 253–289.

Havelaar, A. H. 1986. Concentration of bacteriophages from large volumes of water. Ph.D Dissertation, University of Utrecht, The Netherlands.

Havelaar, A. H. 1987. Bacteriophages as model organisms in water treatment. *Microbiol. Sci.* **4:**362–364.

Havelaar, A. H., and Hogeboom, W. M. 1983. Factors affecting the enumeration of coliphages in sewage and sewage-polluted waters. *Antonie van Leeuwenhoek J. Microbiol.* **49:**387–397.

Havelaar, A. H., and Hogeboom, W. M. 1984. A method for the enumeration of male-specific bacteriophages in sewage. *J. Appl. Bacteriol.* **56:**439–447.

Havelaar, A. H., and Nieuwstad, Th. J. 1985. Bacteriophages and faecal bacteria as indicators of chlorination efficiency of biologically treated wastewater. *J. Water Pollut. Control Fed.* **57:**1084–1088.

Havelaar, A. H., and Pot-Hogeboom, W. M. 1988. F-specific RNA-bacteriophages as model viruses in water hygiene: ecological aspects. *Water Sci. Tech.* **20:**399–407.

Havelaar, A. H., Hogeboom, W. M., and Pot, R. 1984. F-specific RNA bacteriophages in sewage: methodology and occurrence. *Water Sci. Tech.* **17:**645–655.

Havelaar, A. H., Furuse, K., and Hogeboom, W. M. 1986. Bacteriophages and indicator bacteria in human and animal faeces. *J. Appl. Bacteriol.* **60:**255–262.

Hazen, T. C. 1988. Fecal coliforms as indicators in tropical waters: a review. *Toxic. Assess.* **3:**461–477.

Heijthuijsen, J. H. F. G., and Hansen, T. A. 1989. Betaine fermentation and oxidation by marine *Desulfuromonas* strains. *Appl. Environ. Microbiol.* **55:**965–969.

Hirsch, C. F., and Christensen, D. L. 1983. Novel method for selective isolation of actinomycetes. *Appl. Environ. Microbiol.* **46:**925–929.

Hobbie, J. E., Daley, R. J., and Jasper, S. 1977. Use of nuclepore filters for counting bacteria by fluorescence microscopy. *Appl. Environ. Microbiol.* **33:**1225–1228.

Holdeman, L. V., Good, I. J., and Moore, W. E. C. 1976. Human fecal flora: variation in bacterial composition within individuals and a possible effect of emotional stress. *Appl. Environ. Microbiol.* **31:**359–375.

Holm, S. E., and Windsor, J. G., Jr. 1986. Chemical monitoring of sewage effluents using saturated hydrocarbons and coprostanol in estuarine waters. Oceans '86 Conference Record: Science-Engineering-Adventure. *Monitor Strateg. Symp.* **3:** 839–844.

Hunt, D. A., and Springer, J. 1974. Preliminary report on a comparison of total coliform and fecal coliform values in shellfish growing area waters and a proposal for a fecal coliform area standard. In: Wilt, D. S. (ed.) *Proceedings 8th National Shellfish Sanitation Workshop.* Food and Drug Administration, Shellfish Sanitation Branch, Washington, D.C., pp. 97–104.

Isenberg, H. D., Goldberg, D., and Sampson, J. 1970. Laboratory studies with a selective enterococcus medium. *Appl. Microbiol.* **20:**433–436.

James, A. 1979. Chapter 1. The value of biological indicators in relation to other parameters of water quality. In: James, A. and Evison, L. (ed.) *Biological Indicators of Water Quality.* John Wiley & Sons, Chichester, pp. 1-1–1-16.

Jannasch, H. W. 1968. Competitive elimination of *Enterobacteriaceae* from seawater. *Appl. Microbiol.* **16:**1616–1618.

Jiang, X., Estes, M. K., and Metcalf, T. G. 1987. Detection of hepatitis A virus by hybridization with single-stranded RNA probes. *Appl. Environ. Microbiol.* **53:**2487–2495.

Jofre, J., Bosch, A., Lucena, F., Girones, R., and Tartera, C. 1986. Evaluation of *Bacteroides fragilis* bacteriophages as indicators of the virological quality of water. *Water Sci. Technol.* **18:**167–177.

Jones, G. E., and Cobet, A. B. 1975. Heavy metal ions as the principal bactericidal agent in Caribbean sea water. In: Gameson, A. L. H. (ed.) *Discharge of Sewage from Sea Outfalls.* Pergamon Press, Oxford, England, pp. 199–207.

Kanazawa, A., and Teshima, S. 1978. The occurrence of coprostanol, an indicator of faecal pollution, in sea water and sediments. *Oceanol. Acta* **1:**39–44.

Kapuscinski, R. B., and Mitchell, R. 1981. Solar radiation induces sublethal injury in *Escherichia coli* in seawater. *Appl. Environ. Microbiol.* **41:**670–674.

Kapuscinski, R. B., and Mitchell, R. 1983. Sunlight-induced mortality of viruses and *Escherichia coli* in coastal seawater. *Environ. Sci. Technol.* **17:**1–6.

Kasweck, K. L., and Fliermans, C. B. 1978. Lactose variability of *Escherichia coli* in thermally stressed reactor effluent waters. *Appl. Environ. Microbiol.* **36:**739–746.

Kehr, R. W., Levine, B. S., Butterfield, C. T., and Miller, A. P. 1941. A report on the public health aspects of clamming in Raritan Bay. Public Health Service Report. Reissued in June 1954 by Division of Sanitary Engineering Services, Public Health Services, Department of Health, Education, and Welfare.

Keller, R., and Traub, N. 1974. The characterization of *Bacteroides fragilis* bacteriophage recovered from animal sera: observations on the nature of *Bacteroides* phage carrier cultures. *J. Gen. Virol.* **24:**179–189.

Kenner, B. A., Clark, H. F., and Kabler, P. W. 1960. Fecal streptococci. II. Quantification of streptococci in feces. *Am. J. Publ. Hlth.* **50:**1553–1559.

Keswick, B. H., Satterwhite, T. K., Johnson, P. C., DuPont, H. L., Secor, S. L., Bitsura, J. A., Gary, G. W., and Hoff, J. C. 1985. Inactivation of Norwalk virus in drinking water by chlorine. *Apple. Environ. Microbiol.* **50:**261–264.

Ketchum, B. H., Ayers, J. C., and Vaccaro, R. F. 1952. Processes contributing to the decrease of coliform bacteria in a tidal estuary. *Ecology* **33:**247–258.

Kilgen, M. B., and Cole, M. T. 1983. Recovery of poliovirus type I from artificial seawater and natural estuarine water at varying salinities using electropositive filters. *Abstr. Ann. Meet. Am. Soc. Microbiol.* 283.

King, G. M. 1984. Metabolism of trimethylamine, choline, and glycine betaine by sulfate-reducing and methanogenic bacteria in marine sediments. *Appl. Environ. Microbiol.* **48:**719–725.

Kjellander, J. 1960. Enteric streptococci as indicators of fecal contamination of water. *Acta Pathol. Microbiol. Scand. Suppl. 136.* **48:**1–124.

Knittel, M. D., Seidler, R. J., Eby, C., and Cabe, L. M. 1977. Colonization of the botanical environment by *Klebsiella* isolates of pathogenic origin. *Appl. Environ. Microbiol.* **34:**557–563.

Koepfler, E. T., and Kator, H. I. 1986. Ecotoxicological effects of creosote contamination on benthic microbial populations in an estuarine environment. *Toxic. Assess.* **1:**465–485.

Kogure, K., Simidu, U., and Taga, N. 1979. A tentative direct microscopic method for counting living bacteria. *Can. J. Microbiol.* **25:**415–420.

Kory, M. M., and Booth, S. J. 1986. Characteristics of *Bacteroides fragilis* bacteriophages and comparison of their DNAs. *Curr. Microbiol.* **14:**199–203.

Kott, Y. 1981. Viruses and bacteriophages. *Sci. Tot. Environ.* **18:**13–23.

Kuritza, A. P., and Salyers, A. A. 1985. Use of a species-specific DNA hybridization probe for enumerating *Bacteroides vulgatus* in human feces. *Appl. Environ. Microbiol.* **50:**958–964.

LaBelle, R. L., and Gerba, C. P. 1979. Influence of pH, salinity, and organic matter on the adsorption of enteric viruses to estuarine sediment. *Appl. Environ. Microbiol.* **39:**588–596.

LaBelle, R. Y., Gerba, C. P., Goyal, S. M., Melnick, J. L., Cech, I., and Bogdan, G. F. 1980. Relationships between environmental factors, bacterial indicators, and the occurrence of enteric viruses in estuarine sediments. *Appl. Environ. Microbiol.* **38:**93–101.

Landry, E. F., Vaughn, J. M., Vicale, T. J., and Mann, R. 1983. Accumulation of sediment-associated viruses in shellfish. *Appl. Environ. Microbiol.* **45:**238–247.

Leonard, D. L., Broutman, M. A., and Harkness, K. E. 1989. The quality of shellfish growing waters on the east coast of he United States. National Oceanic and Atmospheric Administration. Ocean Assessments Division. Rockville, Maryland, 54 pp.

Le Rudulier, D., and Bouillard, L. 1983. Glycine betaine, an osmotic effector in *Klebsiella pneumoniae* and other members of the *Enterobacteriaceae*. *Appl. Environ. Microbiol.* **46:**152–159.

Lessard, E. J., and Sieburth, J. McN. 1983. Survival of natural sewage populations of enteric bacteria in diffusion and batch chambers in the marine environment. *Appl. Environ. Microbiol.* **45:**950–959.

Levin, M. A. 1977. Bifidobacteria as water quality indicators. In: Hoadley, A. W., and Dutka, B. J. (eds.) *Bacterial Indicators/Health Hazards Associated with Water.* Special Technical Publication 635, American Society for Testing and Materials, Philadelphia, pp. 131–138.

Levin, M. A., and Resnick, I. G. 1981. Bifidobacterium. In: Dutka, B. J. (ed.) *Membrane Filtration: Applications, Techniques, and Problems.* Marcel Dekker, New York, pp. 129–159.

Levin, M. A., Fischer, J. R., and Cabelli, V. J. 1975. Membrane filter technique for enumeration of enterococci in marine waters. *Appl. Microbiol.* **30:**66–71.

Li, W. K. W., and Dickie, P. M. 1985. Growth of bacteria in seawater filtered through 0.2 um Nuclepore membranes: implications for dilution experiments. *Mar. Ecol. Prog. Ser.* **26:**245–252.

Littel, K. J., and Hartman, P. A. 1983. Fluorogenic selective and differential medium for isolation of fecal streptococci. *Appl. Environ. Microbiol.* **45:**622–627.

Logan, K. B., Rees, G. E., Seeley, N. D., and Primrose, S. B. 1980. Rapid concentration of bacteriophages from large volumes freshwater: evaluation of positively charged, microporous filters. *J. Virol. Methods* **1:**87–97.

Lopez-Torres, A. J., Hazen, T. C., and Toranzos, G. A. 1987. Distribution and in situ survival and activity of *Klebsiella pneumoniae* and *Escherichia coli* in a tropical rain forest watershed. *Curr. Microbiol.* **15**:213–218.

Ludwig, W., Seewaldt, E., Kilpper-Balz, A., Schleifer, K. H., Magrum, L., Woese, C. R., Fox, G. E., and Stackebrandt, E. 1985. The phylogenetic position of *Streptococcus* and *Enterococcus*. *J. Gen. Microbiol.* **131**:543–551.

Mackey, B. M., and Derrick, C. M. 1986. Peroxide sensitivity of cold-shocked *Salmonella typhimurium* and *Escherichia coli* and its relationship to minimal medium recovery. *J. Appl. Bacteriol.* **60**:501–511.

Macy, J. M. 1981. Chapter 117. Nonpathogenic members of the genus *Bacteroides*. In: Starr, M. P., Stolp, H., Truper, H. G., Balows, A., and Schlegel, H. G. (eds.) *The Prokaryotes, Vol. II*. Springer-Verlag, New York, pp. 1450–1463.

Mara, D. D. and Oragui, J. I. 1981. Occurrence of *Rhodococcus coprophilus* and associated actinomycetes in feces, sewage and freshwater. *Appl. Environ. Microbiol.* **42**:1037–1042.

Mara, D. D., and Oragui, J. I. 1983. Sorbitol-fermenting bifidobacteria as specific indicators of human faecal pollution. *J. Appl. Bacteriol.* **55**:349–357.

Marshall, K. C. 1985. Mechanisms of bacterial adhesion at solid-water interfaces. In: Savage, D. C. and Fletcher, M. (eds.) *Bacterial Adhesion*. Plenum Press, New York, pp. 133–161.

Martinez, J., Garcia-Lara, J., and Vives-Rego, J. 1989. Estimation of *Escherichia coli* mortality in seawater by the decrease in ^{3}H-label and electron transport system activity. *Microb. Ecol.* **17**:219–225.

Matches, J. R., and Liston, J. 1974. Mesophilic clostridia in Puget Sound. *Can. J. Microbiol.* **20**:1–7.

McCambridge, J., and McMeekin, T. A. 1979. Protozoan predation of *Escherichia coli* in estuarine waters. *Water Res.* **13**:659–663.

McCambridge, J., and McMeekin, T. A. 1980. Relative effects of bacterial and protozoan predators on survival of *Escherichia coli* in estuarine water samples. *Appl. Environ. Microbiol.* **40**:907–911.

McCambridge, J., and McKeekin, T. A. 1981. Effect of solar radiation and predacious microorganisms on survival of fecal coliforms and other bacteria. *Appl. Environ. Microbiol.* **41**:1083–1087.

McFeters, G. A., Bissonnette, G. K., Jezeski, J. J., Thomson, C. A., and Stuart, D. G. 1974. Comparative survival of indicator bacteria and enteric pathogens in well water. *Appl. Microbiol.* **27**:823–829.

Metcalf, T. G., and Jiang, X. 1988. Detection of hepatitis A virus in estuarine samples by gene probe assay. *Microbiol. Sci.* **5**:296–300.

Metcalf, T. G., Slanetz, L. W., and Bartley, C. H. 1973. Enteric pathogens in estuary waters and shellfish. In: Chichester, C. O., and Graham, H. D. (eds.) *Microbial Safety of Fishery Products*. Academic Press, New York, pp. 215–234.

Miescier, J. J., and Cabelli, V. J. 1982. Enterococci and other microbial indicators in municipal wastewater effluents. *J. Water Pollut. Control Fed.* **54**:1599–1606.

Miescier, J. J., Peeler, J. T., Clem, J. D., Read, R. B., Jr., and Furfari, S. A. 1985. Final report on a comparison of two methods for recovery of *Escherichia coli* Type 1 and fecal coliforms from oysters. Food and Drug Administration, Divison of Cooperative Programs, Washington, D.C., 18 pp.

Milne, D. P., Curran, J. C., and Wilson, L. 1986. Effects of sedimentation on removal of faecal coliform bacteria from effluents in estuarine water. *Water Res.* **20**:1493–1496.

Mitchell, R., and Chamberlin, C. 1975. Factors influencing the survival of enteric microorganisms in the sea: an overview. In: Gameson, A. L. H. (ed.) *Discharge of Sewage from Sea Outfalls*. Pergamon Press, Oxford, pp. 237–251.

Mitchell, R., Yankafsky, S., and Jannasch, H. W. 1967. Lysis of *Escherichia coli* by marine micro-organisms. *Nature* **215**:891–892.

Molitoris, E., McKinley, G., Krichevsky, M. I., and Fagerberg, D. J. 1985. Comparison of conventional and miniaturized biochemical techniques for identification of animal streptococcal isolates. *Microb. Ecol.* **11**:81–90.

Moore, W. E. C., and Holdeman, L. V. 1974. Human fecal flora: the normal flora of 20 Japanese-Hawaiians. *Appl. Microbiol.* **27**:961–979.

Morotomi, M., Ohno, T., and Mutai, M. 1988. Rapid and correct identification of intestinal *Bacteroides* spp. with chromosomal DNA probes by whole-cell dot blot hybridization. *Appl. Environ. Microbiol.* **54**:1158–1162.

Mundt, J. O. 1963. Occurrence of enterococci on plants in a wild environment. *Appl. Microbiol.* **11**:141–144.

Mundt, J. O., 1982. Occurrence of enterococci on plants in a wild environment. *Appl. Microbiol.* **11**:141–144.

Mundt, J. O., Johnson, A. H., and Khatchikian, R. 1958. Incidence and nature of enterococci on plant materials. *Food Res.* **23**:186–193.

Mundt, J. O., Coogin, J. H., Jr., and Johnson, L. F. 1962. Growth of *Streptococcus faecalis* var. *liquefaciens* on plants. *Appl. Microbiol.* **10**:552–555.

Munoa, F. J., and Pares, R. 1988. Selective medium for isolation and enumeration of *Bifidobacterium* spp. *Appl. Environ. Microbiol.* **54**:1715–1718.

Munro, P. M., Gauthier, M. J., and Laumond, F. M. 1987. Changes in *Escherichia coli* cells starved in seawater or grown in seawater-wastewater mixtures. *Appl. Environ. Microbiol.* **53**:1476–1481.

Munro, P. M., Gauthier, M. J., Breittmayer, V. A., and Bongiovanni, J. 1989. Influence of osmoregulation processes on starvation survival of *Escherichia coli* in seawater. *Appl. Environ. Microbiol.* **55**:2017–2024.

Murtaugh, J. J. and Bunch, R. L. 1967. Sterols as a measure of fecal pollution. *J. Water Pollut. Control Fed.* **39**:404–409.

Nikaido, H., and Vaara, M. 1987. Outer membrane. In: Neidhardt, F. C. (ed.) *Escherichia coli and Salmonella typhimurium Cellular and Molecular Biology. Vol. 1.* American Society for Microbiology, Washington, D.C., pp. 7–22.

Nishimura, M. 1982. 5Beta-isomers of stanols and stanones as potential markers of sedimentary organic quality and depositional paleoenvironments. *Geochim. Cosmochim. Acta* **46**:423–432.

Nishimura, M., and Koyama, T. 1977. The occurrence of stanols in various living organisms and the behavior of sterols in contemporary sediments. *Geochim. Cosmochim. Acta* **41**:379–385.

O'Malley, M. L., Lear, D. W., Adams, W. N., Gaines, J., Sawyer, T. K., and Lewis, E. J. 1982. Microbial contamination of continental shelf sediments by wastewater. *J. Water Pollut. Control Fed.* **54**:1311–1317.

Oragui, J. I., and Mara, D. D. 1981. A selective medium for the enumeration of *Streptococcus bovis* by membrane filtration. *J. Appl. Bacteriol.* **51**:85–93.

Oragui, J. I., and Mara, D. D. 1983. Investigation of the survival characteristics of *Rhodococcus coprophilus* and certain fecal indicator bacteria. *Appl. Environ. Microbiol.* **46**:356–360.

Oragui, J. I., and Mara, D. D. 1984. A note on a modified membrane-Bovis agar for the enumeration of *Streptococcus bovis* by membrane filtration. *J. Appl. Bacteriol.* **56**:179–181.

Orlob, G. T. 1956. Viability of sewage bacteria in seawater. *Sewage Ind. Wastes.* **28**:1147–1167.

Pagel, J. E., and Hardy, G. M. 1980. Comparison of selective media for the enumeration and identification of fecal streptococci from natural sources. *Can. J. Microbiol.* **26**:1320–1327.

Palmer, L. M., Baya, A. M., Grimes, D. J., and Colwell, R. R. 1984. Molecular genetic and phenotypic alteration of *Escherichia coli* in natural water microcosms containing toxic chemicals. *FEMS Microbiol. Lett.* **21**:169–173.

Perez-Rosas, N. and Hazen, T. C. 1988. In situ survival of *Vibrio cholerae* and *Escherichia coli* in tropical coral reefs. *Appl. Environ. Microbiol.* **54**:1–9.

Pettibone, G. W., and Cooney, J. J. 1986. Effect of organotins on fecal pollution indicator organisms. *Appl. Environ. Microbiol.* **52**:562–566.

Pierce, R. H. and Brown, R. C. 1984. Coprostanol distribution from sewage discharge into Sarasota Bay, Florida. *Bull. Environ. Contam. Toxicol.* **32**:75–79.

Pisano, M. A., Sommer, M. J., and Lopez, M. M. 1986. Application of pretreatments for the isolation of bioactive actinomycetes from marine sediments. *Appl. Microbiol. Biotechnol.* **25**:285–288.

Pocklington, R., Leonard, J. D., and Crewe, N. F. 1987. Le coprostanol comme indicateur de la contamination fecale dans l'eau de mer et les sediments marins. *Oceanol. Acta* **10**:83–89.

Post, F. J., Allen, A. D., and Reid, T. C. 1967. Simple medium for the selective isolation of *Bacteroides* and related organisms, and their occurrence in sewage. *Appl. Microbiol.* **15**:213–218.

Postgate, J. R. 1967. Viability measurements and the survival of microbes under minimum stress. In: Rose, A. H., and Wilkinson, J. R. (ed.) *Advances in Microbial Physiology*. Academic Press, London, pp. 1–23.

Prauser, H. 1984. Phage host ranges in the classification and identification of gram-positive branched and related bacteria. In: Ortiz-Ortiz, L., Bojalil, L. F., and Yakoleff, V. (eds.) *Biological, Biochemical and Biomedical Aspects of Actinomycetes*. Academic Press, New York, pp. 617–633.

Puglsey, A. P. and Evison, L. M. 1975. A fluorescent antibody technique for the enumeration of faecal streptococci in water. *J. Appl. Bact.* **38**:63–65.

Purdy, R. N., Dancer, B. N., Day, M. J., and Stickler D. J. 1984. A novel technique for the enumeration of bacteriophage from water. *Microbiol. Lett.* **21**:89–92.

Purdy, R. N., Dancer, B. N., Day, M. J., and Stickler D. J. 1985. A note on a membrane filtration method for the concentration and enumeration of bacteriophages from water. *J. Apple. Bacteriol.* **58**:231–233.

Rao, V. C., Seidel, K. M., Goyal, S. M., Metcalf, T. G., and Melnick, J. L. 1984. Isolation of enteroviruses from water. suspended solids, and sediments from Galveston Bay: survival of poliovirus and rotavirus adsorbed to sediments. *Appl. Environ. Microbiol.* **48**:404–409.

Rhodes, M. W., and Kator, H. 1988. Survival of *Escherichia coli* and *Salmonella* spp. in estuarine environments. *Appl. Environ. Microbiol.* **54**:2902–2907.

Rhodes, M. W., and Kator, H. I. 1990. Effects of sunlight and authochthomnous microbiota on *Escherichia coli* survival in an estuarine environment. *Curr. Microbiol.* **21**:65–73.

Rhodes, M. W., Anderson, I. C., and Kator, H. I. 1983. In situ development of sublethal stress in *Escherichia coli:* effects on enumeration. *Appl. Environ. Microbiol.* **45**:1870–1876.

Richards, G. P. 1985. Outbreaks of shellfish-associated enteric virus illness in the United States: requisite for development of viral guidelines. *J. Food Protect.* **48**:815–823.

Roberts, M. C., Moncla, B., and Kenny, G. E. 1987. Chromosomal DNA probes for the identification of *Bacteroides* species. *J. Gen. Microbiol.* **133**:1423–1430.

Roper, M. M., and Marshall, K. C. 1974. Modification of the interaction between *Escherichia coli* and bacteriophage in saline sediment. *Microb. Ecol.* **1**:1–13.

Roper, M. M., and Marshall, K. C. 1977. Lysis of *Escherichia coli* by a marine myxobacter. *Microb. Ecol.* **3**:167–171.

Roper, M. M., and Marshall, K. C. 1978. Biological control agents of sewage bacteria in marine habitats. *Aust. J. Marine Freshwater Res.* **29**:335–343.

Roper, M. M., and Marshall, K. C. 1979. Effects of salinity on sedimentation and of particulates on survival of bacteria in estuarine habitats. *Geomicrobiol. J.* **1**:103–116.

Roszak, D. B., and Colwell, R. R. 1987. Survival strategies of bacteria in the natural environment. *Microbiol. Rev.* **51**:365–379.

Roth, W. G., Leckie, M. P., and Dietzler, D. N. 1988. Restoration of colony-forming activity in osmotically stressed *Escherichia coli* by betaine. *Appl. Environ. Microbiol.* **54**:3142–3146.

Rowbotham, T. J., and Cross, T. 1977a. *Rhodococcus coprophilus* sp. nov.: an aerobic nocardioform actinomycete belonging to the "rhodochrous" complex. *J. Gen. Microbiol.* **100**:123–138.

Rowbotham, T. J., and Cross, T. 1977b. Ecology of *Rhodococcus coprophilus* and associated actinomycetes in fresh water and agricultural habitats. *J. Gen. Microbiol.* **100**:231–240.

Rubentschik, L., Roisin, M. B., and Bieljansky, F. M. 1936. Adsorption of bacteria in salt lakes. *J. Bacteriol.* **32**:11–31.

Salyers, A. A. 1984. *Bacteroides* of the human lower intestinal tract. *Ann. Rev. Microbiol.* **38**:293–313.

Sawyer, T. K. 1980. Marine amoebae from clean and stressed bottom sediments of the Atlanctic Ocean and Gulf of Mexico. *J. Protozool.* **27**:13–32.

Sawyer, T. K., Visesvara, G. S., and Harke, B. A. 1977. Pathogenic amoebas from brackish and ocean sediments, with a description of *Acanthamoeba hatchetti,* n. sp. *Science* **196**:1324–1325.

Sawyer, T. K., Lewis, E. J., Galass, M., Lear, D. W., O'Malley, M. L., Adams, W. N., and Gaines, J. 1982. Pathogenic amoebae in ocean sediments near wastewater sludge disposal sites. *J. Water Pollut. Control Fed.* **54**:1318–1323.

Sawyer, T. K., Nerad, T. A., Daggett, P.-M., and Bodammer, S. M. 1987. Potentially pathogenic protozoa in sediments from oceanic sewage-disposal sites. In: Capuzzo, J. M., and Kester, D. R. (eds.) *Oceanic Processes in Marine Pollution, Vol. 1. Biological Processes and Wastes in the Ocean.* Robert E. Krieger, Malaar, Florida, pp. 183–194

Scardovi, V. 1981. Chapter 149. The genus *Bifidobacterium.* In: Starr, M. P., Stolp, H., Truper, H. G., Balows, A., and Schlegel H. G. (eds.) *The Prokaryotes, Vol. II.* Springer-Verlag, New York, pp. 1951–1961.

Scarpino, P. V. 1975. Human enteric viruses and bacteriophages as indicators of sewage pollution. In: Gameson, A. L. H. (ed.) *Discharge of Sewage from Sea Outfalls.* Pergamon Press, Oxford, pp. 49–61.

Schleifer, K. H., and Kilpper-Balz, R. 1984. Transfer of *Streptococcus faecalis* and *Streptococcus faecium* to the genus *Enterococcus* nom. rev. as *Enterococcus faecalis* comb. nov. and *Enterococcus faecium* com. nov. *Int. J. Syst. Bacteriol.* **34**:31–34.

Seeley, N. D. and Primrose, S. B. 1980. The effect of temperature on the ecology of aquatic bacteriophages. *J. Gen. Virol.* **46**:87–95.

Shiaris, M. P., Rex, A. C., Pettibone, G. W., Keay, K., McManus, P., Rex, M. A., Ebersole, J., and Gallagher, E. 1987. Distribution of indicator bacteria and *Vibrio parahaemolyticus* in sewage-polluted intertidal sediments. *Appl. Environ. Microbiol.* **53**:1756–1761.

Sieburth, J. M., and Pratt, D. M. 1962. Anticoliform activity of sea water associated with the termination of *Skeletonema costatum* blooms. *Trans. NY Acad. Sci.* **24**:498–501.

Singh, S. N., and Gerba, C. P. 1983. Concentration of coliphage from water and sewage with charge-modified filter aid. *Appl. Environ. Microbiol.* **45:**232–237.

Slanetz, L. W., and Bartley, C. H. 1965. Survival of fecal streptococci in sea water. *Hlth. Lab. Sci.* **2:**142–148.

Smith, L. DS. 1975. Common mesophilic anaerobes, including *Clostridium botulinum* and *Clostridium tetani,* in 21 soil specimens. *Appl. Microbiol.* **29:**590–594.

Smith, E. M., Gerba, C. P., Melnick, J. L. 1978. Role of sediment in the persistence of enteroviruses in the estuarine environment. *Appl. Environ. Microbiol.* **35:**685–689.

Sobsey, M. D. 1987. Methods for recovering viruses from shellfish, seawater, and sediments. In: Berg, G. (ed.) *Methods for Recovering Viruses from the Environment.* CRC Press, Boca Raton, Florida, pp. 77–108.

Sodergren, A. 1989. Biological effects of bleached pulp mill effluents. National Swedish Environmental Protection Board, Report 3558. Solna, Sweden.

Strange, R. E. 1976. *Microbial Response to Mild Stress.* Meadowfield Press, Durham, England.

Switzer, R. E., and Evans, J. B. 1974. Evaluation of selective media for enumeration of group D streptococci in bovine feces. *Appl. Microbiol.* **28:**1086–1087.

Tartera, C., and Jofre, J. 1987. Bacteriophages active against *Bacteroides fragilis* in sewage-polluted waters. *Appl. Environ. Microbiol.* **53:**1632–1637.

Tartera, C., Lucena, F., and Jofre, J. 1989. Human origin of *Bacteroides fragilis* bacteriophages present in the environment. *J. Appl. Microbiol.* **55:**2696–2701.

Teshima, S., and Kanazawa, A. 1978. Conversion of cholesterol to coprostanol and cholestanol in the estuary sediment. *Mem. Fac. Fish Kagoshima Univ.* **27:**41–47.

Vasconcelos, G. J., and Swartz, R. G. 1976. Survival of bacteria in seawater using a diffusion chamber apparatus in situ. *Appl. Environ. Microbiol.* **31:**913–920.

Vaughn, J. M., and Metcalf, T. G. 1975. Coliphages as indicators of enteric viruses in shellfish and shellfish raising estuarine waters. *Water Res.* **9:**613–616.

Vaughn, J. M., Landry, E. F., Thomas, M. Z., Vicale, T. J., and Penello, W. F. 1979. Survey of human enterovirus occurrence in fresh and marine surface waters on Long Island. *Appl. Environ. Microbiol.* **38:**290–296.

Venkatesan, M. I., Ruth, E., and Kaplan, I. R. 1986. Coprostanols in Antarctica marine sediments: a biomarker for marine mammals and not human pollution. *Marine Pollut. Bull.* **17:**554–557.

Verstraete, W., and Voets, J. P. 1976. Comparative study of *E. coli* survival in two aquatic ecosystems. *Water Res.* **10:**129–136.

Volterra, L., Bonadonna, L., and Aulicino, F. A. 1985. Comparison of methods to detect fecal streptococci in marine waters. *Water, Air, Soil Pollut.* **26:**201–210.

Wade, T. L., Oertel, G. F., and Brown, R. C. 1983. Particulate hydrocarbon and coprostanol concentrations in shelf waters adjacent to Chesapeake Bay. *Can. J. Fish. Aquat. Sci.* **40:**34–40.

Walker, R. W., Wun, C. K., and Litsky, W. 1982. Coprostanol as an indicator of fecal pollution. *CRC Crit. Rev. Envir. Control* **10:**91–112.

Weiss, C. M. 1951. Adsorption of *E. coli* on river and estuarine silts. *Sewage Ind. Wastes* **23:**227–237.

Wheater, D. W. F., Mara, D. D., and Oragui, J. 1979. Indicator systems to distinguish sewage from stormwater run-off and human from animal faecal material. In: James, A., and Evison, L. (eds.) *Biological Indicators of Water Quality.* John Wiley & Sons, Chichester, pp. 21-1–21.25.

Wheater, D. W. F., Mara, D., Opara, A., and Singleton, P. 1980. Anaerobic bacteria as indicators of faecal pollution. *Proc. R. Soc. Edinburgh* **78B:**s161–s169.

Wright, R. T., and Coffin, R. B. 1984. Measuring microzooplankton grazing on planktonic marine bacteria by its impact on bacterial production. *Microb. Ecol.* **10:**137–149.

Xu, H. S., Roberts, N., Singleton, F. L., Attwell, R. W., Grimes, D. J., and Colwell, R. R. 1982. Survival and viability of nonculturable *Escherichia coli* and *Vibrio cholerae* in the estuarine and marine environment. *Microb. Ecol.* **8:**313–323.

Yde, M., De Wulf, E., De Maeyer-Cleempoel, S., and Quaghebeur, D. 1982. Coprostanol and bacterial indicators of faecal pollution in the Scheldt Estuary. *Bull. Environ. Contam. Toxicol.* **28:**129–134.

Yoshpe-Purer, Y. 1989. Evaluation of media for monitoring fecal streptococci in seawater. *Appl. Environ. Microbiol.* **55:**2041–2045.

8

Viruses in Seafoods

Marilyn B. Kilgen and Mary T. Cole

More than 100 enteric viruses can be found in human feces (Melnick et al. 1978). Families of enteric viral pathogens include picornaviruses, reoviruses, adenoviruses, caliciviruses, astroviruses, and unclassified viruses such as Norwalk and Norwalk-like viruses, Snow Mountain agent, small round viruses, and non-A–non-B hepatitis virus (NANB).

Of the enteric viruses, only hepatitis Type A (HAV), Norwalk virus, Snow Mountain agent, caliciviruses, astroviruses, NANB enteral hepatitis, and unspecified hepatitis have been documented to cause seafood-associated illness. With the exception of HAV contamination from infected food handlers, all reported cases of seafood-associated viral infections have been from the consumption of raw or improperly cooked molluscan shellfish (Bryan 1986; CDC 1989; Cliver 1988; Gerba 1988; Richards 1985, 1987; Rippey and Verber 1988).

Human enteric viruses, like all viruses, are inert in food systems and are active only inside the host. They are very species specific and even receptor specific for certain cells. They are usually transmitted by the fecal-oral route which includes contamination from human sewage. Infected persons shed viruses in their feces, which may contaminate seafoods through sewage pollution of marine waters, or through poor personal hygiene habits if they are food handlers. Most of the reported outbreaks of viral illness associated with seafood have involved fecally contaminated raw molluscan shellfish.

Once fecally associated human enteric viruses are released into the marine environment, their survival and persistence are based on many factors including temperature, salinity, ultraviolet inactivation from sunlight, and presence of organic solids or sediments. Of these factors, the most important are temperature <50°F (<10°C) and the protective action of organic material (Gerba 1988). Human enteric viruses have been isolated in field studies only from blue crabs taken from a sludge dump in the North Atlantic and from molluscan shellfish. Other marine animals including lobsters, sandworms, detrital feeding fish, and conch have been shown to take up enteric viruses when seeded into marine waters experimentally, but this has not been reported in field studies. Only molluscan shellfish have been reported in association with enteric virus transmission from contaminated estuarine waters (Gerba 1988).

Enteric viruses also have the potential to contaminate seafood during processing and preparation by infected food handlers. In seafoods, only hepatitis Type A has been documented to be transmitted by infected food handlers. There are no documented cases of transmission of other human enteric viruses from seafood products contaminated at the processing, distribution, or food handling level (Bryan 1986; CDC 1989; Rippey and Verber 1988). Viruses do not replicate in seafood products, so time and temperature are not factors. However, handling of products by a person who is a carrier or who is infected by enteric viruses could result in transmission by the fecal-oral route if poor personal hygiene is allowed (Cliver 1988; Matches and Abeyta 1983).

FAMILIES OF HUMAN ENTERIC VIRUSES ASSOCIATED WITH POLLUTION OF MARINE WATERS WITH HUMAN FECES

Picornaviruses

Picornaviruses make up the largest of all virus families, with nearly 200 host-specific picornaviruses having been identified in man. Of these, 69 types of enteroviruses inhabit the enteric tract (Gerba 1988; White and Fenner 1986).

	Enteroviruses
Polioviruses	(Types 1, 2, and 3)
Echoviruses	(Types 1–34; no 10 or 28)
Coxsackieviruses	(Types A1–A24 and B1–B6; no A23) (subtotal of 67)
Enteroviruses	(Types 68–71)
Enteroviruses	[Type 72–**Hepatitis Type A (HAV)**]

These enteroviruses all have a naked icosahedral capsid 25–30 nm in diameter, appear as smooth and round in outline, are constructed from 60 protomers, and replicate in the cytoplasm. Each protomer is comprised of a single molecule of four polypeptides, VP 1, 2, 3, and 4, or 1D, 1B, 1C, and 1A, respectively. The genome is a single stranded RNA linear molecule of positive polarity with a MW of 2.5×10^6. The molecule is polyadenylated at its 3' end with the protein VP_g covalently linked to its 5' end.

Enteroviruses are resistant to the acidic pH, proteolytic enzymes, and bile salts in the gut. Hepatitis Type A (enterovirus Type 72) is less acid stable than other enteroviruses, but more heat stable, surviving 140°F (60°C) for 4 hours (White and Fenner 1986).

These enteroviruses have been subdivided into the species groups polioviruses (PV), coxsackieviruses, echoviruses, and enteroviruses. With the exception of PV and HAV, the clinical features, epidemiology, and control of these viruses will be classified by syndromes.

Poliovirus (PV1, PV2, and PV3). The three wild types of polioviruses are no longer endemic in this country. **No cases of seafood-associated polio have been reported in this country, and there is virtually no risk of infection with wild-type PV from fecally contaminated seafood, or from seafood processing and handling.** However, vaccine strains of PV are generally the most common enteroviruses isolated from estuarine waters and molluscan shellfish, and have even been suggested as possible indicators of pathogenic enteric viruses in shellfish and estuarine waters (Cole et al. 1986; Richards 1985; Sobsey et al. 1980).

The majority of infections from PV before the use of the Salk inactivated polio vaccine (IPV) and the Sabin oral polio vaccine (OPV) were minor. Epidemiological evidence indicates that only 1 in 100 persons infected with poliovirus shows clinical symptoms of the disease. This virus is spread mainly via the fecal-oral route by direct contact with contaminated hands, food, fomites, etc. In this country, foodborne poliomyelitis was mainly due to raw milk contaminated by milk handlers, but this foodborne route was actually eliminated by pasteurization before the vaccine was developed. Epidemics may still occur in underdeveloped countries by contamination of water supplies or food with human sewage. Poliovirus is endemic in the tropics, and prior to extensive vaccination programs in developed countries, summer epidemics occurred in temperate countries (Cliver 1988; White and Fenner 1986).

MICROBIOLOGY/CLINICAL FEATURES. The symptoms of poliomyelitis include malaise, sore throat, and the possibility of headache and vomiting. The latter two symptoms indicate aseptic meningitis. One percent of these cases progress to secondary central nervous system (CNS) involvement resulting in muscle pain, stiffness, and rapid development of paralysis. Motor function usually improves for a few months with any remaining paralysis considered permanent. The incubation period usually lasts from 10 to 15 days with the outer limits of 3 days to 1 month. The virus is shed abundantly in the feces during the incubation period. Once the virus is ingested it multiplies in the pharynx and small intestine, then spreads to regional lymph nodes resulting in viremia and dissemination of the virus throughout the body. If the CNS becomes involved, the virus is carried to the anterior horn cells of the spinal cord and motor cortex of the brain resulting in either spinal (most common) or bulbar poliomyelitis. Mortalities in the bulbar form are due to cardiac or respiratory motor failure (White and Fenner 1986).

DISEASE CONTROL. Development of IPV and OPV has virtually eliminated poliomyelitis from developed countries. Continued immunization programs are imperative to avoid reintroduction of the disease. Reported cases of poliomyelitis in this country are generally due to failure of OPV, inadequate attenuation of one of the virus types, or immunosuppression of the vaccine recipients. Infection of unimmunized immigrants and unimmunized religious sects such as the Amish community in 1978 has also occurred. Many third world countries continue to have outbreaks of poliomyelitis due to little or no vaccine programs (White and Fenner 1986).

Hepatitis Type A (Enterovirus Type 72). HAV is spread by the fecal-oral route. It is hyperendemic in countries that are overcrowded, and have inadequate sanitation and poor hygiene. Most infections in these communities occur in childhood and are subclinical. In more developed countries the disease is seen most often between the ages of 15 and 30 (Cliver 1988; White and Fenner 1986).

Contaminated food and water, and person-to-person contact are the main routes of transmission of HAV. Each year 20,000–30,000 cases are reported to the Centers for Disease Control (CDC). Of these cases, approximately 140 are due to foods (0.5% of the total). Most of these foodborne outbreaks are due to mishandling of foods by infected individuals (Cliver 1988). Outbreaks can also occur due to inadequate cooking of contaminated foods and by human fecal contamination of drinking water supplies, swimming waters, and shellfish growing waters.

In the 1950s the first documented case of shellfish-associated HAV occurred in Sweden (Cliver 1988). The first case was documented in the U.S. in the 1960s (Cliver 1988; Gerba 1988; Richards 1985). Richards (1985) reported approximately 1,400 cases of molluscan shellfish-associated HAV from 1961 to 1984. Between 1973 and 1987, the CDC foodborne surveillance system (CDC 1989) reported 11 outbreaks involving 437 cases of seafood-related HAV (Table 8.1). However, the total illnesses reported were mainly due to 1 outbreak involving 285 cases in 1973. Two of the outbreaks involving 92 cases were due to contamination from food handlers. The U.S. Food and Drug Administration's New England Technical Services Unit (NETSU) (Rippey and Verber 1988) reported approximately 356 cases of shellfish-associated HAV in the same time period (Table 8.2); but again, these total HAV illnesses were mainly due to 1 outbreak involving 294 cases in 1973. Since the 1973 outbreaks, the incidence of HAV reported by the CDC foodborne surveillance system and the NETSU is small (Tables 8.1 and 8.2). Richards (1985) noted that the incidence of molluscan shellfish-associated HAV has decreased in the last decade.

MICROBIOLOGY/CLINICAL FEATURES. HAV is ingested and multiplies primarily in the intestinal epithelium. Secondary infection of the parenchymal cells of the liver is through the bloodstream. The virus is found in the feces 1 week prior to the clinical signs. It may also be found in the blood 1 week prior to the appearance of the main clinical sign of dark urine. It disappears after serum transaminase levels reach their peak. The incubation period is normally 4 weeks, with limits from 2 to 6 weeks.

Table 8.1 Outbreaks and Cases of Reported Pathogens in Shellfish and Other Seafoods by Year: Centers for Disease Control Summary Data, 1973–1987 (CDC 1980, 1982, 1989)

| | Hepatitis Type A | | | | Other Viral | | | | Unknown Etiologies[b] | | | |
| | Shellfish | | Finfish | | Shellfish | | Finfish | | Shellfish | | Finfish | |
	OB	CA	OB	CA	OB	CA	OB	CA	OB	CA	OB	CA
73	1	285	0	0	0	0	0	0	0	0	1	122
74	0	0	0	0	0	0	0	0	7	34	6	23
75	0	0	0	0	0	0	0	0	8	74	11	77
76	0	0	0	0	0	0	0	0	4	52	8	25
77	1	17	0	0	0	0	0	0	7	93	2	32
78	1	8	0	0	0	0	0	0	10	229	2	11
79	2	14	0	0	1	22	0	0	4	28	0	0
80	1	3	1[a]	7[a]	0	0	0	0	7	81	2	6
81	0	0	0	0	0	0	0	0	4	254	3	76
82	3	8	1[a]	85[a]	0	0	0	0	59	1,312	1	28
83	0	0	0	0	1	20	0	0	19	536	3	44
84	0	0	0	0	0	0	0	0	9	623	1	2
85	0	0	0	0	0	0	0	0	3	128	3	20
86	0	0	0	0	0	0	0	0	2	76	1	16
87	0	0	0	0	0	0	0	0	1	4	0	0
Total	9	335	2[a]	92[a]	2	42	0	0	114	3,524	44	482

Source: CDC 1980, 1982, 1989.
OB, outbreaks; CA, cases.
[a]Food handler + for HAV.
[b]Unknown etiologies, probably not all microbiological pathogens.

The onset of HAV is associated with the clinical symptoms of fever, malaise, anorexia, nausea, and lethargy. Symptoms also include dark urine, jaundice, and an enlarged, tender, palpable liver. Most infections that occur in children are anicteric. The severity of the disease increases with age (Bryan 1986; Cliver 1988; Overby et al. 1983; White and Fenner 1986).

DISEASE CONTROL. Prevention and control of HAV can be accomplished at several levels. Municipal sewage systems should be properly functioning to prevent contamination of public water supplies and shellfish producing waters.

Food handlers should be carefully instructed and watched for good hygiene practices, particularly hand washing and sanitizing following defecation.

There is currently no valid indicator of human enteric viruses such as HAV in shellfish growing waters (Elliot et al. 1985; Kilgen et al. 1988; Cole et al. 1986). Proper classification of shellfish growing areas, based on a valid virus indicator, and restricting shellfish harvest only to approved areas would be very important in preventing HAV contamination from untreated human sewage.

Table 8.2 Cases of Reported Pathogens in Shellfish by Year: USFDA, NETSU Data, 1973–1987

	HAV	*Norwalk*	*Snow Mountain*	*NANB*	*Unspecified Hepatitis[a]*	*Unknown Etiologies[b]*
73	294	0	0	0	0	0
74	0	0	0	0	0	10
75	0	0	0	0	0	52
76	0	0	0	0	1,391	48
77	17	0	0	0	1,443	134
78	0	0	0	0	0	83
79	18	0	0	0	1,645	0
80	1	6	0	0	0	127
81	9	0	0	0	0	210
82	11	0	0	0	0	1,836
83	5	0	71	0	0	2,123
84	0	0	0	0	0	404
85	1	5	0	1	0	195
86	0	0	0	0	0	120
87	0	0	0	0	0	0
Total	356	11	71	1	4,479	5,342

Source: Rippey and Verber 1988.
[a]Etiological agent(s) unknown.
[b]The unknown etiologies are probably not all microbiological pathogens.

Infection with HAV results in permanent immunity; thus a vaccine could eradicate the disease. There is a live trial vaccine for HAV currently being tested by the U.S. Army. However, no licensed vaccine is available to the public at this time. Passive immunization with gamma-globulin following exposure to HAV is the main method of control and prevention (Cliver 1988; White and Fenner 1986).

Other Enteroviruses (Coxsackie, Echo, and Enteroviruses). Although most of these other enterovirus infections are subclinical, they can cause several clinical syndromes including neurological disease, cardiac and muscular disease, enanthems and exanthems, respiratory disease, ocular disease, and abdominal disease (White and Fenner 1986). These enteroviruses grow in both the throat and the intestinal tract. They are shed for longer periods of time in the feces than in the respiratory secretions, and are disseminated via the bloodstream to target organs. Because they are shed in the feces, hygiene is of great importance in the spread of these viruses. However, *none of these other enteroviruses or their syndromes have been shown to be seafood-associated* (CDC 1989; Cliver 1988; Richards 1985; Rippey and Verber 1988).

Reoviruses—Rotaviruses

The rotaviruses, which are members of the family of reoviruses, are the most important cause of infantile gastroenteritis in the world, and cause many deaths in underdeveloped countries. These viruses are shed in large numbers in the feces for 3–4 days; thus spread of the disease is mainly by the fecal route. Nosocomial outbreaks in nurseries are very common (White and Fenner 1986). **No seafood-associated cases of rotavirus infection have been documented in this country.**

MICROBIOLOGY/CLINICAL FEATURES. The rotavirus has a smooth and round outline with two concentric icosahedral shells giving the appearance of a wheel. They have a diameter of 70 nm, a double-stranded RNA 11 molecule segmented genome, and MW of 12×10^6. There are five human serotypes identified. The incubation period for rotavirus is 1–3 days. The virus multiplies in the differentiated columnar epithelial cells of the small intestine. In neonates, older children, and adults infection with rotavirus is usually asymptomatic. However, between the ages of 6 and 24 months clinical signs appear. These symptoms include vomiting and diarrhea for 4–5 days. If untreated, this can result in severe dehydration and death in underdeveloped tropical countries (Cliver 1988; White and Fenner 1986).

DISEASE CONTROL. Improving the nutrition and hygiene of third-world populations would be the greatest aid in control of rotavirus-related mortalities. There is no human vaccine available at this time. Research to develop a live oral vaccine containing all of the human serotypes is ongoing (Cliver 1988; White and Fenner 1986).

Adenoviruses

Adenoviruses originally isolated from human adenoids have recently been associated with gastroenteritis. Adenoviruses are enteric viruses and are shed in human feces. This means that they are spread by the fecal-oral route, by direct contact, cantaminated food and water, and contaminated swimming waters. However, **no seafood-associated cases of adenovirus infections have been documented in this country.**

MICROBIOLOGY/CLINICAL FEATURES. These viruses are icosahedrons, 80 nm in diameter, with two main types of capsomers. The genome is a single linear molecule of double-stranded DNA with an MW of $20–25 \times 10^6$. These viruses replicate in the nucleus of the host genome.

Adenoviruses Types 40 and 41 have been shown to cause 10% of infantile gastroenteritis. However, the virus is found in the feces of some asymptomatic as well as symptomatic patients. The virus multiplies in the small intestines, and most of these enteric infections are subclinical. The incubation period is 5–7 days (White and Fenner 1986).

DISEASE CONTROL. There are no vaccinations for the enteric adenoviruses Types 40 and 41. Control is by personal hygiene and chlorination of drinking water supplies and swimming pools (White and Fenner 1986).

Caliciviruses and Astroviruses

Human caliciviruses are responsible for many cases of gastroenteritis. The Norwalk virus resembles caliciviruses and is considered a possible member of this family of viruses by some scientists (White and Fenner 1986). However, Norwalk will be discussed with the unclassified viruses. The transmission routes of these viruses are the same as that discussed for Norwalk virus and the other enteric viruses. These viruses have been recently reported to be involved in shellfish-associated outbreaks (Gerba 1988). None of these outbreaks has been officially reported to the CDC or to the NETSU.

MICROBIOLOGY/CLINICAL FEATURES. Caliciviruses are slightly larger than picornaviruses and the icosahedral capsid is made up of only one polypeptide. The capsid is made up of 32 cup-shaped (calix) depressions. They are resistant to heat and acid.

The astroviruses are 28-nm particles detected in the feces of infants with IEM. A star-shaped unstained center gives them their name (White and Fenner 1986).

DISEASE CONTROL. Good sanitation, personal hygiene, and proper classification of shellfish growing waters are the most effective measures in preventing any enteric virus infection which is mainly transmitted by the fecal-oral route.

Unclassified Viruses

These include the nonspecific agents of gastroenteritis including Norwalk and Norwalk-like agents, Snow Mountain agent, and small round viruses (SRVs). Non-A–non-B (NANB) enteral hepatitis is also an unclassified virus (Gerba 1988; White and Fenner 1986).

Norwalk, Norwalk-like agents, Snow Mountain Agent, and SRVs.

Outbreaks of viral gastroenteritis due to the Norwalk agent have been associated with swimming in waters contaminated with human sewage, fecal contamination of food or drinking water, and consumption of uncooked or partially cooked shellfish harvested from estuaries contaminated with human fecal material. The first documented shellfish-associated outbreak of gastroenteritis involving Norwalk virus was in 1979 in Australia, where more than 2,000 people were involved (Gerba 1988). Since this time, there have been many reported outbreaks of Norwalk or Norwalk-like viral gastroenteritis in the U.S. (CDC 1989; Morse et al. 1986; Richards 1987; Rippey and Verber 1988). Norwalk virus illness associated with shellfish is a continuing problem and has increased in the last decade (Richards 1985). Between 1973 and 1987, the NETSU (Rippey and Verber 1988) reported 11 shellfish-related cases of Norwalk gastroenteritis and 71 cases of Snow Mountain agent (Table 8.2). The CDC (1989) reported 42 shellfish-associated cases of Norwalk and related viruses (Table 8.1). Richard (1985) reported an outbreak of shellfish-associated gastroenteritis involving 472 cases in Louisiana. Norwalk was suspected but not serologically documented. Richards (1985) also reported over 6,000 cases of shellfish-associated gastroenteritis over the past 50 years. It is probable that many of these are of viral etiology, possibly Norwalk or Norwalk-related agents. In contrast to hepatitis, only a few specialized laboratories are able to serologically diagnose infections with Norwalk and related viruses. Although the CDC and NETSU data bases report relatively few cases of confirmed Norwalk or related virus infections, recent evidence suggests that shellfish-associated infections with these agents occur more frequently than identified (Morse et al. 1986), and it is likely that they may ultimately be the most common shellfish-associated pathogen. The largest number of reported seafood-associated illnesses have **unknown etiologies** (CDC 1989; Rippey and Verber 1988). Although these are probably not all of microbiological etiology, many have onset periods and symptoms consistent with infection with Norwalk and related gastroenteritis agents. Over 75% of these cases have been reported since 1980, which shows increased awareness and reporting practices in regards to shellfish illnesses (Tables 8.1 and 8.2).

MICROBIOLOGY/CLINICAL FEATURES. The Norwalk virus is 25–32 nm in diameter, whereas the SRVs are 27–40 nm. The SRVs have been identified in the feces of infants with diarrhea using immune electron microscopy (IEM) (White and Fenner 1986).

The incubation period for the Norwalk agent is 24–72 hours. Infection results in the sloughing of intestinal villi followed by rapid regeneration. Clinical symptoms include diarrhea, nausea, vomiting, abdominal cramps, and in some cases, headache, myalgia, and low-grade fever. Symptoms are more serious in adults. Immunity following an infection with Norwalk virus is only temporary, lasting approximately 1 year. This may be one of the reasons for the high attack rate of 50–90% in many individuals (Cliver 1988).

DISEASE CONTROL. Good personal hygiene and good manufacturing practices, proper classification of recreational and shellfishing waters, and prevention of sewage contamination in drinking, swimming, or shellfish growing waters are the most effective preventive measures for the Norwalk and related gastroenteritis viruses because they are found only in human feces.

NANB Enteral Hepatitis. Enteral NANB hepatitis is transmitted mainly by sewage contaminated water. It is also transmitted sporadically by person-to-person contact. In the Middle East and Africa, it appears to be endemic (Overby et al. 1983). Cliver (1988) noted water-associated outbreaks have been reported for years from India, Africa, the U.S.S.R., and most recently, Mexico, and that there is no reason for it not to be found in the U.S. Gerba (1988) reported enteral NANB hepatitis cases in the U.S. associated with consumption of raw shellfish. The NETSU reported 1 case of shellfish associated NANB and 4,479 cases of shellfish associated nonspecified hepatitis (with 36 associated fatalities) between 1973 and 1987 (Table 8.2). The etiological agents were unknown for the nonspecified hepatitis. Diagnostic serology was either not done or not conclusive, and the association of the hepatitis with shellfish consumption was probably assumed from post infection case history information only (Rippey and Verber 1988). Hepatitis has a long incubation period, and this makes taking accurate case history information difficult. Also, because shellfish consumption is commonly associated with possible hepatitis A infection, it is certain that patients would be questioned about having eaten shellfish 4–6 weeks prior to the onset of illness.

MICROBIOLOGICAL/CLINICAL FEATURES.   The etiological agent of enteral or epidemic NANB hepatitis is unknown. Virus-like particles 27–30 nm in diameter have been recovered from patients with waterborne NANB hepatitis. These agents have been transmitted to primates and HAV immune human volunteers, and subsequently recovered from their feces (White and Fenner 1986). It has been suggested that the NANB agent could be an enterovirus—possibly a different serotype of HAV (Overby et al. 1983).

Enteral NANB hepatitis can be more severe than HAV with a high incidence of cholestasis. The mean incubation period ranges from 15 to 40 days. The highest attack rate is in young adults, especially pregnant women in the third trimester. The overall mortality rate for NANB hepatitis is only slightly higher than for HAV, but significant mortalities in excess of 10% can occur in pregnant women. Poor nutrition in underdeveloped countries is probably a factor, but NANB is overall more severe than HAV in pregnant women. The incidence of chronic active hepatitis is also extremely low in NANB hepatitis (Overby et al. 1983; White and Fenner 1986).

Unspecified hepatitis is a clinical syndrome that can be caused by several other viral or even some bacterial infections that may spread secondarily to the liver from the blood (i.e., infectious mononucleosis caused by Epstein–Barr virus) (White and Fenner 1986). Unspecified hepatitis may also be caused by one of the hepatitis viruses (Type A, Type B, NANB, or delta) that is not serologically identified and reported as such. If an unspecified type of hepatitis is accurately reported as associated with seafood consumption, it is most likely that it would be hepatitis A or enteral NANB. However, as previously stated, accurate case histories for an infection with such a long incubation (4–6 weeks) are usually difficult to obtain.

DISEASE CONTROL.   The same sanitary measures discussed previously with respect to prevention and control of enterically transmitted disease apply to enteral NANB hepatitis. There is no vaccine available at this time. The etiological agent is not even definitely known.

HUMAN ENTERIC VIRAL PATHOGENS ASSOCIATED WITH PROCESSING, DISTRIBUTION, AND PREPARATION OF SEAFOODS

Hepatitis A

Although hepatitis A virus (HAV) is a pathogen most commonly transmitted by infected food handlers, most of the seafood-associated cases reported to the CDC and NETSU between 1973 and 1987 were in 1973, and were associated with consumption of raw or inadequately cooked molluscan shellfish (Tables 8.1 and 8.2) (CDC 1989; Rippey and Verber 1988).

Each year 20,000–30,000 cases of hepatitis A are reported to the CDC. Of these cases, approximately 140 are directly attributable to contaminated foods (0.5% of the total). Most of these foodborne outbreaks are due to mishandling of foods by infected individuals (Cliver 1988). Between 1973 and 1987, CDC reported 1 outbreak of 7 cases of HAV in 1980 due to tuna served in the home. The food handler was positive for HAV, which indicated contamination in preparation of the seafood. During this time, CDC also reported another outbreak involving 85 cases of HAV from tuna salad in 1982 in a New York restaurant. The food handler was also HAV-positive in these cases (Table 8.1) (CDC 1980, 1982, 1989).

MICROBIOLOGY/CLINICAL FEATURES. The microbiology and clinical features of HAV were discussed previously as a viral pathogen associated with human fecal pollution of the marine environment.

DISEASE CONTROL. To prevent and control HAV in processing, distribution, and preparation of foods, food handlers should be carefully instructed and watched for good hygienic practices, particularly hand washing and sanitizing following defecation. Infection with HAV results in permanent immunity; thus a vaccine could eradicate the disease. A live HAV vaccine currently is being tested by the U.S. Army; however no licensed vaccine is currently available to the public. Passive immunization with gamma-globulin following exposure to HAV is the main method of postexposure control and prevention (Cliver 1988; White and Fenner 1986).

Other Viral Agents

There are no documented cases of transmission of other human enteric viruses from seafood products contaminated at the processing, distribution, or food handling level (Bryan 1986; CDC 1989; Rippey and Verber 1988). Viruses do not replicate in seafood products, so time and temperature are not factors. However, handling of products by a person who is a carrier or who is infected by enteric viruses could result in transmission by the fecal-oral route if poor personal hygiene is allowed (Cliver 1988; Matches and Abeyta 1983).

SUMMARY OF REPORTED SEAFOOD-ASSOCIATED VIRAL DISEASE OUTBREAKS 1973–1987

Centers for Disease Control Data 1973–1987

A summary of the seafood-associated outbreaks and cases of hepatitis A and other viral agents reported to the CDC from 1973 to 1987 is found in Table 8.1.

NETSU Data 1973–1987

A summary of seafood-associated cases of HAV, Norwalk, Snow Mountain Agent, NANA, and unspecified hepatitis reported to the NETSU in the same time period is in Table 8.2.

REFERENCES

Bryan, F. L. 1986. Seafood-transmitted infections and intoxications in recent years. In: Kramer, D. E., and Liston, J. (eds.) *Seafood Quality Determination*. Elsevier Science Publishers B. V., Amsterdam, pp. 319–337.

Centers for Disease Control. Foodborne Disease Outbreaks, Annual Summary 1980, Issued February 1983. USDHHS Publication No. (CDC) 83-8185.

Centers for Disease Control. Foodborne Disease Outbreaks, Annual Summary 1982, Issued September 1985. USDHHS Publication No. (CDC) 85-8185.

Centers for Disease Control. Foodborne Surveillance Data for all Pathogens in Fish/Shellfish for years 1973–1987. Issued December, 1989.

Cliver, D. O. 1988. Virus transmission via foods. A Scientific Status Summary by the Institute of Food Technologists' Expert Panel on Food Safety and Nutrition. *Food Technol.* **42**:241–247.

Cole, M. T., Kilgen, M. B., Reily, L. A., and Hackney, C. R. 1986. Detection of enteroviruses and bacterial indicators and pathogens in Louisiana oysters and their overlying waters. *J. Food Protect.* **49**:596–601.

Elliot, E. L., and Colwell, R. R. 1985. Indicator organisms for estuarine and marine waters. *FEMS Microbiol. Rev.* **32**:61–79.

Gerba, C. P. 1988. Viral disease transmission by seafoods. *Food Technol.* **42**:99–103.

Kilgen, M. B., Cole, M. T., and Hackney, C. R. 1988. Shellfish sanitation studies in Louisiana. *J. Shellfish Res.* **7**:527–530.

Matches, J. R., and Abeyta, C. 1983. Indicator organisms in fish and shellfish. *J. Food Protect.* **37**:114–117.

Melnick, J. L., Gerba, C. P., and Wallis, C. 1978. Viruses in water. *Bull. Wld. Hlth. Org.* **56**:499–504.

Morse, D. L., Gugewich, J. J., Hanrahan, J. P., Stricof, R., Shayegani, M., Deibel, R., Grabau, F. C., Nowak, N. A., Herrmann, J. E., Cukor, G., and Blacklow, N. R. 1986. Widespread outbreaks of clam- and oyster-associated gastroenteritis. Role of Norwalk virus. *N. Engl. J. Med.* **314**:678–681.

Overby, L. R., Deinhardt, F., and Deinhardt, J. (eds) 1983. *Viral Hepatitis: Second International Max von Pettenkofer Symposium.* Marcel Dekker, New York.

Richards, G. P. 1985. Outbreaks of shellfish-associated enteric virus illness in the United States: requisite for development of viral guidelines. *J. Food Protect.* **48**:815–823.

Richards, G. P. 1987. Shellfish-associated enteric virus illness in the United States, 1934–1984. *Estuaries* **10**:84–85.

Rippey, S. R., and Verber, J. L. 1988. Shellfish borne disease outbreaks. Department of Health and Human Services, Public Health Service, Food and Drug Administration, Shellfish Sanitation Branch. NETSU, Davisville, Rhode Island.

Sobsey, M. D., Hackney, C. R., Carrick, R. J., Ray, B., and Speck, M. C. 1980. Occurrence of enteric bacteria and viruses in oysters. *J. Food Protect.* **43**:111–128.

White, D. O., and Fenner, F. 1986. *Medical Virology,* third edition. Academic Press, Orlando, Florida.

9

Seafood-Transmitted Zoonoses in the United States: The Fishes, the Dishes, and the Worms

Thomas L. Deardorff and
Robin M. Overstreet

We live in a parasite-filled world. Nearly every animal—if not parasitic itself—hosts one or more parasitic organisms. Most parasites live in harmony with their hosts and remain of little concern with regard to economics or public health. On the other hand, some parasitic relationships involving humans and domestic animals result in diseases that have a serious impact on the host populations. Examples of such diseases include malaria, African sleeping sickness, and hookworm disease. In areas with heavy infections of such parasites, both human health and the corresponding economic climates are affected.

Parasitic diseases transmitted from wild or domestic animals to humans are called zoonoses. Over 100 zoonoses have been reported worldwide. In the U.S., toxoplasmosis, trichinosis, and echinococcosis are commonly recognized examples of serious zoonotic diseases. Marine animals also vector zoonotic diseases, diseases that represent an increasing area of concern to the U.S. Food and Drug Administration (FDA).

Marine zoonotic infections in humans result from consumption of contaminated edible tissues or products of seafood or, to a lesser extent, from physical contact with contaminated seafood. No protozoan parasites of marine fishes have been demonstrated to infect humans, although some protozoans may be a risk in the future and others such as the myxosporidian *Henneguya salminicola* in Pacific salmon (Figs. 9.1 and 9.2) give the edible tissue a displeasing appearance and significantly alter its texture. In fact, *Kudoa thyrsitis,* infecting Atlantic salmon cultured in Puget Sound, Washington, produces a proteolytic enzyme that digests surrounding muscles (Harrell and Scott 1985). In the Pacific hake, the myxosporidian's natural host, a defense mechanism has apparently been established against the spore mass. On the other hand, *K. paniformis* is associated with enzymatic digestion of the fillet of the Pacific hake, and other species, such as *K. musculoliquefaciens* in the swordfish, liquefy the flesh of their hosts. The combination of potential or perceived health risks to consumers coupled with the negative consequences of some aesthetically disturbing protozoan infections seriously impact the commercial seafood industry.

Additional observations and comparative studies will be necessary to establish that a marine protozoan can infect humans. The possibility of such an

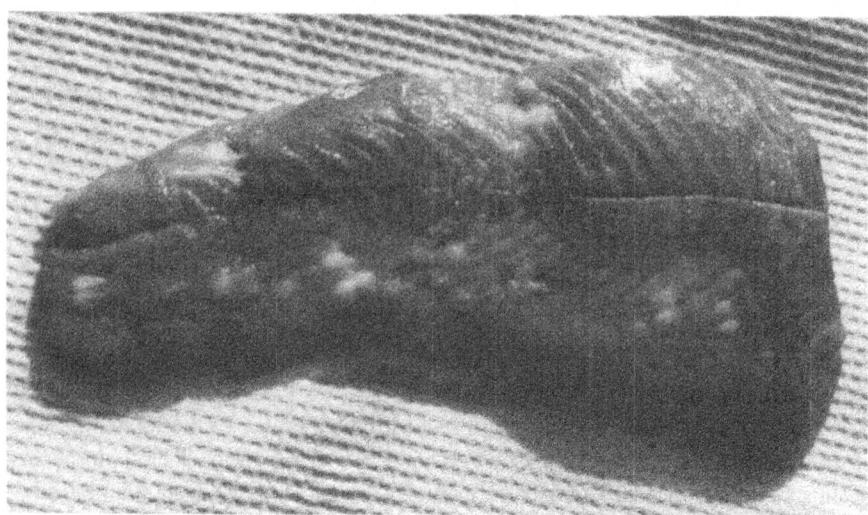

Figure 9.1 Salmon fillet with "tapioca disease," which results from an infection with the myxosporidian *Henneguya salminicola*. Gross view showing conspicuous white myxosporidian nodules. (Original provided by L. W. Harrell.)

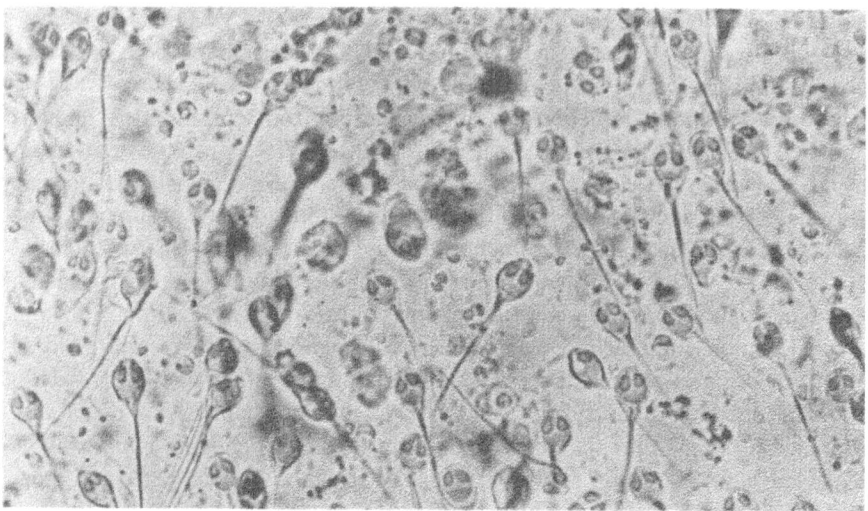

Figure 9.2 Salmon filet with "tapioca disease" which results from an infection with the myxosporidian *Henneguya salminicola*. Light microscopy view of protozoan cysts showing tail-like extensions of the two valves. (Original provided by L. W. Harrell.)

infection exists, especially considering the potential of immunocompromised individuals such as patients on immunosuppressive therapy or those with acquired immunodeficiency syndrome (AIDS) to acquire microsporidosis. For example, microsporidians have been found in the eye, and different fulminating infections with other unidentified species of microsporidians have been reported in other tissues (e.g., Cali and Owen 1988; Canning and Lom 1986). Because fishes serve as normal hosts for numerous species of microsporidians that are sometimes consumed raw, some fish microsporidians might be opportunistic pathogens of humans. One case specifically implicates a piscine species (Fig. 9.3). Spores and developing stages of an unidentified species of *Pleistophora* that fit into a complex of species restricted to fishes were obtained from biopsies of various skeletal muscles of an immunocompromised man in Florida who developed fever, generalized lymphadenopathy, and progressive generalized muscle weakness. Those signs were followed by intense inflammation of infected tissue (Cali and Owen 1988; Ledford et al. 1985). That patient also

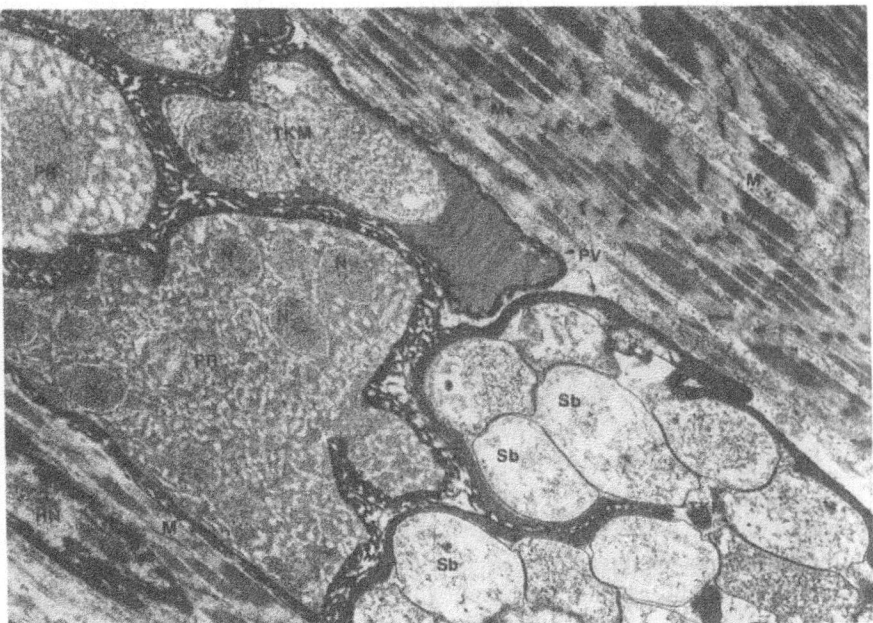

Figure 9.3 Electron micrograph of the microsporidan *Pleistophora* sp. in the skeletal muscle of an immunocompromised patient. Both proliferative stages *(PR)* and sporoblasts *(Sb)*, stages that develop into spores, occur within a thick-walled *(TKM)* parasitophorous vacuole *(PV)*. (From Cali and Owen 1988.)

developed testicular atrophy and gynecomastia, features not reported for AIDS patients, but he refused to allow a testicular biopsy (Ledford et al. 1985).

In contrast to protozoans, over 50 species of helminth parasites from fishes, crabs, crayfishes, snails, and bivalves are known to produce human infections. Most helminth zoonoses are rare and invoke only slight to moderate injury; however, some are more prevalent and pose serious potential health hazards. Extensive reviews of most of these helminth zoonoses have been published by Sprent (1969) and by Williams and Jones (1976).

Most seafood zoonoses occur in regions where seafood constitutes a major portion of the protein intake for the residing population. Historically, populations closely associated with the marine environment have been at greater risk for such zoonoses. In Japan, for example, where seafood constitutes a dietary staple often consumed raw, the annual incidence of the fishborne parasitic disease anisakiasis has been estimated to exceed 1,000 symptomatic episodes per year (Oshima and Kliks 1986). In 1984, 3,141 cases of stomach anisakiasis were reported (Oshima 1987). In contrast, Americans consume much less raw seafood than do Japanese, and only about 50 cases of this zoonotic disease have been confirmed or suspected in the U.S. since 1958.

Within the U.S., most seafood zoonoses occur along coastal regions where seafood products are commonly consumed. Continuing improvements in transportation, technology, and food handling, however, allow fresh seafoods to be shipped from abroad and throughout the U.S.; thus, the potential for acquisition of parasitic infections from marine products is not limited to coastal populations. Meals served both at home and in restaurants have been involved in the transmission of these infections.

Compared with the enormous number of cases of zoonotic parasitic infections occurring throughout the world, relatively few cases have been reported in the U.S. Reasons for such a paucity could be the fewer necessary intermediate hosts required to complete the corresponding parasitic life cycles, better sanitary methods, refined food handling procedures, and relatively meager consumption of raw or undercooked seafood. Generally, most fishery products consumed in the U.S. are effectively processed or cooked.

Some parasitic diseases transmitted by seafoods, however, occur in the U.S., and the number of new cases as well as additional unrecognized parasitic diseases continue to increase (Deardorff et al. 1987). The increasing exploitation of the marine environment by Americans, along with their changing dietary habits incorporating ethnic and "natural" foods and their tendency to reduce cooking times for seafoods, all increase chances to become infected with parasites. Some common raw fish dishes (some described later in the text) known to transmit infective forms of parasites to humans are Japanese sushi and sashimi, Latin American cebiche, Dutch green herring, Scandinavian gravlax, Hawaiian lomi salmon (salmon with chopped tomatoes and green peppers), palu (meat from a fish head and visceral organs that have been allowed to ripen in a closed container), tako poki (Japanese and Hawaiian cephalopod dish), Philippine bagoong (uncooked fish viscera, often in a deteriorated state), Pacific Island poisson cru (fish marinated in coconut juice), and taioro (described later). These and other dishes, some discussed in this chapter, have caused parasitic infections in humans worldwide. Increased consumption of raw or "lightly cooked" seafoods in the U.S. closely corresponds with a notable increase in the incidence of case reports of parasitic diseases.

Several reviews have been published concerning foodborne parasites transmitted to humans from fishes and other aquatic foods with a global emphasis (Higashi 1985; Olson 1987; Williams and Jones 1976). Unlike those reviews, we focus specifically on a few parasitic diseases that have been transmitted by local seafood products to consumers in the U.S.; however, we also mention several other potentially important agents. Emphasis is placed on fin*fishes* that harbor the infective stages of the helminths, *dishes* that are known to transmit the diseases, and *worms* that cause the diseases.

CESTODES

Human infections with the broad fish tapeworm, *Diphyllobothrium latum,* are acquired by ingesting raw fishes that harbor with the plerocercoid metacestode. The patient may harbor one or more worms. The presence of more than one worm usually reflects repeated consumption of infected products.

Although little pain is generally associated with diphyllobothriasis, clinical manifestations may involve complaints of abdominal discomfort, fatigue, dizziness, and alternating bouts of diarrhea and constipation. One of us with a year-long infection noted some of these symptoms several times but only during a short period prior to discharge of senile proglottids. Vomiting may be associated with intestinal obstruction. In some cases, infections have been associated with vitamin B_{12} deficiency and, therefore, pernicious anemia. Diagnosis of diphyllobothriasis is accomplished by finding large characteristic eggs in the feces (Fig. 9.4).

Humans serve as principal definitive hosts of *D. latum* along with a few other fish-eating mammals. The human host is primarily responsible for propagating the species in endemic areas (Beaver et al. 1984). The life cycle (Fig. 9.5) normally involves three hosts: a copepod, a fish, and a mammal. The tapeworm eggs, which pass out with the human feces (Fig. 9.4), are introduced into the aquatic environment after improper sewage treatment or by primitive disposal of feces. Within the egg, the hexacanth embryo, or oncosphere, develops, eventually hatching as a free-swimming ciliated coracidium. If eaten by a suitable calanoid or cyclopoid copepod, it sheds its cilia, and the coracidium penetrates into that copepod's hemocoel and becomes transformed into a procercoid metacestode. When any of the appropriate fishes, including different salmons,

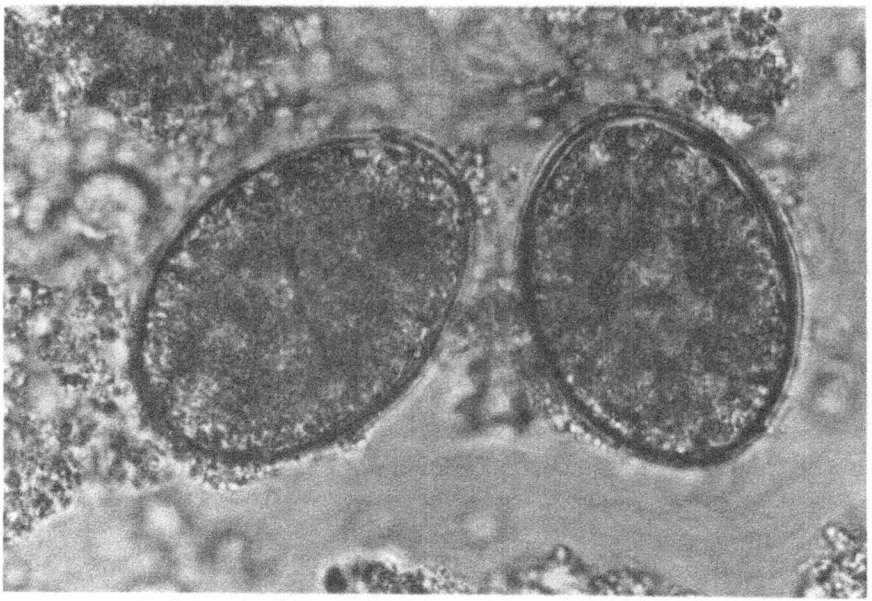

Figure 9.4 Characteristic eggs of *Diphyllobothrium latum* from formalin-fixed human stool. The larger specimen measured 67 μm in length.

eat the infected copepod, the procercoid—freed by digestion in the stomach of the host—migrates onto the viscera or between muscle fibers, where it grows into a plerocercoid metacestode. The plerocercoid represents the infective stage to humans. When contaminated fish products are eaten, the plerocercoid rapidly develops into a mature adult tapeworm up to 10 m long. The scolex of the worm most frequently attaches to the wall of the ileum, occasionally to the jejunum with resulting anemia, or more rarely to the colon. It even has been found in the gallbladder. Gravid proglottids of the mature cestode periodically release eggs into the human gastrointestinal tract to continue the life cycle again. A single worm may discharge as many as 1 million ova per day.

Immigrants from Europe, particularly those from Baltic regions, have been considered as primarily responsible for the introduction of *D. latum* into the waters of North America. Because many of the lake regions of the central U.S. already contained suitable intermediate hosts, establishment of the tapeworm occurred readily. The first patient with diphyllobothriasis in North America has been cited as a Swedish immigrant in 1879. Ward (1930), however, documented that specimens of the cestode were observed from Europeans in the U.S. at least

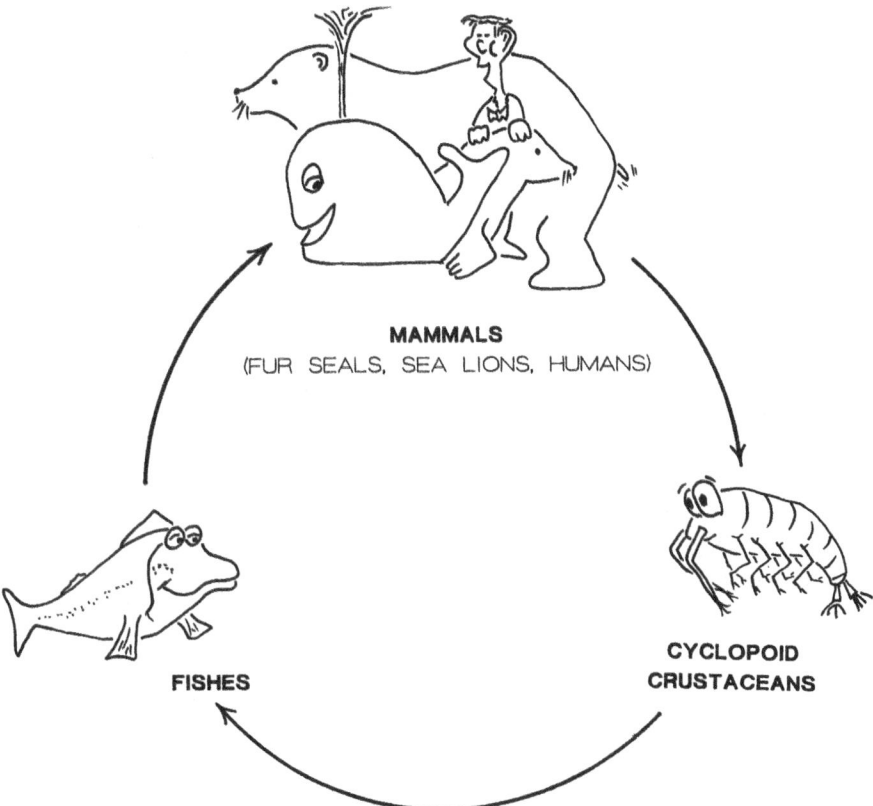

Figure 9.5 Generalized life cycle of *Diphyllobothrium* spp. The adult stage develops in a variety of mammals, including humans. Eggs pass out of proper definitive host in feces, they hatch, and the larvae are eaten by an appropriate copepod crustacean which in turn is eaten by a fish secondary intermediate host. The final host acquires the infection by ingesting a viable metacestode within the fish.

as early as 1858. It was not until 1901 that the first case was reported in someone born in America, although a locally acquired identified infection probably occurred about 1896 (von Bonsdorff 1977; Ward 1930).

Because of the Jewish appetite for "gefüllte fish" and minced fish spread, the Jews in the U.S. demonstrated an unusually high prevalence for the disease. Ironically, properly prepared "gefüllte fish" needs to be thoroughly cooked. Transmission of this disease as well as of heterophyiasis (discussed later) occurred during the preparation stages of the dish. Jewish housewives routinely taste the minced fish meat as various spices are added to ensure proper seasoning. According to de Carneri and Vita (1973), the estimated number of people transmitting *D. latum* in North America was fewer than 100,000.

Although most of human infections of *D. latum* in the U.S. are vectored by freshwater fishes such as pike, burbot, and perch, others are transmitted by salmonid fishes. Pacific salmons (*Oncorhynchus* spp.), which are initially infected in fresh water but spend most of their time at sea still harboring those infective metacestodes, are common vectors of diphyllobothriasis (Ruttenber et al. 1984; Turner et al. 1981). Human infections attributed to *D. latum* from *O. masou* and *O. gorbuscha* in Japan have been determined to represent a separate species, *D. nihonkaiense* (see Yamane et al. 1986).

Whether the incidence of diphyllobothriasis in the U.S. is increasing remains uncertain because the exact number of cases remains unknown (Deardorff and Kent 1989). During the last 4 years that Centers for Disease Control (CDC) monitored this disease (indirectly based on the number of requests for the drug niclosamide), the trend was up (Fig. 9.6). Although recommendations to make diphyllobothriasis reportable to CDC have been suggested, the disease still does not have to be reported.

Attempts to interrupt the tapeworm's life cycle to control diphyllobothriasis

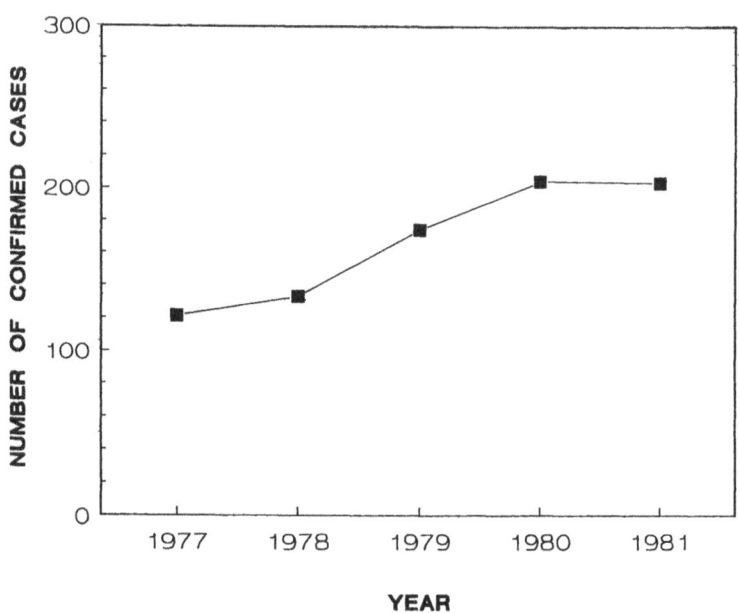

Figure 9.6 Diphyllobothriasis in the United States from 1977 to 1981 as documented by requests to the Centers for Disease Control for the drug niclosamide. (After Ruttenber et al. 1984.)

in the U.S. were initiated in 1923. Methods for preventing infections in fish were also investigated. "Brining" fish so that the concentration of salt in the flesh constitutes about 12% of the weight of the fish fillets and maintaining those fillets in the brine for at least 5 days is effective (Bylund 1976). Infective plerocercoids will die when heated at 129 to 132.8°F (54 to 56°C) for up to 5 minutes or cooled to 15.8–17.6°F (–8 to –9°C) for 8–72 hours. The amount of cooking or freezing necessary so that the internal temperature of the fish reaches those temperatures varies with the size of fillet or whole fish. Anthelmintics such as niclosamide, bithionol, and praziquantel have been found to eliminate most human infections effectively.

Diphyllobothrium latum is discussed as the representative pseudophyllidean tapeworm because it infects a large number of people, even though only a small but growing portion of those cases result from eating raw salmon. Several strictly marine pseudophyllideans also cause human infections as do other species of *Diphyllobothrium* that occur in freshwater and brackish water fishes. For the latter situation, we mention some cases infecting humans in Alaska. Margolis et al. (1973) reported *D. ursi,* a species that, unlike *D. latum,* infects the serosal surface of stomach in sockeye salmon rather than the fillet. Rausch and Hilliard (1970) reported that *D. dendriticum, D. lanceolatum, D. ursi, D. dalliae,* and *D. alascense* also infect humans in Alaska.

Humans abroad serve as hosts for pseudophyllidean species other than *D. latum* that normally infect marine mammals. Many of those infections are obtained from raw and inadequately cooked marine fishes. For example, marine fishes, including several sciaenids, are commonly used to prepare cebiche, also spelled "ceviche," which consists of raw seafish steeped in lime juice with peppers. In Peru where the South American sea lion heavily infected with *D. pacificum* occurs along the coast in abundance, seafood consumers eat ceviche in restaurants, on the street from hand carts, and in homes. Many of those people contract infections from cebiche made from sciaenids or possibly some other fishes (Baer 1969; Baer et al. 1967). On the other hand, Panamanians and people from localities along the Gulf of Mexico who enjoy the same dish prepared with similar fishes that never come in contact with the sea lion do not acquire the infection. Importation of fresh infected fish from Peru or Chile for raw dishes, however, would undoubtedly result in human infection from the metacestodes. The same cestode infects the fur seal in the northern hemisphere and has caused at least one infection in Japan (Kamo et al. 1986).

Several pseudophyllidean cestodes of marine mammals in addition to those indicated above can infect humans and can be expected to cause infections in the future. In Japan, where critical attention has been directed toward identifying the agents of human cases, the species include *Diplogonoporus grandis* (perhaps a junior synonym of *D. balaenopterae;* see Kamo et al. 1971),

D. fukuokaensis (see Kamo and Miyazaki 1970), an unidentified species of *Diplogonoporus* (Kamo and Miyazaki 1971), *Diphyllobothrium yonagoensis*, *D. cameroni*, and species conspecific or similar to *D. hians* and *D. elegans* (see Kamo et al. 1986).

DIGENEANS

Along with turbellarians and monogeneans, which do not infect humans, and the cestodes discussed above, which include members that do infect them, digeneans form the phylum Platyhelminthes, commonly known as flatworms. There are many families of these digeneans, also referred to as trematodes or flukes, and several of them have members that infect humans. These soft-bodied flukes come in a wide variety of sizes and shapes, and as adults they inhabit most tissues in vertebrates. Almost all digeneans use a mollusc as the first intermediate host in which they undergo asexual reproduction, but only a relatively small portion of them uses fishes as second intermediate hosts. Of those that encyst or become encapsulated in fish, few have been reported from humans. The stage (metacercaria) in fishes infective to humans cannot be transferred from an intermediate fish host to a transport fish as occurs for a few tapeworms and for many nematodes and acanthocephalans (a phenomenon called paratenesis for which the infective worm does not develop after entering an acceptable host). Several additional examples are discussed below.

Most of those digenean species that have a metacercaria in fishes and infect people in the U.S. belong to the family Heterophyidae. Nearly all of the heterophyid members have a snail as first intermediate host from which minute cercariae are released. These cercariae penetrate or less often are eaten by specific fishes or groups of fishes in which they encyst in rather specific sites. When the proper avian or mammalian host eats the prey containing the developed encysted metacercariae, the worm matures in the vertebrate's intestine and deposits a relatively small operculate egg. Most species do not reach over 1–2 mm long, and each can infect several species of definitive hosts, usually more than the number of fishes serving as second intermediate hosts. Because of this ability for most heterophyids to infect a variety of final hosts, humans probably serve as final hosts for numerous different heterophyid species. Africa and co-workers reported that a few of these heterophyids produce eggs that filter through the human intestinal wall, gain access to the mesenteric lymphatic system, and accumulate in the cardiac valve and myocardium or in the central nervous system including the brain (e.g., Africa et al. 1935, 1936, 1937). Occasionally, adult specimens also end up in those tissues, such as encapsulated ones in the brain (Deschiens et al. 1958), and, according to Kean and Breslau (1964), who reexamined autopsy tissues for digenean eggs, 15% of those cases from fatal heart attacks in the Philippines may have resulted from heterophyid myocarditis.

Although members of the genus *Heterophyes* apparently do not occur naturally in the U.S., other heterophyids probably affect certain people in a similar manner. Normally, moderate infections produce mild irritation, colicky pain, mucous diarrhea, and superficial necrosis of the mucous coat. Reports of transmission of heterophyes-like intestinal flukes from sushi are reported in the U.S. (Adams et al. 1986). In Hawaii, Alicata and Schattenburg (1938) determined that *Stellantchasmus falcatus,* a fluke from mullet, infected a human who frequently consumed "Japanese salad," a dish consisting of raw fishes (sometimes mullet), fresh lettuce, and soya sauce. A subsequent survey of people in Honolulu revealed about 7–8% of the individuals of Filipino, part-Hawaiian, and Hawaiian ancestry plus two cases of those of Chinese and Japanese ancestry had infections (Ching 1961). Alicata also identified the heterophyid *Diorchitrema pseudocirratum* in native Hawaiians who had eaten local raw mullet (Beaver et al. 1984). Apparently, the mullet constitutes the primary marine fish having public health significance with regard to heterophyiasis in Hawaii. *Mugil cephalus* in Hawaii also hosts *Centrocestus formosanus* and *Haplorchis taichui,* heterophyids known to infect humans in the Far East (Alicata 1964).

Mullets also present potential public health problems in the southeastern U.S. In fact, Welberry and Pacetti (1954) found heterophyid eggs in a child from South Florida with a history of eating smoked mullet and assumed the infection represented *Heterophyes heterophyes* transmitted by mullet that had transferred worms to Florida by short relays. Actually, mullet in that region contain the native *Phagicola longa* (Fig. 9.7; also see Hutton and Sogandares-Bernal 1959; Overstreet 1978). The eggs in *P. longa* are similar but somewhat larger than those found in the child. Nevertheless, there exists in the region at least one other and probably a few other heterophyids (Paperna and Overstreet 1981), all with similar small operculated eggs and many potentially capable of infecting humans. A large number of heterophyid metacercariae can infect mullet. Paperna and Overstreet (1981) mentioned 6,000 metacercariae per gram of flesh as the highest recorded. Keeping in mind that some metacercariae can survive methods of preparation using "minimal amounts of heating" (Paperna and Overstreet 1981), even a small percentage of cysts could cause disease in humans, and a periodic diet of such mullet could maintain a heavy infection. A careful study of those coastal residents of the southeastern U.S. who eat raw, minimally cooked, especially thick-bodied, or cold smoked fish, would lead us to suspect that many would harbor a variety of heterophyids as well as a few other digeneans.

Font et al. (1984) considered *Phagicola nana* to be a potential public health problem because the metacercariae encyst in the somatic musculature of bass and sunfishes. These infected centrarchid fishes occur in coastal regions of the

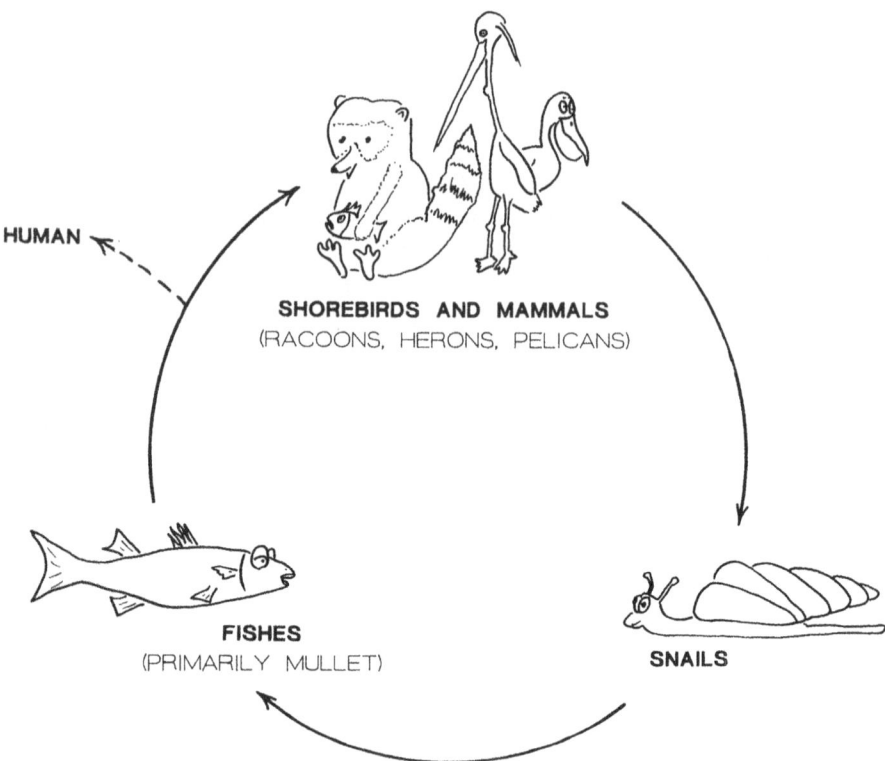

Figure 9.7 Generalized life cycle of *Phagicola longa*. Adult flukes infect several species of shorebirds and mammals. The eggs are introduced into the aquatic environment, and are ingested by snails, which are in turn eaten by a suitable fish. Humans are infected by eating fishes contaminated with variable metacercariae encysted in the edible tissues.

eastern and southern U.S., and seafood consumers could get infected by eating the product raw or inadequately prepared. The authors speculated that some infections with *P. nana* probably have occurred, but had gone undetected because of possible absence of symptoms (Fig. 9.8), failure of individuals to consult physicians, or short life span of adult worms.

Digenean infections also occur in Eskimos in Alaska and in portions of Canada. These have not been identified, but at least *Metorchis* sp., *Cryptocotyle* sp., and a species with smaller eggs have been reported (e.g., Freeman and Jamieson 1976).

The troglotrematid *Nanophyetus salmincola* has long been treated in textbooks as the vector of the rickettsial agent for "salmon poisoning" disease of

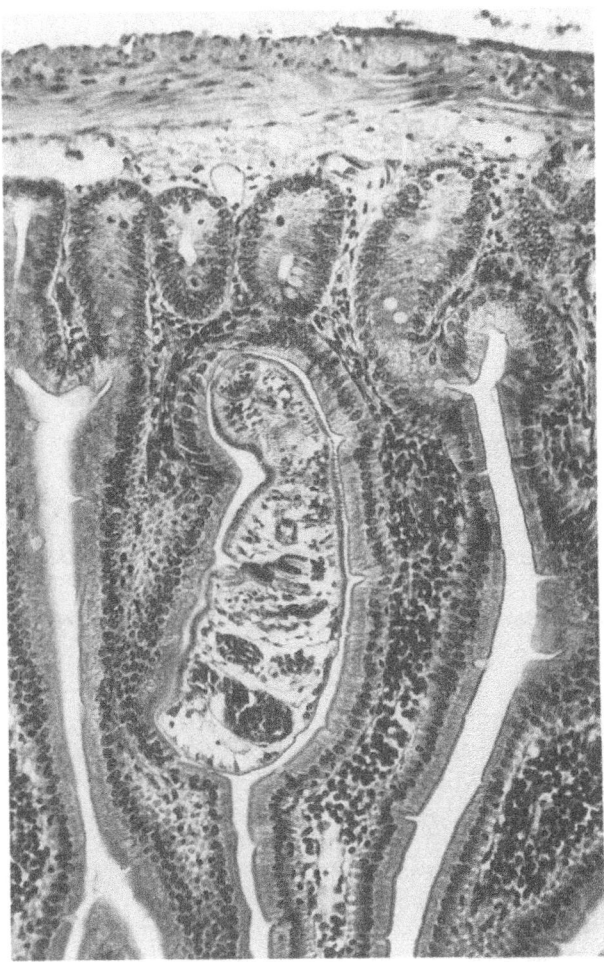

Figure 9.8 Histological section showing adult heterophyid digenean *Phagicola nana* loosely embedded in intestine of hamster. Note lack of pathological response in this experimental host. (From Font et al. 1984.)

dogs and other canines; it now is also known for causing human disease in the U.S. Most canines infected with this parasite, if untreated, die from the accompanying rickettsial infection. The human disease is probably neither fatal nor associated with the rickettsia like in dogs. A generalized life cycle of *N. salmincola* is shown in Figure 9.9.

Small in size like most heterophyids (Fig. 9.10, about 1 mm long), *N.*

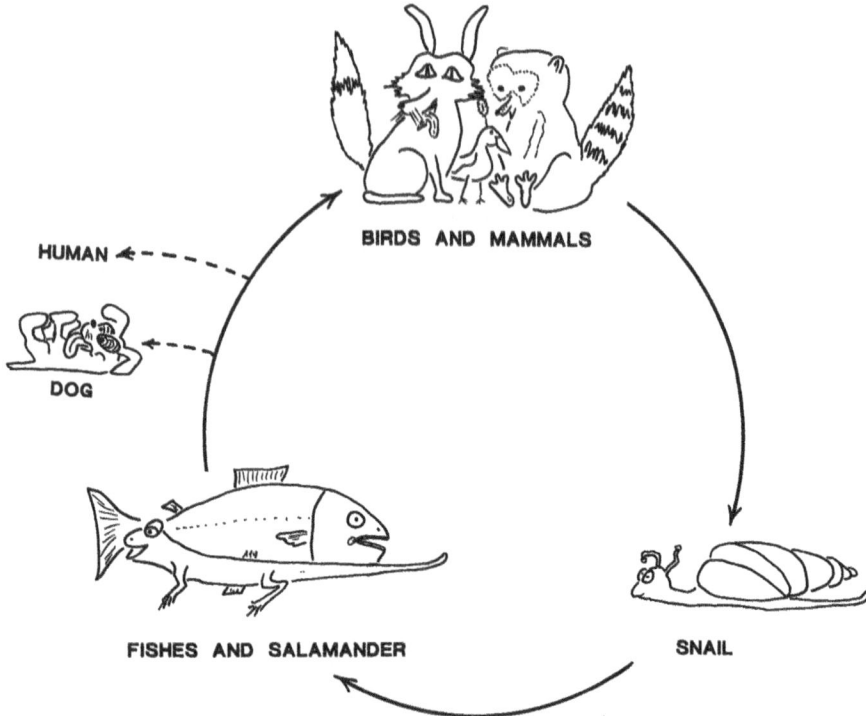

Figure 9.9 Generalized life cycle of *Nanophyetus salmincola*. Adults are found in the intestine of birds and mammals. Eggs are deposited in the water and hatch a miracidium, which actively penetrates the specific molluscan intermediate host [*Juga plicifera* (= *Oxytrema silicula* and *Goniobasis plicifera* var. *silicula*)]. Eventually, xiphidiocercariae are released from the snail into the water and penetrate into fishes and the Pacific giant salamander. This stage subsequently develops in the tissues into the infective metacercariae (see Figure 9.11). Humans are infected by ingesting flesh or eggs with the metacercariae. If canines ingest metacercariae that are naturally hyperparasitized with a rickettsiae *(Neorickettsiae helminthoeca),* they contract "salmon poisoning disease" and often die.

salmincola can produce infections with diarrhea, peripheral blood eosinophilia, abdominal discomfort, nausea, vomiting, weight loss, and fatigue (Eastburn et al. 1987). Symptoms of human disease in eastern Siberia, where prevalence values for infections now reach up to 98% of the inhabitants in some villages, are thought to necessitate a minimal threshold infection of about 500 or more individual worms (Sinovich 1959). The reason for the sudden appearance of infections of *N. salmincola* in the U.S. is the increased consumption of raw, incompletely cooked, and smoked salmon and steelhead trout that contain the metacercaria (Fig. 9.11). Infections may also be caused

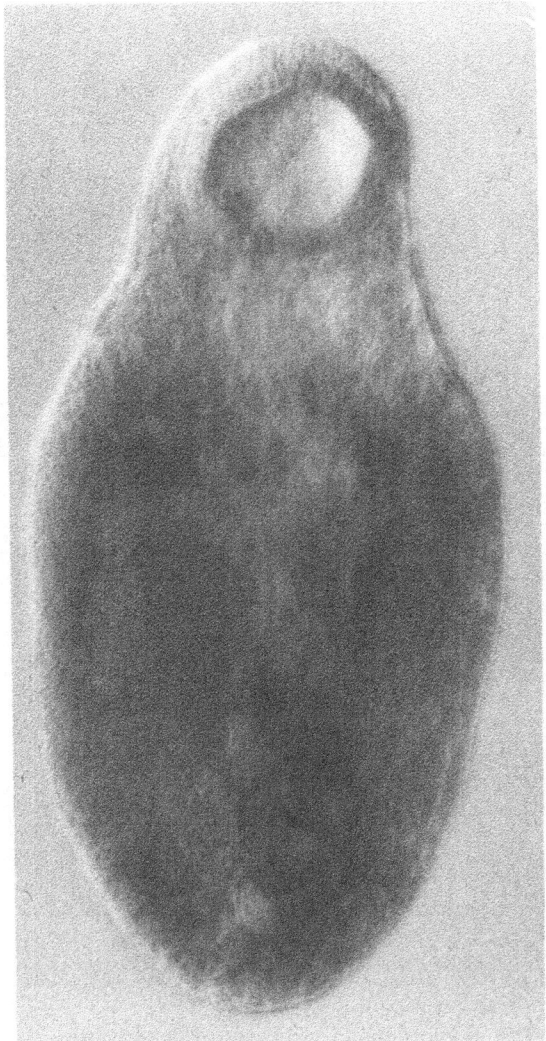

Figure 9.10 The troglotrematid digenean *Nanophyetus salmincola:* Adult stage, about 1 mm long. (Original provided by T. R. Fritsche.)

by eating eggs of those fishes. Praziquantel (20 mg/kg three times a day for 1 day) appears efficacious (Fritsche et al. 1987) in treating infections; however, niclosamide and bithionol also proved effective, and symptoms disappeared slowly over several months without any treatment (Eastburn et al. 1987).

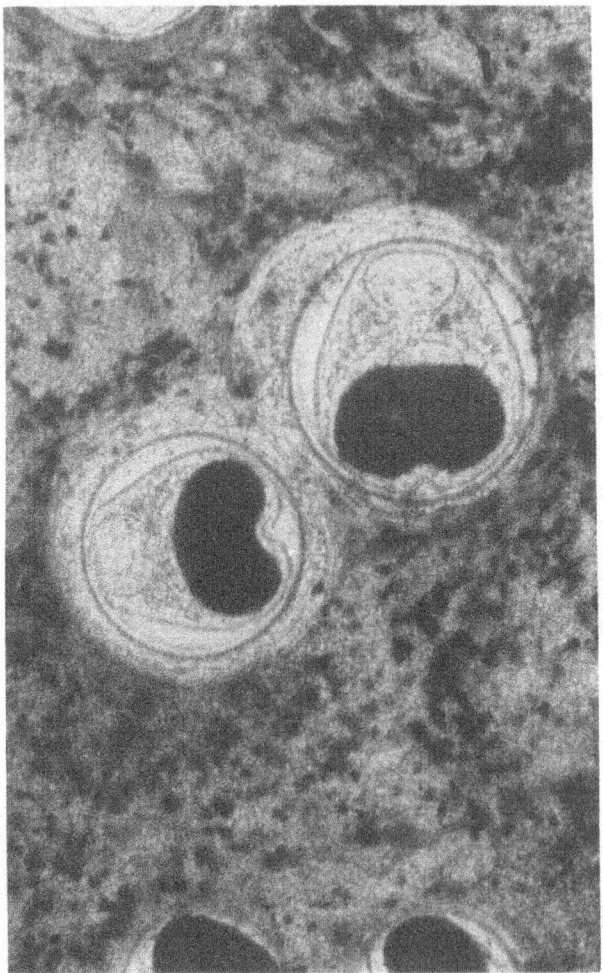

Figure 9.11   The troglotrematid digenean *Nanophyetus salmincola:* infective metacercariae in fish. (Original provided by T. R. Fritsche.)

NEMATODES

Of the seafood-associated zoonoses discussed in this chapter, none has received more media attention in recent months than human anisakiasis. The disease refers to infection by the third-stage juvenile of a few different anisakid nematodes, or "roundworms." Humans may become infected with the parasites by consuming raw or inadequately prepared seafoods. Ingested juveniles may penetrate into (Fig. 9.12) or through the tissues of the gastrointestinal tract of the

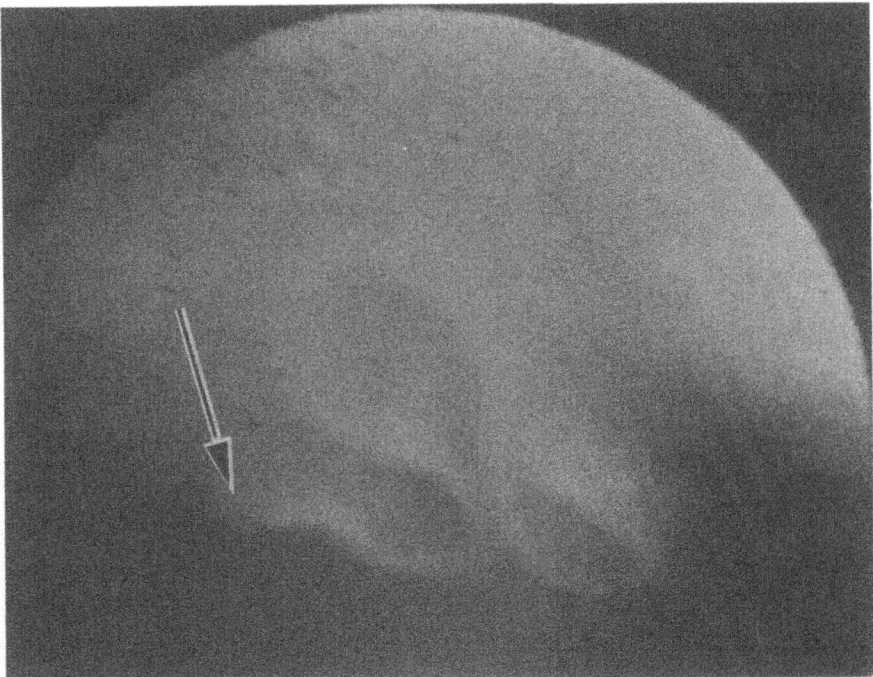

Figure 9.12 Endoscopic photograph of a juvenile *Anisakis simplex* at the greater curvature of patient's stomach; arrow indicates erosion associated with attached anterior end of worm. Bar = 0.5 cm. (After Deardorff et al. 1986.)

host. Although some juveniles have been recovered from the small and large intestine, pancreas, mesentery, and other tissues (Yokogawa and Yoshimura 1967), the majority of juveniles remain within the mucosa and submucosa of the alimentary canal.

In the U.S., the third-stage juveniles of two species, *Anisakis simplex* and *Pseudoterranova decipiens* (= *Phocanema d.*), account for all the reported cases. Over 50 suspected or confirmed cases of anisakiasis have been reported in the U.S. and the number of cases has been increasing (Fig. 9.13). Because anisakiasis is not considered a reportable disease by CDC, we may be looking at the tip of the iceberg. Simply relying on reports of outbreaks in the scientific literature as a barometer for anisakiasis may give a gross misrepresentation of actual occurrences. We doubt that reliable data exist for the incidence of this zoonosis.

Of the two species implicated in anisakiasis, infections with juvenile *Anisakis* are most numerous and, comparatively, more serious. (The name of this disease is derived from the genus *Anisakis*.) The common name for infections with *A. simplex* is "herringworm disease," which refers to an intermediate host of

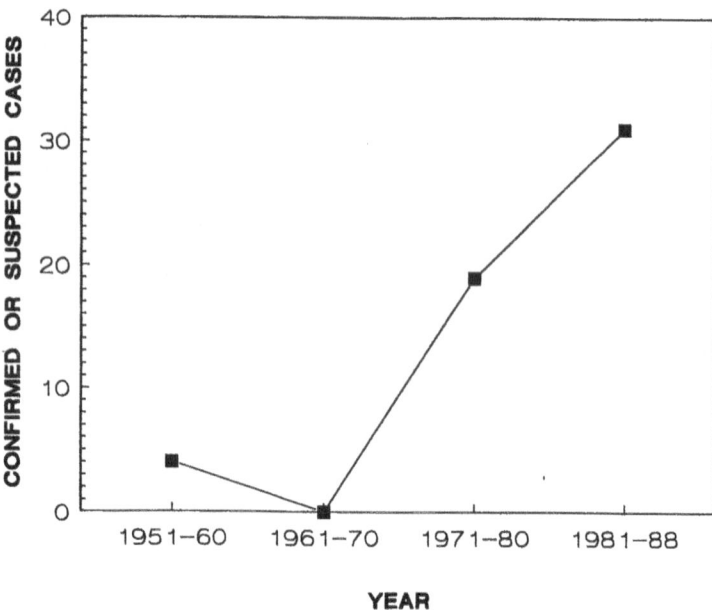

Figure 9.13 Confirmed or suspected cases of anisakiasis in the United States from 1951 to 1988.

the worm in Europe and elsewhere. In the U.S., transmission of the disease has been linked to Pacific salmon (*Oncorhynchus* spp.), rockfishes (*Sebastes* spp.), yellowfin tuna *(Thunnus albacores)*, skipjack tuna (*Euthynnus pelamis* [= *Katsuwonus p.*]), and imported unidentified squid. The majority of cases of human anisakiasis in the U.S. result from home-prepared meals; however, restaurants have been confirmed to be involved in the transmission process (see Deardorff et al. 1987). The juveniles are cream colored and relatively small and therefore rather difficult to detect in tissues.

The clinical features evoked by these juveniles generally involve a sudden and severe onset of epigastric pain, often associated with nausea and vomiting. Symptoms first occur 1–12 hours after eating raw seafood. As the disease moves from the acute to the chronic stage, the epigastric distress becomes vague, occult blood may be seen in the gastric juice or stool, and hematological findings usually reveal a high peripheral eosinophil count. Infective juveniles may evoke a wide variety of clinical manifestations, making infections difficult to diagnose. Also, juvenile *A. simplex* have been diagnostically confused with and clinically associated with malignant cells (Hayakawa et al. 1970; Tsutsumi and Fujimoto 1983).

Endoscopic identification and removal of worms is the most reliable method of diagnosis and treatment for worms in the stomach (see Deardorff et al. 1986a). The endoscope also is useful in those regions of the intestine accessible to the

equipment (e.g., duodenum and descending colon). Should the juvenile penetrate into inaccessible areas of the intestines, the need for laparotomy may be
indicated (e.g., Sakanari et al. 1988).

The signs and symptoms of infections with juvenile *Pseudoterranova decipiens*, commonly called "codworm disease" (because of its occurrence in fillets
of Atlantic cod) or "sealworm disease" (because seals serve as definitive hosts),
are not as profound as with juvenile *A. simplex*. If swallowed, this worm is rarely
invasive in humans, and serious consequences are avoided when specimens are
expectorated by the consumer. Several cases report patients that experience a
"tingle" in their throat, followed by coughing up a worm (e.g., Deardorff et al.
1987; Chitwood 1975; Juels et al. 1975; Kates et al. 1973; Kliks 1983; Lichtenfels and Brancato 1976; Olson et al. 1983). The size and strength of the juvenile
probably account for its ability to migrate up the esophagus and irritate the
throat. This phenomenon is termed the "tingling throat syndrome." Cod worms,
although found in a small portion of fish sold to the public, grow large, often
appear red, and are noticed periodically by potential consumers (Fig. 9.14). The
unsightly appearance encourages consumers to reject infected fillets, to refrain
from purchasing any seafood product, to tell other consumers of the experience,
and to complain to retailers and other involved personnel.

The cod industry tried to reduce the number of worms in cod fillets by
careful visual inspection of fillets, followed by discarding infected areas. No
matter how carefully inspected, some worms avoid detection (Fig. 9.14). Candling fillets, a method of visual inspection commonly used by the cod industry,
involves shining a bright light through the fillets to detect the worms. A study on
the efficacy of candling in Canadian fish plants demonstrated that under commercial working conditions the technique was only about 50% effective in
detecting cod worms (Honans and MacFarlene 1957).

The life cycle of anisakid nematodes involves marine mammals (Fig. 9.15).
Basically, eggs are expelled by a mature female nematode into the gastrointestinal tract of the mammalian host (e.g., whales, seals, and dolphins) and passed
out with its feces into the water. Development of the embryo to first-stage
juvenile and then to the second stage occurs within the egg; the second-stage
juvenile emerges from the egg within a few days. This free-swimming juvenile
may be eaten by an intermediate host or by a crustacean transport host in which
no juvenile development occurs. These transport crustaceans such as copepods
and shrimp transport the worm to its true intermediate host. When the crustacean
is ingested by that acceptable invertebrate or fish intermediate host, the worm
migrates to the hemocoel or mesentery where it develops into a third-stage
juvenile. The third-stage juvenile can infect marine mammals and humans. It
will molt and mature if consumed by an acceptable definitive host. The human
interrupts the cycle by eating the intermediate host, but does not serve as a
definitive host for any anisakiasis-causing parasite.

The relationship between the presence of marine mammals and human
infections is evident. Fishes are more infected in areas where mammalian

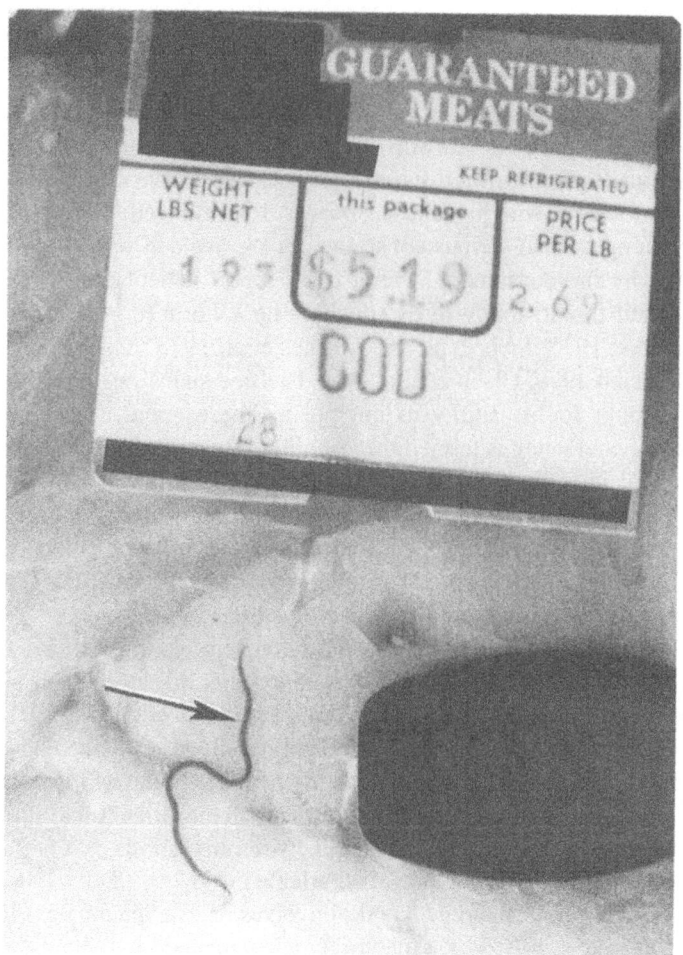

Figure 9.14 Viable juvenile cod worm *(Pseudoterranova decipiens)* in commercially packaged cod. (Original provided by L. A. Jensen.)

definitive hosts are abundant. Surveys for helminths of marine fishes and invertebrates have been conducted for the Atlantic coast (Cheng 1976; Jackson et al. 1978), the Gulf coast (Deardorff and Overstreet 1981; Norris and Overstreet 1976), the Pacific coast (Myers 1979), and the waters near the Hawaiian Islands (Deardorff et al. 1982). Fishes caught off the Pacific coast have a greater worm burden than fishes caught in the other survey areas. When examining these regions for abundance of marine mammals, observers find most along the Pacific coast. Where geographical information was known, 86% of human infections with *Anisakis* juveniles occurred in the western U.S., and 14% occurred in the eastern U.S. (Deardorff et al. 1987).

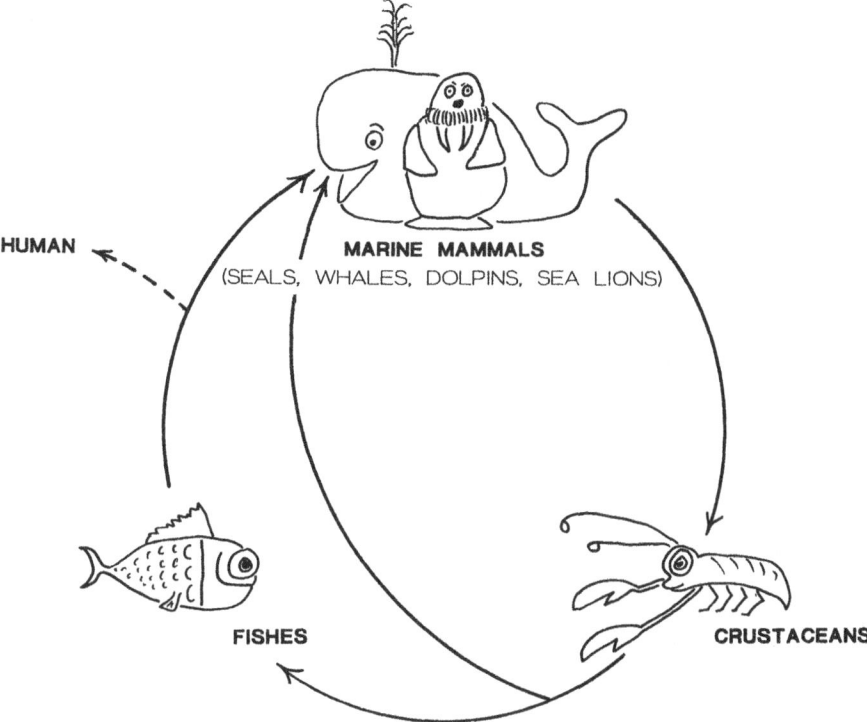

Figure 9.15 Generalized life cycle of a marine ascaridoid. This large group of nematodes has adults that infect the stomach or intestine of a variety of marine mammals and juveniles that encyst in fishes or crustaceans. The juvenile stage that encapsulates in fishes has been confirmed to infect humans.

Since the enactment of the Marine Mammal Protection Act of 1972, the population of marine mammals (e.g., California sea lions, elephant seals, and harbor seals) has increased to numbers not seen since the 19th century (McKer-row et al. 1988). Associated with those increases in seal populations was a substantial increase in both the prevalence and intensity of juvenile nematodes, especially *P. decipiens,* in fillets of cod (Chandra and Khan 1988). This scenario of increasing numbers of mammals and the corresponding increase in number of infected fishes suggests that U.S. seafood consumers could be increasingly endangered.

Evisceration of some infected fishes after capture will lessen their worm burden because most infective ascaridoid worms reside in the viscera. Table 9.1 shows that 75% of the anisakid juveniles were encapsulated in the viscera of a rockfish; therefore, removing viscera from this fish would eliminate the majority of those helminths. Salmons, unfortunately, are the exception; they usually have more worms in the edible musculature than in the viscera (Table 9.1).

Table 9.1   Number of Specimens of *Anisakis simplex* Located in Musculature and in Viscera of Whole Fishes

Type	Muscle	Viscera	Total
Salmon[a]	554	180	736
Rockfish[b]	682	1,618	2,300

[a]*Oncorhynchus nerka*, sockeye salmon, 32 specimens.
[b]*Sebastes pinniger*, canary rockfish, 32 specimens.
Adapted from Deardorff and Throm (1988).

Van Thiel et al. (1960) originally hypothesized that juveniles encapsulated in the viscera migrated to edible portions of the fish after the fish was caught. In an extensive epidemiological study, the authors noted that before 1955 no cases of anisakiasis had been reported in the Netherlands. During that time period, commercial herring fishermen gutted and cured herring at sea. The occurrence of anisakiasis dramatically increased in 1955, when herring began being gutted and cured ashore (probably a result of advanced refrigeration technology used on the boats). Figure 9.16 documents proven cases of anisakiasis from 1955 when immediate gutting was terminated to 1967 when legislative actions required

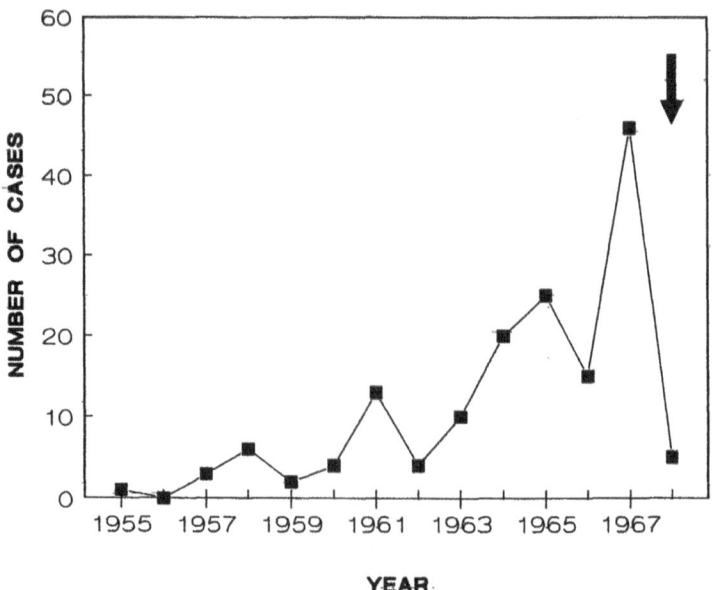

Figure 9.16   Number of confirmed cases of anisakiasis in the Netherlands from 1955 through 1968. Arrow indicates the year legislative action (called the "Green Herring Laws") against this zoonosis was instituted. (After Ruitenberg 1970.)

freezing to kill the worms. Van Thiel et al. (1960) concluded that the worms encapsulated principally in the herring viscera and that the practice of immediate evisceration removed most parasites. Viscera not immediately removed began to decompose, and the worms migrated into the flesh. Therefore, the consumer was exposed to a substantially greater number of worms. This migratory phenomena for anisakid nematode juveniles has been well documented (Deardoff et al. 1984; Hauck 1977; Smith and Wootten 1975; van Thiel et al. 1960; Vik 1966). Because evisceration reduces the number of anisakids available to infect fillets and therefore reduces the public health hazard, we recommend eviscerating all fishes—including salmonids—immediately after capture.

Until recently, only those ascaridoids that mature in marine mammals have been considered to be a public health risk. It is probable, however, that at least members of two genera of ascaridoids that mature in marine fishes may infect humans. In the U.S., *Hysterothylacium* Type MB from a wide range of commercially important fishes and penaeid shrimps occurring from at least Texas to Florida in the Gulf of Mexico (Deardorff and Overstreet 1981) produces a pathological response in both rodents and the rhesus monkey (Norris and Overstreet 1976; Overstreet and Meyer 1981). Although the juvenile stage can probably produce hemorrhaging and an eosinophilic response in humans, the primary danger for cases of moderately intense infections of those individuals not immunocompromised or hypersensitive to ascaridoid infections would probably involve misidentification and treatment of the disease. Also, a juvenile reported as *Terranova* Type HA from Hawaiian bony fishes by Deardorff et al. (1983), which may mature in elasmobranch fishes, induced a histopathological response in rats (Fig. 9.17). Problems involving human hypersensitivity responses to exposure to a variety of ascaridoids and their secretory and excretory products require additional research.

When a parasite invades tissues and fluids, excretory or secretory products (ESP) are released. Those ESP may stimulate or modulate the immune responses of the infected host. The effects of ESP are extremely diverse. For example, ESP may protect the host by stimulating immune responses against the worm or protect the worm by allowing it to evade the host response, stimulate chemotaxis of neutrophils or eosinophils, induce cytotoxic or cytostatic effects on cells, or contribute to the resulting immunopathogenesis. Whatever the bioactivity of a parasite's ESPs, the antigens appear to have sufficient parasite specificity to serve as a valuable starting point for the development of immunodiagnostic tests and vaccinations against helminth infections. Of the seafood transmitted zoonoses that occur in the U.S., the bioactivity of the ESPs of juvenile *Anisakis simplex* has been the subject of many investigations, such as inhibitory effects (Raybourne et al. 1983, 1986), protease activity (Matthews 1982; Sakanari and McKerrow 1988), and serodiagnostics (Bier and Raybourne 1988; Deardorff et al. 1986a; Desowitz et al 1985; Sakanari et al. 1988), and are viewed by Bier et al. (1987).

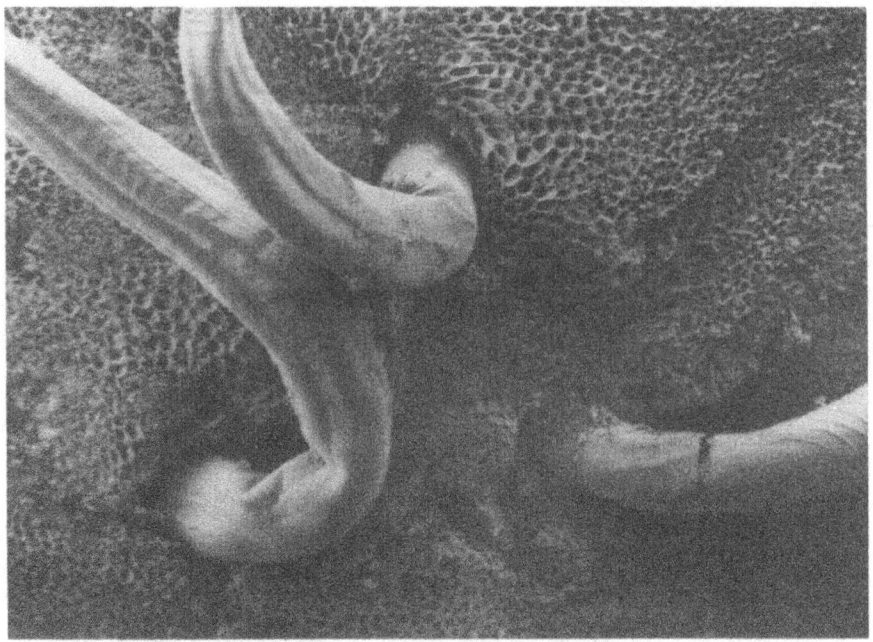

Figure 9.17   Scanning electron micrograph showing three juvenile specimens of *Terranova* Type HA from Hawaiian marine fishes producing a pathological response in a rat stomach 3 days after being gavaged into the host. (From Deardorff et al. 1983.)

For additional information on anisakiasis with a more global prospective, readers are referred to reviews by van Thiel et al. (1960), Ruitenberg (1970), Oshima (1972, 1987), Smith and Wootten (1978), Margolis (1977), Bier et al. (1987), Hafsteinsson and Rizvi (1987), and Sakanari and McKerrow (1989).

As with some of the ascaridoids, some species of other nematode groups also use fishes as paratenic hosts. As indicated above, those nematodes in a piscine paratenic host do not require the fish to develop into the stage infective to the definitive host. That parasite may infect a human, who might be a definitive host, another paratenic host in which no development occurs, or an accidental host in which the worm does not necessarily survive. Actually, populations of many parasites are dependent on paratenic hosts for their perpetuation in nature. Examples of juvenile nematodes other than ascaridoids that humans have acquired by eating paratenic fish hosts are *Dioctophyma renale, Eustrongylides* spp., *Dracunculus insignis, Angiostrongylus cantonensis,* and agents for gnathostomiasis and capillariasis.

Americans have been infected with *Dioctophyma renale* (Figs. 9.18 and 9.19) and *Eustrongylides* spp. These worms have similar life cycles (Fig. 9.20),

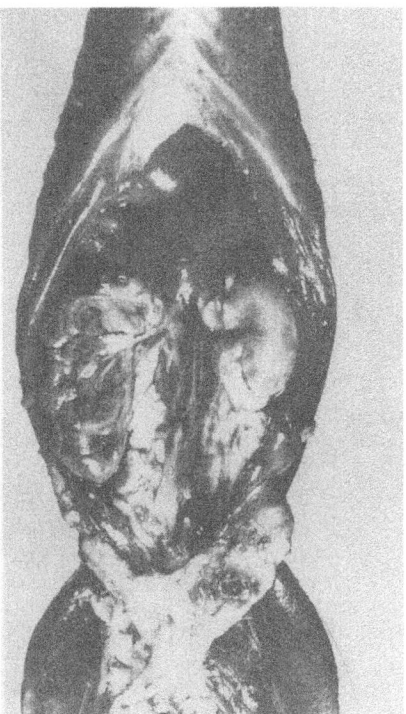

Figure 9.18   Adult stage of the giant kidney worm, *Dioctophyma renale,* in a mink from a brackish marsh in southern Louisiana. Photograph shows greatly enlarged right kidney containing male and female worms. (From Overstreet 1978.)

all of which necessitate oligochaete intermediate hosts (Karmanova 1968; Mace and Anderson 1975; Measures 1988a). At least some fishes, reptiles, and amphibians can serve as paratenic hosts. For at least *E. tubifex* but not *D. renale,* the fish is a necessary true (second) intermediate host in which the juvenile develops into its fourth stage before becoming infective to birds. Whereas all members of this group initially develop in the ventral blood vessel of select aquatic oligochaetes, the juveniles infective for their mammalian or avian definitive hosts are usually much larger in size. They typically grow from about 1 mm to as much as 12 mm after one or no molts, depending on the species. The unusually large reddish third- or fourth-stage juveniles (or larvae, as they are often but incorrectly termed) makes the worms quite conspicuous and therefore probably prevents many humans from consuming them and acquiring infections. When *D. renale* occurs in humans, which is rare, it typically matures in the kidney like in the mink *(Mustela vison)* and other carnivores. In one case, however, what appeared to be a juvenile of this species occurred in a subcutaneous nodule in the chest of a 26-year-old man in California (Beaver and

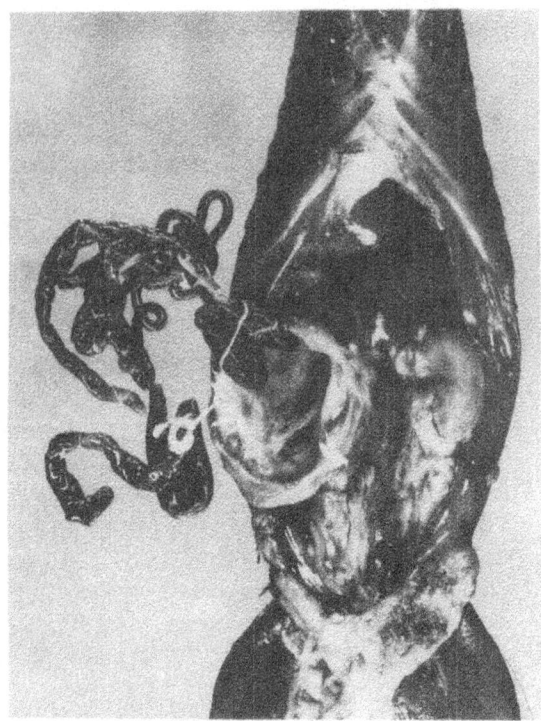

Figure 9.19   Adult stage of the giant kidney worm, *Dioctophyma renale,* in a mink from a brackish marsh in southern Louisiana. Photograph shows same mink as in Figure 9.18 with kidney cut open to reveal worms, kidney devoid of any parenchymal tissue, and bony spicules lining the empty capsule. (From Overstreet 1978.)

Theis 1979). The three or more species (Measures 1988b) of *Eustrongylides* each mature in two to four orders of piscivorous birds, rather than in mammals. Fig. 9.20 depicts a generalized life cycle. Worms typically mature entwined within the stomach wall, primarily in serosal tissue, but occasionally elsewhere in the peritoneal cavity or in the lumen of the alimentary tract. In some instances, heavy infections that possibly may be associated with additional bacterial involvement have caused mass mortalities of the normal bird hosts (e.g., Roffe 1988). Juvenile worms occur in a variety of fish hosts. According to an Associated Press news release, when clients of a tavern in Maryland swallowed live baitfish while drinking beer, reminiscent of the Roaring 1920s practice of gulping goldfish, three of those clients developed peritonitis from the juvenile worms and two required emergency abdominal surgery. Guerin et al. (1982) and Gunby (1982) reported on what may have been the same cases, identifying the patients as fishermen who swallowed their live bait minnows when fish were not

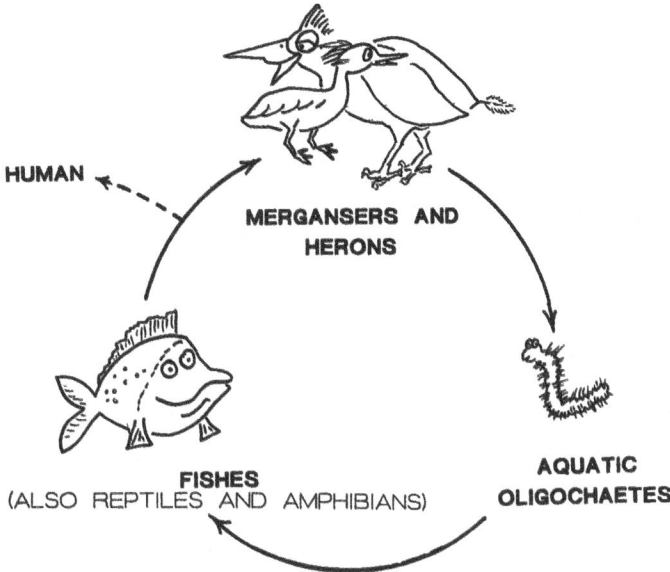

Figure 9.20 Generalized life cycle of *Eustrongylides* spp. Adult worms infect piscivorous birds, and juveniles infect aquatic oligochaetes and both freshwater and estuarine fishes. Fishes, reptiles, and amphibians may serve as paratenic hosts. Humans become infected by eating contaminated fishes.

biting. A case similar to the above infections involving fishermen has occurred in New Jersey (Eberhard 1989). We certainly do not recommend that fishermen, tavern patrons, or others eat live or fresh "bait-fish" during periods of boredom or hunger. One other report refers to a college student who became infected with a juvenile *Eustrongylides* after eating home-prepared sushi in New York City (Schantz 1989; Wittner et al. 1989).

All the above dioctophymatids have a negative impact on the corresponding fisheries because infected fish may appear lumpy, may have the reddish juveniles embedded in or penetrating through the flesh, may have holes or other lesions in the flesh depicting previous paths or sites of the worms, or may be encapsulated in the body cavity. Because the worms are large and reddish, they are conspicuous. We have seen these infections in both freshwater and marine fishes, including specimens in natural habitats, in ponds (e.g., striped bass and catfish), and in aquaria in which inhabitants had been fed wild gambusia and killifishes. To eliminate infections of *Eustrongylides* spp. from ponds in the southeastern U.S., the green-backed heron and mergansers especially should be prevented from defecating into the ponds [accomplished by using bird nets over water (see Fig. 9.29)]. Small wild fishes or oligochaetes should not be fed live to fish in aquaria or raceways.

Wild fishes that harbor large juvenile specimens of *Eustrongylides* spp. can present an aesthetic obstacle. Cooper et al. (1978) determined that recreational and commercial fishermen did not have to become concerned about those juvenile worms migrating through the fillet portion of the catch if they kept the fish cold with ice of other means. In any event, infected fishes should not be eaten raw.

The other nematodes indicated above as being potentially infective to humans in the U.S. and occurring in fish paratenic hosts are likely to infect consumers now that new cuisines are being tried, a larger number of Oriental immigrants have moved into the U.S., and more people are eating more seafood. On the other hand, local cases have not been reported. The guinea worm, *Dracunculus medinensis,* infects large numbers of people and other mammals in the Old World who drink water containing infected cyclopoid copepods. In the U.S. and Canada, *Dracunculus insignis,* which has been considered by some as a synonym of *D. medinensis,* infects primarily the racoon and mink. These infected mammals occur in both fresh and brackish waters of the U.S., and rather than infections originating from feeding on copepods, which is possible, infections are probably acquired from paratenic hosts such as frogs and fish (Crichton and Beverley-Burton 1977).

Angiostrongylus cantonensis, a lungworm of rats, causes eosinophilic meningitis and occasionally eye disease in humans. Nearly all patients develop a severe headache, and many exhibit convulsions, weakness of arms or legs, paresthesia, vomiting, facial paralysis, stiffness of the neck, and fever (Beaver et al. 1984). The disease can be fatal to humans and can be transmitted by a variety of paratenic hosts; however, infections transmitted by marine of estuarine fishes have not been confirmed. Before the causative agent was established, Rosen et al. (1961) hypothesized that the disease was caused by a helminth parasite in skipjack tuna or some other pelagic fish that was commonly eaten raw. Intermediate hosts actually include a variety of slugs, snails, and other gastropods. Nevertheless, some shrimp and possibly fishes (discussed later) serve as paratenic or carrier hosts. Human infections from Tahiti have been traced to eating dishes with raw prawns (*Macrobrachium* sp.), "taioro," and related recipes (Alicata and Brown 1962). Taioro consists of the hepatopancreas (digestive gland) and surrounding tissues of the prawn ground up in fresh water and mixed with grated coconut. The nematode has become established in Hawaii, which is one of several localities in the Pacific region where human infections have been reported (Alicata 1962; Beaver et al. 1984). Since 1981, the worm has been recognized in the western hemisphere, first in Cuba, then in Puerto Rico, and most recently in *Rattus norvegicus* in New Orleans, Louisiana (Campbell and Little 1988). Other species in the genus also occur in the U.S., and one of those, *A. costaricensis,* has caused a zoonotic intestinal infection in children in Costa Rica and abdominal infections in Mexico and South America. The species ranges at least into Texas in a natural host (the cotton rat, *Sigmodon hispidus*).

Where seafood products are concerned, health workers should be aware of a potential for the presence of infective juveniles of these rat metastrongyles in pond-reared finfishes and shrimps. When employees of culture facilities distribute commercial chow to fish, excess food usually remains spread around the ponds. Rats in the area may be infected because fish farms offer proper conditions for establishment of the terrestrial gastropod intermediate hosts. The excess food attracts the rats, which if infected can infect gastropods near the ponds, which in turn can release infective juveniles capable of being washed into the ponds and accidentally or intentionally ingested and then accumulated by the fishes, shrimps, or other cultured stock.

Humans who eat undercooked fishes containing third-stage juveniles of *Gnathostoma* spp. may become infected. The juveniles usually migrate to the superficial layers of the skin; however, they have been reported in muscles, the eye, and central nervous system. Bennish et al. (1986) reported a point source outbreak of gnathostomiasis in the U.S. by individuals attending a dinner.

Other potential nematode agents such as *Capillaria* spp. may also be transmitted to humans through fishes. As different cuisines incorporate different fish products from different localities, we expect to encounter new zoonotic infections with nematodes as well as with other parasites in the U.S.

ACANTHOCEPHALANS

Two reports of juvenile acanthocephalans infecting Alaskans have been published. All spiny-headed worms, as members of this phylum are commonly called, use crustaceans as first or only intermediate hosts. Some species incorporate piscine paratenic hosts in their cycles, and some cycles require that host to ensure its transmission to the appropriate cold- or warm-blooded vertebrate definitive host. Figure 9.21 depicts a generalized life cycle. Schmidt (1971) reported a single juvenile specimen of *Corynosoma strumosum* from the stool of a male Eskimo from Chevak, and he speculated that the patient became infected by eating raw fish containing an encysted worm. Margolis (1982) stated that, because the juveniles of several species of *Corynosoma* occur in salmon, probably members of this genus can temporarily infect humans. Margolis added that because of the scarcity of this juvenile acanthocephalan in Pacific salmon and because of its usual location in the viscera rather than in the flesh, the potential for its transmission to humans was insignificant. However, Schmidt (1971) suggested that accidental parasitism of Eskimos is probably fairly common. The adult stage of *C. strumosum* occurs in pinnipeds. Juvenile worms have been reported from dogs, otters, birds, and other fish-eating animals.

In the second reported human case, specimens of *Acanthoscephalus rauschi* invaded the peritoneal cavity of an Alaskan Eskimo (Golvan 1969). Raw fishes, probably Pacific salmon, were suspected in the transmission process. Schmidt

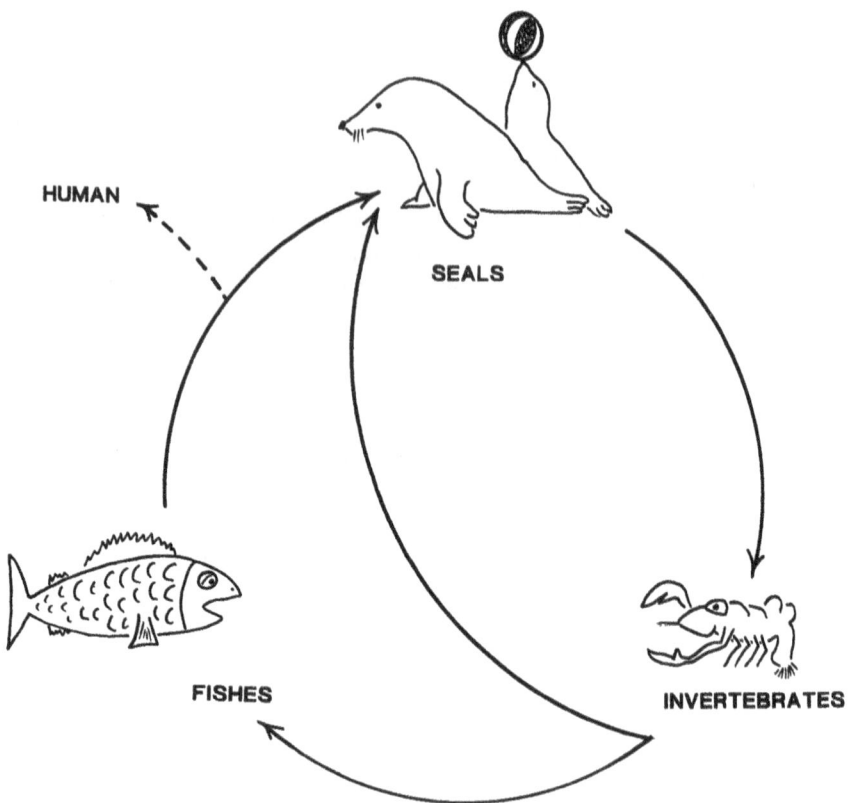

Figure 9.21 Generalized diagram of the life cycle of a marine acanthocephalan. Infection of seals as well as humans occurs by ingestion of invertebrates or fishes.

and Roberts (1981) concluded that acanthocephalans were much more important as parasites of wild and captive animals than as a threat to human health. In confined stocks of hosts, sudden epizootics have been known to kill a great number of individuals in a short time.

Nevertheless, other acanthocephalan infections in Americans can be expected. For example, fishes serve as paratenic hosts of species of *Bulbosoma* that mature in marine mammals. Two separate reports describe human infections of an unidentified species in this genus of acanthocephalan in the peritoneal cavity of two Japanese. One occurred in an eosinophilic granulomatous mass on the serosa of the ileum (Beaver et al. 1983) and one in a purulent mass on the jejunum (Tada et al. 1983). We have seen fishes from American waters infected with a member of this as well as other genera that contain mammalian-infecting species. Moreover, infected fishes could be inadvertently imported.

OTHER METAZOAN AGENTS

New dishes, new activities, newly introduced parasites, and new immigrants as well as an increased interest on the part of the medical and scientific community provide the potential for additional zoonotic diseases. Perhaps a variety of parasitic groups not discussed above will be implicated. One example is the pentastomid *Sebekia mississippiensis* which matures in the American alligator. Because the nymphs of this parasite, a species that superficially resembles an insect larva, utilize fish hosts and are known to elicit a pathological response in mammals, it is a potential health hazard for humans (Overstreet et al. 1985). In addition to infecting a large number of brackish and freshwater fishes in the Southeast, the parasite also infects turtles, snakes, the Virginia opossum, and the nearctic river otter as paratenic hosts (Overstreet et al. 1985). Perhaps other groups of marine and estuarine parasites that have no record of infecting humans will also induce some reactions in people in the U.S. in the future. We have examined specimens encountered by FDA inspectors of an unidentified pentastome in routine inspection of "eel meat chunks" from Thailand.

Additional parasites belonging to some of the same groups already discussed as well as to different ones will also be acquired by seafood consumers eating raw or minimally cooked crustaceans (Overstreet 1987). One important deviation from heterophyid digeneans in finfishes is microphallid digeneans in crustaceans. One species from prawns has been implicated in adverse and fatal involvement of the heart, spinal cord, and other vital organs of Filipinos, and related species occur in wild and cultured palaemonid and penaeid products in the U.S. (Heard and Overstreet 1983).

RISKS

Risks from Imports

In 1987, 65% of the seafood consumed in the U.S. was imported. These products were imported from 141 countries having different requirements for safety and quality of food. With the introduction of sophisticated systems for processing and delivering food, importation of fresh seafood has increased. Much of that seafood originates from a variety of places where parasites potentially harmful to humans exist. The increasing dependence of the U.S. on imports creates potential public health problems that involve a variety of parasitic and other disease conditions that should not be ignored by those seeking to improve the safety and quality of seafoods sold in the U.S. A shipment of imported scallops was recently detained and found to be infected with juveniles of *Echinocephalus sinensis*, a gnathostome nematode (Figs. 9.22 and 9.23). This species, not present in U.S. waters, can cause pathogenic lesions in experimentally infected

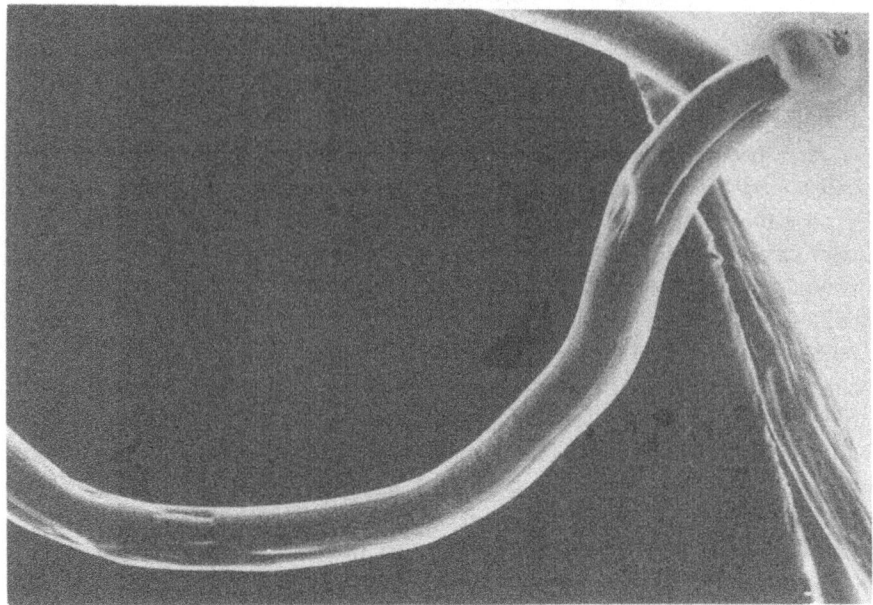

Figure 9.22 Scanning electron micrograph of adult stage of the gnathostome nematode *Echinocephalus sinensis*. General aspect of anterior portion of worm. (From Deardorff and Ko 1983.)

monkeys and other mammals and probably could produce similar effects in humans (Ko 1976). Another shipment, one consisting of live crabs from Asia, was refused entrance into the U.S. by the FDA after specimens were determined to be infected with metacercariae of the Oriental lung fluke, *Paragonimus westermani*. Apparently, the same shipment was refused entrance into Canada for the same reason. Paragonimiasis, a serious human disease with lung, glandular, and cerebral involvement, is chiefly confined to people in the Far East. United States reports of infections with *P. westermani* occur predominantly among refugees, but several other species of *Paragonimus* potentially infective to humans occur in crabs, crayfish, and other crustaceans in both the Eastern and Western Hemispheres, including the U.S. and adjacent countries (Beaver et al. 1984).

A parasite population may become established in the U.S. by introduction of one or more required stages of that parasite within any of several animals, including humans, that constitute hosts in its life cycle. We already discussed how the broad fish tapeworm became established by means of immigrants in the U.S. Williams and Ballard (1982) reported that artificial stocking of fishes in several countries in South America resulted in the introduction of the broad fish tapeworm disease in areas where it was not previously recognized. Within the last few years, *Angiostrongylus cantonensis* has become established in the New

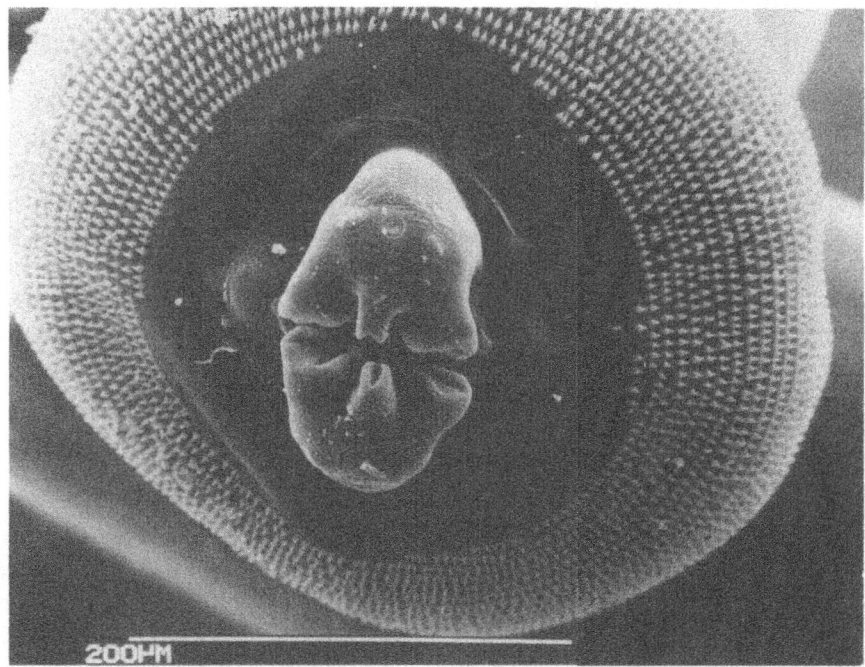

Figure 9.23 Scanning electron micrograph of adult stage of the gnathostome nematode *Echinocephalus sinensis*. Close-up of cephalic bulb showing rows of spines. (From Deardorff and Ko 1983.)

Orleans, Louisiana, area (Campbell and Little 1988). From 1962 to 1976, members of the Department of Tropical Medicine at Tulane University, including one of the authors of this review, examined *Rattus norvegicus* and other rats yearly from areas near the wharfs as part of several different projects and never encountered *A. cantonensis* in any of the rodents. Campbell and Little (1988) found the parasites only in *R. norvegicus,* but the prevalence of 21.4% in 94 specimens collected from April 1986 to February 1987 from the same locations and the ease with which they experimentally infected a variety of local gastropod intermediate hosts suggest that the infection had only recently arrived, probably from rats aboard a ship docking in the area. The abundance of rat, gastropod, and paratenic hosts in the southeastern U.S. should enable the worm to spread extensively within a few years.

Risks to Seafood Handlers

Not all human infections from marine products result from ingestion of the infective form of the parasites. Direct contact with uncooked, infected fish may also expose food handlers to parasitic infections. Deardorff et al. (1986b) reported an accidental infection with an adult marine parasitic nematode. The philometrid penetrated an open lesion of the hand of a fisherman while he filleted the infected fish (Fig. 9.24).

The life cycle of this philometrid nematode is unknown but probably is similar to that of other filarial nematodes. The adult female worm is essentially a large bag full of infective juveniles (Fig. 9.25). The juveniles are released into the seawater (e.g., a portion of the mature female worm protrudes through an ulcer in the fish's skin, releasing the juveniles by way of the fish's reproductive system, and rupturing of the female worm on death of the fish) where they are eaten by the appropriate crustacean intermediate host. The juvenile will develop into the infective stage within the hemocoel of the intermediate host. On ingestion of the intermediate host by a suitable definitive host (the jack fish), the juvenile is released by digestive enzymes into the gastrointestinal tract and begins to migrate to its appropriate site of maturation (Fig. 9.26). Once the male and female worms mate, the male dies and the female begins to swell with juveniles.

Parasitic isopods in the family Aegidae and, less often, Cymothodidae, can leave their fish hosts, attach to human skin, and produce small lesions and urticaria. Also, we have seen an acute case of urticaria in a scuba diver from Hawaii resulting from an amphipod. Hundreds of this unidentified amphipod attached to the exposed skin of the diver when he was cleaning the bottom of a ship's hull. Although these crustaceans are not known to cause serious conditions, they can certainly be annoying.

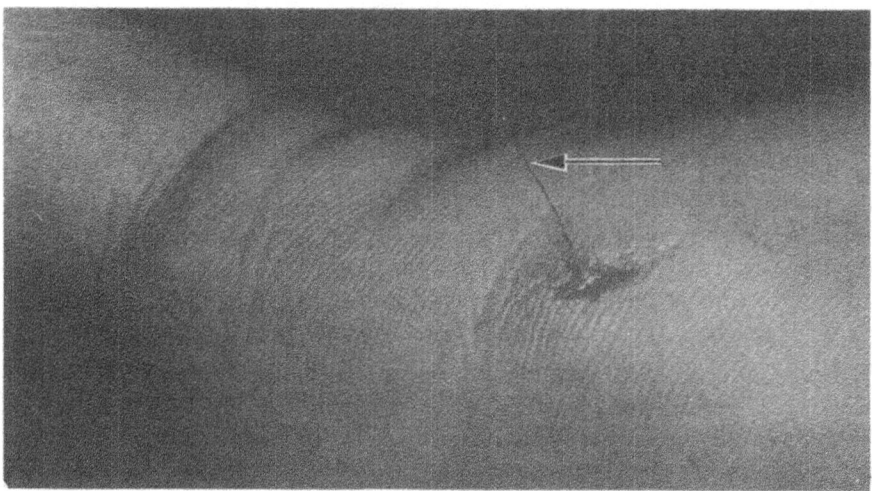

Figure 9.24 *Philometra* sp. infection in human. Gross view of left hand with puncture wound and shriveled, exposed portion of implanted adult nematode (arrow) that entered an open lesion. (From Deardorff et al. 1986.)

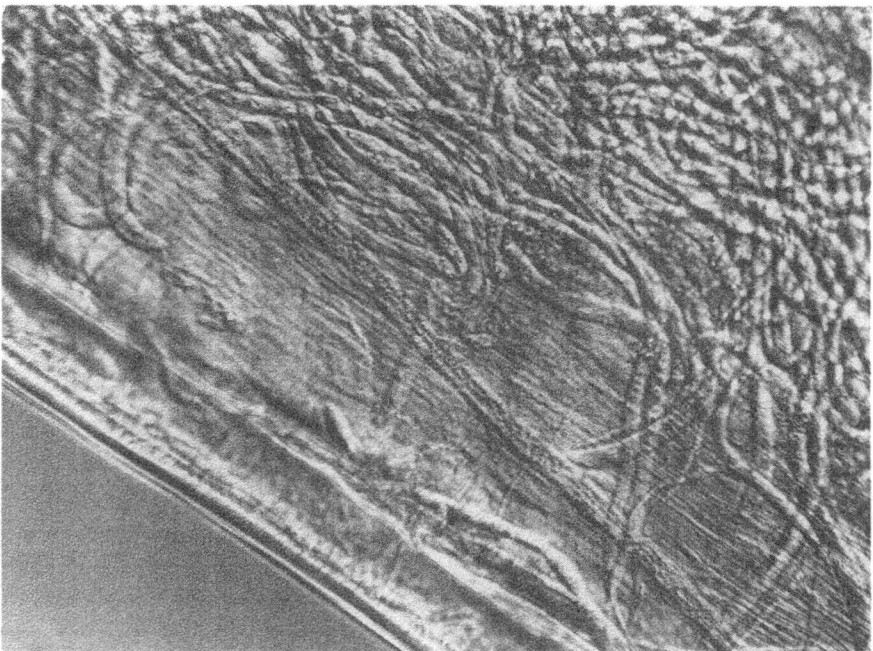

Figure 9.25 *Philometra* sp. infection in human. Cleared wholemount of female specimen showing developing first-stage juveniles in uterus. (From Deardorff et al. 1986.)

Recently, a seafood handler was infected with the digenean *Nanophyetus salmincola* (see Harrell and Deardorff, 1990). This report represents the first U.S. case of human nanophyetiasis that did not involve ingestion of raw or undercooked salmonid fishes. In brief, contamination with the infectious metacercariae of the digenetic trematode *N. salmincola* occurred during handling of fresh killed, juvenile coho salmon *(Oncorhynchus kisutch)*. Figure 9.27 shows an infected fish. Clinical findings included chronic diarrhea, nausea, abdominal discomfort, and a peripheral blood eosinophilia of 43%. Positive diagnosis was based on finding characteristic bipolar eggs in a stool specimen. The patient's occupation involved working with highly infected coho salmon, but he rarely ate seafood products and never ate raw or cold-smoked fish dishes. The patient probably introduced the metacercariae into his mouth accidentally, perhaps while smoking. He responded favorably to praziquantel and was asymptomatic after therapy. Praziquantel, although beneficial in the treatment of the adult trematode human infections, was not effective against metacercariae when in salmon (Foreyt and Gorham 1988). The life cycle of this parasite is shown in Figure 9.9.

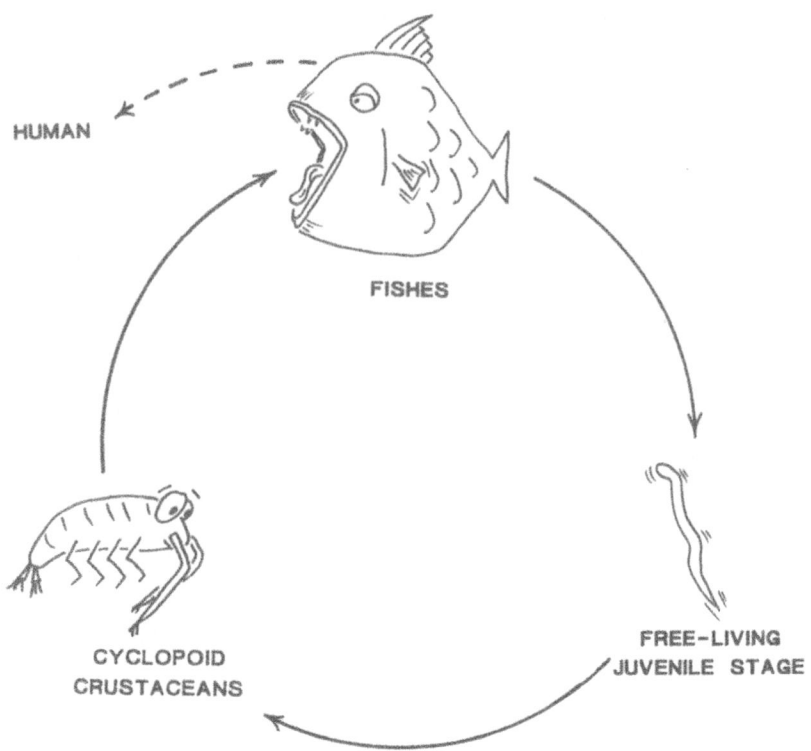

Figure 9.26 Probable life cycle of *Philometra* sp. photographed in Figure 9.24. The adult worm releases the free-living juvenile stage into the water where a specific copepod host eats it. Within the copepod, the juvenile develops and molts into the stage infective for fishes.

Figure 9.27 Salmon smolt showing ulcerative area (arrows) on skin adjacent to encystment of metacercariae of *Nanophyetus salmincola.* (Original provided by L. W. Harrell.)

AQUACULTURE AND PARASITES

Recent advances in aquacultural technology has led to explosive growth in fish production. Culture of channel catfish *(Ictalurus punctatus),* for example, represents a classic example for success of aquacultural fishery which clearly overshadows the once predominant U.S. wild catfish fishery. Like catfish culture, salmon farming has become a viable and rapidly developing industry. On the other hand, production of domestic salmon in culture probably will never equal or replace the commercial wild salmon fishery. The rearing of salmon in seawater net-pens for the commercial market has been widely practiced in Europe and quickly is gaining acceptance in North America. Producers practice net-pen culture in protected coastal temperate waters such as bays and sounds in the Pacific Northwest (Figs. 9.28 and 9.29) and New England. A typical farm is comprised of several pens, each one approximately 6 m long by 12 m wide. Nets forming the pens are attached to floats or to rigidly anchored floating walkways, and those nets extend down about 6–12 m. If net-pens are properly located, oceanic currents ensure adequate flushing with high quality sea water and dispersal of contaminating detrital materials that would otherwise accumulate under the pens. Ponds and raceways are also used to culture a variety of

Figure 9.28 Salmon aquaculture in Puget Sound, Washington. Areal view of salmon pens. (Original provided by M. L. Kent.)

freshwater, estuarine, and marine fishes. Regardless of the species of fish being reared and the facilities used, the fishes will be prone to disease conditions, some of them posing public health risks. None of the facilities eliminate all disease agents that harm reared fishes. On the other hand, most of the helminths and protozoans that could also infect humans can be kept out of the facilities with some effort. The potential for establishing those agents depends on several factors, including the source of water and the ability for prey organisms other than the fish being reared to survive in the system.

Public health aspects of aquaculture depend on a variety of circumstances. For instance, the type of life cycle (i.e., direct or indirect) of the parasite in question limits the species and abundance of individuals, as does how the way fish are fed and the amount of water, the restrictions on the water, and on the animals being introduced into the system.

Helminth parasites that present a public health risk have indirect cycles with two or more hosts. Consequently, depending on the presence of definitive, intermediate, or paratenic hosts, infections can be avoided or can be much more abundant than in the wild counterpart. If an essential host can be eliminated or

Figure 9.29 Salmon aquaculture in Puget Sound, Washington. Closer view of Figure 9.28 showing individual pens with bird netting on top. (Original provided by T. L. Deardorff.)

restricted from a system, most infections should be eliminated. Intermediate hosts can be introduced into a system in a variety of ways.

Net-pens are in direct contact with the surrounding water, and we have already discussed their importance in establishing a thriving salmon culture industry. A study by McGladdery et al. (1990) provides a good example of the importance of pen location and the possibility of atypical facets of infections, even though the agent in question was not pathogenic to humans. Rainbow trout in shallow water pens were exposed to cercariae of *Stephanostomum tenue* from the dog whelk, *Nassarius obsoletus*. This association with the gastropod was not considered important because the digenean's metacercaria in that geographical area infected the peritoneal lining of *Fundulus heteroclitus* and did not infect salmonids. However, the penned trout acquired infections of that worm that were restricted to the pericardial sac where they caused fibrosis and led to mortality of the stock when excess food or other conditions reduced the oxygen concentration in the pen and caused stress in the fish.

Ingestion by salmonids of accidentally introduced animals in brick raceways has been linked the acquisition of the metacestodes of *Diphyllobothrium ditremum* and *D. dendriticum* in cultured stocks in Scotland. Wootten and Smith (1979) found the plerocercoids in farmed rainbow trout, *Oncorhynchus mykiss* (= *Salmo gairdneri*), and suggested that the infections "must have resulted from the ingestion of infected copepods entering the farm via the unfiltered water supply." The infective stage of numerous different parasites can confront salmonids or any other fishes in hosts that are allowed to enter either ponds, raceways, or pens. Many but not all of these agents enter within the intermediate or paratenic host rather than free in the water. Others can make use of hosts, either large or small, already established in the system.

Ponds provide optimal conditions for some intermediate hosts to become established and thrive. For example, the snail hosts for heterophyid digeneans often become established, typically after production of the first crop of fish. The snail *Pirenella conica* flourished in ponds used to culture mullet in Dor, Israel, and that snail was the host for more than one heterophyid infective to humans (Paperna and Overstreet 1981; Paperna and Overstreet unpublished observations). With an increase in estuarine culture in the U.S., we expect similar examples to be commonplace. We have seen extensive heterophyid infections in pond-reared baitfish *(Fundulus grandis)* in the Southeast.

In some cases, infective hosts are introduced into a system by way of food for the fish. Commercially prepared diets of farmed salmon usually consist of fish meal, ground fish, and fish offal. Consequently, such farmed fishes have less opportunity to become infected with anisakid nematodes and other helminths than their wild counterparts unless fed fresh wild fish. Deardorff and Kent (1989) compared infections in 237 market-ready, pen-reared salmon with those in 50 wild-caught salmon and demonstrated that *A. simplex*, *Diphyllobothrium* sp., and *N. salmincola* occurred only in the wild fishes. Table 9.2 demonstrates this difference for the anisakid. Farmers who supplement the diet of their fish with fresh herring *(Clupea harengus)* especially increase the potential for anisakid infections in the net-penned fishes. Deardorff and Kent (1989) also noted that small herrring as well as a variety of invertebrates such as amphipods, copepods, and tunicates covering the surface of the nets all represent potential vectors for disease. Farming the salmon provides an effective means of interrupting the anisakid nematode life cycle. For those who prefer raw salmon dishes, these farmed products accommodate the tastes of that growing number of consumers while increasing their margin of safety from parasitic infections.

Cultured fish are not necessarily parasite-free, even if in covered raceways with well-filtered water. Usually, the parasites that become established in those conditions are those with direct life cycles not requiring intermediate hosts. Confinement and intense crowding, conditions necessary for culture, permit many different parasites with direct cycles to thrive. Diseases caused by these agents may result in significant mortalities. Causative agents include some helminths (e.g., monogeneans and leeches), other metazoans (e.g., copepods and argulids), and a wide variety of protozoans (e.g., amoebas, flagellates,

Table 9.2 Prevalence of Third-Stage Juveniles of *Anisakis simplex* from Pen-Reared Salmons from Puget Sound, Washington, in October 1987 and from Commercially Harvested Salmons

Source	Salmon Species	Number Examined	Fishes Infected (%)	Worms in Muscle (%)	Year
Pen-Reared					
Farm A	coho	200	0	0	1987[a]
Farm B	chinook	31	0	0	1987[a]
Farm B	Atlantic	6	0	0	1987[a]
Totals		237	0	0	
Wild-Caught					
	Sockeye	50	100	100	1987[a]
	Sockeye	32	100	75	1986[b]
	Sockeye	159	100	100	1981[c]
	Pink	30	97	97	1981[c]
	Chinook	19	79	79	1981[c]
	Coho	88	98	98	1981[c]
	Coho	55	36	100	1974–76[d]
	Chum	1	100	100	1974–76[d]

[a]Adapted from Deardorff and Kent (1989).
[b]Adapted from Deardorff and Throm (1988).
[c]Adapted from Bier, Schwien, and Sellers (1982).
[d]Adapted from Myers (1979).

ciliates, coccidians, microsporidians, and myxosporans). Certain other members of those same groups do not necessarily kill the fish, but they adversely affect the marketable product (e.g., "jellied meat" caused by the tissue protozoan *Kudoa thyrsitis* in salmon and *K. histolytica* in mackerel). Both categories of parasites can have a profound impact on the economics of the industry without involving public health.

Infectious diseases (Kent and Elston 1987), waterborne toxicants (Kent et al. 1988), difficulties in obtaining permits, lack of availability of good sites for fish pens, negative environmental and aesthetic impact of the culture facilities, and competitive pricing with wild-caught products represent major hurdles to the projected growth of the salmon and other cultured fish industries in the U.S. Nevertheless, the salmon industry is confronting these problems, and the public now consumes both domestic and foreign pen-reared salmon from markets and in restaurants (Deardorff and Kent 1989).

PREVENTION OF HUMAN INFECTIONS

Antiparasitic drugs are in some instances available and effective for some zoonotic diseases. Because different governments tend to vary in their approval procedures for drugs and their suggested drugs of choice, physicians may have difficulty in determining appropriate treatment regimens. For example, praziquantel and niclosamide, readily available in the U.S. by prescription, are efficacious for tapeworm infections. However, these drugs are classified as emergency drugs in Canada. Examples of differences between the Canadian and U.S. recommendations have been published (Canadian Infectious Disease Society 1988).

In instances where no antiparasitic drugs are available, methods of prevention focus on attempts to interrupt the parasite's life cycle to prevent the food product from becoming adulterated (see Aquaculture and Parasites section) or to kill the helminths in infected seafoods.

Of the measures that we examined to reduce zoonotic infections, temperature extremes appear to be most effective in neutralizing the infective potential of worms. Adequate freezing or heat from thorough cooking of seafoods kills parasites. Freezing provides the more promising preventive measure because some consumers do not want a thoroughly cooked product (Deardorff and Throm 1988; Deardorff et al. 1987) and cooking decreases the market value of most seafood products. Numerous studies have assessed various aspects of temperature extremes for specific groups of parasites (e.g., anisakiasis: Bier et al. 1987; *Hysterothylacium* Type MB: Norris and Overstreet 1976; *Eustrongylides tubifex:* Cooper et al. 1978; *Angiostrongylus cantonensis:* Alicata 1967; *Diphyllobothrium* spp. and related cestodes: Hilliard 1959; Bylund 1982; heterophyids: Paperna and Overstreet 1981).

In the United States

The commercial fishing industry commonly employs a process called "blast-freezing," which rapidly freezes the fishes down to –40°F (–40°C). Deardorff and Throm (1988) demonstrated that commercial blast-freezing was extremely effective in killing juvenile nematodes in fishes (Table 9.3). The study concluded that blast-freezing could effectively prevent anisakiasis when thoroughly cooked seafood was not acceptable. Not only did blast-freezing kill parasites, it was cost-effective, caused little change in the flavor or texture of the products, and had practical importance because the commercial fishing industry currently uses the process.

With the risk of infection from parasites in fishery products and our increased knowledge of the effects of temperature extremes on parasites, the U.S. Food and Drug Administration (FDA) on 21 August 1987 released the following interpretation: "Fishery products which are not cooked throughout to 140°F (60°C) or above, must have been or must, before service or sale in ready-to-eat form, be blast frozen to –31°F (–35°C) or below for 15 hours or regularly frozen to –10°F (–23°C) or below for 168 hours (7 days). Records that establish that fishery products were appropriately frozen on-site must be retained by the operator for 90 days."

The above is a federal recommendation to state and local regulatory agencies that have a primary responsibility for inspecting retail food establishments and was based on current scientific knowledge of parasites and on modern technological practices. The interpretation reflected substantial input from both regulatory agencies and trade associations. The "fishery products" affected by this federal recommendation are defined as finfishes, bivalve molluscs (scallops), gastropods (abalone), cephalopods (octopus and squids), and crustaceans (lobsters and crabs) from both fresh and salt water. The interpretation specifically excluded clams, mussels, and oysters because of the controls

Table 9.3 Viability of Third-Stage Juvenile of *Anisakis simplex* Following up to 7 Days of Commercial Blast-Freezing

Sample	No. of Fishes Examined	No. of Worms Recovered Live[a]	Dead
Salmon[b]	32	2	1,243
Rockfish[c]	32	4	2,296

[a]All numbers in this column represent viable specimens following 1 hour of freezing.
[b]*Oncorhynchus nerka.*
[c]*Sebastes pinniger.*
Adapted from Deardorff and Throm (1988).

imposed by states under the provisions of the National Shellfish Sanitation Program.

Based on reported cases of anisakiasis, chances of a consumer getting infected are significantly greater from a fish dish prepared at home than from one obtained at a sushi bar. This difference may be attributed to a variety of reasons, such as more fish are consumed at home than at raw fish bars, fish prepared at home are likely to be whole fillets hiding the worms rather than thinly sliced portions exposing them to the sushi chef, and lack of experience of the home preparer compared with that of the the trained sushi chef. Also, most sushi bars freeze their fish, according to the Japanese Restaurant Association of Marin County, California (1989).

In addition to the above interpretation, the FDA addresses the degree of exposure that seafood consumers have to parasites not infective to humans. The FDA has established food "defect action levels" (DAL) for some parasites. The DAL are set generally on the basis of having no hazard to human health. Examples of DAL concerned with adulteration of fishes by parasites are listed in Table 9.4.

Although misrepresentation of seafood products is not a major problem or practice within the industry, it does occur. Examples of misrepresentation may involve substituting different products (e.g., "punched out" stingray wings for scallops) or substituting closely related species (e.g., green sturgeon for the gastronomically superior and more expensive white sturgeon, oreo dory for orange roughy). Requirements for wholesale and retail establishments to label fishes correctly by their specific name may reduce the risk from parasitic infections. When a supermarket sells "red snapper" fillets, as indicated on its package label, and the product is a species of the Pacific rockfishes (*Sebastes* spp.) rather than *Lutjanus campechanus,* does the supermarket chain place its customer at risk? Clearly, labeling a fish as "red snapper" for a species other than

Table 9.4 Defect Action Levels for Parasites in Finfishes Established by USFDA

Product	*Defect*	*Action Level*
Tullibees, ciscoes inconnus, chubs, and white fish	Cysts	50 cysts/100 pounds (whole or fillets), provided that 20% of the fish examined are infested
Blue fin and other fresh water herring	Cysts	60 cysts/100 fish (fish 1 pound or less) or 100 pounds of fish (fish averaging over 1 pound), provided that 20% of the fish examined are infested
Red fish and ocean perch	Copepods	3% of the fillets examined contain 1 or more copepods accompanied by pus pockets

From HHS Publication No. (FDA)85-2199.

L. campechanus constitutes misbranding in violation of the Federal Food, Drug, and Cosmetic Act (FDA Compliance Policy Guide No. 7108.21). In the case of intentionally misbranding a rockfish as red snapper, however, the supermarket is also substituting a species commonly infected with potentially invasive nematodes for a species rarely infected and never reported to have worms occurring in edible tissues. Consequently, the supermarket probably increases the consumer's risk of parasitic infection.

The commercial liability for purveying infected fishes has already been tested in U.S. courts. A patient brought legal action against a raw fish restaurant, stating that the restaurant was responsible for transmitting anisakiasis to him. The suit was dismissed because the actual source of the infection could not conclusively be determined. Defense attorneys successfully argued that the patient, a Japanese tourist, could have been infected in Japan and not expressed the signs or symptoms of the disease until his arrival in the U.S. A brief history of this case is covered by Kliks (1983).

Unlike the meat and poultry industry, the seafood industry is presently not subject to mandatory 100% product inspection by the U.S. Federal Government. Although the FDA has taken several steps to ensure seafood product safety and improve its coverage of imported seafood products, the responsibility of insuring seafood products lies with the seafood industry. The role of the FDA is not to act as a quality control system for the seafood industry; rather, it functions as a regulatory agency monitoring the efforts of the industry to produce seafood in compliance with the applicable laws and regulations.

Infections can also be eliminated or controlled by interrupting the life cycle of a parasite in such a way that the seafood product never or seldom comes in contact with the infectious agent; however, numerous problems and methods of handling them exist.

Birds that feed on cultured fish or defecate into the water supply can represent more of a problem to fish farmers than those that just eat a portion of the stock. Birds serve as a source of parasites infective to fishes (e.g., see discussion of *Eustrongylides* spp.). Many aquaculture facilities have nets placed across their pens, ponds, and tanks to discourage birds from feeding and defecating into the holding facilities (see Fig. 9.29). Similarly, we have seen sea lions basking on walkways of floating salmon net-pens. Their contaminated feces could initiate ascaridoid infections in the fish (e.g., see discussions on *Anisakis* and on Aquaculture and Parasites).

Life cycles can be interrupted at various stages. As illustrated in several of the life cycle figures in this chapter, snails and oligochaetes represent potential vectors of parasitic infection and should be controlled when possible. Catfish farmers, whose net-pens were suspended over areas with high concentrations of snails, have experienced a high infection rate in the tissues of their catfish with the yellow grub, *Clinostomum* sp. Changing salinities can rid fishes of some ectoparasites, such as monogeneans and crustaceans, and intermediate hosts of other parasites. Culturists often try to eliminate intermediate hosts with toxins. Occasionally treatments to control infections with toxic chemicals are effective, but usually success necessitates small restricted enclosures. Von Bonsdorff and Bylund (1982) discuss problems associated with people trying to use chemical agents to control *D. latum,* and in those cases concluded that control was more successfully achieved by improving sanitation measures.

Worms and protozoans can be mechanically removed from their hosts after they are located in fillets by candling or other means or by assuming specific parts of all members of a given species have infections and consequently removing those portions without an examination.

Infections can be reduced in some cases by extensively fishing specific members of a stock. These members may be individuals of a certain year-class or individuals occupying specific areas. That catch can be used or treated different-ly from other stocks so that it will not endanger public health or marketability, leaving the rest of the population less infected. Consequently, the infective link can be decreased and the fish products from the remaining stock can be of higher quality.

In Foreign Countries

The U.S. was not the first country to address parasite problems in an attempt to protect its citizens. The Netherlands established legislative regulations concern-ing consumption of raw herring. The regulations, known as "The Green Herring Laws," state that fresh herring should be frozen to a temperature of at least −4°F (−20°C) within 12 hours and should be stored for 24 hours before being released to the public. Lightly salted herring must be similarly treated. The regulations also address marination and hot and cold smoking procedures. Before im-plementation of that legislation, Ruitenberg (1970) reported 149 confirmed cases of anisakiasis from 1955 to 1967. As a direct result of the Netherlands' legisla-tive actions, the number of cases of parasitic infections in humans was dramati-cally reduced from 46 cases in 1967 to merely five cases in 1968 (Fig. 9.16). According to the World Health Organization (WHO 1988), the Netherlands' "Green Herring Laws" have resulted in the disappearance of human anisakiasis from a country that previously had almost 300 cases per annum.

The European Economic Community (EEC) excluding Portugal recently voted for a directive to control infections involving live nematodes in fish (Eurofish Report 1987). According to the WHO (1988), the measures in the EEC draft regulation that are currently being studied propose (1) that fishes be cleared of ascaridoid juveniles by eviscerating the fish as soon as possible after they are caught, washing them carefully, and examining the fillets for worms and (2) that the juveniles present in the processed fish be destroyed by (a) heating the product to at least 149°F (65°C), (b) salting in brine, (c) pickling, or (d) freezing at −4°F (−20°C) for 24 hours. This directive must be applied and adhered to in intracommunity trade and apparently is being enforced. In the summer of 1987, the Italian government detained and rejected shipments of mackerel from Denmark, France, and the Netherlands because of the nematode parasites (WHO 1988).

The WHO (1988, p. 314) concluded, "Furthermore, the proposed EEC regulations should completely obviate the risk of contracting anisakiasis from fish caught by professionals; the prevention of risk in domestic preparation calls for education of consumers, while those with a predilection for raw or marinated fish should examine the fillets carefully to ensure they contain no larvae."

The Japanese have a high regard for quality seafood, and Japan has no official policy regarding anisakiasis or regulation requiring freezing. Oshima (1972) stated that because there were so many species of fish in Japanese waters that could transmit disease, legislative regulations such as freezing were "almost impossible." On the other hand, Japanese cooks reportedly freeze Pacific salmonids (*Oncorhynchus* spp.) before serving raw dishes to kill potentially invasive helminths. Atlantic salmon are considered by the Japanese to be free of these worms.

PUBLIC EDUCATION

Health conscious consumers are aware of the alleged health benefits of seafoods. Many products provide a source of protein low in saturated fats and high in omega-3 fatty acids; however, some of those same consumers remain unaware of the possible health risks associated with marine fish products. The importance of consumer awareness of the possible health hazards, including parasitic diseases that can result from eating raw or poorly prepared fishery products, cannot be overemphasized.

Public education is not totally supported by fishery organizations in the U.S. They fear a consumer backlash like the one that recently occurred in West Germany. Following a television program that focused on the risk of parasitic infection from eating raw herring (a product commonly consumed in Europe), West German fish sales plunged nearly 70% in 1 day. Ironically, the consumer backlash involved not only fresh fish, but also frozen and preserved products. To combat these serious repercussions, the veterinary services of the Federal Republic of Germany were compelled to reassure public opinion by imposing import control measures that were deemed discriminatory by the other member states of the EEC (WHO 1988). Sales eventually recovered.

In contrast with the response that occurred in West Germany, consumer backlash was minor in France when a French television station aired the same German program. *Seafood Leader* (Anonymous 1988, p. 21) attributed the lack of "parasite paranoia" by the French consumers to the fact the "French suppliers had time to educate their customers." Furthermore, following the recent broadcasts in the U.S. and Canada of several similar television programs on potential parasite problems, no serious repercussions such as witnessed in West Germany were reported. There are examples, however, where public education creates additional problems. When some Americans realize that a product is infected by a harmless parasite such as the cestode *Poeciancystrium caryophylum,* they will not eat that seafood product regardless of the amount of education they receive (Overstreet 1983).

Fish parasites have recently been involved in "blackmail," according to the Canadian Television Network (CTN). The incident clearly underscores the close relationship between informed consumers and the fishing industry. In 1988, the Canadian government wanted to reduce the number of gray seals for a variety of reasons; one principle reason was that the adult stage of *Pseudoterranova decipiens* in the seals was infecting commerically important fishes (especially cod) with the juvenile stage of the parasite(s) capable of infecting humans (see previous section on anisakiasis). Because the number of seals offshore of eastern Canada had rapidly increased, the number of worms capable of causing anisakiasis (in the broad sense) in cod and other fishes had also increased. According to a program broadcast by the CTN, the increased abundance of juvenile cod worms in commerically important Canadian fishes was costing the fishing industry and the consumer a considerable amount of money. The industry estimated that it spent nearly 50 million dollars a year to keep worms out of fishes. Environmental groups exerted heavy pressure on the government. According to the televised program, one group, in effect, held the government and industry hostage. To prevent the culling of seals, that group threatened to show worldwide audiences graphic pictures of worm infections in cod. The obvious intent was to create devastating repercussions to the fishing industry as in West Germany by publicizing pictures of large nematodes in fish, thereby protecting the seals. In January 1988, the Canadian government announced that it would not favor killing seals; rather, it extended the 4-year-old ban on culling gray seals. Hafsteinsson and Rizvi (1987) reviewed actions taken by other governments to control the growth of seal populations.

If accurate, responsible, and continuous reporting on ascaridoid parasites is offered, a consumer backlash in the U.S. like the one in West Germany appears unlikely. The American consumer has been educated about other harmful parasites dwelling in other types of foods. For example, most Americans probably realize that they can contract parasitic infections from raw pork (pork tapeworm and trichinosis) and raw beef such as steak tartare (beef tapeworm and toxoplasmosis). Consumers who choose to eat those meats raw or rare have probably considered the risks. Most consumers of seafood products, however, remain uninformed about potential parasite hazards and, depending on the product, may be at risk.

Based on the historical example of health education, the eradication of seafood-transmitted parasitic diseases in the U.S. solely by public education will be difficult. Ingrained habits such as the use of tobacco and alcohol are difficult to break. Public awareness requires continued dispersement of information about the known safety risks because of the increased consumption of "natural" seafood dishes. The seafood industry also must be aware of these problems and how to address them properly. Introduction of new species of marine animals as food items and changing production methods will require constant monitoring of seafood for potential hazards.

Parasitic zoonoses, no doubt, will continue to plague humans. The economic and public health significance of parasites in seafood has provided an impetus for study of methods to prevent zoonoses, and some encouraging advances have been made. Criteria for freezing and heating have been established to kill worms in foods. The potential advantages of commercially fed fishes held in aquaculture facilities shows promise. Further, our knowledge of parasitic diseases has increased rapidly in recent years and has led to the development of more effective antiparasitic drugs and immunological procedures for reliable diagnosis and practical treatment of parasitic infections.

Informed consumers have an advantage from a public health viewpoint: They can evaluate the consequences of eating certain raw seafoods, because they know more about *the fishes, the dishes, and the worms.*

ACKNOWLEDGMENTS

We are grateful to Kyle Headley and Helen Gill (Gulf Coast Research Laboratory) for technical assistance; Dennis Farley (FDA, Los Angeles District) for providing parasites encountered during food inspections; and Lois Tomlinson, George Hoskin, Mary Snyder, Thomas Schwarz, and George Jackson (all of FDA, Washington, D.C.) for comments and suggestions. This study was supported in part by U.S. Department of Commerce, National Marine Fisheries Service Grants, Nos. NA89WC-D-IJ084 and NA90AA-D-IJ217 and by U.S. Department of Agriculture CSRS Grant No. 88-38808-3319 to RMO.

REFERENCES

Adams, K. O., Jungkind, D. L., Bergquist, E. J., and Wirts, C. W. 1986. Intestinal fluke infections as a result of eating sushi. *Am. J. Clin. Pathol.* **86**:688–689.

Africa, C. M., Garcia, E. Y., and de Leon, W. 1935. Intestinal heterophyidiasis with cardiac involvement: a contribution to the etiology of heart failures. *Philippine J. Publ. Hlth.* **2**:1–22.

Africa, C. M., de Leon, W., and Garcia, E. Y. 1936. Heterophyidiasis, IV: lesions found in the myocardium of eleven infested hearts including three cases with valvular involvement. *Philippine J. Publ. Hlth.* **3**:1–27.

Africa, C. M., de Leon, W., and Garcia, E. Y. 1937. Heterophyidiasis, V: ova in the spinal cord of man. *Philppine J. Sci.* **62**:393–403.

Alicata, J. E. 1962. *Angiostrongylus cantonensis* (Nematoda: Metastrongylidae) as a causative agent of eosinophilic meningoencephalitis of man in Hawaii and Tahiti. *Can. J. Zool.* **40**:5–8.

Alicata, J. E. 1964. Parasitic infections of man and animals in Hawaii. *Hawaii Agric. Exp. Stat. Coll. Trop. Agric. Tech. Bull.* **61**:1–138.

Alicata, J. E. 1967. Effect of freezing and boiling on the infectivity of third-stage larvae of *Angiostrongylus cantonensis* present in land snails and freshwater prawns. *J. Parasitol.* **53**:1064–1066.

Alicata, J. E., and Brown, R. W. 1962. Observations on the method of human infection with *Angiostrongylus cantonensis* in Tahiti. *Can. J. Zool.* **40**:755–760.

Alicata, J. E., and Schattenburg, O. L. 1938. A case of intestinal heterophyidiasis of man in Hawaii. *JAMA* **110**:1100–1101.

Anonymous. 1988. Europe. *Seafood Leader.* Spring, p. 21.

Baer, J. G. 1969. *Diphyllobothrium pacificum,* a tapeworm from sea lions endemic in man along the coastal area of Peru. *J. Fish. Res. Board Can.* **26**:717–723.

Baer, J. G., Miranda, C. H., Fernandez, R. W., and Medina, T. J. 1967. Human diphyllobothriasis in Peru. I. Identification of the species. *Z. Parasit.* **28**:277–289.

Beaver, P. C., and Theis, J. H. 1979. Dioctophymatid larval nematode in a subcutaneous nodule from man in California. *Am. J. Trop. Med. Hyg.* **28**:206–212.

Beaver, P. C., Otsuji, T., Otsuji, A., Yoshimura, H., Uchikawa, R., and Sato, A. 1983. Acanthocephalan, probably *Bolbosoma,* from the peritoneal cavity of man in Japan. *Am. J. Trop. Med. Hyg.* **32**:1016–1018.

Beaver, P. C., Jung, R. C., and Cupp, E. W. 1984. *Clinical Parasitology,* ninth edition. Lea and Febiger, Philadelphia.

Bennish, M. L., Sullivan, C., Michelson, S., et al. 1986. A point source outbreak of gnathostomiasis at a diplomatic dinner. Paper read at the 28th Interscience Conference on Antimicrobial Agents and Chemotherapy, Los Angeles, California, 23–26 October 1986.

Bier, J. W., and Raybourne, R. B. 1988. *Anisakis simplex* (Nematoda: Ascaridoidea): formation of immunogenic attachment caps in pigs. *Proc. Helminth. Soc. Wash.* **55**:91–94.

Bier, J. W., Deardorff, T. L., Jackson, G. J., and Raybourne, R. B. 1987. Human anisakiasis. In: Zbigniew S. and Pawlowski (eds.), *Bailliere's Clinical Tropical Medicine and Communicable Diseases, Vol. 2.* Harcourt Brace Jovanovich, London. pp. 723–733.

Bier, J., Schwien, W. G., and Sellers, R. L., Jr. 1982. A metazoan parasite survey of fresh salmon from United States markets. Paper read at the Fifth International Congress of Parasitology, Toronto, Canada, 7–14 August 1982, p. 315.

Bylund, B. G. 1976. The epidemiology and control of the broad fish tapeworm. *Duodecim* **92**:646–648.

Bylund, B. G. 1982. Diphyllobothriasis. In: Jacobs L., and Arambulo P., III (eds.) *CRC Handbook Series in Zoonoses: Parasitic Zoonoses,* CRC Press, Boca Raton, Florida, pp. 217–225.

Cali, A., and Owen, R. L. 1988. Microsporidiosis. In: Balows, A., Hausler, W. J., Jr., Lennette, E. H., Halonen, P., Murphy, F. A., Ohashi, M., and Turano, A. (eds.) *Laboratory Diagnosis of Infectious Diseases Principles and Practice, Vol. I: Bacterial, Mycotic, and Parasitic Diseases, and Vol. II: Viral, Rickettsial, and Chlamydial Diseases:* Springer-Verlag, New York, pp. 929–950.

Campbell, B. G., and Little, M. D. 1988. The finding of *Angiostrongylus cantonensis* in rats in New Orleans. *Am. J. Trop. Med. Hyg.* **38**:568–573.

Canadian Infectious Disease Society. 1988. Treatment of parasitic infections: Canadian versus US recommendations. *Can. Med. Assoc. J.* **139**:849–851.

Canning, E. U., and Lom, J. 1986. *The Microsporidia of Vertebrates*. Academic Press, Orlando, Florida.

de Carneri, J., and Vita, G. 1973. Drugs used in cestode diseases. In: Cavier, R. (ed.) *Chemotherapy of Helminthiasis, Vol. 1, Section 64. International Encyclopedia of Pharmacology and Therapeutics Series*. Pergamon Press, Elmsford, New York, pp. 145–160.

Chandra, C. V., and Khan, R. A. 1988. Nematode infestation of fillets from Atlantic cod, *Gadus morhua*, of eastern Canada. *J. Parasitol.* **74**:1038–1040.

Cheng, T. C. 1976. The natural history of anisakiasis in animals. *J. Milk Food Technol.* **39**:32–46.

Ching, H. L. 1961. Internal parasites of man in Hawaii with special reference to heterophyid flukes. *Hawaii Med. J.* **20**:442–445.

Chitwood, M. 1975. *Phocanema*-type larval nematode coughed up by a boy in California. *Am. J. Trop. Med. Hyg.* **24**:710–711.

Cooper, L. C., Crites, J. L., and Sprinkle-Fastkie, D. J. 1978. Population biology and behavior of larval *Eustrongylides tubifex* (Nematoda: Dioctophymatida) in poikilothermous hosts. *J. Parasitol.* **64**:102–107.

Crichton, V. F. J., and Beverley-Burton, M. 1977. Observations on the seasonal prevalence, pathology and transmission of *Dracunculus insignis* (Nematoda: Dracunculoidea) in the raccoon [(*Procyon lotor* (L.)] in Ontario. *J. Wldlf. Dis.* **13**:273–280.

Deardorff, T. L., and Kent, M. L. 1989. Prevalence of larval *Anisakis simplex* in pen-reared and wild-caught salmon (Salmonidae) from Puget Sound, Washington. *J. Wldlf. Dis.* **25**:416–419.

Deardorff, T. L., and Ko, R. C. 1983. *Echinocephalus overstreeti* sp. n. (Nematoda: Gnathostomatidae) in the stingray, *Taeniura melanopilos* Bleeker, from the Marquesas Islands, with comments on *E. sinensis* Ko, 1975. *Proc. Helminthol. Soc. Wash.* **50**:285–293.

Deardorff, T. L., and Overstreet, R. M. 1981. Larval *Hysterothylacium* (= *Thynnascaris*) (Nematoda: Anisakidae) from fishes and invertebrates in the Gulf of Mexico. *Proc. Helminthol. Soc. Wash.* **48**:113–126.

Deardorff, T. L., and Throm, R. 1988. Commercial blast-freezing of third-stage *Anisakis simplex* larvae encapsulated in salmon and rockfish. *J. Parasitol.* **74**:600–603.

Deardorff, T. L., Kliks, M. M., Rosenfeld, M. E., Rychlinski, R. A., and Desowitz, R. S. 1982. Larval ascaridoid nematodes from fishes near the Hawaiian Islands, with comments on pathogenicity experiments. *Pacif. Sci.* **36**:187–201.

Deardorff, T. L., Kliks, M. M., and Desowitz, R. S. 1983. Histopathology induced by larval *Terranova* (Type HA) (Nematoda: Anisakinae) in experimentally infected rats. *J. Parasitol.* **69**:191–195.

Deardorff, T. L., Raybourne, R. B., and Desowitz, R. S. 1984. Behavior and viability of third-stage larvae of *Terranova* sp. (Type HA) and *Anisakis simplex* (Type I) under coolant conditions. *J. Food Protect.* **47**:49–52.

Deardorff, T. L., Fukumura, and Raybourne, R. B. 1986a. Invasive anisakiasis: a case report from Hawaii. *Gastroenterology* **90**:1047–1050.

Deardorff, T. L., Overstreet, R. M., Okihiro, M., and Tam, R. 1986b. Piscine adult nematode invading an open lesion in a human hand. *Am. J. Trop. Med. Hyg.* **35**:827–830.

Deardorff, T. L., Altman, J., and Nolan, C. M. 1987. Human anisakiasis: two case reports from the State of Washington. *Proc. Helminthol. Soc. Wash.* **54**:274–275.

Deschiens, R., Collomb, H., and Demarchi, J. 1958. Distomastose cerebrale a *Heterophyes*. In: *Abstracts of the Sixth International Congress of Tropical Medicine and Malaria*.

Desowitz, R. S., Raybourne, R. B., Ishikura, H., and Kliks, M. M. 1985. The radioallergosorbent test (RAST) for the serological diagnosis of human anisakiasis. *Trans. R. Soc. Trop. Med. Hyg.* **79:**256–259.

Eastburn, R. L., Fritsche, T. R., and Terhune, C. A., Jr. 1987. Human intestinal infection with *Nanophyetus salmincola* from salmonid fishes. *Am. J. Trop. Med. Hyg.* **36:**586–591.

Eberhard, M. L. 1989. Intestinal perforation caused by larval *Eustrongylides* (Nematode: Dioctophymatoidae) in New Jersey. *Am. J. Trop. Med. Hyg.* **40:**648–650.

Eurofish Report. 1987. E.E.C. directive to control anisakiasis. November 12. Cited in Fish Inspector, p. 4.

Font, W. F., Overstreet, R. M., and Heard, R. W. 1984. Taxonomy and biology of *Phagicola nana* (Digenea: Heterophyidae). *Trans. Am. Microscop. Soc.* **103:**408–422.

Foreyt, W. J., and Gorham, J. R. 1988. Preliminary evaluation of praziquantel against metacercariae of *Nanophyetus salmincola* in chinook salmon *(Oncorhynchus tshawytscha). J. Wldlf. Dis.* **24:**551–554.

Freeman, R. S., and Jamieson, J. 1976. Parasites of eskimos at Igloolik and Hall Beach, Northwest Territories. In: Shephard, R. J., and Iton, S. (eds.) *Proceedings of the Third International Symposium, Yellowknife, N.W.T.*, University of Toronto Press, Toronto, pp. 306–315.

Fritsche, T. R., Eastburn, R. L., Wiggins, L. H., and Terhune, C. A. 1987. Treatment of North American nanophyetiasis with praziquantel. Paper read at The 36th Annual Meeting of The American Society of Tropical Medicine and Hygiene, 29 November–3 December, at Los Angeles Hilton and Towers, Los Angeles, California.

Golvan, Y. J. 1969. Systématique des acanthocéphales (Acanthocephala Rudolphi 1801). Première partie. L'ordre des Palaeacathocephala Meyer 1931. Premier fascicule. La super-famille des Echinorhynchoidea (Cobbold 1876) Golvan et Houin 1963. *Mém. Mus. Natl. Hist. Natur.* **47:**1–373.

Guerin, P. F., Marapendi, S., McGrail, L., Moravec, C. L., and Schiller, E. L. 1982. Intestinal perforation caused by larval *Eustrongylides*-Maryland. *Cent. Dis. Cont. Morbid. Mortal. Wk. Rep.* **31:**383–389.

Gunby, P. 1982. One worm in the minnow equals too many in the gut. *JAMA* **248:**163.

Hafsteinsson, H., and Rizvi, S. S. H. 1987. A review of the seal worm problem: biology, implications and solutions. *J. Food Protect.* **50:**70–84.

Harrell, L. W., and Deardorff, T. L. (1990). Human nanophyetiasis: transmission by handling naturally infected coho salmon *(Oncorhynchus kisutch). J. Infect. Dis.* **161:**146–148.

Harrell, L. W., and Scott, T. M. 1985. *Kudoa thyrsitis* (Gilchrist) (Myxosporea: Multivalvulida) in Atlantic salmon, *Salmo salar* L. *J. Fish Dis.* **8:**329–332.

Hauck, A. K. 1977. Occurrence and survival of the larval nematoda *Anisakis* sp. in the flesh of fresh, frozen, brined, and smoked Pacific herring, *Clupea harengus* Pallasi. *J. Parasitol.* **63:**515–519.

Hayakawa, M., Suzuki, K., Maeda, T., Oikawa, K., and Ishidate, T. 1970. A case of IIa type early gastric cancer associated with eosinophilic granuloma. *Stom. Intest.* **5:**223–227.

Heard, R. W., and Overstreet, R. M. 1983. Taxonomy and life histories of two North American species of "*Carneophallus*" (= *Microphallus*) (Digenea: Microphallidae). *Proc. Helminthol. Soc. Wash.* **50:**170–174

Higashi, G. I. 1985. Foodborne parasites transmitted to man from fish and other aquatic foods. *Food Technol.* **39:**69–74, 111.

Hilliard, D. K. 1959. The effects of low temperatures on larval cestodes and other helminths in fish. *J. Parasitol.* **45:**291–294.

Honans, R. E. S., and MacFarlene, A. S. 1957. Preliminary report on the occurrence of codworms in flounders of the maritimes and on the efficiency of present candling methods in fish plants. Pamphlet H. Inspection Branch, Department of Fisheries, Maritimes Area, Halifax, Canada.

Hutton, R. F., and Sogandares-Bernal, F. 1959. Further notes on Trematoda encysted in Florida mullets. Q.J. *Flor. Acad. Sci.* **21:**329–334.

Jackson, G. J., Bier, J. W., Payne, W. L., Gerding, T. A., and Knollenberg, W. G. 1978. Nematodes in fresh market fish of the Washington, D. C. area. *J. Food Protect.* **41:**613–620.

Japanese Restaurant Association of Marin County, California. 1989. Is it safe to eat sushi? *Edell Hlth. Lett.* **8**

Juels, C. W., Butler, W., Bier, J. W., and Jackson, G. J. 1975. Temporary human infection with a *Phocanema* sp. larva. *Am. J. Trop. Med. Hyg.* **24:**942–944.

Kamo, H., and Miyazaki, I. 1970. *Diplogonoporus fukuokaensis* sp. nov. (Cestoda: Diphyllobothriidae) from a girl in Japan. *Jpn. J. Parasitol.* **19:**635–644.

Kamo, H., and Miyazaki, I. 1971. A case of human infection with unknown species of *Diplogonoporus* in Japan. *Yonago Acta Med.* **15:**55–60.

Kamo, H., Hatsushika, R, and Yamane, Y. 1971. Diplogonoporiasis and diplogonadic cestodes in Japan. *Yonago Acta Med.* **15:**234–246.

Kamo, H., Yazaki, S., Fukumoto, S., Maejima, J., and Sakaguchi, Y. 1986. Two unknown marine species of the genus *Diphyllobothrium* from human cases. *J. Trop. Med. Hyg.* **14:**79–86.

Karmanova, E. M. 1968. Dioctophymidea of animals and man and diseases caused by them. In: *Fundamentals of Nematology, Vol. 20. Academy of Sciences of the USSR.* Translated and published for the U.S. Department of Agriculture. Amerind, New Delhi, p. 383.

Kates, S., Wright, K. A., and Wright, R. 1973. A case of human infection with the cod nematode, *Phocanema* sp. *Am. J. Trop. Med. Hyg.* **22:**606–608.

Kean, B. H., and Breslau, R. C. 1964. Cardiac heterophyidiasis. In: *Parasites of the Human Heart,* Grune & Stratton, New York, pp. 95–103.

Kent, M. L., and Elston, R. A. 1987. Diseases of saltwater reared salmon in Washington State. *Int. Assoc. Aquat. Anim. Med. Proc.* **18:**33–38.

Kent, M. L., Myers, M. S., Hinton, D. E., Eaton, W. D., and Elston, R. A. 1988. Suspected toxicopathic hepatic necrosis and megalocytosis in pen-reared Atlantic salmon *(Salmo salar)* in Puget Sound, Washington, U.S.A. *Dis. Aquat. Org.* **4:**91–100.

Kliks, M. M. 1983. Anisakiasis in the western United States: four new case reports from California. *Am. J. Trop. Med. Hyg.* **32:**526–532.

Ko, R. C. 1976. Experimental infections of mammals with larval *Echinocephalus sinensis* (Nematoda: Gnathostomatidae) from oysters *(Crassostrea gigas). Can. J. Zool.* **54:**597–609.

Ledford, D. K., Overman, M. D., Gonzalvo, A., Cali, A., Mester, S. W., and Lockey, R. F. 1985. Microsporidiosis myositis in patient with the acquired immunodeficiency syndrome. *Ann. Intern. Med.* **102:**628–630.

Lichtenfels, J. R., and Brancato, F. P. 1976. Anisakid larva from the throat of an Alaskan Eskimo. *Am. J. Trop. Med. Hyg.* **25:**691–693.

Little, M. D., and Most, H. 1973. Anisakid larva from the throat of a woman in New York. *Am. J. Trop. Med. Hyg.* **22:**609–612.

Mace, T. F., and Anderson, R. C. 1975. Development of the giant kidney worm, *Dioctophyma renale* (Goeze, 1782) (Nematoda: Dioctophymatoidea). *Can. J. Zool.* **53:**1552–1568.

Margolis, L. 1977. Public health aspects of "codworm" infection: a review. *J. Fish. Res. Board Can.* **34:**887–898.

Margolis, L. 1982. Parasitology of pacific salmon—an overview. In: Meerovitch, E. (ed.) *Aspects of Parasitology*. McGill University, Montreal, pp. 135–226.

Margolis, L., Rausch, R. L., and Robertson, E. 1973. *Diphyllobothrium lobothrium ursi* from man in British Columbia—first report of this tapeworm in Canada. *Can. J. Publ. Hlth.* **64**:588–589.

Matthews, B. E. 1982. Behavior and enzyme release by *Anisakis* sp. larvae (Nematoda: Ascaridida). *J. Helminthol.* **56**:177–183.

McGladdery, S. E., Murphy, L., Hicks, B. D., and Wagner, S. K. 1990 . The effects of *Stephanostomum tenue* (*Digenea: Acanthocolpidae*) on marine aquaculture of the rainbow trout *Salmo gairdneri*. In: Perkins, F. O., and Cheng, T. C. (eds.). *Pathology in Marine Science*. Academic Press, San Diego, California, pp. 305–315.

McKerrow, J. H., Sakanari, J., and Deardorff, T. L. 1988. Anisakiasis: revenge of the sushi parasite. *N. Engl. J. Med.* **319**:1228–1229.

Measures, L. N. 1988a. The development of *Eustrongylides tubifex* (Nematoda: Dioctophymatoidea) in oligochaetes. *J. Parasitol.* **74**:294–304.

Measures, L. N. 1988b. Revision of the genus *Eustrongylides* Jägerskiöld, 1909 (Nematoda: Dioctophymatoidea) of piscivorous birds. *Can. J. Zool.* **66**:885–895.

Myers, B. J. 1979. Anisakine nematodes in fresh commercial fish from waters along the Washington, Oregon and California coasts. *J. Food Protect.* **42**:380–384.

Norris, D. E., and Overstreet, R. M. 1976. The public health implications of larval *Thynnascaris* nematodes from shellfish. *J. Milk Food Technol.* **39**:47–54.

Olson, R. E. 1987. Marine fish parasites of public health importance. In: Kramer, D. E., and Liston, J. (eds.) *Seafood Quality Determination*. Elsevier, The Netherlands, pp. 339–355.

Olson, A. C., Jr., Lewis, M. D., and Hauser, M. L. 1983. Proper identification of anisakine worms. *Am. Soc. Med. Technol.* **49**:111–114.

Oshima, T. 1972. *Anisakis* and anisakiasis in Japan and adjacent area(s). In: Morishita, N., Komiya Y., and Matsubayashi, H. (eds.) *Progress of Medical Parasitology in Japna, Vol. 4*. Meguro Parasitological Museum, Tokyo, pp. 303–393.

Oshima, T. 1987. Anisakiasis—is the sushi bar guilty? *Parasitol. Today* **3**:44–48.

Oshima, T., and Kliks, M. 1986. Effects of marine mammal parasites on human health. *Int. J. Parasitol.* **17**:415–421.

Overstreet, R. M. 1978. Marine Maladies? Worms, Germs, and Other Symbionts from the Northern Gulf of Mexico. *Mississippi-Alabama Sea Grant Consortium* MASGP-78-021. p. 140.

Overstreet, R. M. 1983. Aspects of the biology of the spotted seatrout, *Cynoscion nebulosus*, in Mississippi. *Gulf Res. Rep. Suppl.* **1**:1–43.

Overstreet, R. M. 1987. Solving parasite-related problems in cultured Crustacea. *Int. J. Parasitol.* **17**:309–318.

Overstreet, R. M., and Meyer, G. W. 1981. Hemorrhagic lesions in stomach of rhesus monkey caused by a piscine ascaridoid nematode. *J. Parasitol.* **67**:226–235.

Overstreet, R. M., Self, J. T., and Vliet, K. A. 1985. The pentastomid *Sebekia mississippi-ensis* sp. n. in the American alligator and other hosts. *Proc. Helminthol. Soc. Wash.* **52**:266–277.

Paperna, I., and Overstreet, R. M. 1981. Parasites and diseases of mullets (Mugilidae). In: Oren, O. H. (ed.) *Aquaculture of Grey Mullets*. International Biological Programme 26, Cambridge University Press, England, pp. 411–493.

Rausch, R. J., and Hilliard, D. K. 1970. Studies on the helminth fauna of Alaska. XLIX. The occurrence of *Diphyllobothrium latum* (Linnaeus, 1758) (Cestoda: Diphyllobothriidae) in Alaska, with notes on other species. *Can. J. Zool.* **48**:1201–1219.

Raybourne, R., Desowitz, R. S., Kliks, M. M., and Deardorff, T. L. 1983. *Anisakis simplex* and *Terranova* sp.: inhibition of larval excretory-secretory products of mitogen-induced rodent lymphoblast proliferation. *Exp. Parasitol.* **55**:289–298.

Raybourne, R., Deardorff, T. L., and Bier, J. 1986. *Anisakis simplex:* larval excretory secretory protein production and cytostatic action in mammalian cell cultures. *Exp. Parasitol.* **62**:92–97.

Roffe, T. J. 1988. *Eustrongylides* sp. epizootic in young common egrets *(Casmerodius albus)*. *Avian Dis.* **32**:143–147.

Rosen, L., Laigret, J., and Bories, S. 1961. Observations on an outbreak of eosinophilic meningitis on Tahiti, French Polynesia. *Am. J. Hyg.* **74**:26–42.

Ruitenberg, E. J. 1970. Anisakiasis: pathogenesis, serodiagnosis and prevention. Published doctoral dissertation. Rijkuniversiteit te Utrecht, The Netherlands, 138 pp.

Ruttenber, A. J., Weniger, B. G., Sorvillo, F., Murray, B. A., and Ford, S. L. 1984. Diphyllobothriasis associated with salmon consumption in Pacific coast states. *Am. J. Trop. Med. Hyg.* **33**:455–459.

Sakanari, J. A., and McKerrow, J. H. 1988. Characterization of the secreted proteases of *Anisakis* sp. Paper read at the 63rd annual meeting of The American Society of Parasitologists Winston-Salem, North Carolina, 31 July–4 August 1988, p. 43.

Sakanari, J. A., and McKerrow, J. H. 1989. A review of anisakiasis. *Clin. Microbiol. Rev.* **2**:278–284.

Sakanari, J. A., Loinaz, H. M., Deardorff, T. L., Raybourne, R. B., McKerrow, J. H., and Frierson, J. G. 1988. Intestinal anisakiasis: a case diagnosed by morphologic and immunologic methods. *Am. J. Clin. Pathol.* **90**:107–113.

Schantz, P. M. 1989. The dangers of eating raw fish. *N. Engl. J. Med.* **320**:1143–1145.

Schmidt, G. D. 1971. Acanthocephalan infections of man, with two new records. *J. Parasitol.* **57**:582–584.

Schmidt, G. D., and Roberts, L. S. 1981. *Foundations of Parasitology,* second edition. C. V. Mosby, St. Louis.

Sinovich. L. I. 1959. Nanophyetosis in the Soviet Far East. *10th Conference of Parasitological Problems and Diseases with Natural Reservoirs*. Academy of Sciences of the U.S.S.R., Moscow and Leningrad. **2**:410–411.

Smith, J. W., and Wootten, R. 1975. Experimental studies on migration of *Anisakis* sp. larvae (Nematoda: Ascaridida) into the flesh of herring, *Clupea harengus* L. *Int. J. Parasitol.* **5**:133–136.

Smith, J. W., and Wootten, R. 1978. *Anisakis* and anisakiasis. *Adv. Parasitol.* **16**:93–148.

Sprent, J. F. A. 1969. Helminth "zoonoses": an analysis. *Helminthol. Abstr.* **38**:333.

Tada, I., Otsuji, Y., Kamiya, H., Mimori, T., Sakaguchi, Y., and Makuzumi, S. 1983. The first case of a human infected with an acanthocephalan parasite, *Bolbosoma* sp. *J. Parasitol.* **69**:205–208.

Tsutsumi, Y., and Fujimoto, Y. 1983. Early gastric cancer superimposed on infestation of an anisakis-like larva: a case report. *Tokai J. Clin. Med.* **8**:265–273.

Turner, J. A., Sorvillo, S. J., Murray, R. A., Chin, J., Middaugh, J. P., Dietrich, P. D., Wiebenga, N. H., Googins, J. A., Allard, J., Ruttenber, A. J., Barr, D. B., Bier, J. W., Shandrack, P., and Swanson, J. W. 1981. Diphyllobothriasis associated with salmon—United States. *Morbid. Mortal. Wkl. Rep.* **30**:331–2, 337.

van Thiel, P. H., Kuipers, F. C., and Roskam, R. T. 1960. A nematode parasitic to herring, causing acute abdominal syndromes in man. *Trop. Geogr. Med.* **2**:97–111.

Vik, R. 1966. *Anisakis* larvae in Norwegian food fishes. In: *Proceedings of the First International Congress of Parasitology* **1**:568–569.

von Bonsdorff, B. 1977. *Diphyllobothriasis in Man*. Academic Press, London.

von Bonsdorff, B., and Bylund, G. 1982. The ecology of *Diphyllobothrium latum*. *Ecol. Dis.* **1**:21–26.

Ward, H. B. 1930. The introduction and spread of the fish tapeworm *(Diphyllobothrium latum)* in the United States. *De Lamar Lect.* 1929–1930:1–36.

Welberry, A. E., and Pacetti, W. 1954. Intestinal fluke infestation in a native negro child. *Bulletin of the Dade County Medical Association,* pp. 34 and 45.

Williams, J. F., and Ballard, M. A. 1982. Epidemiology of tropical zoonoses. In: Mettrick, D. F., and Desser, S. S. (eds.) *Parasites—Their World and Ours.* Elsevier, The Netherlands. Biomedical Press, pp. 366–368.

Williams, H. H., and Jones, A. 1976. Marine helminths and human health. In: *CIH Miscellaneous Publications* No. 3. Farnham Royal, Slough, U.K. Commonwealth Agricultural Bureaux, pp. 1–47.

Wittner, M., Turner, J. W., Jacquette, G., Ash, L. R., Salgo, M. P., and Tanowitz, H. B. 1989. Eustrongylidiasis—a parasitic infection acquired by eating sushi. *N. Engl. J. Med.* **320:**1124–1126.

Wootten, R., and Smith, J. W. 1979. The occurrence of plerocercoids of *Diphyllobothrium* spp. in wild and cultured salmonids from the Loch Awe area. *Scott. Fish. Res. Rep.* **13:**1–8.

World Health Organization. 1988. Parasitic diseases. Anisakiasis. *Wkl. Epidemiol. Rec.* **63:**311–314.

Yamane, Y., Kamo, H., Bylund, G. and Wikgren, B. P. 1986. *Diphyllobothrium nihonkaiense* sp. nov. (Cestoda: Diphyllobothriidae)—revised identification of Japanese broad tapeworm. *Shimane J. Med. Sci.* **10:**29–48.

Yokogawa, M., and Yoshimura, H. 1967. Clinicopathologic studies on larval anisakiasis in Japan. *Am. J. Trop. Med. Hyg.* **16:**723–728.

10

Nonindigenous Bacterial Pathogens

John E. Kvenberg

Seafood has the potential to pose a wide spectrum of public health problems from common yet harmful bacteria through contamination during production and distribution from the point of harvest to final preparation. Seafood-borne disease organisms can be divided into several groups based on the source of contamination. The focus of this chapter is on bacterial pathogens that, though often present in seafood, are not common to the marine environment. These pathogens are rather the result of direct fecal transmission from human or animal reservoir or a result of poor general sanitation.

Much human illness caused by foodborne bacterial pathogens results from fecal contamination of food either directly by ill or asymptomatic carriers who are food handlers or by infected domestic animals raised for food. Man, as a carrier of common bacteria such as *Salmonella* spp. and *Shigella* spp., can contribute to human illness by direct fecal–oral transmission of these types of pathogens by human vectors. In this case, seafood acts as a fomite and the level of contamination need not be great because an infectious dose can often be quite low if no final heat process is applied after preparation prior to consumption. This first category of risk includes species of fish eaten raw (sushi) and other aquatic food animals such as molluscan shellfish. Seafoods harvested from polluted waters or raised in aquaculture systems could be considered as vectors rather than fomites because of the close contact of the living product with man and his agricultural livestock in closed or confined harvest areas. Bacterial contamination also enters the seafood supply through a generalized pollution of open harvest areas in the aquatic environment from human sewage or animal waste runoff.

By far, the major opportunity for adulteration of seafood with enteric pathogens is caused by insanitary practices during product handling. Contamination and product temperature abuse can occur during all phases of the human food chain after harvest. This abuse offers the greatest potential chance of contaminated products to reach the consumer through introduction of filth to the product during processing or distribution of products. Products that are fully cooked or otherwise processed are subject to subsequent cross-contamination. Amplification of the numbers of these organisms by time–temperature abuse often coincides with insanitary handling, increasing the potential for disease. Presence of pathogens of terrestrial origin in these cases is not a certainty, but rather the opportunity for infection is provided by man's mishandling of product. Such organisms as all forms of *Salmonella, Campylobacter,* and *Listeria monocytogenes* and pathogenic forms of *Escherichia coli* and *Yersinia enterocolitica* are common in man's environment and the chance of human infection from seafood is increased with product abuse. These niches may not harbor common enteric pathogens common to the alimentary tract of mammals and birds.

This second group of contaminants also includes those found in the natural pristine marine or fresh water environment. These include *Vibrio* spp. which are discussed in Chapter 11 and *Clostridium botulinum* which exist in the water column or sediments of the harvest area.

The potential for contamination from the two sources discussed above may well grow in the future as a result of confounding factors such as the increase in consumer demand for seafood, the advent of massive global quantities of aquaculture products, and the loss of available harvest areas not contaminated by man. In addition, new processing and distribution techniques such as modified and vacuum packaging and sous vide processing may change the risks of potential disease outbreaks because microaerophilic or anaerobic conditions created by these technologies will allow bacterial growth patterns that are different from those normally encountered.

OPEN WATER HARVEST AND POLLUTION

The incidence of common enteric bacteria in freshly harvested seafood is largely dependent on the quality of water from which these products are harvested. An excellent example of control of these conditions can be found in the U.S. molluscan shellfish industry which relies chiefly on certification of safe harvest areas for oysters, mussels, and clams by classification of growing areas that are free from pollution as measured by the index of fecal coliform bacteria. The Interstate Shellfish Shippers Conference (ISSC) is administered by state agencies that classify growing areas as approved for harvest. "Bootlegging" or illegal harvest of closed areas is a continuous enforcement problem, but for the most part the program is successful in limiting disease outbreaks through control of harvest sites. Imported raw molluscan shellfish is likewise afforded a degree of protection by the existence of formal and audited Memoranda of Understanding between the U.S. Food and Drug Administration (FDA) and foreign governments. When uncertified imports of shellfish are encountered the FDA is required to notify ISSC member agencies or to take action to prevent distribution and sale of these products.

It has been suggested that the bacterial flora of all fish is a reflection of the aquatic environment from which they are harvested (Shewan and Hobbs 1967). One obvious implication of this theory is that the microbiological quality of the growing waters affects mobile or migratory species as well as sedentary shellfish. Contamination by enteric bacteria in polluted harvest areas is sporadic and difficult to control but must be considered. It has been suggested further that although the species of fish harvested may play a minor role in the occurrence of certain microbial flora in seafood, environmental factors appear to predominate (Ward 1989). These factors include the presence of human and animal sources of enteric bacteria in the ocean environment. Some comfort can be taken in this idea because there are more than 250 commercial species of seafood harvested from U.S. waters and over 3,000 species world wide (Otwell 1989). Considering each species separately seems to be an insurmountable task.

POLLUTION AND AQUACULTURE SYSTEMS

Growing fish in ponds purposefully contaminated with waste water has been a long standing practice in some parts of the world. This is done to provide more available nutrients and increase yield. It has been demonstrated that fish grown in these ponds accumulate fecal bacteria from effluent and that at levels in water above 10^4/ml these bacteria become detectable in muscle tissue (Ward 1989). The possibility of fish so raised becoming reservoirs of human infection has been suggested (Janssen and Meyers 1968).

Today some 12% of the American fish consumer market is supplied from aquaculture sources. On a global scale, the seafood supply is estimated to be 90 million metric tons annually and 10 million tons of this total is from aquaculture production (Redmayne 1989). This forebodes the potential for new microbiological problems from seafood because of the closeness of man and the product during the growing phase of the food cycle as well as contact during production, distribution, and sale of products.

RELATIVE INCIDENCE OF ENTERICS IN SEAFOOD ILLNESS

The number of outbreaks and illnesses reported to the Centers for Disease Control (CDC) is quite minor when compared to the billions of pounds of seafoods consumed. The data, however, put in perspective the relative importance of enteric pathogens with other causes of disease. For the period 1973–1987, CDC data can be summarized according to two major classifications. These are shellfish and finfish/other species. A comparison of bacteria versus other causes is listed in Tables 10.1 and 10.2. Among shellfish outbreaks

Table 10.1 CDC Reported Outbreaks and Illness
from Shellfish 1973–1987

Vehicle	Outbreaks	Illness
Enterics	13	205
C. perfringens	2	28
Salmonella	3	80
Shigella	4	77
S. aureus	2	14
B. cereus	2	6
Vibrio		
V. parahaemolyticus	18	298
V. cholerae	3	16
V. cholerae (non-0-1)	2	11
Virus	11	377
Hepatitis A	9	335
Other virus	2	42
Shellfish Toxins	21	160
Paralytic (PSP)	19	155
Neurotoxic	2	5
Other Chemical	1	57
Unknown	144	?

Table 10.2 CDC Reported Outbreaks and Illness from Finfish
and Other Species Excluding Shellfish 1973–1987

Vehicle	Outbreaks	Illnesses
Enterics	51	320
C. botulinum	35	77
C. perfringens	3	114
Salmonella	4	96
S. typhi	1	25
Shigella	3	60
S. aureus	3	32
Enterococcus	1	12
B. cereus	1	4
Vibrio	3	130
V. parahaemolyticus	1	122
V. cholerae	2	8
Parasites	6	46
Hepatitis A	1	46
Scombroid poisoning	199	1,200
Ciguatera	232	1,048
Other Chemical	4	24
Unknown	44	?

enteric pathogens represented 13 of 213 incidents and accounted for 205 of 1,124 total cases. By comparison *Vibrio parahaemolyticus* alone accounted for 18 outbreaks and 298 cases. Hepatitis A virus was recorded in 9 outbreaks with 335 cases. Among finfish outbreaks, 51 of 540 were due to enterics and 320 cases of illness were associated with this class of seafood.

CLOSTRIDIUM BOTULINUM

The organism that evokes the most reaction among seafood safety professionals is *Clostridium botulinum* because very small amounts of botulinal toxin can cause human paralyses and in some cases death. The clostridia are anaerobic, Gram-positive sporeformer organisms that can grow between pH 4.6 and 8.5. These characteristics allow *C. botulinum* to survive boiling temperatures and multiply in the absence of oxygen. It is of particular concern to the seafood industry because spores are present in the intestinal tract of finfish, the gills, and viscera of crabs and other shellfish.

Incidence and Symptoms of Botulism

There were 35 outbreaks of botulism from finfish and 77 attendant illnesses reported to CDC over the period 1973–1987. These numbers do not however reflect a level of concern over commercially produced seafood as many of the reported cases were among the native Alaskan population who often ferment various fish in anerobic environments and thus do not reflect industry practices.

Botulism is a rare form of foodborne disease but its occurrence causes great concern because of its life-threatening nature. The onset of symptoms usually occurs 18–36 hours after the toxic food is eaten but may not appear for up to 8 days. The early indications of botulism intoxication are lassitude, weakness, and vertigo followed by double vision and paralysis of the neck muscles which cause difficulty in speech and swallowing. Symptoms become more severe and progress downward through the body. Eventually, the victim has difficulty breathing because of paralysis of the diaphragm and chest muscles and must be ventilated to breathe or death from asphyxia will result. The toxic dose of botulism toxin is very small. A few nanograms of toxin are sufficient to cause symptoms. Another concern is the long-term effects of intoxication. Botulinal toxin causes severe and prolonged neuromotor impairment that may be permanent. The toxin acts by permanently attaching to the myoneural junction, blocking motor nerve impulses and causing paralysis. It is for this reason that antitoxin should be administered quickly to neutralize the effect.

There are seven serologically identifiable toxins of *C. botulinum* designated as Types A through G. Proteolytic strains (all Type A, some Type B and F) require temperatures above 50°F (10°C) for growth and subsequent toxin formation (Sperber 1982). Nonproteolytic strains on the other hand (all Type E, some Type B, and F) can both grow and produce toxin at temperatures as low as 37.4°F (3°C) (Abrahamsson et al. 1966; Solomon et al. 1982). Salt concentrations of 5% inhibit growth of Type E but not Types A and B, which tolerate up to 10% or an A_w of 0.93 (Sakaguchi 1979).

Clostridium botulinum is rather ubiquitous in the natural environment and can be found in ocean sediments as well as in animal, bird, and fish intestines. Type E of *C. botulinum* has been frequently shown as a contaminant of seafood and is primarily of marine origin. This type has been isolated from a number of coastal areas and from various seafood species (Eklund and Poysky 1965, 1967; Nickerson et al. 1967).

In addition to the presence of botulinal spores, other factors must be present for the seafood product to become toxic. The product must receive a treatment that is lethal or at least inhibitory to vegetative bacteria cells. This allows successful competition, growth, and toxin production. Inadequate processing is the chief cause of this selective process and can happen in several ways. Low heat treatments that are intentional or accidental are the primary cause of this failure. A fermentation or acidification process that operates at a pH range above 4.6 or a brining process with low salt levels also can allow survival and growth.

With these conditions in place, the surviving spores then germinate and multiply under the proper circumstances, ultimately making the product toxic. Seafood held under refrigeration temperatures above 38°F (3.3°C) can support the growth of Type E organisms (Schmidt 1961). Because Type E can grow at low temperatures, refrigerated storage is not a barrier to outgrowth and may in fact enhance the competitive capability of this strain. Therefore current refrigeration practices can be effective only against some types of *C. botulinum* and frozen storage is the optimal preservation system (USDHEW 1963).

The advent of vacuum and modified atmosphere packaging increases the anaerobic environment necessary for the growth of *C. botulinum*. Because several outbreaks of botulism have been attributed to vacuum packaged fish, concern is voiced over this practice. Vacuum packaging is not a requirement for toxin formation but does contribute significantly. Spores adjacent to the intestine or under the skin of fish have anaerobic conditions in all instances, but vacuum packaging allows spores on the surface to germinate and thus produce toxin (Bryan 1973).

Finally, for an intoxication to develop, the product must be served without an adequate preparatory cooking step before eating because the preformed toxin is inactivated at 140°F (60°C) within 5 minutes (Sakaguchi 1979).

Botulism from Smoked Fish

Smoked fish from the Great Lakes became a significant source of botulism in the 1960s. During that time, three recorded outbreaks of human botulism were caused by vacuum packaged smoked ciscos, vacuum packaged smoked chubs, and smoked whitefish. Ten deaths were attributed to these outbreaks (USDHEW 1970). It is significant to note that in two of these instances vacuum packaging was involved. In all smoked fish outbreaks the organism was Type E; evidently the process provided did not inactivate the Type E spores (Eklund 1982).

One outbreak that occurred in the 1960s is worthy of mention because it typifies inadequate food sanitation practices that lead to botulism. Whitefish that were harvested from Lake Michigan caused 15 cases of Type E botulism several weeks after processing. Of the 15 cases reported, five individuals died. All had eaten product that was plastic wrapped and temperature abused. The fish were eviscerated and iced on the catch boat 1 day prior to smoking at a temperature of 180°F (82.2°C) as measured by air temperature, for a processing time of 5 hours. The product was then chilled and vacuum packaged the following day. Product was shipped in an unrefrigerated van and was in transit for several days. Some product remained after the outbreak was discovered and approximately one-third of the unsold packages was the fish that contained Type E toxin (Bryan 1973).

Another seafood that caused cases of botulism was salted, uneviscerated whitefish. "Kapchunka" is an ethnic food that is intended for consumption without heat treatment. There have been three outbreaks of botulism associated with kapchunka in recent years, leading to a ban on this product for public health reasons. A classic example of dose–response was demonstrated in 1987 when eight cases of Type E botulism were attributed to Kapchunka. The fish that caused the outbreak contained high levels of Type E toxin. All eight victims developed symptoms within 36 hours, one died, and two required ventilation on a respirator. Three were treated with antitoxin and the rest recovered spontaneously (Conner et al. 1989).

LISTERIA MONOCYTOGENES

Listeria monocytogenes causes the disease listeriosis and is of great concern to special at-risk groups. The susceptible groups include pregnant women and their fetuses, cancer patients and others undergoing immunosuppressive therapy, as well as diabetics and cirrhotics and the elderly. Although the risk of contracting listeriosis is less for normal, healthy individuals, they may also contract the disease. Typical *Listeria* infections result in septicemia, meningitis, and encephalitis, although enteritis has been reported (Janssen and Meyers 1968). Mortality is high among those infected as evidenced by a 29% death rate among patients in a New England outbreak associated with fluid milk (Fleming et al. 1985).

Incidence and Symptoms of Listeriosis

There have been no cases of listeriosis reported in the U.S. that were linked to the consumption of seafood. *Listeria monocytogenes* is an interesting pathogen because it is a facultative intracellular parasite. The organism enters the body through the intestine and has a variable incubation period that may be as short as 1 day to a month or longer. The ingested cells enter the body through ileal villi cells and are subsequently taken up by macrophage cells in the bloodstream. Instead of being digested, the engulfed cells multiply inside the host cell until the macrophage bursts and liberates the *L. monocytogenes* cells to repeat the process. This causes the transitory flu-like symptoms often reported at the initial stage of disease. The enteric phase of the disease is not consistent; some report upset stomach and diarrhea whereas other victims do not display these symptoms. The actual disease known as listeriosis does not occur until a severe form of septicemia, encephalitis, lesions, or meningitis develops. All of these forms of listeriosis may accompany infection of those who are not immunocompetent (Lovett 1989).

Listeria spp. have had little documented association with seafood until very recently. It has long been known that *Listeria* is widely distributed in the environment, having been isolated from soil, water, humans, and a variety of animals (Gray and Killinger 1966). Of particular interest, the association of seagulls as vectors for *L. monocytogenes* is worthy of note (Fenlon 1985).

Seafood as a Cause of Human Listeriosis

There is little known association between seafood and listeriosis. An outbreak of perinatal listeriosis was reported from three obstetric hospitals in Auckland, New Zealand. Most of the 22 cases were due to strain 1b. The cause of the outbreak was not discovered but consumption of raw shellfish and raw fish may have played a role (Lennon et al. 1984).

Food surveillance for *L. monocytogenes* has increased since 1987 after the organism was found in refrigerated and frozen crabmeat.

Listeria are relatively heat resistant but usually are not present in foods receiving an adequate heat treatment. The presence of *L. monocytogenes* in cooked seafood indicates cross-contamination of the product or underprocessing. Several characteristics of *L. monocytogenes* are worth mentioning in relationship to seafood. The *Listeria* are aerobic under most circumstances but can be facultatively anaerobic and thus grow well under reduced levels of oxygen in packaged products. *Listeria* can also survive and grow under adequate refrigeration temperature. These bacteria also survive freezing well, thus making adequate cooking and prevention of recontamination extremely important.

Identification of *Listeria* Strains

Listeria are commonly identified by serotyping. Types 1–7 are known, with Types 1/a, 1/b, and 4b predominating as both environmental and clinical isolates. The serotyping scheme is based on both somatic and flagellar antigens. Phage typing has also been employed as a method of further identifying isolated strains. There are currently 27 phages used in the typing system of Audurier (Mclaughlin et al. 1986).

STAPHYLOCOCCUS AUREUS

Seafoods, regardless of species source, can support the growth of *Staphylococcus aureus* when contaminated and subsequently subjected to temperature abuse because they are high in protein content and fulfill the nutritional needs of the microorganism. Typical outbreaks of staphylococcal seafood poisoning occur in cooked products such as smoked fish or crab because heating destroys competitive microorganisms, allowing staphylococci to predominate. Another factor that allows competitive advantage is the presence of minimal salt levels in cured products. *Staphylococcus aureus* has a very high level of salt tolerance. Products such as salt cured fish successfully inhibit most microorganisms but support the growth of staphylococci with an A_w above 0.86 and allow the production of staphylococcal enterotoxin (Tatini 1973).

Staphylococcus aureus is a spherical or coccus-shaped Gram-positive organism that appears under the microscope in pairs, short chains, or grapelike clusters. The overwhelming majority of strains of *S. aureus* that produce enterotoxin also produce coagulase which has the ability to clot blood plasma. Although some coagulase-negative strains produce enterotoxin, these traits are so closely linked that a coagulation test is used to predict toxigenic potential. A second predictive test is the presence of thermonuclease. There are five serological types of staphylococcal enterotoxin designated A–E. Most food outbreaks involve types A or D (Sperber 1977).

Incidence and Symptoms of Enterotoxicosis

There have been three outbreaks of staphylococcal enterotoxicosis and 32 cases reported to CDC over the period 1973–1987 associated with finfish. Over the same period two outbreaks and 14 cases were reported as associated with shellfish.

It has been estimated that there must be about 1,000,000 organisms/g in food to produce food poisoning symptoms but the organism itself need not be present if the toxin has been preformed. From a study of 16 incidents of foods implicated in outbreaks, levels of <0.01–0.25 $\mu g/g$ of enterotoxin were detected (Gilbert and Wieneke 1973).

The onset of symptoms of enterotoxin poisoning are rapid and acute. Common reactions include nausea followed by vomiting and prostration. In more severe cases transient changes in blood pressure and pulse rate occur. Death from staphylococcal food poisoning is very rare. Although the possibility of death from staphylococcal enterotoxin illness is not frequently mentioned in the literature, it should be noted that the elderly and infants are severely debilitated and are at greatest risk.

Of greatest concern to those producing seafood is the source of *S. aureus* and what can be done to control its presence. Human and domestic animals are the primary reservoirs. Staphylococci are present on the skin, hair, nasal passages, and throats of 50% or more of healthy individuals and carriers or infected food handlers can easily transmit staphylococci to seafood (Sneath et al. 1986). Although food handlers are usually the main source of food contamination, equipment and food contact surfaces can often be fomites to spread contamination.

Enterotoxins are heat resistant. Temperatures of 176°F (80°C) for 3 minutes and 212°F (100°C) are needed to cause loss of serological detection of staphylococcal enterotoxin A (Bergdoll 1979). It appears that inactivation of enterotoxin is dependent on the level present (Deny et al. 1966).

Some degree of controversy exists on the ultimate level of the survival of the biological activity of staphylococcal enterotoxin in retorted or pasteurized foods. Biological activity measured by cat emetic response was lost at 11 minutes at a temperature of 482°F (250°C) (Deny et al. 1966, 1971). The experiments of Bennett and Berry indicate the retention of biological activity through the retorting process in some foods (Bennett and Berry 1987). Caution should be taken with the presence of *S. aureus* in canned seafood products to avoid the possibility of some active preformed toxin remaining.

SALMONELLA

The occurrence of *Salmonella* in fish and shellfish, either in fresh or marine waters, has normally been associated with fecal contamination of the area from which they were harvested (Buttaiux 1962). Concern over *Salmonella* contaminated seafood is not new. In the early years of this century the shellfish industry was plagued with *S. typhi* as a primary pathogen found in raw shellfish harvested from polluted waters. Reports as early as 1954 have indicated that *Salmonella* also could be consistently isolated from fin fish harvested in highly polluted waters (Floyd and James 1954). There is some indication that both fresh and marine species exposed to *Salmonella* in polluted waters remain positive for up to 30 days after exposure of fin fish and shrimp (Lewis 1975).

The salmonellae are rod-shaped, non-spore-forming, Gram-negative bacteria. There are over 2,000 serotypes of salmonellae recognized at this time and more are added to the list every year. The species name for *Salmonella* is determined by unique serotype determination of somatic and flagellar antigens.

The primary location of *Salmonella* spp. is the alimentary tract of mammals, birds, amphibians, and reptiles. Salmonellae are not endemic to the intestinal tracts of finfish, crustaceans, or molluscs. Because this organism is fecal in origin, it can be found in processed seafood through pollution of the water environment or by contamination of the seafood after catch.

Incidence and Symptoms
of Salmonellosis

Data for both finfish and shellfish for the period 1973–1987 show that there were four outbreaks and 96 cases of salmonellosis from finfish. One outbreak of *S. typhi* with 25 cases and three outbreaks with 80 cases of other *Salmonella* spp. from shellfish were reported to CDC.

There are many cases of salmonellosis each year in the U.S. The author believes 2 million cases of human salmonellosis is a realistic estimate. Although salmonellosis is a reportable disease, only a fraction of a percent of the actual disease is reported to the CDC in Atlanta, Georgia, each year. Symptoms of acute salmonellosis are nausea, vomiting, abdominal cramps, diarrhea, fever, and headache. The onset time is usually 6–48 hours but may be no longer with low doses of virulent strains. The infective dose for *Salmonella,* which is quite variable and dependent on serotype and other factors, may range from only a few organisms to over 10^5 to cause illness. The general susceptibility of the person exposed is as important as other factors in illness. The infectious dose may be as low as 15–20 cells, depending on varying virulence factors, age, and general health status of the victim (D'Aoust and Pivnick 1976). Persons of all ages are susceptible to *Salmonella* infections but symptoms are more severe and prolonged among the elderly, infants, and people with underlying illnesses. The acute symptoms may be short in duration (1–2 days) or extended (1 week or more). The duration of illness seems to be affected by the same factors as infectious dose. Several complications can arise from acute salmonellosis. These are septicemia, local lesions, and reactive or sterile site arthritic conditions.

Salmonellosis (Acute Enteritis)

The most common disease associated with *Salmonella* is gastroenteritis, commonly referred to as salmonellosis. The symptoms of intestinal illness occur with an onset time of 6–48 hours. Symptoms include nausea, vomiting, abdominal cramps, diarrhea, fever, and headache. These symptoms normally persist for several days and the illness is self-limiting. The victim, however, may continue to shed *Salmonella* for periods of several weeks or months. These individuals are known as carriers and account for many cases of salmonellosis through person-to-person contact and food preparation activities. The *Salmonella* carrier state is asymptomatic and the host may shed up to 10^8 organisms/g for as long as 6 months after an episode of illness (Frishl et al. 1986).

Typhoid and Paratyphoid Fever

The most serious forms of salmonellosis are typhoid and paratyphoid fever caused by *S. typhi* and the paratyphoid serotypes of *Salmonella*. The septicemia and secondary sequelae caused by these organisms can be life threatening if not successfully treated. The septicemia and organ lesions caused by *S. typhi* and *S. paratyphi* A, B, and C are the cause of prolonged clinical illness of great severity. The fatality rate for typhoid fever is 10% compared to <1% for most serotypes of salmonellae. There are other species that have a percentage of human septicemia associated with infection. *Salmonella dublin*, for instance, has a 15% mortality rate associated with it among the elderly when septicemia develops. *Salmonella enteritidis* has shown a 3.6% mortality rate among hospital and nursing home outbreaks (D'Aoust and Pivnick 1976).

The presence of any type of *Salmonella* contamination of food should be taken seriously because of these grave consequences to enteritis. *Salmonella* septicemia has been associated with subsequent infection of all body organs and a sterile arthritis has been reported as associated with a small percentage of foodborne disease victims. In addition to reactive arthritis, Reiter's syndrome, sometimes referred to as Reiter's triad, occurs. In this situation joint inflammation is accompanied by conjunctivitis and urethritis (Archer 1988).

Reports as early as 1954 have indicated that *Salmonella* could be consistently isolated from fish harvested in highly polluted waters (Floyd and Jones 1954). There is some indication that both fresh and marine species exposed to *Salmonella* in polluted waters remain positive for up to 30 days after exposure of finfish and shrimp (Lewis 1975).

Salmonella in Aquaculture

Seafood can be contaminated with *Salmonella* in aquaculture systems from many sources including farm runoff and direct fecal contamination from livestock or feed. As an example, domestically produced, pond reared catfish carry *Salmonella* at an apparently low level of contamination. A survey of retail samples of raw catfish fillets conducted from product grown in nine southeastern states indicated an overall incidence of 5% (Andrews et al. 1977). The flow of *Salmonella* in the food chain can be complex. This is especially true in aquaculture systems. The role of fish meal, for instance, can ultimately affect the food supply of other commodities. *Salmonella* organisms can survive drying and thus be a problem in fish meal that is fed to livestock and poultry. Concern for the prevalence of *S. agona* in fish meal products became a national concern in the 1960s (Garrett 1973).

Salmonella in Frog Legs

The fact that frogs used for food frequently harbor salmonellae is well established. The reason for this association is that the growing environment is often heavily polluted and subsequent contamination of frog legs occurs during slaughter from the animal's intestinal contents or by cross-contamination during handling. As a consequence the level and kinds of *Salmonella* are high. It is not unusual to recover several types of *Salmonella* from a single lot. Although frog legs have not been implicated as a direct cause of foodborne illness, the contamination of raw product may contribute to dissemination of *Salmonella* in the food chain (WHO 1989).

Salmonella in Smoked Fish

Smoked fish have been a traditional source of salmonellosis outbreaks and minor epidemics over the early and mid-century decades in the U.S. A total of 33 common source outbreaks have occurred between 1940 and 1965 (Bryan 1973).

Typical of these outbreaks was an instance involving *S. newport* where the cause of three separate outbreaks traced to one fish processing plant that had many elements enabled the outbreak to occur. Proper hygiene was not observed and the fish were temperature abused subsequent to their sale. The operation was producing fish under extremely poor sanitary conditions. A human carrier who packed the smoked fish and had been previously ill with salmonellosis was identified and product temperature abuse also contributed to this series of outbreaks (Olitzki et al. 1956).

A second epidemic from *S. java* contaminated white fish illustrates similar points. More than 300 people became ill from contaminated white fish produced in a plant that had been damaged by fire. In addition to poor general sanitation conditions that existed at the time of these outbreaks, water used to wash fish was taken from waters where coliform counts were high (Gangarosa et al. 1968).

Salmonella in Molluscan Shellfish

Molluscan shellfish species require special sanitary practices from harvest to consumption because they are frequently consumed raw. The early public health safeguards put in place for shellfish were developed to protect the public from outbreaks of typhoid fever from oysters and clams infected with *S. typhi*. In the case of *S. typhi* infection, the source of infection is either from sewage pollution or a carrier of the organism. An outbreak of typhoid fever occurred in the southeastern U.S. because a carrier working as an oyster tonger contaminated the shell stock (Bond et al. 1962).

SHIGELLA

The *Shigella* are a genera closely related to *Escherichia* that have long been known as foodborne pathogens. The early known cause of bacillary dysentery was the organism now classified as *S. dysenteriae*. It predominated as a food-borne disease in the early 20th century because of poor sanitary control of human waste. This species is little known in developed countries today, being replaced by the species *S. flexneri* and *S. sonnei* which are usually implicated when a human carrier infects another person (Smith 1987). These transmissions can be either waterborne or foodborne via the fecal–oral route. A lesser encountered species, *S. boydii,* is also a human pathogen.

Shigella spp. are principally thought of as waterborne pathogens. They are however also of concern in foods. The association with both food and water underscores the importance of these pathogens. The shigellae are rod-shaped, nonmotile, non-spore-forming Gram-negative bacteria. Heat shock proteins or other related stress-induced proteins have been linked to virulence in some pathogens, including the shigellae.

Incidence and Symptoms of Shigellosis

According to CDC data there were three outbreaks of shigellosis, with 60 cases reported from finfish and four outbreaks with 77 cases reported from shellfish for the period 1973–1987.

The disease shigellosis is usually spread person-to-person but seafood can serve as a fomite for infection. This disease is rare among domestic animals, being principally a disease of man. Symptoms include abdominal pain, cramps, diarrhea, fever, vomiting, and blood, pus, or mucus in stools. The destructive nature of the disease is due to the ability of this organism to attach, invade, and destroy epithelial cells of the intestine. The infectious dose of the shigellae is quite low; only a few organisms need be ingested to cause disease. Infants, the elderly, and those of compromised health are more severely affected by this disease but all people are susceptible. Complications after infection include ulceration of the mucosal membrane, bleeding, and severe dehydration. Death can occur in as many as 10–15% of cases in some outbreaks. Of great importance is the fact that *Shigella* spp. have a low infectious dose; <100 organisms can cause disease. Recently the spread of a strain with multiple antibiotic resistance characteristics has been discovered in *S. sonnei* which limits the means to treat the disease effectively.

Of the 72 foodborne outbreaks of shigellosis reported to CDC between 1961 and 1975, five involved shrimp or tuna salads and one was caused by raw oysters contaminated from polluted waters (Black et al. 1978). This trend continues at low levels because *Shigella* spp. is an incidental foodborne contaminant that is directly linked to man.

Shigella in Shrimp and Oysters

Shrimp-borne shigellosis has become a major concern after the 1983 outbreak of *S. flexneri* in the Netherlands where cooked product caused illness and death (Bijkerk and van Os 1984). In a recently reported outbreak of *S. sonnei* from raw oysters, 24 persons became ill from a supply traced to a single boat. Investigations led to the discovery that 5-gallon pails were used as toilets aboard oysters boats and that they were dumped overboard after use (Reeve et al. 1989).

The essentials of preventing transmission of seafood-borne shigellosis are sanitation and good personal hygiene habits. Outbreaks occur because a human carrier of *Shigella* contaminates his hands and subsequently the product with fecal material.

CONCLUSIONS

Seafood is a commodity that has a good history of food safety but will always possess the potential to cause major if sporadic outbreaks of disease. An understanding of the major foodborne pathogens and the seafood associated problems they have presented in the past allows for focused attention to control measures that can effectively eliminate or greatly reduce the risk of human illness from these products. Food sanitation and quality control measures that consider the safety of seafood from harvest to consumption must be developed in the future to ensure a continued high level of safety.

REFERENCES

Abrahamsson, K., Gullmar, B., and Molin, N. 1966. The effect of temperature on toxin formation and toxin stability of *Clostridium botulinum* type E in different environments. *Can. J. Microbiol.* **12**:385–394.

Andrews, W. H., Wilson, C. R., Poelma, P. L., and Romero, A. 1977. Bacteriological survey of the channel catfish *(Ictalurus punctatus)* at the retail level. *J. Food Sci.* **42**:359.

Archer, P. L. 1988. The true impact of foodborne infections, *Food Technol.* **42**(7):53–58.

Bennett, R. W., and Berry, M. R. 1987. Serological reactivity and in vivo toxicity of *Staphylococcus aureus* enterotoxins A and D in selected canned foods. *J. Food Sci.* **52**:416–418.

Bergdoll, M. S. 1979. Staphylococcal intoxications. In: Reimann E. R., and Bryan F. L. (eds) *Foodborne Infections and Intoxications*. Academic Press, New York, pp. 443–494.

Bijkerk, H., and van Os, M. 1984 Bacillaire dysenterie *(Shigella flexneri* type 2). door garnalen. *Ned. T.V. Geneesk.* **128**:431.

Black, R. E., Craun, G. F., and Blake, P. A. 1978. Epidemiology of common-source outbreaks of shigellosis in the United States 1961–1975. *Am. J. Epidemiol.* **108**:47–52.

Bond, J. O., Murphey, W. J., Smith, W. H. Y., and Kooman, J. 1962. Typhoid fever related to raw oyster consumption—southeastern. *Weekly Morbid. Mortal. Rep.* **11**:98–99, 104.

Bryan, F. L. 1973. Activities of the Center for Disease Control in public health problems related to the consumption of fish and fishery products. In: Chitchester, C. O., and Grahm, H. D. (eds) *Microbial Safety of Fishery Products.* Academic Press, New York, pp. 273–307.

Buttaiux, R. 1962. Salmonella problems in the sea. In: Borgstrom, G. (ed) *Fish as Food, Vol. 2.* Academic Press, New York, p. 503.

Conner, D. E., Scott, V. N., Bernard, D. T., and Kautter, D. A. 1989. Potential *Clostridium botulinum* hazards associated with extended shelf-life refrigerated foods: A review. *J. Food Safety.* 10:131–153.

D'Aoust, J. Y. and Pivnick, H. 1976. Small infectious dose of *Salmonella. Lancet* i: 866.

Deny, C. B., Tan, P. L., and Borher, C. W. 1966. Heat inactivation of staphylococcal enterotoxin. *J. Food Sci.* **31**:762–767.

Deny, C. B., Humber, J. Y., and Boher, C. W. 1971. Effect of toxin concentration an heat inactivation of staphylococcal enterotoxin A in beef bouillon and in phosphate buffer. *Appl. Microbiol.* **21**:1064–1066.

Eklund M. W. 1982. Significance of *Clostridium botulinum* in fishery products preserved short of sterilization. *Food Technol.* **36**(12):107–112.

Eklund, M. W., and Poysky, F. T. 1965. *Clostridium botulinum* type E from marine sediments. *Science* **149**:306.

Eklund, M. W., and Poysky, F. T. 1967. Incidence of *C. botulinum* type E from the Pacific coast of the United States. In: Ingram, M. and Roberts, T. A. (eds) *Botulism 1966.* Chapman and Hall, London, pp. 49–55.

Fenlon, D. A. 1985. Wild birds and silage as reservoir of *Listeria* in the agriculture environment. *J. Appl. Bacteriol.* **59**:537.

Fleming, D. W., Cochi, S. L. MacDonald, K. L., Brondum, J., Hayes, P. S., Plikaytis, B. D., Holmes, M. B., Audurier, A., Broome, C. V., and Reingold, A. L. 1985. Pasteurized milk as a vehicle of infection in an outbreak of listeriosis. *N. Engl. J. Med.* **312**:404–407.

Floyd, T. M., and Jones, G. B. 1954. Isolation of *Shigella* and *Salmonella* organisms from Nile fish. *Am. J. Trop. Med. Hyg.* 3:475.

Frishl, M. A., Dickinson, G. M., Sinare, C., Pitchenick, A. E. and Cleary, T. J. 1986. *Salmonella* bacteremia as a manifestation of acquired immunodeficiency syndrome. *Arch. Intern. Med.* **146**:113–115.

Gangarosa, E. J., Bisno, A. L., Eichner, E. R., Treger, M. D., Goldfield, M., DeWitt, W. E., Fodor, T., Fish, S. M., Dougherty, W. J., Murphy, J. B., Feldman, J., and Vogle, H. 1968. Epidemic of febrile gastroenteritis due to *Salmonella java* traced to smoked whitefish. *Am. J. Publ. Hlth.* **58**:114–121.

Garrett, S. E. 1973. U.S. salmonella control program relating to fishmeal. In: Chitchester, C. O., and Graham, H. D. (eds) *Microbial Safety of Fishery Products.* Academic Press, New York, pp. 131–135.

Gilbert, R. J., and Wieneke, A. A. 1973. Staphylococcal food poisoning with special reference to the detection of enterotoxin in foods. In: Hobbs B. C., and Christian J. H. B. (eds) *The Microbiological Safety of Food.* Academic Press, New York, pp. 273–285.

Gray, M. L., and Killinger, H. A. 1966. *Listeria monocytogenes* and *Listeria* infections. *Bacteriol. Rev.* **30**:309–382.

Ho, J. L., Shands, K. N., Friedland, G., Ecklind, P., and Fraser, D. W. 1986. An outbreak of type 4b *Listeria monocytogenes* infection involving patients from eight Boston hospitals. *Arch. Intern. Med.* **146**:520–524.

Janssen, W. A., and Meyers, C. D. 1968. Fish: serologic evidence of infection with human pathogens. *Science* **159**:547–548.

Lennon, D., Lewis, B., Mantell, C., Becroft, D., Dove, B., Farmer, K., Tonkin, S., Yeates, N., Stamp, R., and Mickleson, K. 1984. Epidemic perinatal listeriosis. *Pediatr. Infect. Dis.* **3**:30–34.

Lewis, D. H. 1975. Retention of *Salmonella typhimurium* by certain species of fish and shrimp. *J. Am Vet. Med. Assoc.* **167**:551.

Lovett, J. 1989. *Listeria monocytogenes:* In: Doyle M. P. (ed) *Foodborne Bacterial Pathogens*. Marcel Dekker, New York, pp. 283–310.

Maurelli, A. T., and Sansonetti, P. J. 1988. Identification of a chromosomal gene controlling temperature-regulated expression of *Shigella* virulence. *Proc. Natl. Acad. Sci. USA* **85**:2820–2824.

Mclaughlin, J., Audurier, A., and Taylor, A. G. 1986. The evaluation of a phage-typing system for *Listeria monocytogenes* for use in epidemiological studies. *J. Med. Microbiol.* **22**:357–365.

Nickerson, J. T. R., Goldblith, S. A., Digioia, G., and Bishop, W. W. 1967. the presence of *Clostridium botulinum* type E in fish and mud taken from the Gulf of Maine. In: Ingram M. and Roberts T. A. (eds) *Botulism 1966*. Chapman and Hall, London, pp. 25–33.

Olitzki I., Perri, A. M., Shiffman, M. A., and Werrin, M. 1956. Smoked fish as a vehicle of salmonellosis. *US Publ. Hlth. Rep.* **71**:773–779.

Otwell, W. S. 1989. Regulatory status of aquacultured products. *Food Technol.* **43**(11):103–105.

Redmayne, P. C. 1989. World aquaculture developments. *Food Technol.* **43**(11):80–81.

Reeve, G., Martin, D. L., Pappas, J., Thompson, R. E., and Green, K. D. 1989. An outbreak of shigellosis associated with the consumption of raw oysters. *N. Engl. J. Med.* **321**:224–227.

Sakaguchi, G. 1979. Botulism. In: Reimann H. (ed) *Food-Borne Infections and Intoxications*. Academic Press, New York, pp. 390–433.

Schmidt, C. F., Lechowich, R. V., and Folinzao, J. J. 1961. Growth and toxin production by Type E *Clostridium botulinum* below 40°F. *J. Food Sci.* **26**:626–630.

Shewan, J. M., and Hobbs, G. 1967. The bacteriology of fish spoilage and preservation. In: Hackenhull D. J. D. (ed) *Progress in Industrial Microbiology*. Iliffe Books, London.

Smith, J. L. 1987. Shigella as a foodborne pathogen. *J. Food Protect.* **50**:788–801.

Sneath, P. H. A., Mair, N. S., Sharpe, M. E., and Holt, J. G. 1986. *Bergey's Manual of Systematic Bacteriology, Vol. 2*. Williams & Wilkins, Baltimore.

Solomon, H. M., Kautter, D. A., and Lynt, R. K. 1982. Effect of low temperature on growth on non-proteolytic *Clostridium botulinum* types B and F and proteolytic type G in crab and broth. *J. Food Protect.* **45**: 516–518.

Sperber, W. H. 1977. The identification of staphylococci in clinical and food microbiology laboratories. *CRC Crit. Rev. Clin. Lab. Sci.* **7**:121–184.

Sperber, W. H. 1982. Requirements of *Clostridium botulinum* for growth and toxin production. *Food Technol.* **36**(12):89–94.

Tatini, S. R. 1973. Influence of food environments on growth of *Staphylococcus aureus* and production of various enterotoxins. *J. Milk Food Technol.* **36**:559–563.

United States Department of Health, Education and Welfare. 1963. Botulism—Smoked fish products—United States. *Morbid. Mortal. Weekly Rep.* **12**:358.

United States Department of Health, Education, and Welfare. 1970. *Botulism in the United States*. National Communicable Disease Center, Atlanta, Georgia.

United States Department of Health and Human Services. 1987. Nationwide dissemination of a multiply resistant *Shigella sonnei* following a common source outbreak. *Morbid. Mortal. Weekly Rep.* **36**:663–664.

Ward, D. R. 1989. Microbiology of aquaculture products. *Food Technol.* **43**(11):82–87.

World Health Organization. 1989. Consultation on microbiological criteria for foods to be further processed including by irradiation. 29 May–2 June 1989. WHO Geneva Switzerland, 30 pp.

11

Indigenous Pathogens: Vibrionaceae

Gary E. Rodrick

The family of Virbrionaceae includes the genera *Aeromonas, Plesiomonas,* and *Vibrio* among others. Species of these genera are Gram negative, non-spore-forming, facultative aerobic rods that have been isolated from fresh water, seawater, and soil as well as various warm-blooded vertebrates. Some of the species are pathogenic to man and others pathogenic to marine invertebrates and vertebrates. During the past 10 years, an increase in the number of infections involving *Aeromonas, Plesiomonas,* and *Vibrio* bacteria has occurred. Reasons for this increase remain unclear, but may involve a combination of factors. This chapter will attempt to review the epidemiology, pathology, ecology, and prevention of selected species in general *Aeromonas, Plesiomonas,* and *Vibrio.*

VIBRIO

With respect to the genus *Vibrio,* historically, *Vibrio cholerae* 01 has been the organism of most concern; all other species and serotypes were referred to as nonagglutinable or noncholera *Vibrios* (Blake 1980; Desmarchelier 1984). However, *Vibrio* is a large genus containing at least 50 species plus numerous biotypes and chemovars (Oliver 1988). At least 11 species of *Vibrio* are recognized as pathogenic or potentially pathogenic to humans. They may cause a variety of diseases including gastroenteritis, wound infections, ear infections, and primary secondary septicemia (Morris and Black 1985). The members of this genus that can affect humans are discussed in alphabetical order.

Vibrio alginolyticus

Vibrio alginolyticus is a ubiquitous, halophilic bacterium that is widely distributed in the marine environment (Stephen et al. 1978). In the past, this organism was considered to be a biotype of *V. parahaemolyticus,* but is now recognized as a human pathogen associated primarily with wound and ear infections. This *Vibrio* is rather noninvasive and its role in such infection is not well defined (Schmidt et al. 1979). Although it has been isolated from the stools of healthy individuals in the summer owing to ingestion of raw seafood, it has not been associated with clinically expressed gastroenteritis.

Vibrio cholerae

Vibrio cholerae is usually divided into two groups, serotype 01 and non-01 *V. cholerae*. Those groups can be further subdivided as toxigenic and nontoxigenic. Toxigenic strains are capable of producing cholera toxin or a very similar toxin. Toxigenic *V. cholerae* 01 is the causative agent of endemic or asiatic cholera. The 01 serotype contains two biotypes: classic and El Tor, both of which may contain toxigenic and nontoxigenic strains. The biotypes are differentiated by sensitivity to polymyxin B and Murkee's group four phage and by the ability to agglutinate chicken red blood cells (Sakazaki 1979). The classic biotype predominated worldwide until the 1960s. The El Tor biotype is currently predominant worldwide and is the biotype with associated with recent cases in the U.S. (Blake et al., 1980; CDC 1986b; Levine 1981; Morris and Black 1985). All other serotypes are referred to as non 01 *V. cholerae*.

The present pandemic (seventh) that began in Hong Kong and Macao has spread into Russia, Africa, the Middle East, Europe, and the U.S. With respect to the U.S. there had not been an outbreak since 1911. But in 1973 the first documented case of cholera occurred in a coastal town in Texas. The source of *V. cholerae* could not be determined but shellfish were suspected (Weissmann et al. 1975). In 1978, several clusters of cholera outbreaks occurred in southwestern Louisiana. Bacteriological examination of the shellfish eaten (both crabs and shrimp) resulted in the isolation of *V. cholerae* (CDC 1978a,b). In 1979, and 1980 outbreaks of cholera occurred in Florida (CDC 1979, 1980). Raw oysters were found to be the source of the organism. The transmission of *V. cholerae* by fish and other seafood outside the U.S. is well known, but these outbreaks were among the first that were traced to shellfish collected in the U.S. In 1981 there were two cases of cholera involving residents of the Texas Gulf Coast and 17 additional cases on an oil rig in the Gulf (Morris and Black 1985). Thirteen cases of domestically acquired cholera occurred in 1986, 12 in Louisiana and one in Florida (CDC 1986b). Inadequate cooking or improper handling of crustaceans seems to have been the vehicle in these outbreaks.

A great deal is known about the eipidemiology of cholera (Blake et al. 1980; Hughes et al. 1978). However, the ecology of the organism needs more study. *Vibrio cholerae* is widely distributed, occurring in environments where there is no human pollution. The organism has been isolated from shellfish taken from Gulf coast waters and the Chesapeake Bay. Moreover, one investigation of *V. cholerae* in the Chesapeake Bay revealed low levels of the organism throughout the year. In contrast, similar studies by Hood et al. (1984) showed high numbers in both the shellfish harvesting waters in warm-temperature months (May–September). In addition to the various reports of the distribution of *V. cholerae*, much of the information indicates that *V. cholerae* is indigenous to many estuaries. Further, *V. cholerae* can survive well in seawater and can readily attach to numerous substrates such as crabs and shrimp exoskeletons, algae, and $CaCO_3$.

One of the more interesting aspects of the ecology of *V. cholerae* is its ability to go into a nonculturable but viable state (NCBV). The NCBV state of *V. cholerae* is capable of transversing a 0.45-μm membrane filter and sometimes a 0.25-μm membrane filter. MacDonell and Hood (1982) were able to isolate bacteria that passed through 0.4- and 0.2-μm membrane filters. Those that were 0.2-μm filterable cells required low levels of nutrients on initial isolation. Characterization of these bacteria revealed that the majority of them were *Vibrio* spp. and several were strain of *V. cholerae* that had adapted to low levels of nutrients.

There is probably a relationship between the minibacteria described by Watson et al. (1977) and the ultramicrobacteria (Torella and Morita 1981) that have been observed in the water column of various aquatic environments. Understanding the nature of the microbacteria may lead to a better understanding of the mechanism by which *V. cholerae* survives in an estuary. However, regardless of the mechanism of survival of *V. cholerae*, it is well known that various bacteria can survive long periods of time in estuarine water and seawater. Thirty years ago, Pallitzer (1959), in his review of cholera, showed data that indicated that *V. cholerae* is capable of surviving very long periods of time in both marine and estuarine waters.

Moreover, it is apparent that *V. cholerae* may exist in the marine environment under adverse conditions, such as low temperature and lack of nutrient availability in a viable but nonculturable state. *Vibrio cholerae* cell undergo dramatic metabolic and morphological changes. These changes are presumed to be an expression of survival strategy mechanism. The genetic events that control this mechanism(s) are unclear.

Questions concerning whether this stage (VBNC) of *V. cholerae* can produce disease are now being asked. The answer to such questions could be dramatic if this stage remains virulent.

The classic virulence of the cholera organism (toxigenic 01 *V. cholerae*) is caused by the attachment of the bacterium to the mucosa of the small bowel, growth at the site of attachment, and the production of an enterotoxin that causes disruptions in the cyclic AMP system, resulting in water and electrolyte loss. Other pathogenic factors have been elucidated in El Tor strains, including a heat-labile toxin or hemolysin that functions in a similar fashion like the heat-stable hemolysin from *V. parahaemolyticus,* a phospholipase, and possibly a lecithinase.

Clinical symptoms of cholera can vary from the presence of mild, watery diarrhea to acute diarrhea (known as rice water stool), abdominal cramps, nausea, vomiting, dehydration, shock, and, with continued fluid and electrolyte loss, death. Serologically, *V. cholerae* is divided into serotype 01 and non-01. The 01 serotypes include two biotypes, classic (generally considered most toxic) and El Tor, the strain most commonly isolated in cholera cases today.

At least 70 other groups of *V. cholerae* are known to exist. They are referred to collectively as non-01 *V. cholerae* or nonagglutinable (NAG) *V. cholerae*. In contrast to classic cholera, outbreaks caused by these organisms are generally less severe. The virulence of the non-01 *V. cholerae* is not well understood. A toxin that resembles, but is not identical to, the cholera enterotoxin has been found in some strains of the non-01 serotypes. Heat-labile and heat-stable toxins may play a role in the pathogenicity of non-01 serotypes as well as in the invasive properties that have been found in some strains. The function of plasmids in these microorganisms is under investigation, but the role of these extra DNA particles has not yet been defined. Only with the completion of these and additional studies will the virulence mechanism of the non-01 *V. cholerae* be understood.

The majority of the strains isolated from seafood and patients are nontoxigenic strains; fewer than 5% of the non-01 strains from human sources in the U.S. produce cholera toxin. The nontoxigenic strains are prinicipally associated with gastrointestinal illness; but in the U.S. about one-third of the human isolates are from extraintestinal sources, including wound infection, ear infection, and primary and secondary septicemia (Morris and Black 1985). Associated symptoms of gastroenteritis have included diarrhea (100% of cases; 25% have bloody stools), abdominal cramps (93%), and fever (71%). Nausea and vomiting occurs in 21% of the victims. The diarrhea may occasionally be severe, with production of as many as 20–30 watery stools per day (Morris and Black 1985). Almost all of the cases of non-01 *V. cholerae* infections in the U.S. have been associated with eating raw oysters. Considering the relative frequency of isolates from seafood, the incidence of illness is very low. There is evidence that victims often have an underlying liver disease, which might be a host factor for the disease. Also, in most cases the disease may not be severe enough to warrant medical attention and therefore the incidence may be unreported.

Vibrio cincinnatiensis

Vibrio cincinnatiensis is the newest species of *Vibrio* that is pathogenic to human. Brayton et al. (1986) described a Gram-negative bacillus with unique biochemical and genetic properties. Bode et al. (1986) described the isolation of a new *Vibrio* from the spinal fluid and blood of an elderly man. The source of infection could not be determined.

Vibrio damsela

Vibrio damsela is a halophilic bacterium that has been isolated from infected wounds (Morris et al. 1982). In the majority of cases the lesions were erythematous and had a purulent discharge. Wound infections with *V. damsela* have and involved exposure to brackish or salt water and fish fins punctures.

Vibrio fluvialis and V. furnissii

Vibrio fluvialis and *V. furnissii,* originally referred to as Group F or EF6 *Vibrio,* are two relatively new species of *Vibrio* linked to human disease (Brenner et al. 1983; Seidler et al. 1980). They have been reported from both the marine and estuarine environments and from shellfish in England (Furniss et al. 1977; Lee et al. 1978). In addition, these vibrios have been reported in patients with diarrhea from Bangladesh, Bahrain, and Jordon (Lee et al. 1978). Information concerning both the epidemiology and clinical features of this group of vibrios is lacking; however, limited information from the International Center of Diarrhoeal Disease in Bangladesh indicates that the clinical symptoms are similar to cholera, except for occasional bloody diarrhea and abdominal pain (Kahn 1979).

Both of these vibrios tested negative for pathogenicity using adrenal cell assay, infant mouse test, and Sereny test. However, others have shown that this group of vibrios produces a heat-stable toxin that causes fluid accumulation in ligated rabbit ileal loops (Blake et al. 1980). In addition, it was reported that these organisms caused mortality in mice when injected intraperitoneally. Special care should be taken in identifying these two species of vibrios because *V. fluvialis* is similar to anaerogenic *Aeromonas* species.

Vibrio hollisae

Vibrio hollisae was described in 1982 and has been infrequently isolated from individuals with diarrhea (Hickman et al. 1982). Recently, nine cases of *V. hollisae* associated diarrhea have been reviewed (Morris et al. 1982). The source of *V. hollisae* infection was raw oysters and clams in six of the cases, raw shrimp in one case, and two were unknown. Only one of the nine had preexisting hepatic disease. Diarrhea and abdominal pain were universal symptoms and over half of the cases had an evaluated leukocyte number and fever. The ecology of *V. hollisae* is not well understood because the organism grows poorly or fails to grow in thiosulfate–citrate–bile salts–sucrose agar (TCBS), the medium most used in isolation of members of the genus *Vibrio.*

Vibrio metschnikovii

Vibrio metschnikovii is widely distributed in both the environment and in humans (Jean-Jacques et al. 1981; Lee et al. 1978). Specifically, *V. metschnikovii* has been found in sewage, estuaries, and rivers (Lee et al. 1978). To date there is direct evidence that *V. metschnikovii* causes disease in humans. However, Miyake et al. (1988) have reported a clinical case of diarrhea caused by *V. metschnikovii* and isolated and purified the cytolysin. They found the cytolysin caused fluid accumulation in infant mice and increased vascular permeability in rabbit skin.

Vibrio mimicus

In 1981, a new species of *Vibrio* that had been formerly classified with atypical *V. cholerae*, mainly because of its inability to ferment sucrose and Vokes–Proskauer negative reaction, was described (Davis et al. 1981). In earlier publications, they were described as *V. cholerae* of the Hieberg Group 5; however DNA homology studies demonstrated that they were a separate species and the name *mimicus* was proposed because of their biochemical similarity to *V. cholerae* (Colwell 1984; Shandera et al. 1983). *Vibrio mimicus* has been isolated from the gut contents in various geographical locales including Guam, Mexico, Bangladesh, the Philippines, and the United States. Shandera et al. in 1983 reported in a follow-up study that 17 of 19 isolates were found in diarrheal patients. Ciufecu et al. (1983) in Romania isolated *V. mimicus* from patients suffering from gastroenteritis. Raw oysters and boiled crawfish (crayfish) have been implicated as vehicles for the organism. Both toxigenic (produces cholera-like toxin) and nontoxigenic strains have been isolated; however, the food poisoning cases have been mostly from the nontoxigenic strains. Symptoms of the illness have included diarrhea in most cases, but approximately 67% of the cases had nausea, vomiting, and abdominal cramps. Diarrhea may be bloody and will last 1–6 days. *V. mimicus* is widely distributed in nature and can be found in fresh as well as brackish waters. It does show seasonal variation, being present in highest numbers in the warmer months (Bockemühl et al. 1986; Colwell 1984).

Vibrio parahaemolyticus

Vibrio parahaemolyticus is a halophilic, motile (polarly flagellated), Gram-negative rod that is widely distributed in marine and estuarine environments. It has been isolated from seawater, sediments, fish and shellfish, and plankton. A study of the ecology of the bacterium in Chesapeake Bay revealed a seasonal and temperature-related cycle (Kaneko and Colwell 1973). During winter months when water temperatures were below 51.8°F (11°C), the organisms were not present in the water column but were overwintering in the sediments. As temperatures rose, the organisms were released from the sediments, attached to zooplankton, and proliferated in the water column.

In Japan, where large amounts of raw seafood are consumed, *V. parahaemolyticus* has been recognized for many years as the causative agent of over 70% of all cases of gastroenteritis. The first documented case of gastroenteritis caused by *V. parahaemolyticus* in the U.S. occurred in Maryland and was traced to improperly cooked crabs (CDC 1973). Since then, a number of cases have been traced to other seafoods such as shrimp, lobster, and even raw oysters

(CDC 1971). In general, *V. parahaemolyticus* outbreaks in the U.S. have been caused by gross mishandling of the seafood and include such factors as improper refrigeration, insufficient cooking, cross-contamination, and recontamination.

The survival of *V. parahaemolyticus* in seafood has been investigated by numerous workers (Ma-Lin and Benchat 1980; Matches et al. 1971) and research has revealed that although the organism is fairly sensitive to heat in shellfish, it is somewhat resistant to cold temperatures. Specifically, it was found that in shrimp tissue at low inoculum sizes (5×10^2 cells/ml) all cells were destroyed after heating to 140°F (60°C), 176°F (80°C), and 212°F (100°C) for 1 minute, whereas at higher inoculum sizes (5×10^5 cells/ml) surviving cells could be recovered after heat treatments of 140–176°F (60–80°C) for 15 minutes. In contrast, the organism could survive for long periods of time (48 days) at 33°F (0.6°C) in fish homogenates. A survey of processed blue crab (cooked, picked, packed, and refrigerated meats) revealed that the organism was both widely distributed in crabs and could survive processing and storage conditions (Fishbein et al. 1970). Although *V. parahaemolyticus* is a common contaminant of seafood, often present in high numbers, almost none of the isolates from seafood are capable of causing gastroenteritis in man (Blake et al. 1980; Fujino et al. 1974; Hackney et al. 1980, 1981). The test most widely used to differentiate between virulent and avirulent strains is the Kanagawa reaction. Isolates from the marine environment and seafood are predominantly Kanagawa negative. Thompson and Vanderzant (1976) reported only 0.18% of the isolates from water, shellfish, and sediments of the Gulf of Mexico were Kanagawa positive. In Japan 99% of the sea and fish isolates are Kanagawa negative (Sakazaki 1979). Food poisoning victims usually only excrete Kanagawa-positive isolates. Studies have demonstrated that isolates do not change in the intestines and that Kanagawa-positive types are probably part of marine *V. parahaemolyticus* populations, but present in low numbers.

Outbreaks involving *V. parahaemolyticus* include symptoms such as abdominal cramps, diarrhea, nausea, and vomiting without bloody stools, although a dysenteric illness may develop. The incubation period ranges from 4 to 96 hours and last approximately 72 hours. Most people with the symptoms recover quickly without treatment.

The mechanism of pathogenicity has been studied but remains unclear. In early studies it was first thought to be related to the production of a single factor, a heat-stable hemolysin. A direct correlation between the virulence of the organism and the production of β-hemolysis on a special blood agar containing high salt concentrations was demonstrated. Furthermore, when the thermostable hemolysin was isolated and purified, it was shown to cause pathogenic responses in mice, guinea pigs, ileal loops, and cultured mouse heart cells (Honda and Finkelstein 1976). In addition, an invasive pathogenic factor has recently been reported, and studies have shown that the organism is capable of adhering to human fetal cells, colonizing the gut, and penetrating into the lamina propria of the lumen (Boutin et al. 1979).

Vibrio vulnificus

Vibrio vulnificus, previously known as *Beneckea vulnificus*, is a lactose-fermenting, halophilic, rod-shaped, Gram-negative bacterium. The geographical distribution of *V. vulnificus* ranges from the warm coastal waters of both the Atlantic and Gulf of Mexico waters, sediments, oysters, clams, crabs, and plankton (Oliver 1988).

Infections involving *V. vulnificus* have occurred in Belgium, Canada, Japan, and the U.S. (Rodrick et al. 1982). At least 16 states within the U.S. have reported *V. vulnificus* infections and the majority of the cases have occurred during the warm months of April to October. In addition, many of those infected were men over 40 years of age. The principle source of infection seems to be seawater and shellfish. Retrospective review of Florida *V. vulnificus* cases from 1981 to 1987 revealed there were significantly more patients with wound infections than with septicemia onset who had been in contact with seawater (Klontz et al. 1988). In contrast, septicemic patients appeared to have been infected by eating raw oysters. Furthermore, all patients who presented with septicemia onset often ate raw oysters. The average incubation period for *V. vulnificus* wound infections and primary septicemia are 24–48 hours, respectively.

Putative virulence factors produced by *V. vulnificus* include extracellular cytotoxin–hemolysin (Gary and Kreger 1985), phospholipase (Testa et al., 1984), and siderophores (Wright et al. 1986) and a surface antigen-(s) that confer resistance to phagocytosis and bacteriocidal activity in normal serum (Kreger and Lockwood 1981). Of these factors, cytolysin has been found to possess vascular-permeability enhancing activity. In addition, *V. vulnificus* protease has been shown to enhance hypodermic vascular-permeability in guinea pigs (Miyoshi et al. 1987).

Vibrio vulnificus associated pathology has been studied using the laboratory mouse. Subcutaneous injection of large numbers of *V. vulnificus* caused no disease.

One of the more interesting aspects of the pathogenesis of *V. vulnificus* is the recent discovery that two morphological colony types exist for this bacterium, but that only cells of the so-called "opaque" colony type are virulent (Yoshida et al. 1985). The difference in colony opacity is apparently due to the presence of an acidic polysaccharide at the cell surface that has antiphagocytic properties (Simpson et al. 1987; Yoshida et al. 1985).

Retrospective studies of the clinical features of *V. vulnificus* illness in the U.S. reveal two distinct clinical presentations: wound infections and primary septicemia (Blake et al. 1980). The preexisting wound infection group developed rapid swelling and erythema around the wound. In several cases, the lesions often extended into adjacent areas, with vesicles, bullae, and necrosis occurring. *Vibrio vulnificus* was isolated from all wounds and the majority of patients had a fever. In contrast, fever, chills, and malaise developed in the primary septicemia group. Several patients also had vomiting and diarrhea. In addition, one-third of the primary septicemia patients had hypertension, over 75% had metastatic cutaneous lesions, and 75% had preexisting hepatic disease. *Vibrio vulnificus* was isolated in over 83% of the septicemic group. The mortality rate was 46%, and was caused by intractable shock secondary to Gram-negative sepsis.

Recently, it has been shown that *V. vulnificus* can enter into a viable but "nonrecoverable" or "nonculturable" state (MacDonell and Hood 1984). This state may be a consequence of sublethal injury to the cell or a strategy allowing the cell to conserve energy and potentially survive.

Early observations by Xu et al. (1982) found that injured *E. coli* and *V. cholerae* cells were not irreversibly injured, but were in fact viable and capable of being "resuscitated." Such "repaired" cells can be analyzed by standard methods to give valid information on the bacteriological quality of the water under study. Unfortunately, knowledge of how to facilitate cellular repair is extremely limited, and pathogens have been recovered after prolonged incubation in estuarine waters during controlled laboratory studies. Thus, although cells such as *V. vulnificus* must be capable of self-repair as water temperatures become warmer (and thus become platable using standard methods), the factors that control their cellular repair are even less understood than those affecting "dieoff." An obvious implication of the presence of *V. vulnificus* in a nonrecoverable state in the marine environment is that routine examinations for its presence may be negative, although viable and potentially virulent cells are in fact present in high numbers. Until recently, it was unknown whether or not pathogens such as *V. vulnificus,* when present in the nonculturable state in coastal waters and shellfish, were able to repair themselves and produce infection. In 1988, Colwell reported that nonrecoverable strains of *V. cholerae,* when ingested by human volunteers, caused clinical symptoms consistent with non-01 *V. cholerae* infection, and that of *V. cholerae* could be recovered from stool specimens. Thus, it is now apparent that the presence of *V. vulnificus* in the marine environment, especially in oysters, may provide considerable cause for public health concern, even when the cells are present in the nonculturable state.

AEROMONAS

The word *Aeromonas* comes from two Greek words (*aero,* air or gas; *monas,* unit or monad) and means gas-producing monad. *Aeromonas* cells are Gram-negative, straight rods with curved ends and motile by polar and usually monotrichous flagella. They break down carbohydrates to acid and gas and ferment glucose, fructose, maltose, and trehalose and are cytochrome oxidase and peroxidase positive. In addition, Aeromonads can produce protease, diastase, lipase, DNase, and lecithinase.

With regard to the culture characteristics, most aeromonads grow well on blood agar and conventional enteric media (i.e., eosin–methylene blue, salmonella–shigella, MacConkey, and triple-sugar iron). However, the optimal media includes alkaline peptone water, trypticase soy broth plus ampicilin agar, inositol–Brillian Green bile salts agar, and dextrin–fuchsin–sulfite agar.

Because *Aeromonas* is commonly confused with members of the Enterobacteriaceae, the oxidase test is useful (use a platinum loop, as an iron loop will give false positives) in distinguishing the two groups. Exceptions have been reported by McGrath et al. (1979) when oxidase variable strains are grown on Gram-negative and/or differential media or media of pH 5.1 or less or media with lactose.

Several species of Aeromonads are of public health importance and include *A. hydrophila, A. sobria,* and *A. cavial.*

Aeromonas hydrophila is cosmopolitan in its distribution (Hazen et al. 1978; Khardori and Fainstein 1988) and has been recovered in rivers, tap water, swimming pool, lakes, estuaries catfish, and shellfish (De Figueredo and Plumb 1977). In Florida, the highest numbers were recovered during warm-water months from both seawater and shellfish meats. Although *A. hydrophila* has been reported in human stools it is not part of the normal fora. *A. hydrophila* seems to be an opportunistic organism that has the ability to infect all human organ systems. The most common site of infection for Aeromonas are the gastrointestinal tract tissue and blood. Certain risk factors exist for infection and include immunosuppression, caused by leukemia or solid tumors, and cirrhosis. Harris et al. (1985) have reviewed 24 cases of bacteremia caused by *Aeromonas* sp. in hospitalized cancer patients. In contrast to cases described by Walff et al. (1980) there was no known exposure to outdoor water of fish among the cases. However, a seasonal variation was noted with the majority of cases occurring during the warm months.

Moreover, it seems that the patterns of isolation of *Aeromonas* in acute diarrhea vary with age of the patient, geographical distribution, and new techniques in isolation and identification of pathogens. There have been many reports of acute diarrhea associated with *Aeromonas* species (Davis et al. 1978; Trust and Chipman 1979); however, it has not been accepted as an enteric pathogen. Burke et al. (1983) conducted a prospective, 12-month study of 975 nonaboriginal children with diarrhea and matched children of the same age and sex without diarrhea in Perth, Australia. Enterotoxigenic Aeromonas species were found in 10.8% of children with diarrhea whereas the nondiarrheal control showed only 0.7%. The majority of *Aeromonas* species were isolated from the children's feces during the summer and the highest number of *Aeromonas*. Environmental studies of *Aeromonas* have revealed high numbers in water [95°F (35°C)] especially in surface and estuarine waters in the summer (Hazen et al. 1978). These facts have led to speculation that *Aeromonas*-associated gastroenteritis might be waterborne.

PLESIOMONAS

Plesiomonas shigelloides is a Gram-negative, oxidase positive, facultative anaerobic bacillus in the family Vibrionaceae that was first described by Ferguson and Henderson in 1947.

This organism has been referred to in the literature variously as Paracolon C27, *Plesiomonas shigelloides, Aeromonas shigelloides,* or *Vibrio shigelloides.* Separation of the genus *Plesiomonas* from *Aeromonas* and *Vibrio* can be done on the basis of different biochemical and morphological characteristics. Serological studies of the organism show that it possess O antigens which are identical or closely related to those of *Shigella sonnei* phase I. Recently, an antigenic scheme characterizing the organism by the presence of O and H antigens has been reported (Shimada and Sakasaki 1978).

Isolation techniques for *P. Shigelloides* have been studied by Miller and Koburger (1985). They recommended the use of two plating medias, inositol–Brillian Green bile salts agar (IBB) and plesiomonas agar (PL) for the isolation of this organism. They recommended the use of both medias to achieve more accurate results. Specifically, IBB is more inhibitory; however, it works well on environmental samples, and PL allows for the growth of injured organisms. A number of enrichment broths have been tested in combination with these plating medias, but have yielded little success.

Plesiomonas shigelloides is found worldwide and is isolated from a variety of sources that include soil, water, food, and wild and domestic animals, especially during the summer months, as well as diarrhetic and asymptomatic humans. The occurrence of *P. shigelloides* has been noted in oysters in the United States and associated with diarrheal illness (Rutala et al. 1982).

Plesiomonas has been isolated from patients with gastroenteritis. However, it is not a proven cause of gastroenteritis because it does not produce positive results in ligated ileal loop test in rabbits, does not produce disease in human volunteers administered the organism orally and rectally, and fails to produce enterotoxin, such as those of *Escherichia coli.* Several well-studied cases of human gastroenteritis have revealed *P. shigelloides* in the feces in large numbers. This been isolated without the presence of other bacterial pathogens tends to lend more support to the hypothesis that *Plesiomonas shigelloides* can cause a gastroenteritis, and should be considered when bacterial gastroenteritis is present with the absence of normally recognized bacterial pathogens. *Plesiomonas shigelloides* has also been reported as a cause of cellulitis, and of fatal septicemia. The organism is sensitive to chloramphenicol, colistin, and tetracycline, with variable sensitivity to ampicillin, gentamicin, and kanamycin.

SUMMARY

Many gaps concerning the ecology, epidemiology, control, and prevention of *Aeromonas, Plesiomonas,* and *Vibrio* still exist. However, it is clear that both seawater and raw or improperly cooked seafoods can serve as the principle vehicles for their transmission.

Current federal and state guidelines attempt to prevent the consumption of contaminated shellfish, expecially oysters. For the most part, such programs are successful and prevent hepatitis, typhoid fever, viral gastroenteritis, and disease associated with contamination by human feces. However, these guidelines do not relate to *Aeromonas* and *Vibrio* because they seem to be part of the normal estuarine microflora.

Recent epidemiological evidence suggests that people who are immunosuppressed or have liver disease (cirrhosis, hepatitis, or leukemia) should avoid eating raw or improperly cooked seafood because of the high risk of *Vibrio* infections. In addition, the risk of infection may be minimized by not eating raw shellfish collected during warm summer and fall months.

The prevention of *Aeromonas, Vibrio,* and *Plesiomonas* wound infections is difficult, because the majority occur during both occupational and recreational exposures to seawater, especially when the seawater is warm. Immediate cleansing and disinfection of wounds is highly recommended for the prevention of disease. Furthermore, swimmers with a history of ear disease may prevent otitis by preventing seawater entering their ears.

When seafoods are properly cooked before eating, prevention of infection is simple. More specifically, the consumption of seafoods is safe if the food is cooked at temperatures adequate to sterilize them and if cross-contamination after cooking is prevented. Also, such cooked seafood should be held at either temperatures too cold [<39.2°F (4°C)] or too hot [>140°F (60°C)] for their multiplication. Traditional methods of cooking are inadequate for the inactivation of many pathogens.

In summary, cases involving species of *Aeromonas, Vibrio,* and *Plesiomonas* have increased sharply within the past 10 years. Increased awareness of epidemiologists, microbiologists, and physicians concerning the existence of these potentially pathogenic microorganisms and their suspected settings will lead to a better understanding of the epidemiology control and prevention of these potential seafood infections in the future.

REFERENCES

Blake, P. A., Weaver, R. E., and Hollis, D. G. 1980. Diseases of humans other than cholera caused by vibrios. *Annu. Rev. Microbiol.* **34:**341–367.

Bockemühl, J., Roch, K., Wohler, B., Aleksia, S., and Wokatsch, R. 1986. Seasonal distribution of facultative enteropathogenic vibrios *(Vibrio cholerae, Vibrio mimicus, Vibrio parahaemolyticus)* in the freshwater of the Elbe river at Hamburg. *J. Appl. Bacteriol.* **60:**435.

Bode, R. B., Brayton, P. R., Colwell, R. R., Russo, F. M., and Bullock, W. E. 1986. A new *Vibrio* species, *Vibrio cincinnatiensis,* causing meningitis: successful treatment in an adult. *Ann. Intern. Med.* **104:**55–56.

Boutin, B. K., Townsend, S. F., Scarpino, P. V., Twedt, R. M. 1979. Demonstration of invasiveness of *Vibrio parahaemolyticus* in adult rabbits by immunofluorescence. *Appl. Environ. Microbiol.* **37:**647–653.

Brayton, P. R., Bode, R. B., Colwell, R. R., MacDonell, M. T., Hall, H. L., Grimes, D. J., West, P. A., and Bryant, T. N. 1986. *Vibrio cincinnatiensis* sp. nov., a new human pathogen. *J. Clin. Microbiol.* **23:**104–108.

Brenner, D. J., Hickman-Brenner, F. W., Lee, J. V., Steiagerwalt, A. G., Fanning, G. R., Hollis, D. G., Farmer, J. J., III, Weaver, R. E., Joseph, S. W., and Seidler, R. J. 1983. *Vibrio furnissii* (formerly aerogenic biogroup of *Vibrio fluvialis*), a new species isolated from human feces and the environment. *J. Clin. Microbiol.* **18:**816–824.

Burke, V., Gracey, M., Robinson, J., Peck, D., and Beaumon, J. 1983. The microbiology of childhood gastroenteritis: Aeromonas species and other infective agents. *J. Infect. Dis.* **148:**68–74.

Center for Disease Control. 1971. *Vibrio parahaemolyticus*—gastroenteritis—Maryland. *Morbid. Mortal. Wk. Rep.* **20:**356.

Center for Disease Control. 1973. Surveillance summary. *Vibrio parahaemolyticus*—gastroenteritis—United States, 1969–1972. *Morbid. Mortal. Wk. Rep.* **22:**231–232.

Center for Disease Control. 1978a. *Vibrio cholerae*—Louisiana. *Morbid. Mortal. Wk. Rep.* **27:**367.

Center for Disease Control. 1978b. Follow-up on *Vibrio cholerae* serotype Inaba infection—Louisiana. *Morbid. Mortal. Wk. Rep.* **27:**388.

Center for Disease Control. 1979. Non-01 *Vibrio cholerae* infections—Florida. *Morbid. Mortal. Wk. Rep.* **28:**571–579.

Center for Disease Control. 1980. Cholera—Florida. *Morbid. Mortal. Wk. Rep.* **29:**601.

Center for Disease Control. 1986a. Cholera in Louisiana—update. MMWR, Department of Heatlh and Human Services, Atlanta, Georgia.

Center for Disease Control. 1986b. Toxigenic *Vibrio cholerae* 01 Infection—Louisiana and Florida. MMWR, Department of Health and Human Services, Atlanta, Georgia.

Ciufecu, C., Nacescu, N., and Israil, A. 1983. The newly described pathogenic species *Vibrio mimicus* isolated from human diarrheal stools and from a sea water sample. *Z. Baklousol. Parasilenkd. Infektionskr. Hyg. Abt. 1 Orig. Rexhe* A254-89-94.

Colwell, R. 1984. *Vibrios in the Environment.* John Wiley & Sons, New York.

Davis, B. R., Fanning, G. R., Madden, J. M., Steigerwalt, A. G., Bradford, H. B., Jr., Smith, H. L., Jr., and Brenner, D. J. 1981. Characterization of biochemically atypical *Vibrio cholerae* strains and designation of a new pathogenic species, *Vibrio mimicus.* *J. Clin. Microbiol.* **14:**631–639.

Davis, W. A., II, Kane, J. G., and Garagusi, V. F. 1978. Human Aeromonas Infections: a review of the literature and a case report of endocarditis. *Med. Balt.* **57:**267–277.

De Figueredo, J. and Plumb, J. A. 1977. Virulence of different isolates of *Aeromonas hydrophila* in channel catfish. *Aquaculture* **11:**349–354.

Desmarchelier, P. 1984. Significance of *Vibrio* spp. in foods. *Food Technol. Aust.* **36:**220.

Ferguson, W. W., and Henderson, N. D. 1947. Description of strain C 27: a motile organism with the major antigen of *Sigella sonnei* I. *J. Bacteriol.* **54:**179–181.

Fishbein, M., I. J. and Pitcher, J. 1970. Isolation of Vibrio *parahaemolyticus* from the processed meat of Chesapeake Bay blue crabs. *Appl. Microbiol.* **20:**176–178.

Fujino, T., Sakaguchi, G., Sakazaki, R., and Takeda, Y. 1974. *International symposium on Vibrio parahaemolyticus.* Saikon, Tokyo, Japan.

Furniss, A. L., Lee, J. V., and Danoan, R. J. 1977. Group F, a new *Vibrio? Lancet* **ii:**565–566.

Gary, L. D., and Kreger, A. S. 1985. Purification and characterization of an extracellular cytolysin produced by *Vibrio vulnificus. Infect. Immun.* **48:**62–72.

Hackney, C. R., Ray, B., and Speck, M. L. 1980. Incidence of *Vibrio parahaemolyticus* in and the microbiology quality of seafodds in North Carolina. *J. Food Protect.* **43:**769.

Hackney, C., Kleeman, E., Ray, B., and Speck, M. 1981. Adherence as a method for differentiating virulent and avirulent strains of *Vibrio parahaemolyticus. Appl. Environ. Microbiol.* **40:**652.

Hazen, T. C., Fliermans, C. B., Hirsch, R. P., and Esch, G. W. 1978. Prevalence and distribution of *Aeromonas hydrophila* in the United States. *Appl. Environ. Microbiol.* **36:**731–738.

Hickman, F. W., Farmer, J. J., III, Hallis, D. G., Fanning, G. R., Steigerwalt, A. G., Weaver, R. E., and Brenner, J. D. 1982. Identification of *Vibrio hollisae* sp. nov from patients with diarrhea. *J. Clin. Microbiol.* **15:**395–401.

Honda, T. and Finkelstein, R. A. 1979. Purification and characterization of a hemolysin produced by *Vibrio cholerae* biotype El Tor: another toxic substance produced by cholera vibrios. *Infect. Immun.* **26:**1020–1027.

Hood, M. A., Ness, G. E., Rodrick, G. E., and Blake, N. J. 1984. The ecology of *Vibrio cholerae* in two Florida estuaries. In: Colwell, R. R. (ed.) *Vibrios in the Environment.* John Wiley & Sons, New York, pp. 399–409.

Hughes, J. M., Hollis, D. G., Gangarosa, E. J., and Weaver, R. E. 1978. Non-cholera vibrio infections in the United States. Clinical, epidemiologic and laboratory features. *Ann. Intern. Med.* **88:**602–606.

Jean-Jacques, W., Rayashekaraiah, R., Farmer, J. J., III, Hickman, F. W., Morris, J. G., and Kallick, C. A. 1981. *Vibrio metschnikovii* bacteremia in a patient with cholerystitis. *J. Clin. Microbiol.* **14:**1594–1599.

Kahn, M. 1979. An epidemiological study of diarrhea epidemic caused by NAG and Group III Nag like organism. In: *15th Jt. Conf. Cholera U.S. Japan Coop. Med. Sci. Prog.* Bethseda, Maryland, p. 70 (Abstr.).

Kaneko, T. and Colewell, R. R. 1973. Ecology of *Vibrio parahaemolyticus* in Chesapeak Bay. *J. Bacteriol.* **113:**24–32.

Khardori, N., and Fainstein, V. 1988. *Aeromonas* and *Plesiomonas* as etiological agents. *Am. Rev. Microbiol.* **42:**395–419.

Klontz, K. C., Lieb, S., Schreiber, M., Janouski, H. T., Boldy, L. M., and Gunn, R. A. 1988. Syndromes of *Vibrio vulnificus* infections. Clinical and epidemiology features in the Florida cases 1981–1987. *Ann. Intern. Med.* **109:**318–323.

Kreger, A., and Lockwood, D. 1981. Detection of extracellular toxin(s) produced by *Vibrio vulnificus. Infect. Immun.* **50:**534–540.

Lee, J. V., Donovan, T. J., and Furniss, A. L. 1978. Characterization, taxonomy and emended description of *Vibrio metschnikovii. Int. J. Syst. Bacteriol.* **28:**99–111.

Lee, J. V., Shread, P., and Furniss, A. L. 1978. The taxonomy of Group F organisms: relationship to *Vibrio* and *Aeromonas. J. Appl. Bacteriol.* **45:**ix, 178.

Levine, M. 1981. Cholera in Louisiana: old problem, new light. *N. Engl. J. Med.* **302:**345.

Linder, K., and Oliver, J. D. 1989. Membrane fatty acid and virulence changes in the viable but non-culturable state of *Vibrio vulnificus. Appl. Environ. Microbiol.* **55:**2837–2842.

MacDonell, M. T., and Hood, M. A. 1982. Isolation and characterization of ultramicrobacteria from a Gulf Coast estuary. *Appl. Environ. Microbiol.* **43:**566–571.

MacDonell, M. T., and Hood, M. A. 1984. Ultramicrovibrios in gulf coast estuarine waters: isolation, characterization and incidence. In: R. R. Colwell, (ed.) *Vibrios in the Environment.* John Wiley Interscience, New York, pp. 551–562.

Ma-Lin, C. F. A., and Benchat, L. R. 1980. Recovery of chill-stressed *Vibrio parahaemolyticus* from oysters with enrichment broths supplemental with magnesium and iron salts. *Appl. Environ. Microbiol.* **39**:179–185.

Matches, J. R., Liston, J., and Daneault, L. P. 1971. Survival of *Vibrio parahaemolyticus* in fish homogenate during storage at low temperature. *Appl. Microbiol.* **21**:951–952.

McGrath, V. A., Overman, S. B., and Overman, T. L. 1979. Media-dependent oxidase reaction in a strain of *Aeromonas hydrophila*. *J. Clin. Microbiol.* **5**:112–113.

Miller, M., and Koburger, J. A. 1985. *Plesiomonas shigelloides:* an opportunistic food and waterborne pathogen. *J. Food Protect.* **48**:449–457.

Miyake, M., Honda, T., and Miwatani, T. 1988. Purification and characterization of *Vibrio metschnikovii* cytolysin. *Infect Immun.* **56**:954–960.

Miyoshi, N., Miyoshi, S., Suguyamia, K., Suzuki, Y., Furuta, H., and Shinoda, S. 1987. Activation of the plasma kallikrein–kinin system by *Vibrio vulnificus* protease. *Infect. Immun.* **55**:1936–1939.

Morris, J., and Black, R. 1985. Cholera and other vibrioses in the United States. *N. Engl. J. Med.* **312**:343–350.

Morris, J. G., Jr., Wilson, R., Hallis, D. G., Weaver, R. E., Miller, H. G., Tacket, C. O., Hickman, F. W., and Blake, P. A. 1982. Illness caused by *Vibrio damsela* and *Vibrio hollisae*. *Lancet* **i**:1294–1296.

Oliver, J. D. 1988. *Vibrio Vulnificus*. In: Doyle, M. (ed.) *Foodborne Bacterial Pathogens*. Marcel-Dekker, New York. pp. 569–600.

Pallitzer, R. 1959. *Cholera*. WHO Monogr. Series No. 43. World Health Organization, Geneva.

Rodrick, G. E., Hood, M. A., and Blake, N. J. 1982. Human vibrio gastroenteritis. *Med., Clin. North Am.* **66**:665–673.

Rutala, W. A., Sarulli, F. A. J., Finch, C. S., MacCormack, J. N., and Steinkraus, G. E. 1982. Oyster-associated outbreak of diarrheal disease possible caused by *Plesiomonas shigelloides*. *Lancet* **1**:739.

Sakazaki, R. 1979. Vibrio infections: In: Rieman, H., and Bryon, F. (eds.) *Food-Borne Infections and Intoxications*. Academic Press, New York.

Schmidt, U., Chmel, H., and Cobbs, C. 1979. *Vibrio alginolyticus* infections in humans. *J. Clin. Microbiol.* **10**:666–668.

Seidler, R. J., Allen, D. A., Colwell, R. R., Joseph, S. W., and Daily, O. P. 1980. Biochemical characteristics and virulence of environmental group bacteria isolated in the United States. *Appl. Environ. Microbiol.* **40**:715–720.

Shandera, W. X., Johnson, J. M., Davis, B. R., and Blake, P. A. 1983. Disease from infection with *Vibrio mimicus*, a newly recognized *Vibrio* species. *Ann. Intern. Med.* **99**:169–171.

Shimada, T., and Sakasaki, R. 1978. On the serology of *Plesiomonas shigelloides*. *Jpn. J. Med. Sci. Biol.* **31**:135–142.

Simpson, L. M., White, V. K., Zane, S. F., and Oliver, J. D. 1987. Correlation between virulence and colony morphology in *Vibrio vulnificus*. *Infect. Immun.* **55**:269–272.

Stephen, S., Vaz, A. L., Chandraahekara, I., Achyutha, Rao, K. N. 1978. Characterization of *Vibrio alginolyticus (Beneckea alginolytica)* isolated from the fauna of Arabian Sea. *Ind. J. Med. Res.* **68**:7–11.

Testa, J., Daniel, L. W., and Kreger, A. S. 1984. Ertracellular phospholipase A_2 and lysophopholipase produced by *Vibrio vulnificus*. *Infect. Immun.* **45**:458–463.

Thompson, C., and Vanderzant, C. 1976. Serological and hemolytic characteristics of *Vibrio parahaemolyticus* from marine sources. *J. Food Sci.* **41**:204

Torrela, F., and Morita, R. Y. 1981. Microcultural study of bacteria size changes and microcolony and ultramicrocolony formation by heterotrophic bacteria in sea water. *Appl. Environ. Microbiol.* **41**:518–527.

Trust, T. J., and Chipman, D. C. 1979. Clinical involvement of *Aeromonas hydrophila*. *Can. Med. Assoc. J.* **120**:942–946.

Watson, S. W., Novitsky, T. J., Quinby, H. L., and Valois, F. W. 1977. Determination of bacterial number and biomass in the marine environment. *Appl. Environ. Microbiol.* **33**:940–946.

Weissman, J. B., DeWitt, W. E., Thompson, J., Mushnick, C. N., Portnoy, B. L., Feeley, J. C., and Gangarosa, E. J. 1975. A case of cholerae in Texas. *Am. J. Epidemiol.* **100**:487–498.

Wolff, R. L., Wiseman, S. L., Kitchens, C. G. 1980. *Aeromonas hydrophila* bacteremia in ambulatory immunocompromised hosts. *Am. J. Med.* **68**:238–242.

Wright, A. C., Simpson, L. M., Richardson, K., Maneual, D. R., Jr., Oliver, J. D., and Morris, J. G. Jr. 1986. Sidephore production and outer membrane proteins of selected *Vibrio vulnificus* strains under conditions of iron limitation. *FEMS Microbiol. Lett.* **35**:255–260.

Yoshida, S., Ogawa, M., and Mizuguchi, Y. 1985. Relation of capsular materials and colony opacity to virulence of *Vibrio vulnificus*. *Infect Immun.* **47**:446–451.

Xu, H., Roberts, N., Singleton, F. L., Attwell, R. W., Grimes, D. J., and Colwell, R. R. 1982. Survival and viability of non-culturable *Escherichia coli* and *Vibrio cholerae* in the estuarine and marine environment. *Microb. Ecol.* **8**:313–323.

12

Natural Toxins

Sherwood Hall

Marine food products are generally wholesome, nutritious, healthful, and desirable. However, they may on occasion contain potent toxins of natural origin that could pose a significant threat to the health of consumers. To enjoy the benefits of seafoods, as good food and profitable business, it is necessary to recognize and deal with these potential problems. Seafood toxicity generally involves one of several fairly well-defined, recognizable syndromes. This chapter first discusses some general considerations and introduces each of the main known syndromes. Management strategies for dealing with these toxins will then be discussed.

It is useful to note the distinctions between seafood toxins and other problems that can occur with seafood products. The seafood toxins are preformed toxins that are present in the product when it is taken from the marine environment. Although some can undergo transformations that can increase their toxicity, they do not propogate like bacterial contaminants; the number of toxin molecules will decrease with time, never increase. Although the various marine toxins are less stable than many other organic substances, they are far more stable to heat than the proteinaceous toxins produced by bacterial seafood contaminants. Therefore heat treatment, which is relied on to eliminate bacterial contaminants and which largely deactivates the proteinaceous toxins produced by bacteria, is not reliable for protection against seafood toxins.

The toxic substances that accumulate in seafoods can be classified as either natural or anthropogenic. Although public concern about food safety has generally focused on the anthropogenic toxins such as pesticides and heavy metals, natural toxins actually present the more challenging problem because they tend to be both more toxic and more difficult to control. Higher potency implies that detection methods for levels relevant to public health must be more sensitive. The control of contamination by anthropogenic toxins is more practical because it results from human activities that are, at least in principle, subject to regulation.

The majority of the known seafood toxin syndromes are associated with bivalves. Clams and other bivalves subsist by filtering particulate food from the water and are thus a valuable means for converting phytoplankton and other microscopic particulates rich in food value into a form suitable for human consumption. By accomplishing such a transformation in a single step, bivalves are much more efficient as converters than are finfish. However, in addition to the primary metabolites (amino acids, fats, sugars) that are thus obtained, bivalves can accumulate a number of other things: pathogenic bacteria, toxic anthropogenic pollutants such as heavy metals, and a variety of natural toxins, most of which originate in dinoflagellates. These include the paralytic shellfish poisons (PSP), diarrhetic shellfish poisons (DSP), neurologic shellfish poisons (NSP), and domoic acid, responsible for amnesic shellfish poisoning (ASP).

Among the natural toxin syndromes associated with finfish, ciguatera results from the accumulation of toxins from benthic organisms, currently believed to also be dinoflagellates, that encrust seaweeds on tropical reefs. Fugu or pufferfish poisoning results from toxins that are produced within the pufferfish, apparently through the activity of endosymbiont bacteria, rather than through accumulation of toxins from a dietary source. Scombroid toxicity results from the accumulation of histamine, or compounds with similar effects on consumers, apparently due to degradation of flesh after the fish are caught. It is therefore principally a handling problem and will not be discussed further here.

It is important to recognize that the various toxin syndromes are entirely natural processes and have probably been around a good deal longer than people have been eating seafood. There is little hope that human efforts can improve the situation through extermination of the source organisms. It should be recognized, however, that some human activities may make matters worse, either through the nutrient enrichment (eutrophication) of coastal waters with sewage and agricultural runoff, or through the dispersion of the resting cysts of toxigenic dinoflagellates by dredging, discharge of ship ballast water, or transplanting of shellfish.

THE KNOWN
SEAFOOD TOXIN SYNDROMES

Paralytic Shellfish Poisoning (PSP)

Paralytic shellfish poisoning is a potentially lethal syndrome resulting from the consumption of mussels, clams, oysters, or other animals that are contaminated with the saxitoxins, saxitoxin and its derivatives, which are also referred to collectively as paralytic shellfish poison. The abbreviation PSP is commonly used for both the syndrome and the toxins.

Vectors   PSP is principally associated with bivalves, the largest number of human deaths being reported from the consumption of mussels (Combe 1828; Gacutan et al. 1985; Grindley 1969; Luthy 1979; Meyer et al. 1928; Quayle 1969).

It appears that the saxitoxins can also be passed through food webs, there being the potential for organisms that are relatively insensitive to the saxitoxins to accumulate them, posing a risk to subsequent consumers that are more sensitive. Thus there have been mass mortalities of seabirds that ate small fish, that ate zooplankters, that had in turn apparently grazed on blooms of toxigenic dinoflagellates, as well as fish kills among herring by a similar path (White 1977, 1981a). Several finfish of major economic importance have been found to be sensitive to the saxitoxins so that, although they themselves may be at risk, there is little chance that they could accumulate the toxins and be a threat to human consumers (White 1981b). On the other hand, carnivorous univalves such as whelks have been found able to accumulate the saxitoxins from their bivalve prey and must be considered a PSP risk.

Some species of crabs from the western Pacific may on occasion contain large amounts of the saxitoxins and have been responsible for some human deaths (Raj et al. 1983). Some evidence indicates that the crabs accumulate the toxins from a coralline red seaweed, *Jania* sp. The situation in this case is not yet clear.

Mackerel on the Atlantic coast of North America have on occassion been found to contain PSP in their viscera, principally in the liver. Although it is likely that this material was accumulated by consumption of zooplankters containing sublethal levels, the source has not yet been identified. There is at this point no indication that there is a risk to human consumers.

Symptoms   The signs and symptoms of PSP generally develop rapidly, within 0.5 to 2 hours of a meal depending on the dose, and in severe cases may progress to death due to respiratory paralysis within a few hours (Sommer and Meyer 1937). Victims who survive more than 12 hours generally recover, promptly and without lasting effect. In the majority of cases, respiratory support will ensure survival and full recovery of the victim. The saxitoxins do, however, appear to have a relatively weak hypotensive effect that may, where the dose has been extremely large, supervene and lead to death through cardiovascular collapse despite respiratory support.

Toxins   The 12 saxitoxins known from gonyaulacoid dinoflagellates are shown in Figure 12.1, and their structural relationships summarized in Figure 12.2 (Hall 1982). They consist of the historical and conceptual parent, saxitoxin, 1, and the 11 derivatives formed by 3 substituents: N-1-hydroxyl, 11-hydroxysulfate, and 21-sulfo (Bordner et al. 1975; Boyer et al. 1978; Hall et al. 1984; Schantz et al. 1975; Shimizu et al. 1978; Wichmann et al. 1981).

As a group, the toxins are small molecules soluble in water and insoluble in most organic solvents. They are poorly adapted to many of the sensitive analytical methods that work well for other compounds because they are highly polar, involatile, and neither fluoresce nor absorb ultraviolet light (UV) in a useful range. Fortunately, they do yield fluorescent products on oxidation. Most

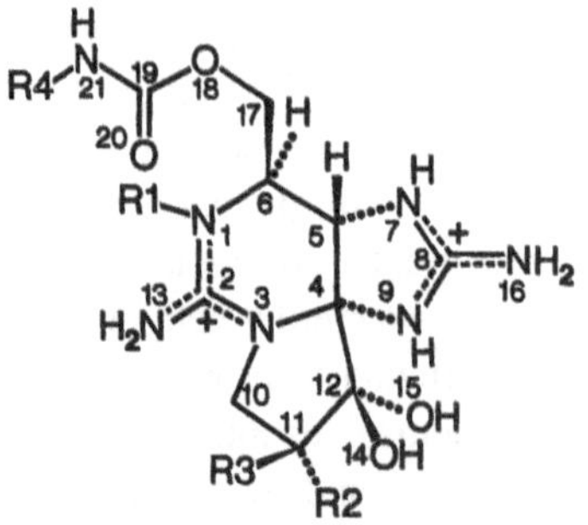

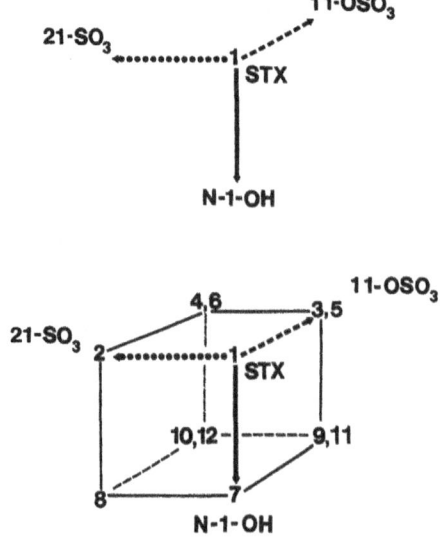

	R1	R2	R3	R4	
1	H	H	H	H	STX
2	H	H	H	SO₃	B1
3	H	OSO₃	H	H	GTX2
4	H	OSO₃	H	SO₃	C1
5	H	H	OSO₃	H	GTX3
6	H	H	OSO₃	SO₃	C2
7	OH	H	H	H	NEO
8	OH	H	H	SO₃	B2
9	OH	OSO₃	H	H	GTX1
10	OH	OSO₃	H	SO₃	C3
11	OH	H	OSO₃	H	GTX4
12	OH	H	OSO₃	SO₃	C4

Figure 12.1   The twelve saxitoxins known from gonyaulacoid dinoflagellates of the genus *Alexandrium*.

Figure 12.2   The structural relationships among the saxitoxins.

successful methods for chemical detection of the saxitoxins are based on this property (Buckley et al. 1976; Sullivan and Iwaoka 1983).

Saxitoxin includes two guanidinium groups that are protonated under normal conditions. One of these, apparently that centered on carbon 2, is strongly basic as normally expected for a guanidinium group, remaining protonated up through about pH 12. The other, however, loses a proton at an anomolously low pH, having a pK_a about 8.5. Saxitoxin is therefore nominally a dication, but may bear an average net charge slightly less than +2 under physiological conditions.

Compounds bearing the N-1-hydroxyl substituent give relatively low yields of fluorescent oxidation products in solution, compared to the N-1-H toxins. This complicates their detection. These compounds also have another dissociable proton with a pK_a around 6.5, and thus under physiological and other conditions with pH around 7 and above will have about 1 unit less net charge.

The presence of the 11-hydroxysulfate substituent, which is fully ionized under all practical conditions, reduces net charge by 1 unit. This substituent can be present in either of two orientations and can invert from one to the other with frustrating ease. Thus a solution of any one of the 11-hydroxysulfate toxins will eventually equilibrate to a mixture of both epimers, with the 11-α-hydroxysulfate epimer predominating. The equilibrium ratios of epimers appear to be in the range 3:1 to 5:1. Analyses of dinoflagellate extracts indicate that dinoflagellates contain only the 11-β epimers, with the 11-α epimers arising through spontaneous conversion once the toxins have been released from the cells (Hall 1982). In contrast, the 11-α epimers tend to be either significant or the predominant components in shellfish extracts, and may serve as a clock to indicate the time elapsed since the exposure of the shellfish to toxic dinoflagellates.

Compounds bearing the 21-sulfo substituent, referred to collectively as sulfamates, have relatively low mammalian toxicity compared to their carbamate counterparts (Hall 1982; Hall et al. 1980). The sulfo group is easily hydrolyzed, converting each sulfamate into the respective carbamate, with increases in potency ranging from $5\times$ to $70\times$ on the basis of the mouse bioassay. The presence of the 21-sulfo group also reduces net charge by 1 unit.

Detection   The mouse bioassay has for many years been the principle means for detection of the saxitoxins and remains the approved regulatory standard (Helrich 1990; Sommer and Meyer 1937). In the Association of Official Analytical Chemists (AOAC) standard method, samples are macerated with dilute acid and heated. Portions of the resulting supernatant or dilutions thereof are injected into mice. The level of toxicity, expressed as the equivalent amount of saxitoxin, is calculated from the mouse death time, mouse weight, dilution, and a calibration factor determined for the test population of mice by standardization with reference standard shellfish poison (Schantz et al. 1958).

The detection limit of the mouse bioassay corresponds to about 0.18 μg of saxitoxin/ml of solution (0.49 μM), equivalent to 36 μg/100 g of shellfish (about 1 μM) prepared according to the standard AOAC protocol.

Source   The principle known marine source of the toxins is a group of dinoflagellates now assigned to the genus *Alexandrium* (Balech 1985), including the species *A. catenellum* (Whedon and Kofoid 1936) and *A. tamarense* (Lebour 1925; Needler 1949). These are motile organisms, about 40 μm in diameter, covered by a theca—a thin shell consisting of polygonal plates apparently made of cellulose. The metabolism of the organisms is plant-like, in that their growth depends on the availability of light and mineral nutrients (carbon dioxide, nitrate, phosphate, etc.) in the water. They grow by division, typically once every 2–5 days. Under suitable conditions they undergo sexual fusion resulting a dormant, immotile cyst stage that can remain viable in the sediment for years and then germinate to produce the actively growing vegetative stage.

The taxonomy of the group has been in a state of flux and has recently been revised. The genus *Alexandrium* now encompasses a group of dinoflagellates, mostly toxic and most previously assigned to the genus *Gonyaulax,* within which they have long been recognized to form a distinct subgroup. Prior to the acceptance of the current revision with the genus name *Alexandrium,* other groupings and names had been suggested, including *Protogonyaulax* and *Gesnerium.*

Two other, entirely distinct dinoflagellates are now recognized as sources of the saxitoxins. The first is *Pyrodinium bahamense* (Steidinger et al. 1980), which has been responsible for many outbreaks of toxicity, in Papua New Guinea (Maclean 1975a,b; 1977), Sabah, Brunei, the Philippine Islands (Gacutan et al. 1985), and the Pacific coast of Central America (Hall and Cahahui, unpublished). These outbreaks have caused well over a thousand illnesses and over 70 deaths. Toxicity has only been observed in Pacific populations of this organism. In the tropical Atlantic, where the organism was first described, its blooms have not yet been associated with toxicity. *Pyrodinium bahamense* is similar in size to *Alexandrium,* but has a rigid, distinctively marked theca.

The second is *Gymnodinium catenatum,* which has been discovered as a source of the saxitoxins in Mexico, Spain, Venezuela (LaBarbera, unpublished), Tasmania (Hallegraeff et al. 1988; Oshima et al. 1987), and Japan (Ikeda et al. 1988). This organism lacks the theca characteristic of *Alexandrium* but is otherwise similar in size and appearance.

Some strains of the freshwater bluegreen alga *Aphanazomenon flos-aquae* contain saxitoxins. Human intoxications have not yet been reported, probably due to the lack of a vector, but dramatic livestock kills have occurred.

Pharmacology The saxitoxins act through a reversible binding to the voltage-sensitive sodium channel, a membrane protein essential for the transmission of signals along most nerves and muscles in humans and most animals. It is possible to observe this action in vitro by incorporating a single molecule of the channel protein into a lipid bilayer, opening it with appropriate techniques so that a current of sodium ions flows through it (Moczydlowski et al. 1984a), and then adding the various saxitoxins (Moczydlowski et al. 1984b). Figure 12.3 shows the results of such experiments for several of the saxitoxins. The current due to

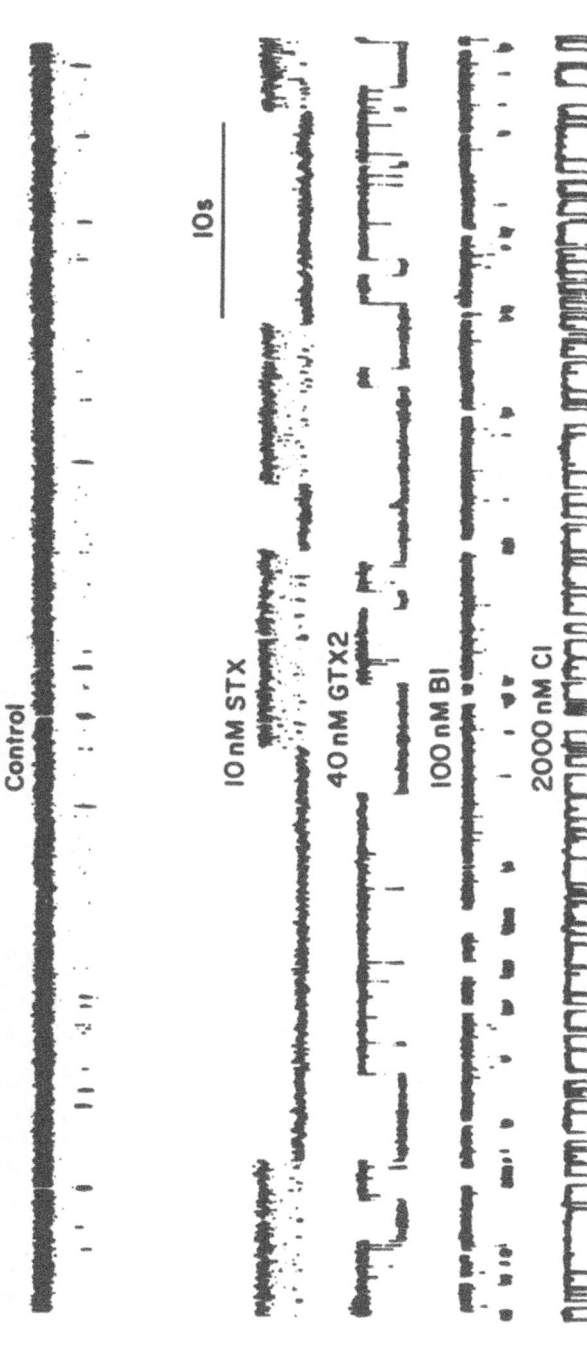

Figure 12.3 The effect of the saxitoxins at the molecular level. For these experiments single voltage-activated sodium channels, prepared from rat muscle, have been incorporated into artificial lipid bilayers under conditions such that they remain open. The top (control) record shows the flow of sodium ions through such a channel without further treatment. Various of the saxitoxins have been added in the other experiments, resulting in interruption of the flow of sodium ions (deflection of the record downward) while the toxin molecule is present at the toxin binding site on the channel. The differences in the rates at which the toxin molecules bind to the channel, and the lengths of time they remain once bound, are the molecular basis for their differences in toxicity. Note particularly how the dwell times of the sulfamates B1 and C1 are much shorter than those of the corresponding carbamates, saxitoxin and gonyautoxin 2.

307

the flow of sodium ions through the channel is blocked (recording deflected downward) when a toxin molecule arrives at the site, and is restored when the toxin molecule leaves. Note how the toxins differ in the concentration required to attain a given rate of arrival events, and how the mean dwell time varies, ranging from about 1 to 60 seconds.

There are two important points here. First, the binding of the toxins is strictly reversible, consistent with clinical experience with PSP victims. Second, the action of all the compounds is qualitatively the same although quantitatively differing over a broad range.

Distribution

SPATIAL   While on the basis of earlier literature accounts it once appeared that PSP was restricted to the temperate/boreal regions, PSP outbreaks have now been reported from equatorial to high latitudes in both hemispheres.

TEMPORAL   Although PSP is to some extent seasonal at higher latitudes, being least frequent during the winter months when there is relatively little phytoplankton growth and most frequent during the months when light and nutrient conditions make blooms of the source dinoflagellate possible, there is not yet any reliable basis for predicting when or whether outbreaks will actually occur. Some species, such as the Alaska butter clam, retain significant levels of toxicity for long periods, such that stocks may remain toxic through the winter. In some areas such as California where experience over several years has led to a routine seasonal closure, severe outbreaks have occurred during what was normally considered the safe period. Still, it has been noted by some investigators in this field that having a carefully planned, well funded study set up to focus on PSP during a given season appears to significantly reduce the probability of a PSP outbreak during the study period.

Characteristics of PSP Relevant to Food Safety

COMPARISON OF PSP WITH BOTULISM   It is instructive to compare and contrast botulism and PSP. Both involve foodborne paralytic agents. In the case of botulism, the agent is a heat labile protein that does long term damage to victims who survive an exposure. PSP involves small molecules that are not effectively deactivated by high food processing temperatures, but that have no lasting effects on survivors. Botulism involves contamination of food with an organism that can propogate in the finished product. In PSP, shellfish accumulate the saxitoxins from the dinoflagellate source.

The management of food production to protect against botulism involves scrupulous control of process parameters to ensure sterility in susceptible product, with the additional safeguard that suitable cooking at the time of use is relatively effective in destroying the toxins. The management of PSP, in contrast, depends entirely on adequate sampling and detection because no process can be relied on to remove or destroy the toxins.

Toxin Chemistry

TOXIN MULTIPLICITY   The task of dealing with the saxitoxins is complicated by the large number of compounds involved, structurally related but quite dissimilar in their potencies and many aspects of their chemistry. For instance, no simple procedure has been found or is likely to be found that can quantitatively enrich all of the toxins from a sample as a cleanup step prior to analysis. Detection methods (discussed below) must cope with a broad range of response factors.

POTENCY   As a group, the toxins are extremely potent. To ensure a safe food product, it is therefore necessary to detect them at low levels. The current regulatory limit, a toxicity equivalent to 80 μm of saxitoxin/100 grams of shellfish meat, corresponds to a saxitoxin concentration of about 2.2 μM. While there are methods for detecting many organic contaminants in foods at this and much lower levels, such determinations generally rely on favorable separation and detection chemistry. Even in favorable cases, they also generally require some sort of sample cleanup and relatively sophisticated equipment for the determination itself.

LATENT TOXICITY   The sulfamate toxins have relatively low potency, but can be easily converted to more toxic compounds. When present, they therefore constitute a reservoir of latent toxicity. The conversion may occur either through removal of the 21-sulfo group, converting sulfamates to carbamates (Hall 1982), or through an enzymatic process that apparently involves removal of the sulfamate side chain to yield the decarbamoyl toxins (Sullivan et al. 1983). The toxicity of a sample could therefore be given as the actual toxicity which would, if the sulfamates were present, include a large component of latent toxicity, or as the potential toxicity—the level of toxicity that would result if these conversions occurred. Since, given the chemistry of the 21-sulfo group, it is impractical to ensure that the conversions do not occur in a food product, it seems prudent to base regulatory decisions on potential toxicity.

This is of particular importance, because the extraction method used in the standard AOAC mouse bioassay fails to completely hydrolyze the 21-sulfo group, and may therefore indicate a toxicity substantially below the potential toxicity (Hall 1982).

Patterns of Toxin Composition

GEOGRAPHIC PATTERNS IN DINOFLAGELLATES   Studies of cultured dinoflagellate clones obtained from several areas indicate that each clone has a characteristic toxin composition, a subset of the array of 12 toxins, which is a conservative property showing little or no variation with growth conditions. Toxin compositions differ dramatically from one area to another, but are remarkably similar for clones isolated from the same area. This implies that the composition of toxins available to shellfish will have a similar geographical pattern of variation (Hall 1982).

It is important to recognize that, in distinct contrast to the uniformity of toxin composition within a population, the total toxin content per cell of gonyaulacoid dinoflagellates varies substantially with growth conditions and during the growth of a bloom. Cells of clones isolated from the northeast Pacific contain far more toxin when grown under conditions approaching those found in their natural environment (low temperature and low nutrient concentrations) than when grown under the relatively high temperature and high nutrient conditions that have often been employed in laboratory studies to optimize growth. The cell concentrations necessary to sustain a given toxin concentration in shellfish are therefore a good deal lower than has been realized. The variation in toxin per cell that occurs through a growth cycle furthermore implies that a close correlation between cell concentration and shellfish toxicity should not be expected (Hall 1982).

COMPOSITIONAL DIAGENESIS IN SHELLFISH   In a given species of shellfish, the various toxins can be expected to act differently, such that toxin composition may start out being close to that of the source dinoflagellate, but evolve with time into something quite different. First, though it is at this point speculative, it seems likely on the basis of laboratory experience that the various toxins would differ in the rates and efficiencies with which they were accumulated and released by a shellfish.

Second, it has been shown that the saxitoxins can be transformed through metabolic processes in the accumulating shellfish. Both the N-1-hydroxyl and the 11-hydroxysulfate substituents can be removed (Shimizu and Yoshioka 1981). The work of Sullivan (Sullivan et al. 1983) has shown that in little-neck clams an enzyme system is present that converts the toxins to metabolites other than the 12 dinoflagellate toxins, apparently the decarbamoyl derivatives. The presence of yet another saxitoxin derivative in scallops from eastern Canada (Koehn 1983) suggests that a different transformation is occurring there.

In some areas butterclams at times contain principally saxitoxin, whereas the resident dinoflagellate populations contain only others of the toxins, implying a substantial amount of metabolic transformation by the clam.

DIFFERENCES AMONG SPECIES OF SHELLFISH   Different shellfish species respond to the toxins in differing ways. As noted above, butterclams, littleneck clams, and scallops each appear capable of performing different transformations.

Although relative patterns of toxin composition have not yet been worked out, oysters are relatively sensitive to the toxins and tend to accumulate relatively low levels, whereas mussels are relatively insensitive to the toxins, accumulate them rapidly to very high levels, and yet lose them relatively quickly, usually after a few weeks. Butterclams remain toxic for much longer periods (Neal 1967; Sribhibhadh 1963). Thus one must expect that the various species of shellfish in a given area may differ in toxin composition.

Neurotoxic Shellfish Poisoning

Vectors   Neurotoxic shellfish poisoning (NSP) was originally recognized and described as a "ciguatera-like" syndrome in people who consumed bivalves that had been exposed to a bloom of *Ptychodiscus brevis* (previously called *Gymnodinium* brevis), the best known "red tide" organism of the Florida and Gulf coast (McFarren 1971).

Symptoms   Victims experience nausea and neurological symptoms (Baden 1983). The symptoms generally pass within a few days and deaths have not been reported among human victims, despite the vast numbers of fish that are killed by the toxins during *P. brevis* blooms. Respiratory irritation is experienced by people in the vicinity of the toxic blooms, due to aerosols containing the toxins (Asai et al. 1982; Baden et al. 1982; Pierce 1986). Reports of such irritation can provide warning that there is a toxic bloom in the area.

Toxins   The toxins involved in this syndrome are referred to collectively as the brevetoxins (Baden et al. 1981; Chou et al. 1985; DiNovi et al. 1983; Golik et al. 1982; Lin et al. 1981; Shimizu et al. 1986). The known brevetoxins, shown in Figure 12.4, are based on two skeletons, both cyclic polyethers. The compounds are lipophilic—relatively insoluble in water, and soluble in nonaqueous solvents. They are structurally related to the ciguatoxins and certain of the DSP toxins.

Source   The sole known source of the brevetoxins is a photosynthetic dinoflagellate originally named *Gymnodinium breve*, at present called *Ptychodiscus brevis* (Baden 1977). This organism is easily ruptured because it lacks the fairly rugged theca (outer coating) found on many other toxic dinoflagellates. As with the saxitoxins in *Alexandrium*, toxin composition varies among clones (Baden and Thomas 1988).

Pharmacology   The brevetoxins also act on the voltage-activated sodium channel in nerve and muscle membranes, but bind to a site distinct from that affected by the saxitoxins, causing prolonged excitation of the channel rather than blocking (Baden et al. 1981; 1988; Huang et al. 1984; Poli, 1985; Poli et al. 1986).

Occurrence   NSP and *P. brevis* blooms are best known from the west coast of Florida. They have been reported along the Texas coast, Mexico, and elsewhere around the Gulf, and have now occurred in North Carolina. The latter episode apparently resulted from bloom material being swept by the Gulf Stream from Southwest Florida to the East and then North, until a bolus from the Gulf Stream impinged on the Atlantic coast South of Cape Hattaras (Tester and Fowler 1990).

Type 1 Skeleton

	R1	R2	R3
Brevetoxin 2	H	CH₂	CHO
3	H	CH₂	CH₂OH
5	COCH₃	CH₂	CHO
6	H	CH₂	CHO (27,28-β-epoxide)
8	H	O	CH₂Cl

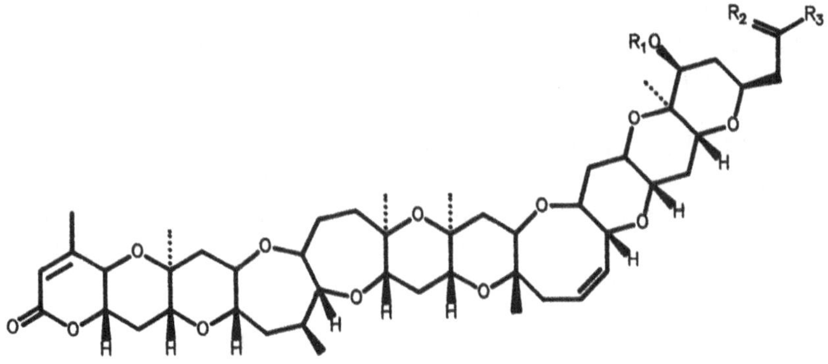

Type 2 Skeleton

	R1	R2	R3
Brevetoxin 1	H	CH₂	CHO
7	H	CH₂	CH₂OH

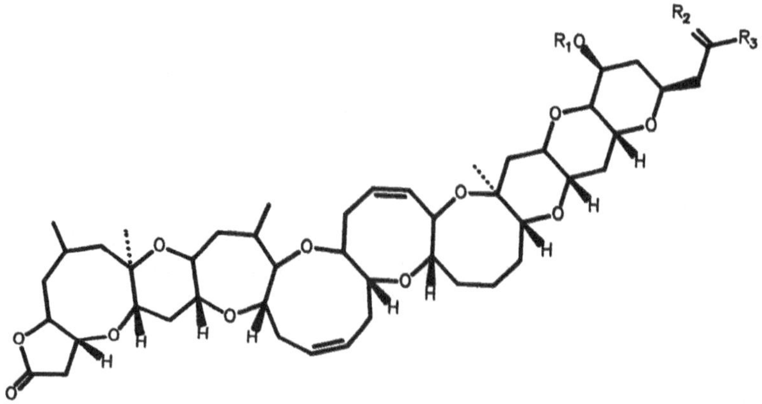

Figure 12.4 The brevetoxins.

312

Diarrhetic Shellfish Poisoning

Vectors   The known vectors of in DSP outbreaks have been bivalves. It is likely that, given suitable analytical methods, the DSP toxins will be found in other seafoods.

Symptoms   The primary symptom of the syndrome is acute diarrhea, with onset more rapid than that typical for bacterial food poisonings (Yasumoto et al. 1978).

Toxins   The DSP toxins fall into three structural classes (Figure 12.5), all being lipophilic compounds, one class closely resembling the brevetoxins (Murata et al. 1986, 1987; Yasumoto et al. 1985). The most widely distributed and apparently most significant DSP toxins are okadaic acid and its derivatives. Okadaic acid itself was previously isolated from a marine sponge (Tachibana et al. 1981), has now been isolated from the dinoflagellates and shellfish responsible for DSP, and is the principal DSP toxin in Northern Europe.

Source   The DSP toxins are apparently accumulated by shellfish from dinoflagellates of the genera *Dinophysis* and *Prorocentrum* (Murakami et al. 1982; Yasumoto et al. 1980). Species of the latter have been cultured and produce primarily okadaic acid. Studies of *Dinophysis* have been limited by the repeated failure of attempts to culture species from the genus. Data on toxin composition are based on analyses of wild material, in some cases following laborious pipet isolation of single cells to provide a pure sample.

Pharmacology   The mode of action of okadaic acid is being studied (Bialojan et al. 1987, 1988; Erdodi et al. 1988). Its acute effects appear to be due to inhibition of protein phosphorylation. In addition to its acute effect on the gastrointestinal tract, okadaic acid has been shown to be a potent tumor promoter (Suganuma et al. 1988). Whether this implies a significant risk for human health remains to be clarified. Some of the toxins associated with the syndrome that have been isolated on the basis of their activity in animal lethality assays may not actually cause diarrhea.

Distribution   DSP was originally recognized in Japan (Yasumoto et al. 1978). There have subsequently been severe outbreaks in Europe (Kumagai et al. 1986). The syndrome is suspected but has yet to be confirmed in North America.

Amnesic Shellfish Poisoning

Vectors   Amnesic shellfish poisoning (ASP) was first recognized in Canada in 1987, and traced to mussels coming from the Eastern shore of Prince Edward

	R1	R2
Okadaic acid	H	H
Dinophysistoxin 1	H	CH₃
Dinophysistoxin 3	acyl	CH₃

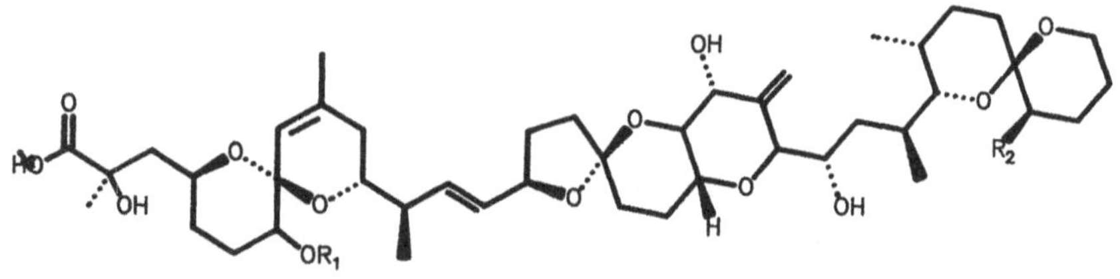

	R
Pectenotoxin 1	CH₂OH
Pectenotoxin 2	CH₃
Pectenotoxin 3	CHO
Pectenotoxin 6	COOH

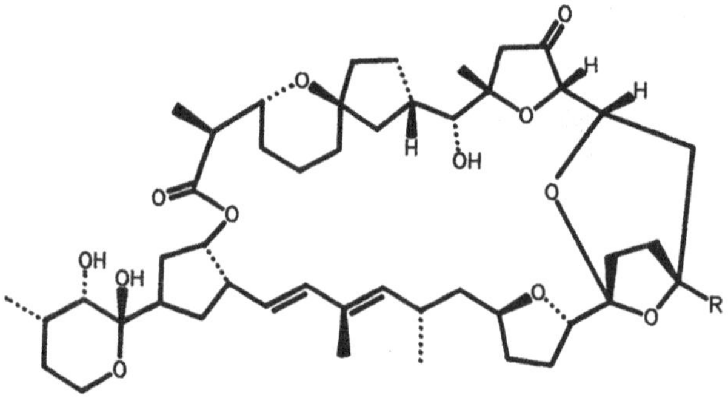

Yessotoxin

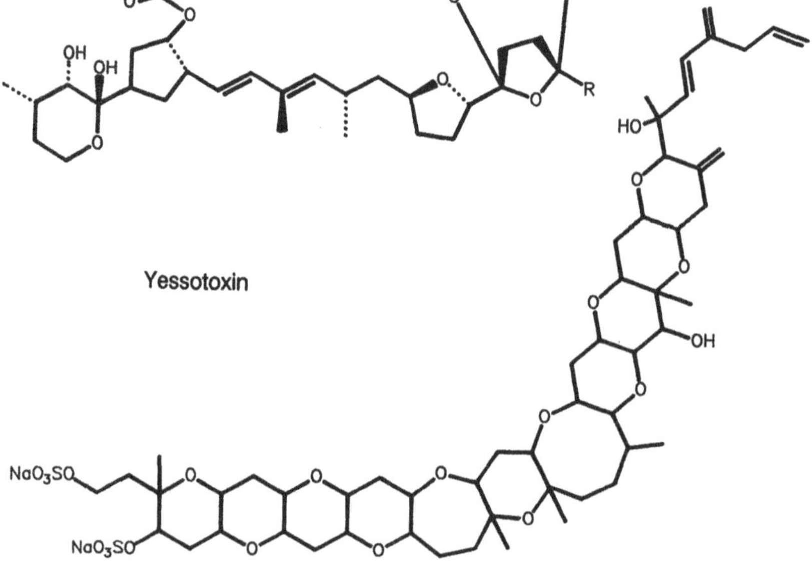

Figure 12.5 The DSP toxins. (a) Okadaic acid and related toxins. (b) Pectinotoxins. (c) Yessotoxin. Note the similarity of yessotoxin to the brevetoxins.

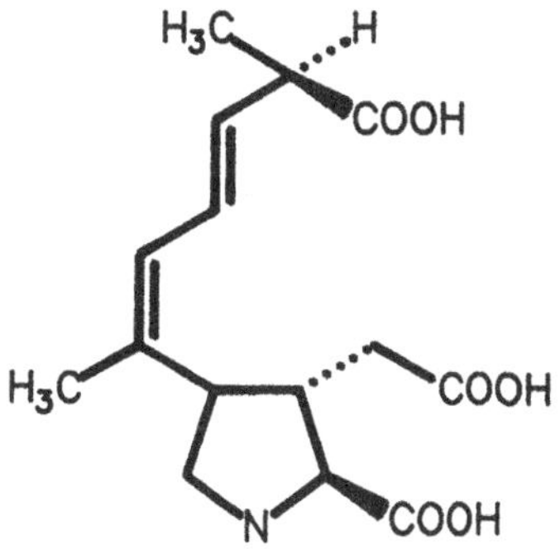

Figure 12.6 Domoic acid, the compound responsible for amnesic shellfish poisoning.

Island (Todd 1990). ASP cases that have been recognized have been due to the consumption of mussels, although other bivalves have been shown to contain the toxin at lower levels.

Symptoms Symptoms include nausea, loss of equilibrium, and central neural deficits including confusion and memory loss. There have been three deaths. Unlike most other shellfish toxin symptoms, the brain damage resulting from ASP appears irreversible.

Toxins ASP is due to the compound domoic acid (Ohfune and Tomita 1982) (Figure 12.6). Although isomers of domoic acid co-occur as minor components, it is not clear what effect they have.

Source Domoic acid, originally isolated and characterized as a constituent of seaweed, is apparently produced and supplied to bivalves from various species of the diatom *Nitschia*.

Pharmacology Domoic acid is a glutamate analog, binding to one class of glutamate receptors in the brain. As such it is an excitotoxic amino acid, capable of exciting a receptive brain cell until it dies from overstimulation. Animal studies suggest that domoic acid passes though the gastrointestinal tract of healthy individuals with little or no absorption but that, when it is absorbed, it is highly potent.

Distribution Significant levels of domoic acid have at this point been detected only in shellfish on the eastern end of Prince Edward Island, with low levels detected in some neighboring areas, principally along the Atlantic coast of Canada.

Ciguatera

Vectors Although the name ciguatera is derived from the Spanish term for a marine snail, the syndrome as it is currently recognized results from the consumption of tropical fish. The risk of ciguatera is highest from the consumption of herbivorous reef fish and the carnivorous fish that feed upon them.

Symptoms Symptoms are nausea and neurological deficits similar to but frequently more severe and much more enduring than those of NSP. Victims of ciguatera occasionally report relapses months or years after the initial episode, in some cases precipitated by consumption of alcoholic beverages. Deaths are rare. The neurological disfunctions include a reversal of the sensations of hot and cold (Capra 1986; Hokama 1988; Hokama and Miyahara 1986; Maharaj et al. 1986; McMillan et al. 1980).

Toxins The characterization of ciguatera toxins has lagged due to the complexity of the molecule and for want of adequate material. Within the last year, however, a collaborative effort (Murata et al. 1989) between chemists in Tahiti, who purified the material, and in Japan, who did spectroscopic studies, has provided the first complete structures (Figure 12.7a). Given the experience with other seafood toxins, it seems likely that other derivatives will be identified. Chromatographic evidence indicates that several other related toxins are present (Legrand et al. 1990).

Detection The development of detection methods for ciguatera toxicity has lagged for the same reasons as, and in part because of, the lack of structural information. Animal assays have been used (Chungue et al. 1984; Hoffman et al. 1983; Sawyer et al. 1984). Immunoassays have been developed (Hokama, 1985; Hokama et al. 1977) but have met with limited success. This may in part be due to the purity of the material to which the antibodies were raised. It is to be hoped that knowledge of the structures of the ciguatera toxins and their availability in pure form will permit the development of more efficient assays.

Source The source of the ciguatera toxins is less clear-cut than for other seafood toxins. The principal source recognized is the benthic dinoflagellate *Gambierdiscus toxicus,* which tends to cluster on the surfaces of tropical seaweeds (Bagnis et al. 1980; Chungue et al. 1979; Durand-Clement 1986; Yasumoto et al. 1979). It is because of this growth habit that it can be efficiently accumulated by herbivorous reef fishes that do not consume phytoplankton per se. One of the structures shown (Figure 12.7b) is of toxin isolated from *Gambierdiscus toxicus* obtained directly from coral reef seaweeds. Cultures of *Gambierdiscus toxicus,* however, have been found to produce primarily a different toxin,

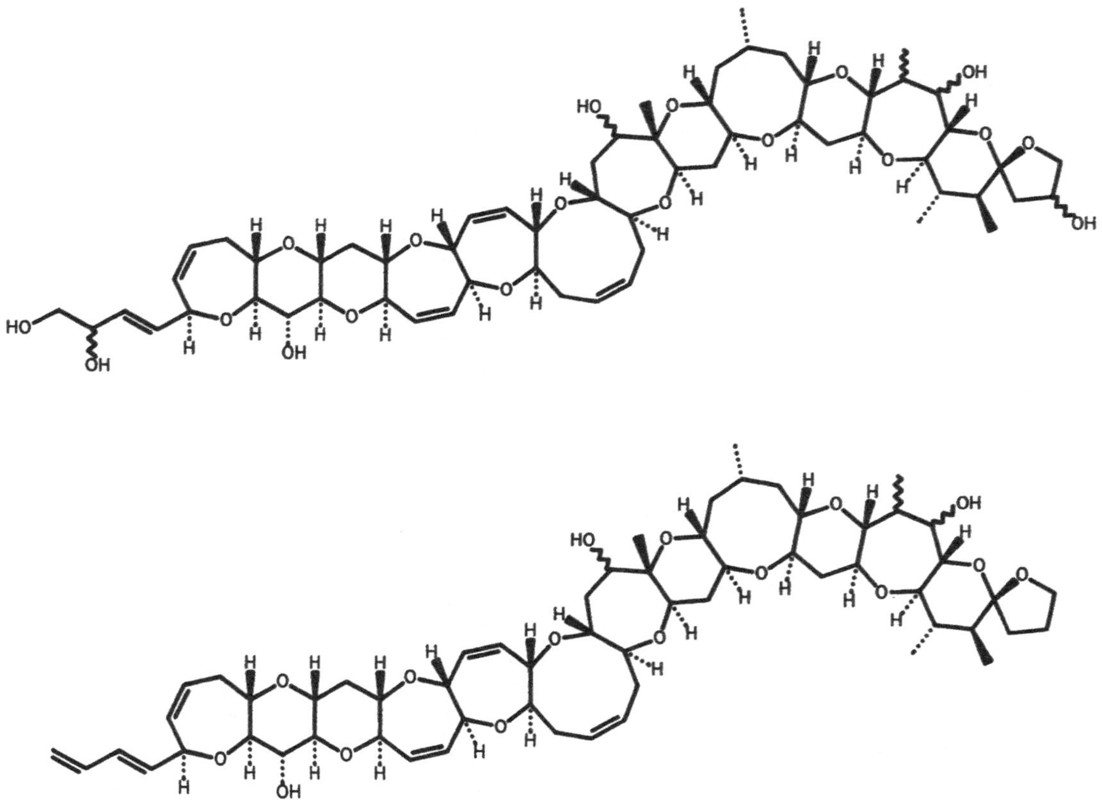

Figure 12.7 Ciguatera toxins. Again, note the similarity to the brevetoxins. The toxin in (a) was isolated from moray eels, in (b) the toxin was isolated from natural populations of the dinoflagellate *Gambierdiscus toxicus*.

maitotoxin, the structure of which is not yet fully known. Other benthic di-noflagellates, with growth habits and distributions similar to *Gambierdiscus toxicus,* may also be involved (Tindall et al. 1990).

Pharmacology   Ciguatoxin appears to act on the voltage-activated sodium channel (Bidard et al. 1984), at the same sodium channel binding site as the brevetoxins (Baden et al. 1985). The known differences between the ciguatoxins and brevetoxins lie in the longer duration of the effects of ciguatoxin in human victims.

Distribution   Ciguatera occurs in tropical fish (Bagnis 1981, 1986) but is becoming a public health problem in other regions as the importation of tropical species becomes more common (Anon. 1984). In general, ciguatera is a more difficult problem than the shellfish toxicities because fish swim. Where shellfish beds can be monitored and their toxicity established, it is more difficult to define and characterize the toxicity of a fish population. On the other hand, in contrast to the shellfish toxin syndromes, the source organism is relatively (though not strictly) benthic, while the dinoflagellates that supply toxins to bivalves move with the water masses. Ciguatera may therefore tend to be more persistent in a given location, in contrast to the ephemeral nature of bivalve toxicity which arises from transient populations of phytoplankton. Near the source, the distribution of ciguatera is patchy, frequently with distinct zones of toxicity that are recognized by local fishermen. Higher carnivores, like barracuda, which both move about and can accumulate high levels of toxicity, present more of a problem.

Pufferfish Poisoning

Vectors   Pufferfish poisoning is due almost entirely to the consumption of pufferfish—tetrodon fish or fugu. Other organisms contain the same family of toxins, however, and have on occasion caused human illness or death (Miyazawa et al. 1985; Maruyama et al. 1983, 1984).

Symptoms   Pufferfish poisoning causes neurological symptoms similar to PSP: tingling in the lips and extremities, paralysis, and death by respiratory arrest and/or cardiovascular collapse. The cardiovascular effects are much more severe than with PSP, and account for the relatively high death rate in pufferfish intoxications.

Toxins   The pufferfish toxins (Nakamura and Yasumoto, 1985; Woodward, 1964) are tetrodotoxin and related compounds (Figure 12.8).

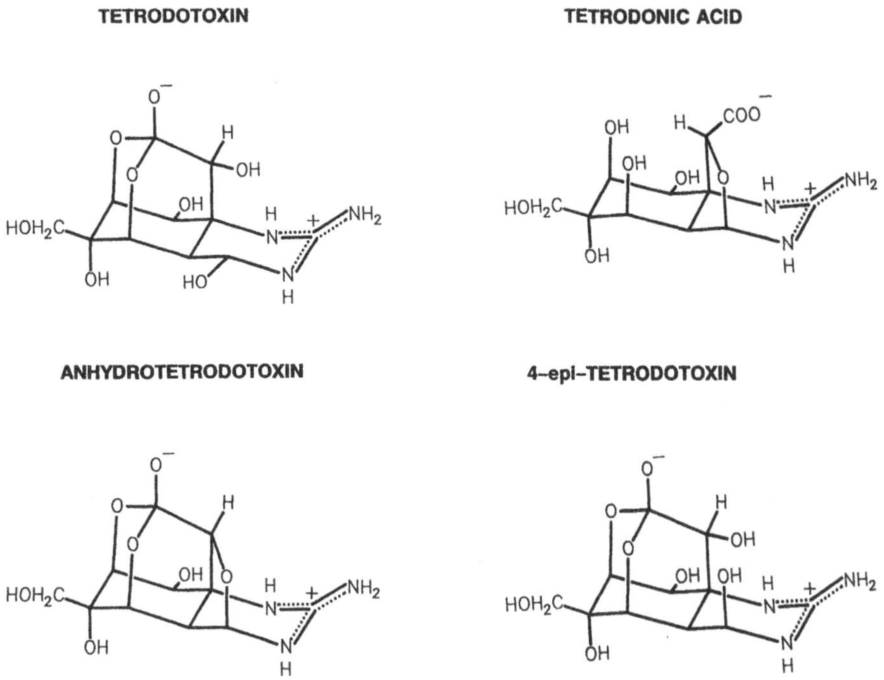

TETRODOTOXIN

TETRODONIC ACID

ANHYDROTETRODOTOXIN

4–epi–TETRODOTOXIN

Figure 12.8 Tetrodotoxins. Tetrodotoxin itself is generally the predominant toxin in pufferfish and is the most toxic compound of the group.

Source Unlike the other seafood toxicities described, the pufferfish toxins do not appear to be accumulated from food, but are apparently produced by symbiont bacteria (Noguchi 1987; Yasumoto et al. 1986; Yotsu et al. 1987). Some such bacteria have been detected and cultured. Research in this area is currently vigorous. There are some indications that the tetrodotoxins can be secreted by pufferfish as a defensive tactic (Kodama et al. 1985).

Pharmacology The pharmacology of the tetrodotoxins is, qualitatively, virtually indistinguishable from that of the saxitoxins. They bind to the same site on the voltage-activated sodium channel, causing the same effect but to differing degrees.

Distribution Tetrodon fishes are broadly distributed. There are marked differences in toxicity among species, some being virtually nontoxic (Lalone et al. 1963). In toxic species, the toxicity is highest in the skin, viscera, and gonads. Although intoxications occur mostly in Japan, where the toxic species are considered a prized delicacy, some intoxications and deaths have been reported in other countries.

MANAGEMENT

Strategies

As outlined above, seafoods of economic importance can accumulate potent toxins from certain varieties of phytoplankton. This is a problem because it is not immediately obvious to the consumer. Products thus contaminated appear wholesome but can cause illness or death. In the case of shellfish, this threat is generally dealt with through monitoring programs that attempt to intercept or quarantine shellfish lots or growing areas. The successful conduct of such programs is particularly challenging due to several factors, including the very sporadic spatial and temporal patterns of occurrence, which demand intensive sampling for reliable performance, the extreme potency of the toxins, requiring that they be detected at very low levels, and the chemical nature of the toxins, which makes it difficult to design suitable detection methods.

Effective management of seafood with respect to natural toxins depends on efficient monitoring such that contaminated product does not reach the consumer. Ideally, contaminated product is not even harvested, but left in place until toxicity has decreased to acceptable levels.

The costs, limitations, and difficulties of this approach make alternatives tempting, but none offer much hope of being practical. Among these possible alternatives are control of toxigenic dinoflagellates, prediction of their occurrence, remote detection of toxicity, and decontamination of toxic product. However, it seems unlikely that it will ever be possible to control phytoplankton composition in growing areas, eliminating toxigenic species. Although useful predictions can be made about the growth of phytoplankton in general, there is no reliable way to forecast when a particular phytoplankter will grow, thus no way to predict the blooming of a toxigenic dinoflagellate. There seems little hope that toxigenic organisms or shellfish toxicity can be detected by means other than frequent, direct sampling.

Removal of toxin, either by depuration of live stock or processing of product, would seem to have some potential. However, depuration is slow, so the economics are marginal even in favorable cases. Transport of contaminated shellfish to another area presents the risk of inadvertently contaminating the receiving area with the resting cysts of the toxigenic dinoflagellate.

Depuration also requires the cooperation of the shellfish. Consider a population of highly toxic shellfish placed in an intensive, short-term treatment system. (Such systems have been proposed, although none has been shown to function.) At the start, the toxin distribution among individuals in this batch permits a reasonable estimate of mean and extreme toxicity to be made from a practical sample. Through the depuration process, sampling can demonstrate a decrease in mean toxicity as the shellfish pump and cleanse themselves. However, a very small number of individuals may decline to open and pump, will therefore retain their original level of toxicity, and will likely go undetected by any practical sampling protocol. Although this hypothetical process would have indeed reduced the mean toxicity of the batch, it would in fact have rendered it more hazardous by increasing the dispersion of toxicities among individual shellfish and making it difficult to detect the few dangerously toxic individuals.

From the known chemistry of the various seafood toxins, it appears that processes that would reliably remove or destroy them from seafoods would render the products unpalatable. Processing does, in some cases, reduce toxicity to some extent. No known process, however, reliably can ensure the reduction of toxicity to an acceptable level while maintaining product quality.

For want of better options, we are therefore left with effective monitoring as the principle management strategy. Monitoring would be easiest if it were possible to sense directly the general level of contamination of shellfish in a bed, as one might scan for radioactivity with a Geiger counter. However, given the chemistry of the toxins, no known or realistically conceived field or laboratory method permits detection of shellfish toxicity without actually obtaining samples of shellfish. Given this requirement, it would then be easiest if a field method permitted direct detection of toxicity in shellfish flesh, without requiring the preparation of an extract. Such a method is conceivable, though none has been reduced to practice. However, the anatomical distribution of toxicity within shellfish is anything but uniform, so it would be difficult to integrate the values thus obtained to obtain a value for the toxicity of a whole shellfish, or other portion relevant to consumer safety. Regrettably, there is therefore no apparent option to obtaining samples, extracting them, and determining the toxicity of the extracts. This should be remembered in the discussions below that dwell on the obviously worthwhile pursuit of improved detection methods, because detection is only one component of the effort and cost of monitoring.

Monitoring: Effective Monitoring Requires a Reliable Sampling Plan and Efficient Means of Detection

The structure of a sampling plan is contingent on the temporal and spatial distribution of toxicity in the populations to be monitored. It is generally necessary to sample at a fairly high temporal and spatial density to ensure safety. Lot sampling, monitoring through sampling of defined lots of harvested or processed product, may largely circumvent temporal variation, but is heavily dependent on an understanding of the dispersion of toxicity among individuals, which corresponds essentially to spatial dispersion of toxicity in the product as grown. Circumventing temporal variation has considerable merit: In the case of PSP in shellfish, particularly mussels, toxicity can increase from negligible to lethal levels in less than 1 week.

The toxicity of individual shellfish varies somewhat within a given growing location. It seems reasonable that the geometry of the growing situation relative to the location of toxigenic dinoflagellates in the water would be a major factor in determining the degree of variation. Experience suggests that this is the case. Thus, shellfish grown in conventional situations—broad flats with tidal or other flow providing some degree of mixing—show enough uniformity that sampling of beds or lots appears to provide adequate protection. Conversely, however, it is possible for the dispersion of toxicity at a location to be much greater when the geometry of the situation does not lead to uniform exposure. Two recent episodes illustrate this. In Funka Bay, Japan, scallops hung on strings became intensely toxic, but only at certain levels, several meters below the surface. These levels were found to correspond to dense, thin strata of the toxigenic dinoflagellate. In southern Taiwan, purple clams were being cultured in small ponds connected by raceways to an estuary. These became intensely toxic, but in only one of several ponds. It is therefore important to recognize that the geometries employed in newer forms of intensive shellfish culture may result in a broad, nonuniform dispersion of individual toxicities, which may compromise an existing sampling plan that has proven satisfactory for conventionally grown product.

Sampling finfish for ciguatera is complicated by the difficulty of knowing where a specific fish came from and what it has eaten. The one step that can be taken is to ensure that the already complex natural situation is not worsened by mixing product in market channels: fish taken from regions where ciguatera is a possibility should be handled in lots that clearly identify their origins.

Detection methods for seafood toxins are efficient insofar as they provide an accurate measure of toxicity, or potential toxicity, to human consumers, and have sensitivity adequate to ensure an ample margin of safety. One of the principal concerns of the Natural Products and Instrumentation Branch, FDA, is the development of more efficient detection methods for foodborne natural toxins such as the seafood toxins.

Detection Methods: Assays and Analyses

It is useful to distinguish between two classes of detection methods which, for want of better terms and for the purposes of this discussion, will be designated assays and analyses. An assay provides a single result for a sample, the net response for all active substances present, whereas an analysis resolves the substances such that each can be determined separately.

Assays: Limitations Developing assays for seafood toxins would be a relatively straightforward matter if each syndrome involved only a single toxin, or if the several toxins associated with each occurred in a mixture of constant composition which would allow a method to treat them as a single entity. However, neither situation can be assumed to generally prevail. Instead, we are faced with mixtures of toxins with a broad range of compositions, varying with time, location, and type of product.

BIOASSAY   In the case of the mouse bioassay for the saxitoxins and tetrodotoxins, the accuracy is basically good, because the assay is based on the response of mammalian sodium channels, likely to have a response similar to the sites affected in humans. However, it may be a long way in physiological terms from the site of administration (intraperitoneal, in this case) to the active site. In mice, such data as are available indicate that the toxins or toxin mixtures that have been tested are about 40-fold less toxic to mice when administered orally than when administered by intraperitoneal injection as normally done for the PSP assay. Given the magnitude of this difference, and the broad range of chemical properties of the toxins, there is reason to suspect that the ratio of human oral potently (HOP) to mouse intraperitoneal potency (MIP) may not be uniform among the toxins.

Although assays based on intraperitonal injection of mice can be employed for ASP, NSP, DSP, and ciguatera, the margins between the detection limits of the assays and the levels at which human symptoms occur are much smaller than with the saxitoxins and tetrodotoxins. Fortunately, there are simple and effective methods for HPLC analysis of the ASP toxin, domoic acid.

In the Netherlands, an assay for DSP has been developed in which rats are fed samples of shellfish and observed for diarrhea. This may prove to be the method of choice, because it proves a direct measure of diarrhetic potency.

IMMUNOASSAYS   The problem in designing an immunoassay, given that several toxins will generally be present in an indeterminate ratio, is to have the response of the assay to the various toxins correspond to the response ratios required to estimate toxicity to humans.

The antibodies developed so far for the saxitoxins lack the response factors necessary for a general use assay; they respond to some toxins but not to others. To be useful for PSP monitoring an immunoassay will have to employ either a single antibody with response factors closer to ideal, or a cocktail of antibodies tailored to give the required response. In the case of DSP, an immunoassay system has been developed that responds to Okadaic acid and dinophysistoxin 1, but not to the other DSP toxins (Usagawa et al. 1989). The assay may have some utility in regions where okadaic acid appears to be the primary DSP toxin present.

HPLC ANALYSES   Although HPLC analyses for marine toxins tend to require expensive equipment and are generally impractical for use in the field, they have the important ability, or at least the potential, to determine the identity and concentration of each toxin present.

In the case of ASP, simple and reliable HPLC methods have been developed using UV detection of domoic acid itself (Quilliam et al. 1990) or, for very low levels, fluorescence, detection of a domoic acid adduct (Pocklington et al. 1990).

Among the DSP toxins, those with carboxylic acid or alcohol functional groups can be determined by HPLC of fluorescent adducts (Lee et al. 1987, 1989). Although HPLC can be used for the purification, characterization, and identification of the breve toxins, and the two known ciguatera toxins, these procedures are currently based on direct detection of UV absorbance by compounds that lack distinct, intense UV chromophores and therefore have limited sensitivity. Because of this their application to seafood samples generally requires extensive enrichment and cleanup, which makes the methods impractical for general use.

The saxitoxins and tetrodotoxins also lack useful UV absorbance or intrinsic fluorescence, but both families of compounds can be treated to yield fluorescent degradation products. HPLC of the saxitoxins, using either gradient (Sullivan 1990) or isocratic (Oshima et al. 1989) conditions, is followed by post-column reaction with a periodate-based oxidant to convert the toxins to fluorescent products which then flow to a fluorescence detector. Sensitivity with these methods is generally on the same order as or somewhat better than that of the mouse bioassay.

Tetrodotoxins are similarly determined (Yotsu and Yasumato 1989) by post-column reaction of the separated toxins with strong base, which transforms the toxins into fluorescent products.

SUMMARY

Natural toxins in seafoods present many challenges, many of which have yet to be satisfactorily addressed. The development and implementation of appropriate measures is and will remain expensive. But the payoff is great, and more than justifies the investment. Seafood can be a wholesome, enjoyable, and affordable source of nutrition and will find an eager market if quality and safety are assured.

ACKNOWLEDGMENTS

I would like to thank Dr. F. Fry and L. Hohneke for assistance in preparing the structure diagrams.

REFERENCES

Alam, M. I., Trieff, N. M., Ray, S. M., and Hudson, J. E. 1975. Isolation and partial characterization of toxins from the dinoflagellate *Gymnodinium breve*. *J. Pharm. Sci.* **64**:865–867.

Anonymous. 1984. Ciguatera poisoning in Canada. *Bull. Pan. Am. Health Organ.* **18**:293–294.

Asai, S., Krzanowski, J. J., Anderson, W. H., Martin, D. F., Polson, J. B., Lockey, R. F., Bukantz, S. C., and Szentlvanyi. A. 1982. Effects of the toxin of red tide, *Ptychodiscus brevis*, on canine tracheal smooth muscle: a possible new asthma-triggering mechanism. *J. Allergy Clin. Immunol.* **69**:418–428.

Baden, D. G. 1977. Metabolism and toxinology of the marine dinoflagellate, *Gymnodinium breve*. Doctoral dissertation, University of Miami, Miami, Florida. 206 pp.

Baden, D. G. 1983. Marine food-borne dinoflagellate toxins. *Int. Rev. Cytol.* **82**:99–150.

Baden, D. G., Mende, T. J., Lichter, W., and Wellham, L. 1981. Crystallization and toxicology of T34: a major toxin from Florida's red tide organism *(Ptychodiscus brevis)*. *Toxicon* **19**:455–462.

Baden, D. G., and Tomas, C. R. 1988. Variations in major toxin composition for six clones of *Ptychodiscus brevis*. *Toxicon* **26**:961–963.

Baden, D. G., Mende, T. J., Bikhazi, G., and Leung, I. 1982. Bronchoconstriction caused by Florida red tide toxins. *Toxicon* **20**:929–932.

Baden, D. G., Mende, T. J., and Brand, L. E. 1985. Cross-reactivity in immunoassays directed against toxins isolated from *Ptychodiscus brevis*, In: Anderson, D. M., White, A. W., and Baden, D. G. (Eds). *Toxic Dinoflagellates, Proceedings of the Third International Conference*. Elsevier, New York. pp. 363–368.

Baden, D. G., Mende, T. J., Szmant, A. M., Trainer, V. L, Edwards, R. A., and Roszell, L. E. 1988. Brevetoxin binding: molecular pharmacology versus immunoassay. *Toxicon* **26**:97–103.

Bagnis, R. 1981. The ciguatera type ichthyosarcotoxism: a complicated phenomenon of marine and human biology. *Oceanol. Acta* **4**:375–387.

Bagnis, R. 1986. Ciguatera, health and development in French Polynesia. *Union Med. Can.* **115**:502–506.

Bagnis, R., Chanteau, S., Chungue, E., Hurtel, J. M., Yasumoto, T., and Inoue, A. 1980. Origins of ciguatera fish poisoning: a new dinoflagellate, *Gambierdiscus toxicus* Adachi and Fukuyo, definitively involved as a causal agent. *Toxicon* **18**:199–208.

Balech, E. 1985. The genus *Alexandrium* or *Gonyaulax* of the *Tamarensis* group. In: Anderson, D. M., White, A. W. and Baden, D. G. (Eds.). *Toxic Dinoflagellates, Proceedings of the Third International Conference*. Elsevier, New York. pp. 33–38.

Bialojan, C., Takai, A., and Ruegg, J. C. 1987. Inhibition of protein phosphatase activity and actin–myosin interaction by black sponge toxin. *Adv. Protein Phosphatases* **4**:253–287.

Bialojan, C., Ruegg, J. C., and Takai, A. 1988. Effects of okadaic acid on isometric tension and myosin phosphorylation of chemically skinned guinea pig taenia coli. *J. Physiol. (Lond.)* **398**:81–95.

Bidard, J. N., Vijverberg, H. P., Frelin, C., Chungue E., Legrand, A. M., Bagnis, R., and Lazdunski, M. 1984. Ciguatoxin is a novel type of Na+ channel toxin. *J. Biol. Chem.* **259**:8353–8357.

Bordner, J., Thiessen, W. E., Bates, H. A., and Rapoport, H. 1975. Structure of a crystalline derivative of saxitoxin. Structure of saxitoxin. *J. Am. Chem. Soc.* **97**:6008–6012.

Boyer, G. L, Schantz, E. J., and Schnoes, H. K. 1978. Characterization of 11-hydroxysaxitoxin sulfate, a major toxin in scallops exposed to blooms of the poisonous dinoflagellate *Gonyaulax tamarensis*. *J. Chem. Soc. Chem. Commun.* **20**:889–890.

Buckley, L. J., Ikawa, M., and Sasner, J. J., Jr. 1976. Isolation of *Gonyaulax tamarensis* toxins from soft shell clams *(Mya arenaria)* and a thin-layer chromatographic-fluorometric method for their detection. *J. Agric. Food Chem.* **24**:107–111.

Capra, M. F. 1986. Ciguatera poisoning. *Proc. Nutr. Soc. Aust.* **11**:63–72.

Chou, H.-N., Shimizu, Y., Van Duyne, G. D., Clardy, J. 1985. Two new polyether toxins of *Gymnodinium breve* (=*Ptychodiscus brevis*). In: Anderson, D. M., White, A. W., and Baden, D. G. (Eds.). *Toxic Dinoflagellates, Proceedings of the Third International Conference.* Elsevier, New York. pp. 305–308.

Chungue, E., Bagnis, R., and Parc, F. 1984. The use of mosquitoes *(Aedes aegypt)* to detect ciguatoxin in surgeon fishes *(Ctenochaetus striatus). Toxicon* 22:161–164.

Chungue, E., Chanteau, S., Hurtel, J. M., and Bagnis, R. 1979. Toxicological study of several species of benthoplankton algae of ciguatera biotopes, cultivated in a nonaxenic artificial medium. *Rev. Int. Oceanogr. Med.* 55:35–40.

Combe, J. S. 1828. On the poisonous effects of the mussel *(Mytllus edulis). Edin. Med. Surg. J.* 29:86–96.

DiNovi, M., Trainor, D. A., Nakanishi, K., Sanduja, R., and Alam, M. 1983. The structure of PB-1, an unusual toxin isolated from the red tide dinoflagellate *Ptychodiscus brevis. Tetrahedron Lett.* 24:855–858.

Durand-Clement, M. 1986. A study of toxin production by *Gambierdiscus toxicus* in culture. *Toxicon* 24:1153–1157.

Erdodi, F., Rokolya, A., Di Salvo, J., Berany, M., and Barany, K. 1988. Effect of okadaic acid on phosphorylation-dephosphorylation of myosin light chain in aortic smooth muscle homogenate. *Biochem. Biophys. Res. Commun.* 153:156–161.

Gacutan, R. Q., Tabbu, M. Y., Aujero, E. J., and Icatio, F. J. 1985. Paralytic shellfish poisoning due to *Pyrodinium bahamense var. compressa* in Matl, Davao Oriental, Philippines. *Marine Biol.* 87:223–227.

Golik, J., James, J. C., Nakanishi, K., and Lin, Y. Y. 1982. The structure of brevetoxin C. *Tetrahedron Lett.* 23:2535–2538.

Grindley, J. R., and Sapieka, N. 1969. The cause of mussel poisoning in South Africa. *S. A. Med. J.* 43:275–279.

Hall, S. 1982. Toxins and toxicity of *Protogonyaulax* from the northeast Pacific. Doctoral dissertation, University of Alaska, Fairbanks, Alaska.

Hall, S., Darling, S. D., Boyer, G. L., Reichardt, P. B., and Liu, H. W. 1984. Dinoflagellate neurotoxins related to saxitoxin: structures of toxins C3 and C4, and confirmation of the structure of neosaxitoxin. *Tetrahedron Lett.* 25:3537–3538.

Hall, S., Reichardt, P. B., and Neve, R. A. (1980). Toxins extracted from an Alaskan isolate of *Protogonyaulax* sp. *Biochem. Biophys. Res. Comm.* 97:649–653.

Hallegraeff, G. M., Steffensen, D. A., and Wetherbee, R. 1988. Three estuarine Australian dinoflagellates that can produce paralytic shellfish toxins. *J. Plant. Res.* 10:533–541.

Helrich, K. 1990. *Official Methods of Analysis of the Association of Official Analytical Chemists,* 14th edition. AOAC, Washington, DC, pp. 881–882.

Hoffman, P., Granade, H. R., and McMillan, J. P. 1983. The mouse ciguatoxin bioassay: a dose–response curve and symptomatology analysis. *Toxicon* 21:363–369.

Hokama, Y. 1985. A rapid simplified enzyme immunoassay stick test for the detection of ciguatoxin and related polyethers from fish tissues. *Toxicon* 23:939–946.

Hokama, Y. 1988. Ciguatera fish poisoning. *J. Clin. Lab. Anal.* 2:44–50.

Hokama, Y., and Miyahara, J. T. 1986. Ciguatera poisoning: clinical and immunological aspects. *J. Toxicol. Toxin Rev.* 5:25–53.

Hokama, Y., Banner, A. H., and Boylan, D. B. 1977. A radioimmunoassay for the detection of ciguatoxin. *Toxicon* 15:317–325.

Huang, J. M. C., Wu, C. H., and Baden, D. G. 1984. Depolarizing action of a red-tide dinoflagellate brevetoxin on axonal membranes. *J. Pharmacol. Exp. Ther.* 229:615–621.

Ikeda, T., Matsuno, S., Sato, S., Ogata, T., Kodama, M., Fukuyo, Y., and Takayama,

H. 1988. First report on paralytic shellfish poisoning caused by *Gymnodinium catenatum* Graham (Dinophyceae) in Japan. In: Okaichi, T., Anderson, D. M., and Nemoto, T. (Eds.) *Red Tides: Biology, Environmental Science, and Toxicology*. Elsevier, New York. pp. 411–414.

Kodama, M., Ogata, T., and Sato, S. 1985. External secretion of tetrodotoxin from puffer fishes stimulated by electric shock. *Marine Biol. (Berl.)* **87**:199–202.

Koehn, F. E. 1983. A chemical investigation of saxitoxin and derivatives. Doctoral dissertation, University of Wisconsin, Madison, Wisconsin. 313 pp.

Kumagai, M., Yanagi, T., Murata, M., Yasumoto, T., Kat, M., Lassus, R, and Rodriguez-Vasquez, J. A. 1986. Okadaic acid as the causative agent of diarrhetic shellfish poisoning in Europe. *Agric. Biol. Chem.* **50**:2853–857.

Lalone, R. C., DeVillez, E. D., and Larson, E. 1963. An assay of the toxicity of the Atlantic puffer fish, *Spheroides maculatus*. *Toxicon* **1**:159–164.

Lawrence, D. N., Enriquez, M. B., Lumish, R. M., and Maceo, A. 1980. Ciguatera fish poisoning in Miami. *JAMA* **244**:254–258.

Lebour, M. V. 1925. *The Dinoflagellates of Northern Seas*. Marine Biology Association of the UK, Plymouth. 250 pp.

Lee, J. S., Yanagi, T., Kenma, R., and Yasumoto, T. 1987. Fluorometric determination of diarrhetic shellfish toxins by high-performance liquid chromatography. *Agric. Biol. Chem.* **51**:877–881.

Lee, J. S., Murata, M., and Yasumoto, Y. 1989. Analytical methods for determination of diarrhetic shellfish toxins. In: Natori, S., Hashimoto, K., and Ueno, Y. (Eds.). *Mycotoxins and Phycotoxins '88*. Elsevier, New York, pp. 327–334.

Legrand, A.-M., Cruchet, P., Bagnis, R., Murata, M., Ishibashi, Y., and Yasumato, T. 1990. Chromatographic and spectral evidence for the presence of multiple ciguatera toxins. In: Graneli, E., Sundstrom, B, Edlar, L., and Anderson, D. M. (Eds.). *Toxic Marine Phytoplankton*. Elsevier, New York, pp. 374–378.

Lin, Y.-Y., Risk, M., Ray, S. M., Van Engen, D., Clardy, J., Golik, J., James, J. C., and Nakanishi, K. 1981. Isolation and structure of brevetoxin B from the "red tide" dinoflagellate *Ptychodiscus brevis (Gymnodinium breve)*. *J. Am. Chem. Soc.* **103**:6773–6775.

Luthy, J. 1979. Epidemic of paralytic shellfish poisoning in western Europe. In: Taylor, D. L. and Sellger, H. H. (Eds.). *Proceedings of the Second International Conference on Toxic Dinoflagellate Blooms*. Elsevier/North Holland, Amsterdam, pp. 15–22.

Maclean, J. L. 1975a. Paralytic shellfish poison in various bivalves, Port Moresby, 1973. *Pac. Sci.* **29**:349–352.

Maclean, J. L. 1975b. Red tide in the Morobe district of Papua New Guinea. *Pac. Sci.* **29**:7–13.

Maclean, J. L. 1977. Observations on *Pyrodinium bahamense* Plate, a toxic dinoflagellate, in Papua, New Guinea. *Limnol. Oceanogr,* **22**:234–254.

Maharaj, S. R., Desai, P., Kulkami, P., and Pinnock, M. A. 1986. Ciguatera fish poisoning in a Jamaican family. *WI Med. J.* **35**:321–323.

Maruyama, J., and Noguchl, T. 1984. Tetrodotoxin—a review with special reference to its distribution in nature. *La Mer* **22**:299–304.

Maruyama, J., Noguchi, T., Jeon, J. K., Yamazaki, K., and Hashimoto, K. 1983. Paralytic poisoning by tetrodotoxin from the trumpet shell *Charonia saullae*. *Shokuhin Eiseigaku Zasshi* **24**:465–468.

McFarren, E. F. 1971. Assay and control of marine biotoxins. *Food Technol.* **25**:38–46.

McMillan, J. P., Granade, H. R., and Hoffman, P. 1980. Ciguatera fish poisoning in the United States Virgin Islands: preliminary studies. *J. Coll. Virgin Is.* **6**:84–107.

Meyer, K. F., Sommer, H., and Schoenholz, P. 1928. Mussel poisoning. *J. Prevent. Med.* **2**:365–394.

Miyazawa, K., Noguchi, T., Maruyama, J., Jeon, J. K., Otsuka, M., and Hashimoto, K. 1985. Occurrence of tetrodotoxin in the starfishes *Astropecten polyacanthus* and *A. scoparlus* in the Seto Inland Sea. *Marine Biol. (Berl.)* **90**:61–64.

Moczydlowski, E., Garber, S. S., and Miller, C. 1984a. Batrachotoxin-activated Na$^+$ channels in planar lipid bilayers. Competition of tetrodotoxin block by Na$^+$. *J. Gen. Physiol.* **84**:665–686.

Moczydlowski, E., Hall, S., Garber, S. S., Strichartz, G. S., and Miller, C. 1984b. Voltage-dependent blockade of muscle sodium channels by guanidinium toxins. Effect of toxin charge. *J. Gen. Physiol.* **84**:687–704.

Murakami, Y., Oshima, Y., and Yasumoto, T. 1982. Identification of okadaic acid as a toxic component of a marine dinoflagellate *Prorocentrum lima*. *Nippon Suisan Gakkaishi* **48**:69–72.

Murata, M., Sano, M., Iwashita, T., Naoki, H., and Yasumoto, T. 1986. The structure of pectinotoxin-3, a new constituent of diarrhetic shellfish toxins. *Agric. Biol. Chem.* **50**:2693–2695.

Murata, M., Kumagai, M., Lee, J. S., and Yasumoto, T. 1987. Isolation and structure of yessotoxin, a novel polyether compound implicated in diarrhetic shellfish poisoning. *Tetrahedron Lett.* **28**:5869–5872.

Murata, M., Legrand, A.-M., Ishibashi, Y., and Yasumoto, T. 1989. Structures of ciguatoxin and its congener. *J. Am. Chem. Soc.* **111**:8929–8931.

Nakamura, M., and Yasumoto, T. 1985. Tetrodotoxin derivatives in puffer fish. *Toxicon* **23**:271–276.

Neal, R. A. 1967. Fluctuations in the levels of paralytic shellfish toxin in four species of lamellibranch mollucs near Ketchikan, Alaska, 1963–1965. Doctoral dissertation, University of Washington, Seattle.

Needler, A. B. 1949. Paralytic shellfish poisoning and *Gonyaulax tamarensis*. *J. Fish. Res. Board Can.* **7**:490–504.

Noguchi, T., Hwang, D. F., Arakawa, O., Sugita, H., Deguchi, Y., Shida, Y., and Hashimoto, K. 1987. *Vibrio alginolyticus*, a tetrodotoxin-producing bacterium, in the intestines of the fish *Fugu vermicularis vermicularis, Marine Biol. (Berl.)* **94**:625–630.

Ohfune, Y., and Tomita, M. 1982. Total synthesis of (–)-domoic acid. A revision of the original structure. *J. Am. Chem. Soc.* **104**:3511–3513.

Oshima, Y., Sugino, K., and Yasumoto, T. 1989. Latest advances in HPLC analysis of paralytic shellfish toxins. In: Natori, S., Hashimoto, K., and Ueno, Y. (Eds.). *Mycotoxins and Phycotoxins '88* Elsevier, New York, pp. 319–326.

Oshima, Y., Hasegawa, M., Yasumoto, T., Haliegraeff, G., and Blackburn, S. 1987. Dinoflagellate *Gymnodinium catenatum* as the source of paralytic shellfish toxins in Tasmanian shellfish. *Toxicon.* **25**:1105–1111.

Pierce, R. H. 1986. Red tide *(Ptychodiscus brevis)* toxin aerosols: a review. *Toxicon.* **24**:955–965.

Pocklington, R., Milley, J. E., Bates, S. S., Bird, C. J., de Freitas, A.S.W., and Quilliam, M. A. 1990. Trace determination of domoic acid in seawater and phytoplankton by high-performance liquid chromatography of the fluorenylmethoxycarbonyl (FMOC) derivative. *Int. J. Environ. Anal. Chem.* **38**:351–368.

Poli, M. A. 1985. Characeterization of the binding of the *Ptychodiscus brevis* neurotoxin T17 to sodium channels in rat brain synaptosomes. Doctoral dissertation, University of Miami, Miami, Florida. 121 pp.

Poli, M. A., Mende, T. J., and Baden, D. G. 1986. Brevetoxins, unique activators of voltage-sensitive sodium channels, bind to specific sites in rat brain synaptosomes. *Mol. Pharmacol.* **30**:129–135.

Quayle, D. B. 1969. Paralytic shellfish poisoning in British Columbia. *Bull. Fish. Res. Board Can.* **268**:1–68.

Quilliam, M. A., Sim, P. G., McCullock, A. W., and McInnus, A. G. 1989. *Int. J. Environ. Anal. Chem.* (in press).

Raj, U., Haq, H., Oshima, Y., and Yasumoto, T. 1983. The occurrence of paralytic shellfish toxins in two species of xanthid crab from Suva barrier reef, Fiji Islands. *Toxicon* **21**:547–551.

Sawyer, P. R., Jollow, D. J., Scheuer, P. J., York, R., McMillan, J. P., Withers, N. W., Fudenberg, H. H., and Higerd, T. B. 1984. The effect of ciguatoxin-associated toxins on body temperature in mice. In: Ragelis, E. P. (Ed.). *Seafood Toxins*, ASC Symposium Series 418. American Chemical Society, Washington, D.C. pp. 321–329.

Schantz, E. J., McFarren, E. F., Schaeffer, M. L., and Lewis, K. H. 1958. Purified shellfish poison for bioassay standardization. *J. Assoc. Off. Agric. Chem.* **41**:160–168.

Schantz, E. J., Ghazarossian, V. E., Schnoes, H. K., Strong, F. M., Springer, J. P., Pezzanite, J. O., and Clardy, J. 1975. Structure of saxitoxin. *J. Am. Chem. Soc.* **97**:1238–1239.

Schimizu, Y., and Yoshioka, M. 1981. Transformation of paralytic shellfish toxins as demonstrated in scallop homogenates. *Science* **212**:547–549.

Shimizu, Y., Hsu, C.-P., Fallon, W. E., Oshima, Y., Miura, I., and Nakanishi, K. 1978. Structure of neosaxitoxin. *J. Am. Chem. Soc.* **100**:6791–6793.

Shimizu, Y., and Yoshioka, M. 1981. Transformation of paralytic shellfish toxins as demonstrated in scallop homogenates. *Science* **212**:547–549.

Shimizu, Y., Chou, H. N., Bando, H., Van Duyne, G., and Clardy, J. 1986. Structure of brevetoxin A (GB-1 toxin), the most potent toxin in the Florida red tide organism *Gymnodinium breve (Ptychodiscus brevis)*. *J. Am. Chem. Soc.* **108**:514–515.

Sommer, H., and Meyer, K. F. 1937. Paralytic shellfish poisoning. *Arch. Pathol.* **24**:560–598.

Sribhibhadh, A. 1963. Seasonal variations of shellfish toxicity in the California mussel, Mytilus californianus Conrad, and the Pacific oyster, *Crassostrea gigas* (Thunberg), along the Strait of Juan de Fuca and in Willapa Bay. Doctoral dissertation, University of Washington, Seattle.

Steidinger, K. A., Tester, L. S., and Taylor, F. J. R. 1980. A redescription of *Pyrodinium bahamense var. compressa* (Bohm) stat. nov, from Pacific red tides. *Phycologia* **19**:329–337.

Suganuma, M., Fujlki, H., Suguri, H., Yoshizawa, S., Hirota, M. N., Ojika, M., Wakamatsu, K., Yamada, K., and Sugimura, T. 1988. Okadaic acid: an additional non-phorbol-12-tetradecanoate-13-acetate-type tumor promoter. *Proc. Natl. Acad. Sci. USA* **85**:1768–1771.

Sullivan, J. J. 1990. High-performance liquid chromatographic method applied to paralytic shellfish poisoning research. In: Hall, S., and Strichartz, G. R. (Eds.). *Marine Toxins: Origin, Structure, and Molecular Pharmacology*. ACS Symp. Series #418. American Chemical Society, Washington, D.C., pp. 66–77.

Sullivan, J. J., and Iwaoka, W. T. 1983. High pressure liquid chromatographic determination of toxins associated with paralytic shellfish poisoning. *J. Assoc. Off. Anal. Chem.* **66**:297–303.

Sullivan, J. J., Iwaoka, W. T., and Liston, J. 1983. Enzymatic transformation of PSP toxins in the littleneck clam *(Protothaca staminea)*. *Biochem. Biophys. Res. Commun.* **114**:465–472.

Tachibana, K., Scheuer, P. J., Tsukitani, Y., Kikuchi, H., Van Engen, D., Clardy, J., Gopichand, Y., and Schmitz, F. J. 1981. Okadaic acid, a cytotoxic polyether from two marine sponges of the genus *Halichondria*. *J. Am. Chem. Soc.* **103**: 2469–2471.

Tester, P. A., and Fowler, P. K. 1990. Brevetoxin contamination of *Mercenaria mercenaria* and *Crassostrea virginica:* A management issue. In: *Toxic Marine Phytoplankton*. Graneli, E., Sundstrom, B., Edler, L., and Anderson, D. M. (Eds.). Elsevier, New York. pp. 499–503.

Tindall, D. R., Miller, D. M., Tindall, P. M. 1990. Toxicity of *ostreiopsis lenticulards* from the British and United States Virgin Islands. In: Graneli, E., Sundstrom, B., Edlar, L., and Anderson, D. M. (Eds.). *Toxic Marine Phytoplankton.*Elsevier, New York, pp. 424–429.

Todd, E. C. D. 1990. Amnesic shellfish poisoning—A new seafood toxin syndrome. In: Graneli, E., Sundstrom, B., Edlar, L., and Anderson, D. M. (Eds.). *Toxic Marine Phytoplankton.*Elsevier, New York, pp. 504–508.

Usagawa, T., Nishimura, M., Itoh, Y., Uda, T., and Yasumoto, T. 1989. Preparation of monoclonal antibodies against okadaic acid prepared from the sponge. *Halichondria okadai*. *Toxicon* **7**:1323–1330.

Whedon, W. F., and Kofoid, C. A. 1936. Dinoflagellates of the San Francisco region. I. On the skeletal morphology of two new species, *Gonyaulax catenella* and *G. acatenalla*. *Univ. Calif. Publ. Zool.* **41**:25–34.

White, A. W. 1977. Dinoflagellate toxins as probable cause of an Atlantic herring *(Clupea harengus harengus)* kill, and pteropods as apparent vector. *J. Fish. Res. Board Can.* **34**:2421–2424. .

White, A. W. 1981a. Marine zooplankton can accumulate and retain dinoflagellate toxins and cause fish kills. *Limnol. Oceanogr.* **26**:103–109.

White, A. W. 1981b. Sensitivity of marine fishes to toxins from the red-tide dinoflagellate *Gonyaulax excavata* and implications for fish kills. *Marine Biol.* **65**:255–260.

Wichmann, C. F., Niemczura, W. P., Schnoes, H. K., Hall, S., Reichardt, P. B., and Darling, S. D. 1981. Structures of two novel toxins from *Protogonyaulax*. *J. Am. Chem. Soc.* **103**:6977–6978.

Withers, N. W. 1982. Ciguatera fish poisoning. *Annu. Rev. Med.* **33**:97–111.

Woodward, R. B. 1964. The structure of tetrodotoxin. *Pure Appl. Chem.* **9**:49–74.

Yasumoto, T., Oshima, Y., and Yamaguchi, M. 1978. Occurrence of a new type of shellfish poisoning in the Tohoku District. *Bull. Jpn. Soc. Sci. Fish.* **44**:1249–1255.

Yasumoto, T., Nakajima, I., Oshima, Y., and Bagnis, R. 1979. A new toxic dinoflagellate found in association with ciguatera. *Dev. Mar. Biol.* **1**:65–70.

Yasumoto, T., Oshima, Y., Sugawara, W., Fukuyo, Y., Oguri, H., Igarashi, T., and Fujita, N. 1980. Identification of *Dinophysis fortii* as the causative organism of diarrhetic shellfish poisoning. *Bull. Jpn. Soc. Sci. Fish.* **46**:1405–1411.

Yasumoto, T., Murata, M., Oshima, Y., Sano, M., Matsumoto, G. K., and Clardy, J. 1985. Diarrhetic shellfish toxins. *Tetrahedron* **41**:1019–1025.

Yasumoto, T., Yasumura, D., Yotsu, M., Michishita, T., Endo, A., and Kotaki, Y. 1986. Bacterial production of tetrodotoxin and anhydrotetrodotoxin. *Agric. Biol. Chem.* **50**:793–795.

Yotsu, M., Yamazaki, T., Meguro, Y., Endo, A., Murata, M., Naoki, H., and Yasumoto, T. 1987. Production of tetrodotoxin and its derivatives by *Pseudomonas* sp. isolated from the skin of a pufferfish. *Toxicon* **25**:225–228.

Yotsu, M., and Yasumoto, T. 1989. An improved tetrodotoxin analyzer. *Agric. Biol. Chem.* **53**:893–895.

13 Scombroid Poisoning

Jayne E. Stratton and
Steve L. Taylor

Toxins produced by microorganisms of marine origin or microorganisms proliferating on marine foods are responsible for several types of foodborne disease. One of these illnesses, histamine poisoning, can result from the ingestion of food containing unusually high levels of histamine. Fish of the families Scombridae and Scomberesocidae are commonly implicated in incidents of histamine poisoning; hence the usage of the term, "scombroid fish poisoning," developed.

Histamine is formed in foods largely from the growth of microorganisms that possess the enzyme histidine decarboxylase. Scombroid fish such as tuna and mackerel have large amounts of free histidine in their muscle tissue which can serve as a substrate for bacterial histidine decarboxylase. The enzyme converts histidine into histamine, thus resulting in the accumulation of histamine in the fish. On consumption of the affected fish, a short incubation period is followed by the onset of symptoms. Because the disease is rather mild and self-limited, it is thought to be greatly underreported. It is difficult to make an estimate of worldwide occurrence because many countries do not keep official records of food poisoning outbreaks. Also, most countries do not have regulatory limits on the allowable levels of histamine in food.

In this chapter, we will review the clinical aspects of histamine poisoning, the epidemiology of fishborne histamine poisoning, the formation and control of histamine, histamine toxicity and metabolism, methods for the analysis of histamine, and the regulatory situation relative to histamine in fish in the U.S. Previous reviews of histamine poisoning may contain additional details on the clinical and toxicological features (Taylor 1986), microbiological formation of histamine in foods (Arnold and Brown 1978; Kimata 1961), and the metabolism of histamine (Beavan 1978).

CLINICAL ASPECTS

Symptomology

Histamine poisoning is usually a rather mild illness with a wide variety of symptoms. The primary symptoms are cutaneous, gastrointestinal, hemodynamic, and neurological. Cutaneous symptoms include rash, urticaria, edema, and localized inflammation, whereas symptoms affecting the gastrointestinal tract include nausea, vomiting, diarrhea, and abdominal cramps. Hemodynamic manifestations include hypotension. Lastly, neurological involvement is manifested by headache, palpitations, tingling, flushing, burning sensations in the mouth, and itching. Bartholomew et al. (1987), in a review of over 250 suspected incidents in Britain from 1976 to 1986, noted that the most common symptoms were rash, diarrhea, flushing and sweating, and headache.

Facial flushing appears to be a frequent occurrence in cases of histamine poisoning (Kasha and Norins 1988; Kim 1979), as does a "metallic" or "peppery" taste. These symptoms help to distinguish histamine poisoning from other types of foodborne intoxications. Individuals may experience a few or several of these symptoms. More serious complications arising from histamine intoxication rarely occur. When they do, they usually involve the cardiac and respiratory systems. Acute symptoms in the elderly or those with underlying cardiac or respiratory problems could prove life-threatening. Russell and Maretic (1986) relate the case of a young child with a history of bronchial asthma who reportedly suffered respiratory collapse after having consumed fish containing excessive amounts of histamine. The severity of the symptoms may also in part be related to the quantity of histamine ingested.

Because histamine poisoning is a chemical intoxication, the incubation period is rather short with the onset of the illness occurring rather rapidly. Symptoms usually appear within minutes to a few hours after ingestion of the toxin food. The "peppery" taste can be noted immediately after placing the food in the mouth, although it is uncertain whether this taste is always associated with the spoiled fish. Other symptoms such as those previously mentioned often occur within 5 minutes after consumption. The duration is typically short, with symptoms subsiding after a few hours even when untreated.

Diagnosis

Diagnosis of histamine poisoning is usually based on symptomology and a history of the foods eaten by the patient just before onset of the illness. If the symptoms occur quite rapidly following the ingestion of a commonly implicated fish and are characteristic of scombroid fish poisoning, then a tentative diagnosis can be made. Confirmation is accomplished by the detection of high amounts of histamine in the implicated food.

Diagnosis based on symptomology alone can be particularly difficult. Those treating patients who develop only gastrointestinal problems have many other causes to consider. Physicians must try to differentiate from other foodborne intoxications by asking about manifestations that may be ignored by the patient. This may include the burning sensation in the mouth, tingling, and flushing. Also, the beneficial response of the patient to antihistamines may assist in the diagnosis of histamine poisoning. Kasha and Norins (1988) offer an excellent example of the diagnosis of two cases of histamine poisoning.

Sometimes histamine poisoning can be confused with food allergies. This is due to the similarity of the symptoms, and because antihistamines are successfully used to treat both conditions. As consumers become more aware of the role of cholesterol in heart disease, more individuals are choosing fish as an alternative to high cholesterol meats. Thus, histamine poisoning may occur more frequently, and the chance for a misdiagnosis to occur likewise becomes greater. Physicians can avoid a misdiagnosis by learning if the patient has a prior history of allergic reactions to the implicated food, and if others eating the same meal also became ill. It is extremely unlikely that more than one individual in a group will have the same type of food allergy so such observations would point toward a diagnosis of histamine poisoning. Finding high levels of histamine in the fish would also strongly indicate scombroid poisoning.

Treatment and Prevention

Most patients respond to antihistamine therapy. Symptoms normally subside rapidly after treatment with an H1 antagonist such as diphenhydramine or chlorpheniramine. However, H2 antagonists such as cimetidine may also be effective in treating the illness (Blakesley 1983). However, because the disease is self-limited and usually mild, drug treatment may not be necessary.

Prevention of histamine intoxication from fish includes avoidance of certain types of fish, removal of contaminated fish from the market, and control of bacterial histamine formation. Obviously, absolute avoidance of certain types of fish may be undesirable or even impossible for some individuals. A more logical approach would be to avoid eating raw fish or fish with obvious signs of decomposition. Removal of contaminated fish is controlled by government regulatory agencies working in conjunction with processing plants or distributors. Quality control programs and frequent inspections may be a part of this aspect of prevention. Because cooking does not destroy the toxin, good hygiene and refrigeration practices by those preparing the fish in either restaurants or markets are needed to control the growth of histamine-producing bacteria. Unfortunately, both of these last preventive measures cannot be controlled by the consumer.

Fish Implicated in Histamine Poisoning

Fish is the most common food implicated in outbreaks of histamine poisoning. As shown in Table 13.1 fish belonging to the families Scombridae and Scomberesocidae are most often linked to histamine poisoning. The scombroid fish group includes the various species of tuna, mackerel, bonito, and saury. Tuna,

Table 13.1   Fish Species Implicated in Histamine Poisoning

Common Name	Scientific Name	
	Family	Genus and Species
	Scombridae	
Yellowfin tuna[a]		*Thunnus albacares*
Blackfin tuna		*T. atlanticus*
Southern bluefin tuna		*T. maccoyii* or *T. thynnus maccoyii*
Big-eye tuna		*T. obesus* or *Parathunnus mebachi*
Atlantic bluefin tuna		*T. thynnus thynnus*
Pacific bluefin tuna		*T. thynnus orientalis*
Longtail tuna		*T. tonggol*
Albacore		*T. alalunga*
Skipjack tuna[a]		*Euthynnus pelamis* or *Katsuwonas pelamis*
Kawakawa		*E. affinis*
Little tunny		*E. alletteratus*
Black skipjack		*E. lineatus*
Slender tuna		*Allothunnus fallai*
Bullet tuna or bullet mackerel		*Auxis rochei*
Frigate tuna, frigate mackerel, or plain bonito		*A. thazard*
Atlantic bonito		*Sarda sarda*
Indo-Pacific or striped bonito		*S. orientalis*
Eastern Pacific bonito		*S. chiliensis*
Australian bonito		*S. australis*
Atlantic mackerel[a]		*Scomber scombrus*
Chub or Pacific mackerel[a]		*S. japonicus*
King mackerel[a]		*Scomberomorus cavalla*
Spanish mackerel[a]		*S. maculatus*
Monterey Spanish mackerel		*S. concolor*
Cero		*S. regalis*
Sierra		*S. sierra*
	Scomberesocidae	
Atlantic saury[a]		*Scomberesox saurus*
Pacific saury or mackerel pike[a]		*Cololabis saira*
	Pomatomidae	
Bluefish[a]		*Pomatomus saltatrix*
	Coryphaenidae	
Dolphin fish, dorado, or mahi-mahi[a]		*Coryphaena hippurus*
	Carangidae	
Horse mackerel[a]		*Trachurus trachurus* or *T. japonicus*
Jack mackerel[a]		*T. symmetricus*
Pacific amberjack		*Seriola colburni*
Yellowtail[a]		*S. dorsalis* or *S. grandis*

Table 13.1 (continued)

	Scientific Name	
Common Name	*Family*	*Genus and Species*
Greater amberjack		*S. dumerili*
	Clupeidae	
Atlantic herring		*Clupea harengus harengus*
Pacific herring		*C. harengus pallasi*
Sprat or brisling		*C. sprattus*
Pacific sardine or pilchard[a]		*Sardinops sagax*
Pilchard or sardine[a]		*Sardina pilchardus*
Golden sardine		*Sardinella aurita*
Spanish sardine		*S. anchovia*
	Engraulidae	
European anchovy		*Engraulis encrasicolus*
Pacific or northern anchovy		*E. mordax*
Anchoveta		*Centengraulis mysticetus*

[a]Species most commonly implicated.

mackerel, and skipjack are most frequently involved, but this partly is due to their high rate of consumption worldwide. Some scombroid fish may be less susceptible to histamine poisoning and are thus only infrequently implicated in outbreaks. Albacore is one such example of a highly consumed scombroid fish with only a few incidents of illness associated with it.

Particular types of nonscombroid fish may also be involved in histamine poisoning. In the U.S., mahi-mahi *(Coryphaena hippurus)* is the most commonly implicated nonscombroid fish. In other countries, pilchards, herring, anchovies, bluefish, and sardines have been involved in a number of cases. In fact, sardines and pilchards have become a major source of histamine poisoning in Great Britain (Bartholomew et al. 1987). Bluefish *(Pomatomus saltatrix)* has been responsible for several outbreaks in the U.S. and has caused at least one outbreak in Australia (Taylor 1985). Outbreaks have also occurred in the U.S. implicating pink salmon, redfish, yellowtail, marlin, and amberjack (Table 13.2). Japan has had an outbreak associated with black marlin (Yamanaka et al. 1987), and anchovies have been implicated in single incidents in Japan, the U.S., and Great Britain.

Whether the fish was imported or not sometimes makes a difference in the profile of the outbreak being reported. With some species, domestic sources seem to be the most common sources for incriminated fish. This would be true for tuna in the U.S., mackerel in England, and most species in Japan. In other cases, imported fish seem to create the problems. This is true for mahi-mahi in the U.S., and sardines, pilchards, and tuna in England.

Table 13.2   Outbreaks, Cases, and Causes of Histamine
Poisoning in the U.S. (1968–1986)

Year	No. of Outbreaks	No. of Cases	Implicated Food (No. of Outbreaks)
1968	3	19	Tuna (2)
			Mahi-mahi (1)
1969	1	3	Mackerel (1)
1970	1	2	Bonito (1)
1972	6	?	Mahi-mahi (2)
			Tuna (1)
			Albacore (1)
			Pork fish (1)
			Marine fish (1)
1973	12	326	Mahi-mahi (6)
			Tuna (5)
			Ulua or jack (1)
1974	10	26	Tuna (6)
			Mahi-mahi (1)
			Spanish mackerel (1)
			Red snapper (1)
			Kumu-red goat fish (1)
1975	6	16	Tuna (3)
			Mahi-mahi (2)
			Skipjack (1)
1976	3	43	Tuna (1)
			Mahi-mahi (1)
			Swiss cheese (1)
1977	13	71	Tuna (5)
			Mahi-mahi (3)
			Bluefish (3)
			Yellow tail (1)
			Anchovies (1)
1978	7	30	Mahi-mahi (7)
1979	12	132	Mahi-mahi (5)
			Tuna (2)
			Other fish (5)
1980	29	153	Mahi-mahi (21)
			Tuna (3)
			Other fish (5)
1981	7	67	Tuna (4)
			Mahi-mahi (2)
			Other fish (1)
1982	18	58	Tuna (3)
			Mahi-mahi (5)
			Other fish (10)
1983	13	27	Mahi-mahi (6)
			Bluefish (4)
			Tuna (3)
1984	13	67	Mahi-mahi (6)
			Tuna (4)

Table 13.2 (continued)

Year	No. of Outbreaks	No. of Cases	Implicated Food (No. of Outbreaks)
			Bluefish (1)
			Marlin (1)
			Unknown (1)
1985	14	56	Tuna (6)
			Bluefish (3)
			Mahi-mahi (1)
			Raw pink salmon (1)
			Amberjack (1)
			Yellowtail (1)
			Unknown (1)
1986	20	60	Tuna (14)
			Mahi-mahi (2)
			Mackerel (1)
			Redfish (1)
			Opelu (1)
			Yellowtail (1)

Compiled from Centers for Disease Control, *Annual Summaries of Foodborne and Waterborne Disease Outbreaks*, Atlanta.

Constraints to Surveillance

Satisfactory statistics on the prevalence of histamine poisoning throughout the world do not exist. Several countries with a high consumption of fish do not report any foodborne illness at all. Thus, it is impossible to know the prevalence of the disease in these countries. Even in those countries such as the U.S., England, and Japan that have adequate reporting programs, several cases are probably missed each year for various reasons.

Histamine poisoning is not often reported because it is an illness of short duration and patients often will not seek medical attention. Physicians may be a source of ignorance by remaining unaware of histamine poisoning and thus do not consider it as a possible diagnosis. Even when a patient seeks medical attention and a correct diagnosis is made, histamine poisoning is not a notifiable illness in some countries and is never reported. Due to these constraints, the true incidence of histamine poisoning remains unknown.

EPIDEMIOLOGY OF SCOMBROID FISH POISONING

Worldwide Occurrence

Histamine poisoning occurs worldwide. Those countries reporting the most incidents are Japan, the U.S., and England. This is probably due to a high rate of consumption of certain types of fish and especially to better reporting of the illness. Incidents have also been documented in Canada and New Zealand. Several other countries, some of which may not keep official records, have published reports on cases of histamine poisoning. Some of those countries include both East and West Germany, France, Norway, Sri Lanka, Czechoslovakia, Sweden, Australia, Indonesia, and South Africa (Taylor 1986). The fact that the disease has been reported on nearly every continent indicates that the occurrence is truly worldwide.

Japanese researchers were the first to recognize histamine poisoning in the early 1950s. Histamine poisoning was a major cause of foodborne disease in that country as evidenced by published epidemiological information during that period. (Kawabata et al. 1955). Over 75% of the 14 incidents recorded from 1951 to 1954 were associated with the ingestion of samma sakuraboshi, a dried seasoned saury *(Cololabis saira)*. The other incidents involved iwashi sakuraboshi or canned mackerel *(Scomber japonicus)*, and frigate tuna *(Auxis thazard)*. All of the incidents involved symptoms typical of histamine poisoning. In more recent years, histamine poisoning in Japan has encompassed a much wider variety of fish, and many of the incidents have involved large numbers of people (Taylor 1986). However, some Japanese scientists feel that the total number of those affected by this illness may be much higher despite the adequate reporting system.

In the U.S., from the period of 1968–1986, a total of 188 outbreaks of histamine poisoning were reported as shown in Table 13.2. In contrast to the situation in Japan, most incidents involved less than five individuals. In 1973 a large outbreak involving commercially canned tuna (Anonymous 1973; Chin et al. 1973; Merson et al. 1974) served to increase the awareness of histamine poisoning in the U.S. Since then numerous incidents have been reported throughout the U.S. (Anonymous 1986, 1988; Etkind et al. 1987; Kasha and Norins 1988; Russell and Maretic 1986). Mahi-mahi, now a popular fish entree, is frequently involved in cases of histamine poisoning. Histamine poisoning is one of the primary foodborne diseases of chemical etiology in terms of the number of reported cases.

In Great Britain, 258 incidents of suspected histamine poisoning occurred between 1976 and 1986 (Bartholomew et al. 1987). Mackerel consumption accounted for most of the cases reported. Cases may not have been reported before 1976 due to lack of recognition of the illness. At the present time, however, reporting of histamine poisoning in Great Britain may be much more complete than other countries because it is one of the few nations where the disease is a notifiable illness.

Reports from other countries can be rather sporadic. Some countries lack an effective system of collecting information on incidents of foodborne disease. Of those countries that do report, Canada, New Zealand, Finland, Norway, Sweden, and Iceland report few incidents despite a high consumption of fish.

The low incidence in Canada, where large amounts of fish commonly associated with scombroid fish poisoning such as mackerel and tuna are consumed, is not readily explained. However, unlike Canada, the northern European countries consume fish not susceptible to histamine formation. This, along with cool climates that results in less opportunity for temperature abuse, may account for the low incidence in these countries.

FORMATION OF HISTAMINE AND ITS CONTROL

Histidine Decarboxylase

Histamine is generated in foods from the amino acid histidine via an enzymatic decarboxylation reaction catalyzed by histidine decarboxylase. Certain fish, especially scombroid fish, possess large amounts of free histidine in their muscle tissues (Hibiki and Simidu 1959; Lukton and Olcott 1958). This free histidine can serve as a substrate for bacterial histidine decarboxylase. Autolysis or bacterial proteolysis may also be important in that it causes the release of free histidine from tissue proteins.

Histamine-Producing Bacteria

Histidine decarboxylase is not commonly found within most bacterial groups. The enzyme is found in certain Enterobacteriaceae, *Clostridium,* and *Lactobacillus* species. Only a few of these species are prolific histamine producers, and these seem to be the ones that cause the problems. Enteric bacteria appear to be the most important histamine-producing bacteria in fish. *Morganella* (formerly *Proteus) morganii* (Kawabata et al. 1956; Sakabe 1973a), *Klebsiella pneumoniae* (Schuurkamp et al. 1987; Taylor et al. 1979), and a few strains of *Hafnia alvei* (Havelka 1967) are the only prolific histamine-producing bacteria that have been isolated from fish implicated in histamine poisoning incidents. Certain strains of *Lactobacillus* that are also prolific histamine producers would probably only be of importance in fermented fish (Taylor 1986). Taylor et al. (1978) provide a list of bacteria that are known to possess histidine decarboxylase. Whereas most strains of *M. morganii* are able to produce histamine, only certain strains of *K. pneumoniae* and *H. alvei* are potent histamine producers (Havelka 1967; Taylor et al. 1978, 1979).

Certain species of luminous bacteria also produce histamine. Okuzumi et al. (1982) isolated species that were both psychrophilic and halophilic, whereas a more recent study isolated luminous halophiles that were mesophilic (Ramesh and Venugopalan 1986). *Photobacterium,* one of the genera isolated by Okuzumi et al. in 1984, may be of importance in relation to the hygiene of fish products. It has been speculated that this organism may play a significant role in histamine formation when scombroid fish are stored at 50°F (10°C) or below where the Enterobacteriaceae are no longer active. Extensive work by Morii et al. (1986, 1988) has shown that one species, *Photobacterium phosphorum,* may be principally responsible for histamine formation in scombroid fish at low temperatures. The halophilic nature of some of these bacteria may be important because fish are often cooled in refrigerated sea water or salt brine.

Control of Histamine Formation

Low-Temperature Storage of Fish. Histamine poisoning can be prevented by a variety of methods. However, the most important and practical method that can be used in any phase of the fishing industry is cold-temperature storage. Histamine-producing bacteria appear to reside in the gills and/or intestine of fish (Frank et al. 1981; Lerke et al. 1978; Taylor and Speckhard 1984). When the fish is stored at higher temperatures for a prolonged period of time, these bacteria are able to invade the muscle tissue and convert histidine to histamine. Current fishing practices for tuna involve either the immediate icing or storage of the fish in refrigerated seawater before freezing. However, if proper conditions are not met almost immediately, a temperature abuse situation may occur, because of the large size of most fish.

Studies on the effect of incubation temperature on bacterial histamine formation have yielded variable results (Behling and Taylor 1982). Arnold et al. (1980) determined that histamine production by *M. morganii* and *P. vulgaris* was delayed and diminished by incubation at 44.6°F (7°C) as compared to 86°F (30°C) or 66.2°F (19°C). Similar results were noted by Behling and Taylor (1982) with *M. morganii, K. pneumoniae,* and *H. alvei.* However, Middlebrooks et al. (1988) found potential histamine-formers in tissues of Spanish mackerel incubated at 32°F (0°C). This contrasts with the previous study where no histamine-producing bacterial isolates were found in fish incubated at 33.8°F (1°C). Although most histamine-producing bacteria are mesophilic and effectively controlled by cold storage temperatures, some are psychroptrophic such as the luminous strains discussed earlier (Okuzumi et al. 1984; Morii et al. 1986, 1988).

Most researchers agree that storage below or at 32°F (0°C) limits histamine formation in fish. However, situations may arise where the fish is exposed to higher temperatures for short periods of time. Klausen and Huss (1987a) demonstrated that fish samples or broth inoculated with *Morganella morganii* will continue to show an increase in histamine production at low temperatures following a short period of exposure to elevated temperatures. According to Baranowski et al. (1985), histidine decarboxylase already produced by bacteria that has ceased growing can still convert histidine to histamine. Thus at lower temperatures, resting cells may continue to produce histamine even though they are biologically inactive. Other investigators have also shown that fish subjected to storage at 68°F (20°C) for a short period (1 day) will yield high levels of histamine following subsequent storage at refrigeration temperatures (Sakabe 1973b; Smith et al. 1980).

Another possible conclusion as to the origin of the histamine produced in fish could lie in the muscle tissue itself. Histamine might be produced by a high histidine decarboxylase activity in the muscle. This could occasionally account for the accumulation of histamine in fish, especially in those instances where bacterial counts are low (Yamanaka et al. 1987). However, more evidence is needed to substantiate this idea.

Hygiene Practices. Histamine-producing bacteria isolated from spoiling fish could most likely be due to postharvest contamination (Taylor and Speckhard 1984) although they are often considered part of the normal microflora of the fish (Yoshinaga and Frank 1982). Research needs to be done on the microflora of freshly caught tuna to determine whether the histamine-producers are on the fish due to postcatch contamination or constitute part of the normal microflora.

Contamination of the fish can occur at any level of processing. They may be contaminated on the fishing boat by human workers or equipment, in the processing plant, or while in the distribution system. Food handlers or consumers may use unhygienic practices to store and/or prepare the fish. In an episode in West Germany, a restaurant was responsible for an outbreak due to the improper handling of canned tuna (Yamani et al. 1981). Proper hygienic practices could control the formation of histamine if postcatch contamination is a major source of these bacteria. By monitoring these practices, the resulting improved sanitation would not only decrease the likelihood of histamine poisoning, but also perhaps that of other foodborne illnesses caused by contaminated fish as well.

HISTAMINE TOXICITY

Evidence for Histamine as the Causative Agent

Although evidence linking histamine to scombroid fish poisoning is mostly circumstantial, there is little doubt that it is the causative agent. High levels of histamine have consistently been found in fish samples implicated in outbreaks. The symptoms involved are similar to those of an allergic reaction, in which histamine plays a major role (Taylor 1985). Also, drug therapy utilizing antihistamines such as diphenhydramine and cimetidine (Blakesley 1983) proves quite effective in the treatment of this illness. Furthermore, several individuals on drugs such as isoniazid that inhibit histamine-detoxifying enzymes have experienced severe cases of histamine poisoning, thus giving further evidence toward the involvement of histamine (Kahana and Todd 1981; Kottegoda and Uragoda 1976; Senanayake and Vyravanathan 1981; Senanayake et al. 1978; Uragoda 1980; Uragoda and Kottegoda 1977).

Toxicological Response to Histamine

Histamine can exert powerful biological responses. It can directly stimulate the heart, cause contraction or relaxation of extravascular smooth muscle, control gastric acid secretion, and stimulate both sensory and motor neurons (Soll and Wollin 1977; Taylor et al. 1984). Although toxic, it is not an unfamiliar substance to the body. Histamine is one of the chemicals released from mast cells during an allergic reaction, and it serves as one of the primary mediators of the allergic response (Beavan 1978). Histamine interacts with specific tissue histamine receptors thereby eliciting many of the symptoms associated with allergies.

Most of the common symptoms of histamine poisoning are due to cardiovascular response. Histamine causes dilatation of the peripheral blood vessels, capillaries, and arteries which can result in hypotension, flushing, and headache (Beavan 1978). Another action, contraction of smooth muscle (Taylor et al. 1984), may account for the abdominal cramps, diarrhea, and vomiting often noted in cases of histamine poisoning. Because histamine would enter the gastrointestinal tract first, contraction of intestinal smooth muscle would be particularly likely. Another function controlled by histamine is gastric acid secretion, which is regulated via receptors on the parietal cells (Soll and Wollin 1977). However, this particular function plays an unknown part in the symptomology of histamine poisoning. Pain and itching associated with the urticarial lesions in scombroid poisoning may be due to the stimulation of sensory and motor neurons by histamine (Taylor 1986).

Metabolism of Histamine

Two enzymes, diamine oxidase (DAO) and histamine N-methyltransferase (HMT), are primarily responsible for the metabolism of histamine in the intestinal tract (Taylor 1986). The existence of these enzymes (in the absence of potentiators) probably explains why orally administered histamine shows no toxic effects (Geiger 1955; Granerus 1968; Motil and Scrimshaw 1979). Diamine oxidase converts histamine into imidazoleacetic acid which can ultimately be conjugated with ribose and excreted. Histamine-N-methyltransferase converts histamine into N-methylhistamine which is then converted by monoamine oxidase into N-imidazoleacetic acid. End-products from both enzymatic pathways are excreted in the urine (Beavan 1978; Maslinski 1975; Schayer 1966; Taylor et al. 1984; Wetterqvist 1978). This detoxification system functions primarily to metabolize endogenously produced histamine, but apparently also functions as a protective mechanism against exogenous histamine.

Possible Potentiation of Histamine Toxicity

Histamine-*N*-methyltransferase has been found to be widespread in human tissues, with high levels of activity in many tissues, especially the liver (Hesterberg et al. 1984). In contrast, the highest activity of DAO exists the small intestine (Hesterberg et al. 1984). When the activity of these enzymes is suppressed by one or more potentiators, the detoxification of histamine is inhibited. (Bjeldanes et al. 1978; Parrot and Nicot 1966). Studies have shown that several substances, including some putrefactive amines, can inhibit rat intestinal HMT and DAO in vitro (Lyons et al. 1983; Taylor and Lieber 1979). Two such putrefactive amines, putrescine and cadaverine, inhibit both of the histamine-metabolizing enzymes (Taylor et al. 1984). Cadaverine/histamine or putrescine/histamine ratios of 5:1 or greater in the duct of rats were required to inhibit histamine detoxification in vivo as assessed by measuring histamine metabolites in the urine (Hui and Taylor 1985).

The ratios of cadaverine to histamine and putrescine to histamine can also reflect the temperature at which the fish decomposed (Middlebrooks et al. 1988). In fish held at 32°F (0°C), maximum cadaverine/histamine and putrescine/histamine ratios of over 20:1 were observed in anterior and posterior fillets, respectively. At 59°F (15°C) both ratios were less than 1:1 in all sections, while at 86°F (30°C) a putrescine/histamine ratio of about 8:1 was found in some samples (Middlebrooks et al. 1988).

The ability to produce putrescine and cadaverine appears to be widespread. Many species within the Enterobacteriaceae group possess either one or both of the necessary enzymes, ornithine decarboxylase or lysine decarboxylase (Taylor and Sumner 1986). However, only a few species also contain histidine decarboxylase. Therefore, the potentiators putrescine and cadaverine are probably formed by bacteria other than those responsible for histamine formation.

Several other foodborne substances are inhibitors of HMT and DAO. The use of drugs such as certain antihistamines, antimalarials, and others may also inhibit histamine-metabolizing enzymes (Barth and Lorenz 1978; Cohn 1965; Hui and Taylor 1985). As previously mentioned, isoniazid has also been implicated as a contributing factor in several cases of histamine poisoning. Isoniazid can inhibit histamine detoxification in vivo (Hui and Taylor 1985). No doubt the possibility exists that many other drugs and foodborne substances not yet discovered may act as potentiators of histamine toxicity. Histamine poisoning could be quite severe in weakened patient undergoing a particular kind of drug therapy. Therefore, more research is needed in this area to eliminate potentially hazardous combinations of foods that are occasionally high in histamine from the diets of patients receiving certain drug treatments.

ANALYSIS OF HISTAMINE IN FISH

AOAC Procedure

The AOAC procedure is the official method of analysis of histamine in foods in U.S. (Williams 1984). The procedure removes interfering compounds, followed by a derivatization of the histamine with o-phthalaldehyde (OPT). The fluorescence of the resulting compound is then measured (Williams 1984). This method was first developed by Staruszkiewicz et al. (1977), and data from studies using the procedure indicate that the method is not only sensitive, but also reproducible from laboratory to laboratory.

According to the AOAC method, marine products are first prepared based on the type and form of the product being analyzed. For example, fresh fish is prepared differently from canned fish, and products packed in oil, brine, or salt need to be drained or rinsed before analysis (Williams 1984). The fish is then homogenized in methanol, heated to 140°F (60°C), filtered, and then run through an anion-exchange column. The material collected from the column is then derivatized with OPT.

The fluorometric detection of histamine following its condensation with OPT was originally developed by Shore et al. (1959) and Shore (1971). The cleanup procedure preceding derivatization has been modified several times. The AOAC method utilizes an anion-exchange resin, but cation-exchange resins have also been used successfully to remove interfering compounds (Lerke and Bell 1976). Cationic exchange resins are part of the accepted procedure in West Germany (Taylor 1985). Another procedure, currently used in the U.K., uses sequential extractions to remove interfering substances prior to condensation with OPT (Taylor 1985). Because fluorometric methods are used worldwide, one of them should be selected as the international standard procedure.

Other Methods

Many enzymatic methods have been developed to detect histamine in blood and other biological tissues (Beavan and Horakova 1978; Dyer et al. 1982; Snyder 1971). These methods are quite sensitive in that they use the enzyme HMT and radioactive S-adenosylmethionine. However, the use of radioisotopes can present a problem for most laboratories analyzing fish samples. Also, the level of sensitivity elicited from these tests is generally much greater than needed to analyze foods for dangerous amounts of histamine.

Numerous chromatographic methods exist that are applicable to histamine analysis. One method, thin-layer chromatography (TLC), offers a distinct advantage. It can be used in virtually any analytical laboratory because no expensive or elaborate equipment is needed to perform the assay. Even though this and related techniques yield only semiquanitative results, this is often all that is needed to detect spoiled or hazardous fish. Lieber and Taylor (1978a) concluded that methanol/ammonia (20:1) and chloroform/methanol/ammonia (2:2:1) were the best solvent combinations, and that Silica gel G was an adequate stationary phase. They also identified *o*-diacetylbenzene and fluorescamine as the most specific spray reagents used to detect histamine on TLC plates (Lieber and Taylor 1978b). Other TLC methods have been developed, but have not been used in food samples (Fleisher and Russell 1975).

More elaborate, quantitative techniques developed specifically for histamine involve high-pressure liquid chromatography (HPLC) and gas–liquid chromatography (GLC). These systems are extremely accurate and can be used to detect other putrefactive amines along with histamine simultaneously (Taylor and Sumner 1986). However, unlike TLC, these techniques require sophisticated equipment that may not be available to many laboratories.

Detection of Histamine-Producing Bacteria

Many methods have been developed to detect histamine-producing bacteria in foods such as fish. Most of these detection systems are based on a specific media that is either selective and/or differential for histamine-producing bacteria. Some methods require the preliminary isolation of bacteria so that individual isolates can be assessed for histamine formation. Obviously, these procedures can be quite laborious. One system utilizes a histidine-fortified trypticase soy broth that is optimized for bacterial histamine production (Taylor and Woychik 1982). Another technique utilizes a tuna fish infusion broth (Omura et al. 1978). In these procedures, the histamine is extracted and measured by one of the fluorometric assays described earlier. Another system for the detection of histamine-producing bacteria in fish is a differential medium containing bromocresol purple. Histamine formation by a colony causes a color change due to a shift in pH (Niven et al. 1981). MRS broth (deMan et al. 1960), fortified with histidine, can be used in the identification of histamine-producing lactobacilli (Sumner et al. 1985). The histamine can be extracted and measured fluorometrically or by a recently developed procedure based on diamine oxidase and leucocrystal violet can be applied directly with the growth medium (Sumner and Taylor 1989).

A recent method developed by Klausen and Huss (1987b) is based on automated conductance measurements in a histidine-containing medium incubated at 77°F (25°C). The method is rapid, with histamine-producing bacteria being detected within 24 hours by means of conductance measurements. Such a method could be used to screen large numbers of fish samples which may be contaminated with significant amounts of histamine-producing bacteria, and thus could be part of a quality assurance program (Klausen and Huss 1987b). However, this method has the disadvantage of requiring expensive instrumentation.

HISTAMINE REGULATION (U.S.)

Most countries do not have any regulatory limits on the allowable levels of histamine in foods. The U.S. is one of the few countries that has regulatory limits for histamine. These regulations are strictly for fish products, specifically tuna, and not for other fish or foods potentially high in histamine such as mahi-mahi, bluefish, cheeses, or wines. Lack of regulation concerning histamine levels in foods may be due to uncertainty regarding the threshold toxic dose for histamine.

The U.S. currently has a two-stage limit of the allowable levels of histamine in tuna (Taylor 1985). The amount constituting a known human health hazard is called the hazard action level, which is now set at 50 mg/100 g for tuna. A defect action level, an amount that signifies some mishandling of the fish, has also been established for tuna alone. This level is set at 10 mg/100 g when odors of decomposition or honeycombing are present, and at 20 mg/100 g if based on histamine levels only. These regulatory limits, based on several years of investigative experience with histamine poisoning, unfortunately ignore the variable levels of histamine within the fish and the existence of suspected potentiators.

Because histamine toxicity is not fully understood, the current regulatory situation probably provides consumers with only minimal protection against histamine poisoning. Even though the health hazard associated with ingesting tuna containing 50 mg/100 g is well established, it is unknown whether similar levels in other fish are also hazardous. Independent evaluation on each type of fish needs to be made before establishing defect action levels. Some of the data from other countries has contrasted with the situation in the U.S. For example, investigative experience in Great Britain seems to imply that the presence of >10 mg/100g not >50 mg/100g in fish is indicative of its potential to cause histamine poisoning (Bartholemew et al. 1987). It is therefore important to consider all data when trying to establish suitable guidelines regarding toxic levels of histamine in foods.

It is essential to gain a more detailed knowledge of the mechanism of histamine toxicity. Moreover, research needs to be done to identify and evaluate potentiators and their respective roles in the degree of toxicity of fish products. By taking into account cumulative data and research from around the world, safe and accurate regulations could be established for those foods that have the potential to cause this disease. However, for some fish and for cheeses and other foods, additional data will need to be gathered before any attempt is made to establish guidelines or standards.

REFERENCES

Anonymous. 1973. Follow-up scombroid fish poisoning in canned tuna fish-United States. *US Morbid. Mortal. Wk. Rep.* **22**:78.

Anonymous. 1986. IV. Formation of histamine and its control. *Food Chem. News* May 5, pp. 60.

Anonymous. 1988. Fish poisoning is new to landlocked New Mexico. *Food Protect. Rep.* March, pp. 4–5.

Arnold, S. H., and Brown, W. D. 1978. Histamine toxicity from fish products. *Adv. Food Res.* **24**:113–154.

Arnold, S. H., Price, R. J., and Brown, W. D. 1980. Histamine formation by bacteria isolated from skipjack tuna, *Katsuwonas pelamis. Bull. Jpn. Soc. Sci. Fish.* **46**:991–995.

Baranowski, J. D., Brust, P. A., and Frank, H. A. 1985. Growth of *Klebsiella pneumonaie* UH-2 and properties of its histidine decarboxylase system in resting cells. *J. Food Biochem.* **9**:349–360.

Barth, H., and Lorenz, W. 1978. Structural requirements of imidazole compounds to be inhibitors or activators of histamine methyltransferase: investigation of histamine analogues and H2-receptor antagonists. *Agents Actions* **8**:359–365.

Bartholomew, B. A., Berry, P. R., Rodhouse, J. C., and Gilbert, R. J. 1987. Scombrotoxic fish poisoning in Britain: features of over 250 suspected incidents from 1976 to 1986. *Epidemiol. Infect.* **99**:775–782.

Beavan, M. A. 1978. Histamine: its role in physiological and pathological processes. *Monogr. Allergy* **13**:1–113.

Beavan, M. A. and Z. Horakova. 1978. The enzymatic isotopic assay of histamine. In: Rocha e Silva M. (ed) *Handbook of Experimental Pharmacology, Vol. 18* (part 2). Berlin, Springer-Verlag, Berlin, pp. 151–173.

Behling, A. R., and Taylor, S. L. 1982. Bacterial histamine production as a function of temperature and time of incubation. *J. Food Sci.:* **47**:1311–14, 1317.

Bjeldanes, L. F., Schutz, D. E., and Morris, M. M. 1978. On the aetiology of scombroid poisoning: cadaverine potentiation of histamine toxicity in the guinea pig. *Food Cosmet. Toxicol.* **167**:157–159.

Blakesley, M. L. 1983. Scombroid poisoning: prompt resolution of symptoms with cimetidine. *Ann. Emerg. Med.* **12**:104–106.

Chin, J., Fleming, D. S., Edwards, M. M., Press, E., Erdmann, R., Westaby, R. S., Hutcheson, R. H., Jr., Simpson, N., Dickerson, M. S., and Skinner, H. G., 1973. Scombroid fish poisoning in canned tuna fish-United States. *US Morbid. Mortal. Wk. Rep.* **22**:69–70.

Cohn, V. H. 1965. Inhibition of histamine methylation by antimalarian drugs. *Biochem. Pharmacol.* **14**:1686–1688.

deMan, J. C., Rogosa, M., and Sharpe, M. E. 1960. A medium for the cultivation of lactobacilli. *J. Appl. Bacteriol.* **23**:130–35.

Dyer, J., Warren, K., Merlin, S., Metcalfe, D. D., and Kaliner, M. 1982. Measurement of plasma histamine: description of an improved method and normal values. *J. Allergy Clin. Immunol.* **70**:82–87.

Etkind, P., Wilson, M. E., Gallagher, K., and Cournoyer, J. 1987. Bluefish-associated scombroid poisoning. An example of the expanding spectrum of food poisoning from seafood. *JAMA* **258**:3409–3410.

Fleisher, J. H., and Russell, D. H. 1975. Estimation of urinary diamines and polyamines by thin-layer chromatography. *J. Chromatogr.* **110**:335–340.

Foo, L. Y. 1975. Scombroid-type poisoning induced by ingestion of smoked kahawai. *N.Z. Med. J.* **81**:476–477.

Frank, H. A., Yoshinaga, D. H., and Nip, W.-K. 1981. Histamine formation and honeycombing during decomposition of skipjack tuna, *Katsuwonas pelamis,* at elevated temperatures. *Marine Fish. Rev.* **43**:9–14.

Geiger, E. 1955. Role of histamine in poisoning with spoiled fish. *Science* **121**:865–866.

Granerus, G. 1968. Effects of oral histamine, histidine and diet on urinary excretion of histamine, methylhistamine, and 1-methyl-4 imidazoleacetic acid in man. *Scand. J. Clin. Lab. Invest.* **22**:49–58.

Havelka, B. 1967. Role of the *Hafnia* bacteria in the rise of histamine in tuna fish meat. *Cesk. Hyg.* **12**:343–52.

Hesterberg, R., Sattler, J., Lorenz, W., Stahlknecht, C. D., Barth, H., Crombach, M., and Weber, D. 1984. Histamine content, diamine oxidase activity and histamine methyl-transferase activity in human tissues: fact or fiction? *Agents Actions* **14**:325–334.

Hibiki, S., and Simidu, W., 1959. Studies on putrefaction of aquatic products. 27. Inhibition of histamine formation in spoiling of cooked fish and histidine content in various fishes. *Bull. Jpn. Soc. Sci. Fish.* **24**:916–919.

Hui, J. Y., and Taylor, S. L. 1985. Inhibition of in vivo histamine metabolism in rats by foodborne and pharmocologic inhibitors of diamine oxidase, histamine N-methyltransferase, and monoamine oxidase. *Toxicol. Appl. Pharmacol.* **81**:241–249.

Kahana, L. M., and Todd, E. 1981. Histamine poisoning in a patient on isoniazid. *Can. Dis. Wk. Rep.* **7**:79–80.

Kasha, E. E., and Norins, A. L. 1988. Scombroid fish poisoning with facial flushing. *J. Am. Acad. Dermatol.* **18**:1363–1365.

Kawabata, T., Ishizaka, K., and Miura, T. 1955. Studies on the allergy-like food poisoning associated with putrefaction of marine products. I. Episodes of allergy-like food poisoning caused by "samma sakuraboshi" (dried seasoned saury) and other kinds of marine products. *Jpn. J. Med. Sci. Biol.* **8**:487–501.

Kawabata, T., Ishizuka, K., Miura, T., and Sasaki, T. 1956. Studies on the food poisoning associated with putrefaction of marine products. VII. An outbreak of allergy-like food poisoning caused by "sashimi" of *Parathunnus mebachi* and the isolation of causative bacteria. *Bull. Jpn. Soc. Sci. Fish.* **22**:41–47.

Kim, R. 1979. Flushing syndrome due to mahi mahi (Scombroid fish) poisoning, *Arch. Dematol.* **115**:963–965.

Kimata, M. 1961. The histamine problem. In: Borgstrom G. (ed) *Fish as Food, Vol. I:* Academic Press, New York, pp. 329–352.

Klausen, N. K., and Huss, H. H. 1987a. Growth of histamine production by *Morganella morganii* under various temperature conditions. *Int. J. Food Microbiol.* **5**:147–156.

Klausen, N. K., and Huss, H. H. 1987b. A rapid method for detection of histamine-producing bacteria. *Int. J. Food Microbiol.* **5**:137–146.

Kottegoda, S. R., and Uragoda, C. G. 1976. Interaction of isoniazid and fish exhibiting high histamine levels. *IRCS Med. Sci.* **4**:370

Lerke, P. A., and Bell, L. D. 1976. A rapid fluorometric method for the determination of histamine in canned tuna. *J. Food Sci.* **41**:1282–1284.

Lerke, P. A., Werner, S. B., Taylor, S. L., and Guthertz, L. S. 1978. Scombroid poisoning. Report of an outbreak. *West J. Med.* **129**:381–386.

Lieber, E. R., and Taylor, S. L. 1978a. Comparison of thin-layer chromatographic detection methods for histamine in tuna fish. *J. Chromatogr.* **153**:143–152.

Lieber, E. R., and Taylor, S. R. 1978b. Comparison of thin-layer chromatographic detection methods for histamine from food extracts. *J. Chromatogr.* **160**:227–237.

Lukton, A. and Olcott, H. S. 1958. Content of free imidazole compounds in the muscle tissue of aquatic animals. *Food Res.* **23**:611–618.

Lyons, D. E., Beery, J. T., Lyons, S. A., and Taylor, S. L. 1983. Cadaverine and aminioquanidine potentiate the uptake of histamine in vitro in perfused intestinal segments of rats. *Toxicol. Appl. Pharmacol.* **70**:445–458.

Maslinski, C. 1975. Histamine and its metabolism in mammals. II. Catabolism of histamine and histamine liberation. *Agents Actions* **5**:183–225.

Merson, M. H., Baine, W. B., Gangarosa, E. J., and Swanson, R. C. 1974. Scombroid fish poisoning. Outbreak traced to commercially canned tuna fish. *JAMA* **228**:1268–1269.

Middlebrooks, B. L., Toom, P. M., Douglas, W. L., Harrison, R. E., and McDowell, S. 1988. Effects of storage time and temperature on the microflora and amine development in Spanish mackerel *(Scomberomorus maculatus)*. *J. Food Sci.* **53**:1024–1029.

Morii, H., Cann, D. C., Taylor, L. Y., and Murray, C. K. 1986. Formation of histamine by luminous bacteria isolated from scombroid fish. *Bull. Jpn. Soc. Sci. Fish.* **52**:2135–2141.

Morii, H., Cann, D. C., and Taylor, L. Y. 1988. Histamine formation by luminous bacteria in mackerel stored at low temperatures. *Nippon Suisan Gakkaishi* **54**:299–305.

Motil, K. J. and Scrimshaw, N. S. 1979. The role of exogenous histamine in scombroid poisoning. *Toxicol. Lett.* **3**:219–223.

Niven, C. F., Jr., Jeffrey, M. B., and Corlett, D. A., Jr. 1981. Differential plating medium for quantitative detection of histamine-producing bacteria. *Appl. Bacteriol.* **23**:130–135.

Okuzumi, M., Okuda, S., and Awano, M. 1982. Occurrence of psychrophiles and halophilic histamine-forming bacteria (N-group bacteria) on/in red meat fish. *Bull. Jpn. Soc. Fish.* **48**:799–804.

Okuzumi, M., Yamanaka, H., and Kubozuda, T. 1984. Occurrence of various histamine-forming bacteria on/in fresh fishes. *Bull. Jpn. Soc. Sci. Fish.* **50**:161–167.

Omura, V., Price, R. J., and Olcott, H. S. 1978. Histamine-forming bacteria isolated from spoiled skipjack tuna and jack mackerel. *J. Food Sci.* **43**:1779–1781.

Parrot, J.-L., and Nicot, G. 1966. Pharmacology of histamine, Absorption de l'appareil digestif: In: Rocha e Silva M. (ed) *Handbook of Experimental Pharmacology, Vol. 18* (part 1). Springer-Verlag, New York, pp. 148–161.

Ramesh, A., and Venugopalan, V. K. 1986. Densities and characteristics of histamine-forming luminous bacteria of marine fish. *Food Microbiol.* **3**:103–105.

Russell, F. E., and Maretic, Z. 1986. Scombroid poisoning: minireview with case histories. *Toxicon* **24**:967–973.

Sakabe, Y. 1973a. Studies on allergylike food poisoning. Histamine production by *Proteus morganii*. *J. Nara Med. Assoc.* **24**:248–256.

Sakabe, Y. 1973b. Studies on allergylike food poisoning. 2. Studies on the relation between storage temperatures and histamine production. *J. Nara Med. Assoc.* **24**:257–264.

Schayer, R. W. 1966. Catabolism of histamine in vivo. In: Rocha e Silva M. (ed) *Handbook of Experimental Pharmacology, Vol. 18* (part 2). Springer-Verlag, Berlin, pp. 131–173.

Schuurkamp, G. J., Sabuin, R. H., and Kereu, R. K. 1987. A report of scombroid fish poisoning from skipjack tuna *(Kutsuwonas pelamis)* at Tabubil, Western Province, Papua New Guinea. *Papua New Guinea Med. J.* **30**:203–206.

Senanayake, N., and Vyravanathan, S. 1981. Histamine reactions due to ingestion of tuna fish *(Thunnus argentivittatus)* in patients on anti-tuberculois therapy. *Toxicon* **19**:184–185.

Senanayake, N., Vyravanathan, S., and Kanagasuriyam, S. 1978. Cerbrovascular accident after a skipjack reaction in a patient taking isoniazid. *Br. Med. J.* **2**:1127–1128.

Shore, P. A. 1971. Fluorometric assay of histamine. *Methods Enzymol.* **17B**:842–845.

Shore, P. A., Burkhalter, A., and Cohn, V. H., Jr. 1959. A method for the fluorometric assay of histamine in tissues. *J. Pharmacol. Exp. Ther.* **127**:182–186.

Smith, J. G. M., Hardy, R., and Young, K. W. 1980. A seasonal study of the storage characteristics of mackerel at chill and ambient temperatures. In: Connell J. J. (ed) *Advances in Fisheries Science Technology* Fishing News Book, Farnham U. K.

Snyder, S. H. 1971. Determination of tissue histamine by an enzymatic isotopic assay. *Methods Enzymol.* **17B**:842–845.

Soll, A. H., and Wollin, A. 1977. The effects of histamine, prostaglandin E2, and secretion on cyclic AMP in separated canine fundic mucosal cells. *Gastroenterology* **72**:1166.

Staruszkiewicz, W. F., Jr., Waldron, E. M., and Bond, J. F. 1977. Fluorometric determination of histamine in tuna: development of method. *J. Assoc. Off. Anal. Chem.* **60**:1125–1130.

Sumner, S. S., and Taylor, S. L. 1989. Detection method for histamine-producing, dairy-related bacteria using diamine oxidase and leucocrystal violet. *J. Food Protect.* **52**:105–108.

Sumner, S. S., Speckhard, M. W., Somers, E. B., and Taylor, S. L. 1985. Isolation of histamine-producing *Lactobacillus buchneri* from Swiss cheese implicated in a food poisoning outbreak. *Appl. Environ. Microbiol.* **50**:1094–1096.

Taylor, S. L. 1985. *Histamine Poisoning Associated with Fish, Cheese, and Other Foods.* World Health Organization, pp. 1–47.

Taylor, S. L. 1986. Histamine food poisoning: toxicology and clinical aspects. *CRC Crit. Rev. Toxicol.* **17**:91–128.

Taylor, S. L., and Lieber, E. R. 1979. In vivo inhibition of rat intestinal histamine-metabolizing enzymes. *Food Cosmet. Toxicol.* **17**:237–240.

Taylor, S. L. and Speckhard, M. W. 1984. Inhibition of bacterial histamine production by sorbate and other antimicrobial agents. *J. Food Protect.* **47**:508–511.

Taylor, S. L. and Sumner, S. S. 1986. Determination of histamine, putrescine, and cadaverine. In: Kramer, D. E., and Liston, J. (eds) *Seafood Quality Determination* Elsevier Science Publishers, Amsterdam, pp. 235–245.

Taylor, S. L., and Woychik, N. A. 1982. Simple medium for assessing quantitative production of histamine by Enterobacteriaceae. *J. Food Protect.* **45**:747–751.

Taylor, S. L., Guthertz, L. S., Leatherwood, M., Tillman, F., and Lieber, E. R. 1978. Histamine production by foodborne bacterial species. *J. Food Safety* **1**:173–187.

Taylor, S. L., Guthertz, L. S., Leatherwood, M., and Leiber, E. R. 1979. Histamine production by *Klebsiella pneumoniae* and an incident of scrombroid fish poisoning. *Appl. Environ. Microbiol.* **37**:274–278.

Taylor, S. L., Hui, J. Y., and Lyons, D. E. 1984. Toxicology of scombroid poisoning. In: Ragelis, E. P. (ed) *Seafood Toxins* American Chemical Society, Washington D.C.; pp. 417–430.

Uragoda, C, G. 1980. Histamine poisoning in tuberculosis patients after ingestion of tuna fish. *Am. Rev. Respir. Dis.* **121**:157–159.

Uragoda, C. G., and Kottegoda, S. R. 1977. Adverse reactions to isoniazid and fish exhibiting high histamine levels. *Tubercle* **58**:83–89.

Wetterqvist, H. 1978. Histamine metabolism and excretion. In: Rocha e Silva M. (ed) *Handbook of Experimental Pharmacology. Vol. 18* (part 2): Springer-Verlag, New York, pp. 131–173.

Williams, S. (ed). 1984. *Official Methods of Analysis of the Association of Official Analytical Chemists,* 14th edition. Association of Official Analytical Chemists, Arlington, Virginia, p. 341.

Yamanaka, H., Shimakura, K., Shiomi, K., Kikuchi, T., and Okuzumi, M. 1987. Occurrence of allergylike food poisoning caused by "mirin"-seasoned meat of dorado *(Coryphaena hippurus). J. Food Hyg. Soc. Jpn.* **28**:354–358.

Yamani, M. I., Dickertmann, D., and Untermann, F. 1981. Histamine formation by *Proteus* species from tuna fish. *Zbl. Bakt. Hyg. I Abt. Orig. B* **173**:478–487.

Yoshinaga, D. H., and Frank, H. A. 1982. Histamine-producing bacteria in decomposing skipjack tuna *(Katsuwonas pelamis). Appl. Environ. Microbiol.* **44**:447–452.

Part 3

Special
Processing and
Packaging

14

Principles of Pasteurization and Minimally Processed Seafoods

Cameron R. Hackney,
Thomas E. Rippen, and
Donn R. Ward

Moderate temperature thermal processing is used to extend the refrigerated shelf life of certain prepackaged seafoods. The relatively mild heating conditions results in color, texture, and flavor characteristics that are similar to those of "fresh" products, but with greatly extended shelf life. Although almost any seafood can be moderately heat processed, only smoked fish, crawfish tail meat, and crabmeat have received significant attention. The U.S. experience in moderate temperature thermal processing of seafoods is based largely on processing of meat from the blue crab *(Calinectes sapidus)*. As a consequence, this chapter focuses primarily on principles associated with pasteurization of crabmeat, especially as they impact the product's microbiology; however, these principles also apply to other minimal processing techniques.

PASTEURIZATION

Pasteurization is a term that is used to refer to a mild heating process, usually $<212°F$ ($100°C$). By definition the term indicates that the product is not sterile, and therefore may continue to harbor microorganisms. Consequently, pasteurized products must be continuously refrigerated so that the surviving microorganisms will not multiply too rapidly and thus shorten the product's anticipated shelf life.

Pasteurization of foods other than crabmeat is often defined in terms of a target organism. When the D-values of the microorganism are known it is easy to determine a pasteurization process at their corresponding temperatures using the formula $F_{T_x}{}^z = D_{T_x}n$ (where T_x = temperature of interest and n = the number of decimal reductions desired). For example, in milk the target organism is *Coxiella burnetti* which has a heat resistance of $D_{150} = 0.60$ minutes and a z-value $9°F$. The above formula can be used to determine a pasteurization process for milk. If a desired pasteurization temperature is $150°F$ ($65.5°C$) the pasteurization process would be $F^9_{150} = D_{150}(n)$ or $0.6 \times 15 = 9$ minutes. When the z-value is known, processes at other temperatures can be calculated using the formulas for equivalent D- and F-values given in the appendix. In this example a 15-D process was selected. This might surprise observers who are used to thinking in terms of a 12-*D* process for canned foods; the larger value is in response to the expected number of microorganism that might be present in the raw product. *Clostridium botulinum* is the organism of concern in most canned products. The number of *C. botulinum* spores encountered in foods is usually very low (an average of less than one per container is assumed); therefore a 12-*D* process provides a very large safety factor. Other target organisms may be used in other preservation systems and if high numbers are expected a greater process is warranted.

There is no target organism for the pasteurization of crabmeat. The process evolved based on shelf life extension. The traditional pasteurization process, which was recommended by the Tri-State Seafood Committee soon after the procedure was released from earlier patent rights, was based on heating one pound cans (401 × 301) of crabmeat to a cold point (slowest heating point) temperature of 185°F (85°C) and holding for 1 minute. Recommendations then called for cooling the product to a cold point temperature of 100°F (37.7°C) within 50 minutes. The heating curve from this process gives an average F^{16}_{185} = 31 minutes; however many processors are achieving processes (F^{16}_{185}) = 60–120 minutes. In view of this new perspective on the dynamics of pasteurization, the Tri-State recommendations were revised in 1984 by the National Blue Crab Industry Association and a National Industry Pasteurization Standard was recommended.

The *z*-value of 16 was picked arbitrarily, because there is no specific target organism. There is debate whether this value is truly appropriate, but from a historical standpoint it has worked and its use will probably continue. Within the range of normal crabmeat pasteurization temperatures, *F*-value calculations based on $z = 16$ produce a reasonable and conservative process. However, caution should be exercised when calculating equivalant *F*-values at minimal processing temperatures [below about 170°F (76.6°C)] for which inoculated pack studies are recommended.

To our knowledge, controlled studies have not been conducted to equate various crabmeat pasteurization schedules with shelf life. However, considerable empirical data have been accumulated in mid-Atlantic commercial blue crab processing facilities which support the observations listed in Table 14.1. Obviously, the actual shelf life will also depend on such factors as the initial microbial load, composition of the microbial population, storage temperature, and container integrity.

Pasteurization has been proposed for other products including shrimp (Lerke and Farber 1971), crawfish, and smoked fish (Eklund et al. 1988). The latter group examined the feasibility of pasteurizing vacuum packaged, hot processed smoked fish. Since the U.S. Food and Drug Administration dropped the Good Manufacturing Practice (GMP) regulations on the processing of hot smoked fish, there has been increasing concern regarding the potential of a botulism outbreak associated with these products. The pasteurization process

Table 14.1 Observed Relationship of Blue Crab Pasteurization Process to Refrigerated Shelf Life

Process (F^{16}_{185}, minutes)	Shelf life, months
10–15	1.5
15–20	2–4
20–25	4–6
25–30	6–9
30–40	9–18
>40	12–36

described by Eklund et al. (1988) has the potential to minimize health safety concerns. Hot smoked fish were vacuum packaged and pasteurized in hot water at various temperatures. Both Type E and nonproteolytic B were used as test organisms. Samples were processed in 185°F (85°C), 193.8°F (89.9°C), and 197.9°F (92.2°C) water baths and required 29, 28.5, and 27.7 minutes, respectively, for the internal temperatures to equilibrate. Type E was the most heat sensitive of the test organisms, but none of the samples processed for 175 minutes at 185°F (85°C), or 55 minutes at 197.9°F (92.2°C) developed toxin after 6 months of refrigerated storage. Unfortunately, F-values for the processes were not published. The sensory qualities of the pasteurized fish were considered unchanged with respect to taste and texture. The color did darken; however, a lighter smoke before pasteurization eliminated the problem. The researchers reported that pasteurization was more effective for smoked fillets and steaks than for dressed fish.

MINIMALLY PROCESSED SEAFOODS (SOUS VIDE SEAFOOD PRODUCTS)

Minimally processed foods, including seafood, are being introduced into the U.S. market. The process was developed in France where the products are called "sous vide," which simply means under vacuum. The products are portion controlled, vacuum packaged in plastic pouches or ridged containers, which are impermeable to oxygen and moisture, and then cooked in either a water bath or high humidity oven. Cooking temperatures are usually far less than those associated with pasteurization of crabmeat. The cooking procedures may involve using temperatures very close to that desired for the internal temperature maximum for the product, therefore requiring long cooking times; or products may be cooked quickly using temperatures considerably above the desired internal maximum. In either case, the principles that apply to pasteurization also apply to this processing technology. The products should be cooked to a desired F-value and cooled quickly. Pasteurization principles related to container size, initial bacterial population, cooling, and survivors also apply to sous vide processing.

Sous vide is an outgrowth of the French cooking method "en papillote" (cooking a product in oiled parchment to lock in flavor). Although the idea of cooking in vacuum-sealed pouches has been around since the early part of this century (it was actually patented by W.R. Grace and Co.), it was not until the mid 1970s that the French chef George Pralus made it a popular cooking method in Europe. Pralus first used sous vide as a preferred cooking method for preparing foie gras. He discovered that by cooking under vacuum the product had less shrinkage, better flavor, and improved color retention. The product also had an extended shelf life when held under refrigeration. Furthermore, vacuum cooking allows the food to cook in its own juices and loss of flavor volatiles is minimized. The products are usually packaged raw and lightly spiced, since flavors are retained in the packages. They are usually produced at central facilities and used at upscale restaurants as a means of enhancing menu selection.

The safety of sous vide products has been questioned because these products are only minimally processed and do not contain preservatives to control microbial growth. Furthermore, the cooking process does not eliminate nonproteolytic types of *C. botulinum* and there is some question as to whether it may allow other organisms such as *Listeria* to survive. Shelf life and safety of these products is dependent solely on refrigeration; therefore it is critical that psychrotrophic pathogens not survive and grow. In the U.S. companies are producing these products as refrigerated, ready-to-eat, heat-and-serve products. They are produced under controlled conditions and caution is exercised during distribution. As of late 1989, the products are being sold only to food service establishments and are not available sold retail. This will probably change in the near future as demand increases. The U.S. Food and Drug Administration has limited the production of sous vide products to approved food processing operations and currently is not allowing production at retail establishments, such as grocery stores. It is important that these products be produced under an approved HACCP (hazard analysis critical control point) program, and that the principles that have been outlined for pasteurization be understood and applied to their production.

EFFECT OF CONTAINER TYPE

As discussed earlier, crabmeat traditionally has been pasteurized to a cold point temperature of 185°F (85°C) for 1 minute and then cooled to 100°F (37.7°C) within 50 minutes. This process has been based on the pasteurization of 16 ounces of crabmeat in 401 × 301 cans. The obvious questions is: What would the effect be of using this same processing parameter on smaller cans? The answer to the question is, quite simply, "under processing." This, of course, assumes that the traditional process in 16 ounce (401 × 301) cans is the reference process. Because the containers are smaller, they require less time to reach 185°F (85°C) at the cold point. Consequently, the total time the product is exposed to the lethal effects of heat is significantly reduced. Hence, the F_{185} values are smaller than the 31 minutes achieved in the traditional 401 × 301 cans. Furthermore, pasteurization of 16 ounces of crabmeat in cans of dimensions other than 401 × 301 will impact the process. Industry interest in new generation packaging and packaging materials has increased in recent years. Specifically, there is growing interest in thin profile containers such as pouches, boil-in-bags, and molded trays and cups, all of which can significantly increase the heating and cooling rates. This type of packaging is often perceived as resulting in products closer to fresh, and does not have the "canned" product connotation of tinplate and aluminum containers. Some designs also accommodate numerous convenience features. They may be microwavable or dual ovenable, easy opening, reusable, and resealable. In actuality, product quality is not substantially different than when processed in conventional containers. Moreover, special care should be taken with any innovative packaging, especially seal integrity. Any new package must be able to withstand the rigorous conditions encountered during commercial filling and processing operations, as well as distribution channels.

COMMON PROBLEMS

Aside from the product safety issue, spoilage is the major concern of most processors. When spoilage problems do occur in pasteurized seafood, they can often be traced to slow cooling, leaky can seams, insufficient heating, and high initial bacteria counts. A fifth cause of spoilage is temperature abuse during distribution or storage. This is a serious problem but, fortunately, is now less common due to the industry's awareness of the risk. However, processors should never take lightly the importance of proper temperature control. This includes advising their customers of proper handling practices.

Cooling

All too frequently, the pasteurization process is regarded merely as the heating of crabmeat to the desired temperature for the appropriate length of time. This, however, is a misconception that has proven costly to many processors. The process consists not only of heating the product, but also of cooling the product to temperatures at which the growth rate of surviving microorganisms is greatly reduced. Moreover, even during the cooling phase heat remaining in the cans contributes to the process lethality. This latter statement does not imply that cooling should be prolonged to take advantage of the lethal impact of residual heat. On the contrary, cooling should be as rapid as possible.

Many different types of microorganisms grow in a wide range of temperatures. Fortunately, those microorganisms that would ordinarily spoil crabmeat are very susceptible to destruction by the heat encountered in the normal pasteurization process. Conversely, those organisms that do survive are typically those that grow well at elevated temperatures but not at refrigerated temperatures, hence the importance of refrigeration after pasteurization. Nonetheless, even if the product has been adequately heated but then is allowed to cool at a slow rate, microorganisms that have survived pasteurization will start to multiply and thus shorten the anticipated shelf life of the product. Therefore, it is imperative that processors reduce the internal temperature of the product as quickly as possible.

Tri-State and the more recent National Blue Crab Industry Association (NBCIA) standards recommend immersion of heated containers in an ice-water bath. Work conducted by Virginia Tech's Seafood Agricultural Experiment Station supports this recommendation. An ice-water bath is the most efficient method of cooling the product. The Tri-State Seafood Committee previously recommended that the heated cans be cooled to 100°F (37.8°C) in an ice-water bath within 50 minutes of processing, then removed to refrigerated storage. The rationale for these recommendations was that some residual heat was needed to evaporate the moisture from the cans to prevent rusting. Rusting is no longer a problem with the vast majority of the cans being used today. The problem of microbial growth is of much greater consequence. The more recent standards (NBCIA) call for the heated cans to remain in the ice-water bath until the cold point temperature reaches 55°F (12.8°C) or cooler before being removed to refrigerated storage at 35°F (1.7°C). This standard is intended to improve both the quality of the product and its safety.

It is very important that the cooling water be chlorinated. There is a certain failure rate in any canned product. When the cans are hot the seals are more easily breached by microorganisms in the cooling water. As the sealant hardens, the cans become impermeable to microorganisms. Break-point chlorination of the cooling water kills organisms that might otherwise cause spoilage or safety problems if they were to get into the food; however, chlorination cannot be expected to compensate for defective seam integrity.

Can Seams

Defective can seams pose a serious threat to the health of consumers. Bacterial pathogens drawn into the can during cooling could possibly establish themselves, especially in the absence of the competitive microflora normally found in fresh crabmeat. Therefore, it is *critical* that seams be routinely checked to ensure that they meet the manufacturer's specifications. Also, can seaming equipment should be inspected frequently by a qualified technician, with adjustments made as needed and any corrective actions recorded.

Underprocessing

Inadequate heat processing often leads to premature spoilage. Both time and temperature are critical elements in any thermal process. A difference of a few degrees in the temperature of the water bath, e.g., 185°F (85°C) versus 189°F (87.2°C), or a few minutes in the processing time may significantly impact the lethality of the process and thus affect shelf life. Moreover, pasteurization systems should be loaded and operated so as to promote uniform circulation of the heating water, thus avoiding "cold spots" that may result in underprocessing.

Product Temperature

An often overlooked factor is the initial temperature of the product before it is pasteurized. Crabmeat that is held in ice overnight before processing may require an additional 15 minutes or more of heating compared to meat packed and processed directly from the picking tables. When the product temperature is unknown, or when both cold and warm crabmeat are processed in the same batch, managers should select the longer processing schedules established for refrigerated meat.

Initial Microbial Population

Pasteurization functions by destroying bacteria, with a certain number being killed by a given process. If the meat has low initial counts, then essentially all may be killed, whereas on another occasion high initial counts may lead to a shorter than normal shelf life. This concept is described in more detail in the discussion on the "Concept of Microbial Survivors in a Batch." A crucial point is that pasteurization cannot be successfully used as a salvage technique for crabmeat of marginal quality.

The initial microbiological quality of crabmeat is affected by the cooking method, the processing environment, and storage conditions. The cooking of crabs in commercial processing is done by either boiling or retorting (pressure steaming), and usually varies by geographical location. Commercial processors in states boarding the Gulf of Mexico often boil crabs, whereas in some Mid- and South Atlantic states it is mandated by state law that crabs be cooked by retorting.

When crabs are cooked by boiling, it is recommended that the water be allowed to return to a rolling boil and that cooking continue for at least an additional 10 minutes. Nonetheless, even following this recommendation, microorganisms may survive the boiling process. Cann (1977) observed that crabs in the middle of the basket did not reach a temperature of 140°F (60°C) during commercial processing, whereas Shultz et al. (1984) reported that crabs boiled for 10 minutes obtained a temperature of only 143°F (61.7°C). On the other hand, Dickerson and Berry (1976) found that temperatures would often approach 190°F (87.8°C). The differences in observations may be partially explained by the time required to return to a rolling boil. In some instances, this time interval can be quite long, with the temperature of the crabs slowly increasing during the "come up" period. The final product temperature achieved during boiling, as well as the length of time at lethal temperatures, affects the microbial load of the crabs. The microflora of live crabs is comprised largely of Gram-negative, heat-sensitive bacteria, which are killed at temperatures achieved by proper boiling. Nonetheless, it is not unusual to observe aerobic plate counts (APCs) of 10^3–10^4 per gram when freshly boiled crabs are tested.

Retorting crabs can result in a product that is essentially free of microorganisms, as sampled directly from the retort. The process obtained from retorting crabs for 10 minutes at 250°F (121.1°C) is equivalent to F^{18}_{250} of 0.7–1.8 minutes, depending on the size of the individual crabs and whether or not female crabs are bearing an egg mass often referred to as the "sponge" (sponge crabs heat much more slowly) (Dickerson and Berry 1976). For reference, an F^{18}_{250} of 2.3–3 minutes is considered adequate for commercial sterilization of canned foods.

Although cooking, especially retorting, can eliminate most of the crab's natural microflora, microorganisms are introduced during subsequent processing steps. Sources of microorganisms include the workers, utensils, contact surfaces, and insects. Since picking of the crabmeat is so labor intensive, the workers are perhaps the greatest source of microbial contamination. In areas of the country where crabs are debacked, washed, and refrigerated before picking (removing the meat), microorganisms are introduced from the workers and the wash water during debacking. In other areas, the customary practice is to cool the crabs in the retort baskets and allow the pickers to deback the crabs immediately prior to picking. This latter processing protocol is superior from a microbiological standpoint.

Microorganisms that can be introduced onto crabmeat from workers include spoilage organisms and pathogens. A partial list of pathogens that can be introduced in this manner includes *Staphylococcus aureus,* salmonellae, *Campylobacter, Listeria* (this is more likely to be an environmental contaminant), and enteropathogenic *Escherichia coli*. Of the pathogens listed, *S. aureus* is the most common contaminant. The organism is naturally present on the skin, hair, and nasal passages of much of the population. Because of the extensive human handling of crabmeat, *S. aureus* is a major concern of regulatory agencies.

Listeria can be found in the processing plant environment, and as a consequence can potentially be introduced into the crabmeat through a variety of routes. Some of the more obvious would be contamination of the cooked whole crabs prior to picking through contact with the floor or contaminated shovels used to load crabs onto the picking tables. Perhaps a less obvious route would be drip from ceiling condensate in cold rooms used to store cooked crabs. This can occur if cooked crabs are not sufficiently air cooled before being placed in refrigerated storage; steam rising from the warm crabs condense on the ceiling of the cooler and contaminate the crabs with a variety of microorganisms, *Listeria* among them. Once on the surface of the cooked crabs, these organisms easily contaminate the meat as it is being removed by pickers. Other pathogens from the environment might include *Vibrio* species and salmonellae. Hackney et al. (1980) demonstrated that *Vibrio* can contaminate crabmeat from the environment. Reservoirs can include waste containers, insects, and dust.

Most spoilage microorganisms and pathogens are heat sensitive and can be destroyed by low to moderate heat. The heat resistance of various non-spore forming microorganisms are summarized in Table 14.2. The D-values of pathogens and spoilage organisms were calculated from heat resistance data in the literature, with milk being the most common heating medium. Heat resistance of microorganisms is significantly affected by the food medium being heated. Crabmeat is low in fat and therefore should provide minimal protection to microorganisms. However, organisms imbedded in hepatopancreas tissue, which is often observed in crabmeat, may have a far greater heat resistance due to higher concentrations of fat.

CONCEPT OF MICROBIAL SURVIVORS IN A BATCH

Occasionally, seafood processors are baffled by the premature spoilage of a few containers from a batch of thermally processed products. Because of the sporadic occurrence, it might be assumed to be a problem of defective container seams. However, this is not the only possible explanation; random microbial survival patterns should also be suspected.

If for a given initial microbial load the absurd were possible, and all of a very large batch of crabmeat were packed in one giant can and processed to an accepted F-value, it would spoil due to growth of survivors. The more cans we pasteurize (i.e., the smaller the can) the more cans are likely to spoil, but also the smaller the percentage of spoiled to unspoiled cans. This effect is simply the result of partitioning and probability of microbial survival. If the initial microbial load is small, or the F-value high, fewer survivors and spoiled cans will result. This simplistic model may not appear to describe pasteurization because significant numbers of organisms may survive in nearly every can. However, the thermal-tolerant bacteria present after pasteurization grow very slowly at

Table 14.2   *D*-Value in Seconds of Nonspore-forming Microorganisms at Either 185°F (85°C) or 150°F (65.5°C)

Organism[a]	D-value, 185°F (85°C) (sec)	D-value, 150°F (65.5°C) (sec)	z-value (°F)	Heating Medium	Reference
V. ch.	0.16	11.7	19	Buffer	Shultz
V. ch.	0.29	93	14	Crabmeat	Shultz
V. p.	0.14	2.8	27	Clam	Delmore
L. m.	0.02	11.2	13	Milk	Bradshaw
L. m.	0.03	13.3	13	Milk	Bradshaw
internalized by phagocytes					
L. m.	0.026	12.8	13	Skim milk	Bradshaw
L. m.	0.04	17.6	13	Cream	Bradshaw
S. au.	0.002	15.0	9.2	Various foods	Stumbo
S. au.	0.04	132.0	10	Various foods	Stumbo
S. ty.	0.002	2.3	11.2	Milk	Bradshaw
S. ty.	0.001	4.2	10	?	Bradshaw
S. s.	0.017	53.6	10	?	Stumbo
Y. en.	0.0007	21.4	10	Milk	Lovett
Sh. dy.	0.0002	3.0	8.4	Milk	Stumbo
C. j.	0.0007	1.03	10.5 −12	Skim milk	Doyle
Most nonspore-forming	0.008–0.01	60–180	8–10	Various foods	Stumbo

spoilage bacteria i.e. (Pseudomonas, Achromobacter, Enterobacter, Micrococcus, Lactobacillus)

[a]V. ch., *Vibrio cholerae;* V. p., *Vibrio parahaemolyticus;* L. m., *Listeria monocytogenes;* S. au., *Staphylococcus aureus;* S. ty., *Salmonella typhimurium;* S. s., *Salmonella senftenberg;* Y. en., *Yersinia enterocolitica;* Sh. dy., *Schigella dysenteria;* C. j., *Campylobacter jejuni*

refrigeration temperatures. Only those that result in spoilage prior to the expected shelf life of the product are of concern.

The initial microbial load is important, but its significance may be poorly understood by manages who view the thermal process as a highly forgiving clean-up step. An *F*-value defines the number of targeted bacteria that are killed. Using a hypothetical situation, if a 1-pound container of crabmeat contains 10,000 thermoduric bacteria that are capable of prematurely spoiling the pasteurized product, then an *F*-value resulting in their destruction (a 4-*D* or four log-cycle reduction) will produce the expected shelf life. Every time this pasteurization process is applied to a container of crabmeat starting with no more than 10,000 of these thermodurics, satisfactory results are achieved. And anytime more than 10,000 are present, there is a high probability the container will spoil prematurely.

In actual production we must kill far more organisms than those found in one container. If 50,000 of the 1-pound containers of crabmeat are pasteurized during a season, then 500,000,000 of the offending organisms must be destroyed. Even a 6-D process would result in 500 survivors and a potential premature spoilage rate of up to 1% (500 cans) of the year's pack. In practice, heat is never uniformly applied throughout the container. The F-value achieved near the sidewalls is likely to be many times that at the center. However, the concept still holds since a process is established based on a desired F-value at the container center, and any survivors there will ultimately spoil the contents of the entire can.

USE OF STEAM TUNNEL PROCESSES TO REDUCE MICROBIAL LEVELS

Consumers have typically demonstrated a preference for fresh crabmeat over meat that is either pasteurized or frozen. However, some potential fresh markets are beyond reach due to shelf life limitations. Based on a preliminary study by North Carolina State University researchers (Gates 1977), an investigation was conducted at Virginia Polytechnic Institute and State University to determine the feasibility of using atmospheric steam to extend refrigerated shelf life (Rippen et al. 1988).

Flake crabmeat was spread on trays and passed through a steam tunnel so that the meat achieved a temperature of 167°F (75°C). It was hand packed, while still warm, into standard plastic containers used for fresh meat and stored on ice. Researchers felt that prechilling or packing with sterile implements would reduce its acceptance by the industry. Results indicated significant reductions in aerobic plate counts, and complete elimination of coliforms, fecal coliforms, and *S. aureus*. No changes were found in color, moisture content, or sensory quality. The crabmeat maintained significantly lower APCs than untreated controls during storage, and sensory shelf-life was extended by 85–100%. A process such as this has merit where the intent is to extend the shelf life of a fresh precooked product. Furthermore, the competitive microflora reduces the risk of botulism because subsequent contamination with a mixed spoilage microflora is expected, and containers are not hermetically sealed. Far more caution is warranted for products that are mildly pasteurized after they are placed in hermetic containers.

CLOSTRIDIUM BOTULINUM TYPE E

As has been noted, seafood pasteurization processes are not based on destruction of *C. botulinum;* instead it is shelf life that is the important consideration. If one looks at the heat resistance of Type E, then we see the process provides a large safety factor. *Clostridium botulinum* type E has a D_{185} value of 0.2–0.32

minutes. Therefore, a process of F_{185} of 31 minutes provides at least a 96-*D* process. However, other types of psychrotrophic nonproteolytic *C. botulinum* are more heat resistant. Nonproteolytic Type B is reported to have *D*-values of 0.45–14.33 minutes. A 31-minute *F*-value only provides a 2.2-*D* process for certain strains of this organism. Type F is reported to have a similar heat resistance.

Despite the safety factor, in any discussion of pasteurized crabmeat and public health, the principal consideration is the potential presence of *C. botulinum* Type E toxin. Although unlikely, the possibility exists for the toxin to be present and therefore merits attention.

Human botulism is relatively rare; however, its control and prevention is one of the most important considerations in food processing. History has shown repeatedly that an outbreak of botulism can cause severe, often ruinous, economic problems for processors. Furthermore, when a problem does arise, a whole segment of the food industry is often affected, not just the processor involved (Eklund 1982; Eyles 1986).

Clostridium botulinum, the etiological agent of botulism, is divided into eight types, based on serological differentiation of the neurotoxin: A, B, C1, C2, D, E, F, and G (Sakaguchi 1979). The types have been divided into four groups according to proteolytic activity (Smith 1977). Group I and II are the most important with respect to human botulism. Group I includes Type A and proteolytic strains of Types B and F. This group is strongly proteolytic and produces putrid, unpleasant odors. This group also produces highly heat-resistant spores and has a minimum growth temperature of about 50°F (10°C). Group II includes all Type E and nonproteolytic strains of B and F. Group II is neither proteolytic nor gelatinolytic and cultures do not produce putrid odors in food. This group can grow at temperatures as low as 37.9°F (3.3°C), and the spores are heat labile.

The symptoms of botulism usually develop in 12–36 hours after ingestion of the food; the range is 2 hours–14 days. In general, the shorter the onset time, the more severe the symptoms. The amount of toxin in food does vary, and death has been reported after merely tasting a small piece of bean pod or asparagus. Food poisoning occurs most frequently in women because they are more likely than men to prepare and taste food before it is adequately heated. Botulism is difficult to diagnose. Gastrointestinal problems are often the first sign (i.e., vomiting, nausea, and sometimes diarrhea). Other early symptoms are weakness, lassitude (weariness), dizziness, and vertigo. These can be followed by eye problems such as blurred vision, diplopia (double vision), dilated and fixed pupils, and impaired reflection to light. Other symptoms are weakness of facial muscles, pharyngo-laryngeal paralysis (difficulty in speech and swallowing), impaired salivation (dryness of the mouth, tongue, and throat), and complaint of great thirst. Abdominal pain is severe and often accompanied by constipation. Muscle weakness occurs in the soft palate, tongue, diaphragm, neck, and extremities, causing difficulty in walking and grip. Fever is absent and mental processes are normal. The major cause of death is respiratory failure and airway obstruction.

Clostridium botulinum is widely distributed in soils and because of run off, all types may be isolated from the aquatic environment (Dolman 1964). Type E is the toxin type most frequently isolated from aquatic environments and is usually implicated in botulism associated with seafood products. The spores of Type E are most often isolated from freshwater and marine sediments in temperate zones (Dolman 1964). The numbers for all types of *C. botulinum* found in waters, seafood, and sediment are usually low, with the highest counts being found in sediment (<100/g). The incidence in marine fish usually follows patterns associated with the bottom sediments. Presnell et al. (1967) examined the incidence in Mobile Bay, Alabama. *Clostridium botulinum* was found in only 4.1% of the sediment samples and 2.7% of the oyster samples. Ward et al. (1967a) surveyed the U.S. Gulf Coast. They found 3–5% of fish, and 5–8% of the sediment sampled to contain *C. botulinum*. In further work, Ward et al. (1967b) found a slightly lower incidence on the Atlantic Coast. Cockey and Tatro (1974) found *C. botulinum* in 21 of 24 crab samples from the Chesapeake Bay. In these surveys, Type E was usually the predominate type. Most outbreaks of botulism associated with fishery products have implicated semipreserved products, i.e., smoked, salted, or fermented products that are eaten without further cooking (Eklund 1982; Lynt et al. 1982). Type E is inhibited by water activity of <0.975 (5% NaCl) and pH of <5.3 (Emodi and Lechowich 1969). The spores are sensitive to heat. Decimal reduction times at 180°F (82.2°C) range from 0.49–6.6 minutes, depending upon the heating medium and the strain (Lynt et al. 1982; Simunovic et al. 1985). The spores are most resistent in tuna packed in oil. For foods not packed in oil a *D*-value of 4.3 minutes at 180°F (82.2°C) is usually considered the maximum. *z*-values range from 4.8°C to 9.6°C (Simunovic et al. 1985). For comparison, other members of Group II produce slightly more heat-resistant spores with *D*-values for nonproteolytic type B ranging from 1.49 to 32.3 minutes at 180°F (82.2°C) (Scott and Bernard 1982). The *D*-values for nonproteolytic Type F are similar to those of nonproteolytic Type B.

APPENDIX: EXPLANATION OF TERMS

Thermal Processing: Thermal processing is the application of heat to a food or container of food so that a certain temperature is achieved at the cold point of the product. The terms *D*-value, *z*-value, and *F*-value are used to define a thermal process, including pasteurization.

Commercial Sterilization: A process that destroys all microbial pathogens, and all other organisms that could lead to spoilage under normal distribution and storage conditions. However, certain nonpathogenic thermophilic sporeformers may survive. Because virtually all psychrotrophic and mesophilic microorganisms have been destroyed, the product need not be refrigerated to achieve the anticipated shelf life.

D-value: Decimal reduction time; the time needed to reduce a population of microorganisms by 90% or one log cycle. *D*-values can be determined from survivor curves where the log population is plotted against time or by the formula:

$D_{\text{reference Temperature}}$ = Time/ $\log_a$ − $\log_b$ where a = the initial population and b = the survivors after a time interval. Care must be taken when determining *D*-values because the curves are not always completely linear.

The heat resistances of bacteria, vegetative cells, or spores vary with the species of bacteria, conditions under which the cells are grown (temperature of incubation, phase of growth, age of the spores) and the inoculum level. The heating menstruum, or medium in which the spores are tested, also affects the heat resistance. Factors affecting bacterial heat resistance include water content, pH, and the chemical composition (fats, salts, proteins, carbohydrates, and minerals) of the heating menstruum. Therefore the *D*-values of microorganisms will vary depending on the conditions mentioned above.

Sometimes it is desirable to be able to calculate *D*-values at other process temperatures. Equivalent *D*-values are calculated by the formula: $\log D_2 = \log D_1 - (T_2 - T_1/z)$ where D_1 = the known reference value; D_2 = the desired value; T_1 = the reference temperature; T_2 = the desired temperature; z = the z-value as described below. It is important that equivalent *D*-values not be calculated for temperatures far greater or less than the reference temperature, because the real *D*-values may be considerably different than the calculated values. Survival curves are not always linear; they often have shoulders and tailing is often observed.

z-value: The number of degrees Fahrenheit or centigrade required for a thermal death time curve to traverse one log cycle. The z-value gives an indication of the heat resistance of an organism, with higher values indicating a greater level of resistance. The z-value is obtained by plotting the logarithms of at least two *D*-values versus temperature. (Several *D*-values are required for greatest accuracy). Conversely, z-values can be used to calculate *D*-values at various temperatures and are essential in process calculations. These conversions are reasonably accurate within the range of normal processing temperatures. The formula for calculating z-values is: $z = T_2 - T_1 /\log D_1 - \log D_2$. (Terms are the same as described above for equivalent *D*-value.)

F-value: The *F*-value is a term that is useful in describing the total heating value of the process. It is the equivalent, in minutes at a given temperature, of accumulated heat with respect to its capacity to destroy spores or vegetative cells of a particular organism. The *F*-value defines a process that is equivalent to that which would result from instantaneously heating a product to a given temperature, holding it for a specified time, and instantaneously cooling it. Of course, instantaneous heating is impossible in real processing. Therefore an *F*-value is calculated to account for heating and cooling rates. Heating and cooling curves tend to appear somewhat bell shaped. The shape of the curve is affected by product type, container size, container shape, method of loading the pasteurizer, pasteurizer style, amount of agitation, temperature of the waterbath (Delta T), etc. The destruction of microorganisms begins at a relatively low temperature and accelerates with increasing temperature. As the internal temperature of the product approaches or exceeds the reference temperature, the destructive impact becomes maximal. Even as the product cools microorganisms continue to die because the heat that remains in the can is contributing—in decreasing proportions—to the lethality of the process.

To determine the *F*-value, the area under the curve must be integrated. This can be approximated by dividing the curve into small sections, calculating the *F*-value for each section, and then adding them together. The formula for each section (time interval) is:

$$F = \log^{-1} [(T_2 - T_{\text{reference}})/z] \times \text{time interval}$$

In this instance T_2 is the calculated midpoint temperature for the time interval. Both the heating and cooling sections of the curve are considered in the calculations. When the time interval is large the curve used to calculate *F*-values has a staircase appearance. However, when the time interval is small, such as when the data are recorded and calculated by computer the steps are virtually eliminated and the determination of accumulated *F*-values becomes nearly accurate.

Sometimes it is desirable to process at a different temperature than the referencew. A reference *F*-value can be converted to an equivalent F value using the formula:

$$F_{\text{temperature desired}} = F_{\text{reference temperature}} \times \log^{-1} (T_1 - T_2)/z.$$

Cold Point: The slowest heating point in a container.

REFERENCES

American Public Health Association. 1985. Speck, M. L. (ed.) *Compendium of Methods for the Microbiological Examination of Foods.* APHA, Washington, DC.

Boutin, B., Bradshaw, J., and Stroup, W. 1986. Heat processing of oysters naturally contaminated with *Vibrio cholerae. J. Food Protect.* **49.**

Bradshaw, J., Peeler, T., Corwin, J., Hunt, J., Tierney, J., and Twedt, R. 1985. Thermal resistance of *Listeria monocytogenes* in milk. *J. Food Protect.* **48:**743.

Bradshaw, J., Peeler, T., Corwin, J., Hunt, J., and Twedt, R. 1987a. Thermal resistance of *Listeria monocytogenes* in dairy products. *J. Food Protect.* **50:**543.

Bradshaw, J., Peeler, T., Corwin, J., Barnett, J., and Twedt, R. 1987b. Thermal resistance of disease-associated *Salmonella typhimurium* in milk. *J. Food Protect.* **50:**95.

Bunning, V., Crawford, R., Bradshaw, J., Peller, J., Tierney, J., and Twedt, R. 1986. Thermal resistance of intracellular *Listeria monocytogenes* suspended in bovine milk. *Appl. Environ. Microbiol.* **52:**1398.

Cann, D. 1977. Bacteriology of shellfish with reference to international trade. In: *Handling, Processing and Marketing of tropical Fish*. Tropical Products Institute, London, p. 377.

Cockey, R., and Tatro, M. 1974. Survival studies with spores of *Clostridium botulinum* Type E in pasteurized meat. *J. Milk Food Technol.*

Delmore, R. P., and Crisley, 1979. Thermal resistance of *Vibrio parahaemolyticus* in clam homogenate. *J. Food Sci.* **41**:899.

Dickerson, R., and Berry, M. 1976. Heating curves during commercial cooking of the blue crab. *J. Milk Food Technol.* **39**:258.

Dolman, C. 1964. Botulism as a world problem. In: Lewis, K., and Cassel, (eds.) *Botulism*. Cincinnati, Ohio. Public Health Series, U.S. Department of Health, Education and Welfare.

Doyle, M., and Roman, D. 1981. Growth and survival of Campylobacter jejuni as a function of temperature and pH. *J. Food Protect.* **44**:596.

Eklund, M. W. 1982. Significance of Clostridium botulinum in fishery products. Preserved short of sterilization. *Food Technol.* **36**:107–112.

Eklund, M., Peterson, M., Paranjpuke, R., and Pelroy, G. 1988. Feasibility of a heat-pasteurization process for the inactivation of non-proteolytic *Clostridium botulinum* Types B and E in vacuum-packaged, hot-process (smoked) fish. *J. Food Protect.* **51**:720.

Emodi, A., and Lechowich, R. 1969. Low temperature growth of type E *Clostridium botulinum* spores. 1. Effect of sodium chloride and pH. *J. Food Sci.* **34**:78.

Eyles, M. 1986. Microbiological hazards associated with fishery products. *CSIRO Food Res. Q.* **46**:8.

Garcia, G., Genigeorgis, C., and Lindroth, S. 1987. Risk of growth and toxin production by *Clostridium botulinum* non-proteolytic types B, E, and F in salmon fillets stored under modified atmospheres at low and abused temperatures. *J. Food Protect.* **50**:330.

Gates, K. 1977. Steaming hand picked crab meat improves quality. Timely tips. North Carolina State University newsletter.

Hackney, C. R., Ray, B., and Speck, M. L. 1980. Incidence of *Vibrio parahaemolyticus* in and the microbiology quality of seafoods in North Carolina. *J. Food Protect.* **43**:769.

Himelbloom, B. H., Rutledge, J. E., and Biede, S. L. 1983. Color changes in blue crabs *(Callinectes sapidus)* during cooking. *J. Food Sci.* **48**:652.

Hobbs, G. 1976. Clostridium botulinum and its importance in fishery products. *Add. Food Res.* **22**:135.

International Commission on Microbiological Specifications for Foods. 1980. *Microbial Ecology of Foods, Vol. 2—Food Commodities*. Academic Press, Orlando, Florida.

Lerke, P., and Farber, L. 1971. Heat pasteurization of crab and shrimp from the Pacific coast of the United States: public health aspects. *J. Food Sci.* **36**:277.

Lovett, J., Bradshaw, J., and Peeler, J. 1982. Thermal inactivation of Yersinia enterocolitica in milk. *Appl. Environ. Microbiol.* **44**:517.

Lynt, R. K., Solomon, H. M., Lilli, T., Jr., and Kautter, D. A. 1977. Thermal death time of Clostridium botulinum Type E in meat of the blue crab. *J. Food Sci.* **42**:1022–1025.

Lynt, R. K., Kautter, D., and Solomon, H. 1982. Differences and similarities among proteolytic and nonproteolytic strains of *Clostridium botulinum* A, B, E, and F: a review. *J. Food Protect.* **45**:466.

McClain, D., and Lee, W. 1987. A method to recover *Listeria monocytogenes* from meats. ASM Abstracts.

Rippen, T., Hackney, C., Mayer, B., Lynn, S., Soul D., and Sanders, L. 1988. Effect of atmospheric steam on the shelf-life of blue crab meat. Institute of Food Technology Abstracts.

Rippen, T., Hackney, C., Mayer, B., and Sutton, H. 1989. Improvement of heat transfer rates and shelf-life of pasteurized blue crab meat. Institute of Food Technology Abstracts.

Sakaguchi, G. 1979. In: Rieman, H., and Bryon, F. (eds.) *Botulism in Foodborne Infection and Intoxication* Academic Press, New York, pp. 389–442.

Scott, V., and Bernard, D. 1982. Heat resistance of spores of nonproteolytic type B *Clostridium botulinum*. *J. Food Protect.* **45**:909.

Shultz, L., Rutledge, J., Grodner, R., and Biede, S. 1984. Determination of the thermal death time of *Vibrio cholerae* in blue crabs *(Callinectes sapidus)*. *J. Food Protect.* **47**:4.

Simunovic, J., Oblinger, J., and Adams, J. 1985. Potential for growth of nonproteolytic types of *Clostridium botulinum* in pasteurized meat products. A review. *J. Food Protect.* **48**:265.

Smith, L. 1977. *Botulism: the Organism, Its Toxins, the Disease*. Charles C Thomas. Springfield, Illinois.

Stumbo, C. 1973. *Thermobacteriology in Food Processing,* second edition. Academic Press, Orlando, Florida.

Tri-state Seafood Committee. 1971. Regulations governing the pasteurization of crabmeat. Maryland, North Carolina, and Virginia state health departments and Land Grand Universities.

Young, F. 1987. Food Safety and FDA's Action Plan Phase II. *Food Technol.* **41**:116.

15

Modified Atmosphere Packaging

L. E. Lampila

Modified atmosphere packaging (MAP) refers to a system in which air is replaced either totally or in part by other gases. Over the course of storage, the atmosphere, although different from air, will be modified as a result of metabolic activity and permeability of the packaging material. Controlled atmosphere packaging (CAP) is similar to MAP; however, gas composition is monitored and regulated to maintain a stable composition during the storage period. In both MAP and CAP, CO_2 is primarily used, but mixtures containing nitrogen, oxygen, and air have also been tested. Coyne (1932, 1933) first reported the bacteriostatic effect to purified bacterial cultures incubated in the presence of CO_2. Results of further studies with CO_2 levels ranging from 20 to 100% (optimum 40–50%) remarkably preserved haddock, cod, sole, whiting, and plaice stored either in the round or as fillets (Coyne 1933). A comprehensive review of early modified atmosphere storage has been published by Parkin and Brown (1982).

PROPOSED MECHANISMS OF ACTION

Four general mechanisms are attributed to the preservative effects of CO_2 in foods. Most likely, the preservative effect is due to a combination of factors. First, the displacement of oxygen by CO_2 inhibits the growth of aerobic microorganisms and results in a delayed growth phase (prolonged lag phase) while there is a shift of dominant flora (Callow 1932) that tend to be antagonistic to spoilage bacteria (Price and Lee 1970). This would be only a partial reason since it has been shown that 100% CO_2 inhibits the growth of anaerobes to a greater extent than 100% N_2. Second, the hydration of CO_2 to carbonic acid may result in acidification of the tissues (Aichin and Thomas 1975; Becker 1933; King and Nagel 1967). It should be noted, however, that the hydration of CO_2 to carbonic acid is estimated to be only 1 (Knoche 1980) to 2% (Brinkman et al. 1933; Quinn and Jones 1936). Carbonic acid dissociates rapidly with the formation of bicarbonate (HCO_3^-) and hydrogen (H^+) ions. Third, either CO_2 or its ions may alter the bacterial cell permeability characteristics (Krough 1919; Sears and Eisenberg 1961) and specifically by increasing fluidity of membrane fatty acids

(Castelli et al. 1969) and/or the carbamination of proteins (Fox 1981). Last, metabolic pathways may be affected by the presence of CO_2 and result in altered bacterial enzymatic activity, e.g., an increased rate of succinate formation by *Bacillus* (Elsden 1938), inhibition of fumaric acid formation by *Rhizopus nigricans* (Foster and Davis 1949), inhibition of isocitrate dehydrogenase and malate dehydrogenase by *Pseudomonas* (King and Nagel 1975), interference with hydrolases and thus autolysis (Mitsuda et al. 1980), and a near complete inhibition of NADH oxidation by the submitochondrial complex I–II–III–IV (Smikahl 1985). This may be the result of either altered solubility of proteins (Mitz 1957, 1978) and/or feedback inhibition mediated by enhanced CO_2 solubility at reduced temperatures (Delente et al. 1969; Krebs and Roughton 1948).

Modified atmosphere storage is more desirable for seafoods, due to a lower myoglobin content, than for red meats. Red meats develop a brown discoloration in the presence of >20% CO_2 (Ledward 1970; Silliker et al. 1977).

MAP STORAGE OF GUTTED FISH

Villemure and Simard (1988) evaluated the effect of either 25% CO_2/75% N_2 or air holding on the quality of fresh, gutted cod *(Gadus morhua)* stored at 35.6°F (2°C) for up to 11 days. A third treatment involved packaging fresh, gutted cod aboard the fishing boat under the aforementioned MAP conditions. After 8 days, the air-held cod had reached a bacterial count of 10^6 colony forming units (cfu)/g whereas the MAP (packaged onshore) had a count of 10^4 cfu/g. It is interesting to note that cod treated at sea had a 24-hour count 1 log unit less (at 10^2) than those fish treated onshore. Similarly by 11 days the air-held and on-shore MAP- and vessel-treated cod had counts of 10^7, 10^4, and 10^3 cfu/g, respectively. Total volatile nitrogen (TVN) production was detected at levels that would parallel the microbial counts and that of both MAP-treated samples was 35% less than that of the air-held control at 15 days. Results of sensory evaluation indicated that the air-held samples were rejected at 8 days and the MAP-treated at 11 days. This indicates that microbial quality may not be a suitable indicator for MAP of seafoods.

Round Pacific cod *(Gadus macrocephalus)*, 9 hours after catch, were transferred into chilling systems consisting of either ice or CO_2 modified refrigerated seawater [(MRSW) (Reppond and Collins 1983)]. Two hundred kilograms of MRSW (3.5% NaCl saturated with CO_2, pH 4.3) and 100 kg of fish were stored at 33.9°F (1.1°C) for 12 days. The pH of the system continued to increase to 6.2 by 8 days despite continued addition of CO_2. The ice was replenished as it melted in the control treatment. The samples were evaluated for total solids, chloride, total volatile acids (TVA), trimethylamine (TMA), and sensory quality. The results indicated that flavor and texture of the ice-held samples deteriorated significantly over the 12-day period and that there were no flavor and texture changes in those samples held in MRSW ($p < 0.05$). The NaCl content of the MRSW samples did increase to 1.0% by 8 days. Development of TVA was similar among samples although slightly inhibited by storage in MRSW. TMA accumulation was not affected by MRSW. Although no microbial counts were determined, the MRSW system appeared to show promise for the shelf life extension of fresh product, although elevated levels of sodium may make these fish less desirable for those on sodium-restricted diets.

As an extension of MRSW studies, Reppond et al. (1985) investigated the effects of holding walleye pollock *(Theragra chalcogramma)* in ice (fish to ice, 0.5:1); slush-ice [fish to ice to 1.75% brine, 8:3:1, 30.1°F (−1.1°C)]; RSW [fish to RSW, 1.5:1, 1.75% NaCl brine, 31.5–35.2°F (−0.3–1.8°C)]; and MSRW [fish to MSRW, 1.5:1, 1.75% NaCl, CO_2 to pH 6.0, 30.2°F (−1.0°C)]. Fish were removed for filleting at 0, 2, 4, 6, and 8 days. Chemical analyses were performed for NaCl, dimethylamine (DMA), trimethylamine (TMA), formaldehyde (FA), volatile acid number (VAN), inosine monophosphate (IMP), hypoxanthine (Hx), viscosity, and emulsifying capacity (EC). Sensory evaluation for flavor, texture, and moistness was also conducted.

The results indicated that there were no differences in fillets at 2 days; however, at 4 days the MRSW fillets were softer with some gaping. RSW-held fish had a strong ammoniacal odor and those held in slush ice were less aromatic by 8 days. The NaCl content of samples in RSW rose to 1.0%, in the MRSW system to 0.5%, in slush ice to 0.3%, and in ice, to 0.2% by 8 days. VAN at 0 hours was 2.0 and rose to 6.0 by 8 days in the MRSW, ice, and slush systems but in RSW rose to 16.0 in the same time period. The TMA content increased from ca. 0.1 mg TMA-N/100 g fish to 7.5 in RSW-held fish by 8 days. The balance of the treatments remained at ca. 1.0 mg TMA-N/100 g flesh. The DMA content increased in all treatments except the MRSW sample which showed no change. Over 12 months storage, the FA content increased linearly with DMA ($r = 0.946$, $p < 0.001$). IMP and Hx were not detectable by the methods used and usually do not provide stable quality indices for fish. Neither emulsifying capacity nor viscosity was a reliable indicator of protein quality since no differences were seen among the treatment means.

Sensory evaluation indicated that fish held in RSW systems were too salty by day 8. The MRSW-treated fillets were firmer at 8 days; however, all samples were judged acceptable. During extended (12 month) storage sensory differences were negated.

STUDIES WITH FRESH FILLETS AND STEAKS

Brown et al. (1980) studied the effect of 20 and 40% CO_2 with or without CO (the balance air) and 100% air (control) over 14 days at 40.1°F (4.5°C) on sensory, microbiological, and selected biochemical parameters of *Sebastes miniatus* (Vermillion rockfish or California red snapper) fillets and *Onchory-nchus kisutch* (coho or silver salmon) steaks. Since these researchers determined in preliminary work that CO_2 could promote surface darkening in highly pigmented species due to the oxidation of oxymyoglobin, 1% CO was introduced to some treatments to form carboxymyoglobin, a form of myoglobin with enhanced resistance to oxidation.

The results indicated that by 7 days there was a visual slime on the surface of the rockfish controls (air) but not on any of the CO_2-treated samples. Under 1% CO, the fillets retained a red color, presumably due to the formation of carboxymyoglobin. Both fresh and air-stored fillets were judged more aromatic than CO_2-treated samples; the presence of 1% CO, however, nullified this difference. By 7 days, there were no differences in odor between samples held either in air or under 20% CO_2 with or without CO. Salmon steaks held for 7 days under 40% CO_2 were determined to smell stronger than fresh steaks but were not as aromatic as the air-held samples ($p < 0.001$). Unlike in rockfish, there was no significant difference ($p < 0.001$) in salmon fillets stored with either CO_2 or CO_2 plus CO. The FDA may not, however, accept the use of CO as it may be viewed as a means to enhance color artificially.

TBA values in all treatments increased over the storage trials and although some significant differences did arise the overall levels remained rather low. Consequently, lipid oxidation was not considered to be exacerbated by MA with either rockfish or salmon. Hx values typically show large variation and were similarly unaffected by the MA treatments. TMA production was inhibited over the entire 14-day storage period by 20% CO_2 in rockfish but only for 3 days at 40% CO_2. As anticipated, TMA levels rose more in the rockfish than the salmonid species, despite the treatment. Microbial counts by 7 days on the air-held controls were 3.3×10^7 whereas on the CO_2- treated they were only 3.0×10^5 cfu/g. Although initial counts were slightly greater in salmon than in rockfish, a 100-fold difference within treatments was also noted in 7 days.

Haard and Lee (1982) stored Atlantic salmon *(Salmo solar)* steaks at 37.4°F (3°C) in air, CO_2 at atmospheric pressure, or in a hypobaric chamber with CO_2 flowing continuously at 100 cc/min. Samples were evaluated every 2 days over a 20-day period for sensory evaluation, aerobic plate count (APC), and TMA content. The results indicated that air-held samples developed an undesirable fishy odor and appearance after 10 days; those in CO_2 quickly had a carbonated, bland flavor and powdery texture; and hypobarically stored samples under CO_2 retained a fresh flavor and texture. In steaks with low (10^2) initial counts, the CO_2 treatments suppressed the lag phase of bacterial growth ca. 4 days, initiation of the log phase being slightly more inhibited by CO_2 at atmospheric pressure. The air-stored (controls) samples reached 10^9 APC by 10 days; however, both CO_2 treated samples reached an APC of 10^6 by 16 days and began to decline. Although salmon steaks with higher (10^5) initial bacterial counts sustained inhibited bacterial growth under continuous flow CO_2, hypobaric storage, it would not be recommended to extend the shelf life of either aged or improperly handled product. TMA development did not begin until 6 days in all treatments, it was suppressed ca. 25% by 14 days under atmospheric pressure with CO_2, and was inhibited ca. 70% of the control by hypobaric storage in combination with CO_2. This indicates that reduced pressure and an altered atmosphere inhibit the expected spoilage microflora.

Lannelongue et al. (1982) studied the effect of MAP on brown shrimp *(Penaeus aztecus)*, 50–60 count, using 100% CO_2, 66% CO_2 plus 34% O_2, 38% CO_2 plus 62% O_2, 65% CO_2 plus 35% N_2, and 35% CO_2 plus 65% N_2. Samples were stored at 39.2°F (4°C) and evaluated at 3, 6, 10, 14, and 19 days.

The initial load was high at 5.8×10^4 cfu/g, 69% of which were Gram-negative organisms, an indication of 3 days holding on ice. All CO_2 treatments extended the lag growth phase, with the 100% CO_2 treatment inhibiting growth to 1.7×10^6 by 14 days. The air-held samples had a population of 10^8 by 14 days. The population of the air-held samples was 89% Gram-negative and 38% in the 100% CO_2-treated sample by 14 days. The short shelf life for shrimp was considered to be due to the type of holding; normally shrimp held on ice causes a washing/dilution effect over time. The TVN accumulation was rapid in air and increased from 4.75 to 9 mg TVN-N/100 g of shrimp. Its development was offset to the greatest extent by the presence of 100% CO_2, and under 66% CO_2/34% O_2 and was one-half to one-third the level of the air-held samples. Nitrogen did not inhibit the formation of TVN, especially at the 35% level of CO_2. The pH of the samples decreased initially but increased over the total storage period which indicated deamination and ammonia accumulation. The pH did rise proportionally to the CO_2 content of each package.

In the air-held samples it was determined by headspace analysis of the gases that CO_2 did constitute 20% of the gases by 10 days. This would be indicative of microbial respiration and enzymatic activity. In all of the MAP treatment samples there was a decrease in CO_2 by 3 days and for the 100% CO_2 treated samples there was sufficient absorption by the shrimp to cause deformation of the packages.

CHARACTERIZATION AND SHIFT OF MICROFLORA

Richter and Banwart (1983) investigated the effect of MAP on 24-hour-old ocean perch *(Sebastes marinus)* and sea trout *(Cynoscion nebulosus)*. Fillets were packed onto plastic trays containing an absorbent pad and placed in either a nylon–Surlyn bag (oxygen transmission rate of 25 cc/mm^2/day at 50% RH, CO_2 transmission rate of 160 cc/mm^2/day at 50% RH) or were overwrapped with polyvinyl chloride film (oxygen transmission rate of 150 cc/mm^2/day at 50% RH, CO_2 transmission rate of 970 cc/mm^2/day at 50% RH; the control was an air atmosphere and the experimental treatment was backflushed with 80% CO_2). The results indicated that under MAP both seafood species showed a 7-day lag in growth of total mesophiles and oxidase-positive mesophiles, and proteolytic and psychrotrophic microorganisms. Although the CO_2-treated samples were considered significantly ($p < 0.05$) more desirable than the air-held treatments, none of the samples were considered acceptable at 14 days. Under CO_2 the pH of the sea trout increased over the 14-day storage period while that of the ocean perch remained static. The authors were in agreement with Wolf (1980), who stated that MAP should be used to provide a better quality product to the consumer but not to extend shelf life.

Scott et al. (1986) packed 48-hour-old snapper *(Chrysophrys auratus)* fillets on a PVC tray overwrapped with an O_2/CO_2 permeable film; a CO_2 flushed pouch or under 40% CO_2/60% N_2 (CN46) in a ratio of 1:5 fish to gas, and stored samples at 30.2°F (-1°C). The APC of the control exceeded 10^8 by 11 days; that of the CO_2 flush pack, 4.3×10^5 by 18 days; and that of the CN46 treated samples, 3×10^6 cfu/g by 11 days. Although the APC of the CN46 pack was 5.6×10^6 cfu/g by 9 days, it was judged unacceptable due to the high level (10^4 cfu/g) of sulfide producing microorganisms. There was also a high level of sulfide producers in the control pack; expression of their deteriorative enzymes may have been inhibited by a rapid outgrowth of *Lactobacillus* spp. Although lactics are not usually associated with aerobic packs, their counts reached 10^5 by day 10 in these studies. Isolation of lactics often depends on the media used and the growth atmospheres (Molin and Stenstrom 1984). In the control and 40% CO_2/60% N_2 treatments, microbial spoilage patterns paralleled acceptability as determined by sensory evaluation. In the CO_2 flush pack, flavor quality declined by 11 days, and the texture became "wet cardboard-like" but sensory quality did not reflect the acceptable bacterial quality indices. For snapper, sensory quality may suffer at the expense of microbial quality. This indicates that MAP may need to be evaluated for each seafood species.

Gray et al. (1983) studied the effect of MAP on 24-hour-old fillets or whole, gilled, gutted, and rinsed perch *(Meronia americanus),* seatrout *(Cynoscion regalis),* croaker *(Micropogon undulatus),* and bluefish *(Pomatomus salatrix)* at either 33.9°F (1.1°C) or 50°F (10°C) for up to 30 days (Phase I). In Phase II of study sea trout were scaled, headed, gutted, and spray washed at a commercial plant 4 days after harvest and handled under one of five conditions: (1) packaged under CO_2 at a 20:1 ratio (20 in^3 CO_2/1 pound flesh), at a 10:1 ratio, or without gas treatment and transported via normal distribution channels; (2) packaged without gas treatment or under CO_2 at a 20:1 ratio with no refrigeration for 6 hours prior to transportation to distribution channels (simulated dock abuse); (3) packaged without gas treatment or under CO_2 at a 20:1 ratio and shipped directly to the researchers' laboratory; (4) packaged without gas treatment or under CO_2 at a 20:1 ratio, held for 4 days at the processing plant, and then transported by customary distribution channels; and (5) packaged without gas treatment or under CO_2 at a $20:1$ ratio, held 4 days at the processing plant, and transported to the researchers' laboratory.

The results of Phase I (temperature studies) indicate that the population of pyschrotrophic organisms increased to 10^7 cfu/g (CO_2) to 10^8 (air) at 33.9°F (1.1°C) by 20 days whereas the mesophilic population for either air-held or MAP product was 10^5 cfu/g in the same time period. Although the species were deemed unacceptable from a sensory view, the mesophilic counts would indicate acceptable product. As anticipated, the MAP treatments showed a lag in the log growth of both psychrotrophs and mesophiles; however, by 10 days the difference was negligible between the mesophilic counts. Among seafood species, the differences in psychrotrophic counts between MAP-stored samples and those of the air-held samples diminished by 20 days. Additionally, spoilage of samples held at 50°F (10°C) was more rapid than that at 33.9°F (1.1°C); by 10 days, bacterial counts for the higher temperature samples were 3 log units greater than those at the low-temperature treatments. This is supported by Ogrydziak and Brown (1982) and many others.

From Phase II studies, it became apparent that between 5 and 11 days the generation time for the MAP samples was 24 hours as opposed to 14 hours for the air-held samples. It was noted that samples abused at the dock, treated with either air or MAP, spoiled at a rate comparable to nonabused fish. The $20:1$ (CO_2/fish) treatment, transported through normal distribution channels, again verified a 6-day lag prior to attainment of air-treated microbial numbers. A reduced $10:1$ ratio was inferior to the $20:1$ ratio and resulted in a spoilage rate similar to that of the air control. While MAP (20 CO_2:1 fish) resulted in a similar lag whether dock-abused, delayed in transport, or held under laboratory conditions, this is no substitute for good manufacturing practices. MAP of abused product could theoretically lead to concentration of toxins such as histamine, which in this case was not assayed. Watts and Brown (1982), however, studied the effects of MAP (80% CO_2/20% air) on the formation of histamine, tyramine, cadaverine, and putrescine in abusively stored Pacific mackerel. After 48 hours at 68°F (20°C), the histamine levels in the MAP-stored mackerel were 62% less than those of the air-held control. Although this difference was eliminated by 72-hour storage, species such as bluefish may be less tolerant to abusive handling.

Molin et al. (1983) studied the effect of air storage, CO_2, and N_2 modified atmospheres on herring fillets and the shift in microflora. Product was considered spoiled when the APC exceeded 10^7 cfu/g which was 3, 6, and 21 days for the air, N_2, and CO_2 treatments, respectively. Although the initial flora were characterized and determined to consist primarily of coryneforms, *Flavobacteria, Pseudomonads, Moraxella,* and like organisms, there was a shift in the microbial populations over the 28-day study. *Pseudomonads, Moraxella,* and like organisms dominated the spoilage flora (9 days) of the air-held samples; *Enterobacteriacea, Vibrios,* and *Lactobacilli* were the major spoilage organisms of the N_2-held samples (14 days); and *Lactobacilli* predominated the flora of the CO_2-treated samples at 28 days. The CO_2 treatment and the resulting outgrowth of competitive lactobacilli successfully inhibited the outgrowth of coliforms, whereas there was comparable lag growth of coliforms in both the air and N_2 environments. Competitive microflora in all three gaseous treatments protected against the outgrowth of *Clostridia*. These researchers concluded that the competition of microflora would protect against the outgrowth of pathogens and that the reduced deteriorative capacity of the predominant microflora (and extended shelf life) would indicate that modified atmospheres held potential for the industry.

Controlled atmosphere packaging was used to study 3-day-old swordfish (*Xiphias gladius*) from the Gulf of Mexico (Oberlender et al. 1983). Steaks were stored at 35.6°F (2°C) under one of six atmospheres (air; 100% CO_2; 70% CO_2 plus 30% O_2; 40% CO_2 plus 60% O_2; 70% CO_2 plus 30% N_2; and 40% CO_2 plus 60% N_2) under a continuous flow of 2 L/hour for 22 days. APC, characterization of the microflora, TVN, and TMA were determined. The results indicated that the initial microflora consisted of *Micrococcus* species, and low percentages of *Bacillus*, coryneforms, *Flavobacteria, Moraxella–Acinetobacter, Citrobacter freundii*, and pseudomonads and the APCs ranged from 3.8×10^2 to $1.4 \times 10^3/cm^2$. Samples stored in air 40% CO_2 plus 60% O_2, 40% CO_2 plus 60% N_2, and 70% CO_2 plus 30% O_2 reached an APC of 10^6 cfu/cm^2 in 6, 14, 20, and 22 days, respectively. Counts of samples stored in 100% CO_2 (2.6×10^4) and 70% CO_2 plus 30% N_2 (4.5×10^3) did not approach the same levels by 22 days at 2°C.

In the air-held (control) sample, pseudomonads constituted 100% of the population. By 22 days, samples held under 100% CO_2 had populations consisting of 89% heterofermentative *Lactobacilli*, and the balance *Pseudomonas* whereas in 40% CO_2 plus 60% N_2 only 17% of the population consisted of heterofermentative *Lactobacilli* and 79% *Brochotrix thermosphacta*. Under 70% CO_2/30% N_2, 19% of the population has heterofermentative *Lactobacilli*, 55% *Brochotrix thermosphacta*, and 26% *Pseudomonas;* under 40% CO_2 plus 60% O_2, 3% heterofermentative *Lactobacilli*, 38% *B. thermosphacta*, and 56% *Pseudomonas;* and under 70% CO_2 plus 30% O_2, 31% heterofermentative *Lactobacilli*, 23% *B. thermosphacta*, and 46% *Pseudomonas*. The intitial pH (pH 5.9) of the flesh rose to pH 6.5, and as anticipated, the CO_2 treatment pH values remained near the 0-day level. In the continuous CO_2 flow, there was presumably continuous hydration to carbonic acid and therefore stability of the pH.

The TVN was similar in all treatments to the 8-day value (11–14 mg TVN/100 g fish) but then rose in the air-held sample to 25 mg TVN/100 g fish by 14 days while there was a lag in the remaining treatments for an additional 4–6 days. It should be noted that guidelines for TVN value for freshness have been recommended as: fresh fish, 12 mg; good quality, 12–20 mg; edible, 20–25 mg; decomposed and inedible, >25 mg TVN-N/100 g tissue (Stansby 1963).

Although none of the values determined exceeded the guidelines for edible flesh, the presence of CO_2 inhibited those organisms responsible for the accumulation of TVN-N.

TMA content increased in all treatments except the 70% CO_2 plus 30% O_2 and the 100% CO_2 in which the levels remained static for 20 days. No fish tested exceeded the generally accepted levels for freshness.

Stenstrom (1985) studied the effect of MAP on day-old cod *(Gadus morhua)* fillets. The gaseous mixtures tested included air; 48% CO_2/52% O_2; 48% CO_2/52% :N_2; 90% CO_2/10% N_2 or 100% CO_2 and each treatment was held at 35.6°F (2°C) for up to 34 days. Total bacterial counts of 10^6 cfu/g were used to determine spoilage and, as anticipated, growth was inhibited at higher concentrations of CO_2. Data indicated that a 10^6 cfu/g population was attained under air, by 3 days; 48% CO_2 plus 52% O_2, 19 days; 48% CO_2 plus 52% N_2, 26 days; 90% CO_2 plus 10% O_2, 34 days; 90% CO_2 plus 10% N_2, 34 days; and 100% CO_2, >34 days. The initial flora were composed of *Flavobacterium* spp. (28%), coryneforms (25%), *Alteromonas putrefaciens* (12%), *Pseudomonas* spp. (non-flourescent, 10%), and *Acinetobacter calcoaceticus* and *Micrococcus* spp. (5% each). In air storage *A. putrefaciens* (62%) and *Pseudomonas* spp. (28%) became dominant, *A. putrefaciens* (75%) in the 48% CO_2 plus 52% N_2 treatment, and homofermentative *Lactobacillus* spp. in the remaining treatments containing CO_2. In only one treatment, 48% CO_2 plus 52% O_2, at 6 days, were >5 cfu/g sulfite-reducing *Clostridia* detected.

Mokhele et al. (1983) investigated the effects of air and MA (80% CO_2, 20% air) storage on the numbers and types of microorganisms on fresh rock cod *(Sebastes* spp.) stored at 39.2°F (4°C). Fillets were analyzed after 0, 7, 14, and 21 days of storage and total plate counts were conducted aerobically, under MA at 39.2°F (4°C), 68°F (20°C), and 95°F (35°C) or anaerobically at 95°F (35°C). The results indicated that the air-stored fillets were spoiled (subjective) by 7 days with an APC of 10^9 cfu/g at incubation temperatures of 39.2°F (4°C) and 68°F (20°C) and 10^7 at 95°F (35°C). Incubated under MA, the air-stored fillets had counts of <10^3, 10^9, and 10^7 cfu/g at 39.2, 68, and 95°F (4, 20, and 35°C), respectively. Plates from the MAP samples incubated in either air or under MA at 68°F (20°C) had very similar counts at each time interval and reached 10^7 cfu/g by 21 days. This indicates that the total numbers would not differ significantly between aerobic and MA incubation of the experimental astmosphere. Anaerobic counts of the air-held fillets reached 10^8 by 14 days; however, MA fillets lagged 3 log cycles behind. The greatest variety of bacteria were isolated from air-stored samples incubated in air at 68°F (20°C) and consisted primarily of *Acinetobacter, Pseudomonas spp., Moraxella spp.,* and *Micrococcus spp.* The flora shifted to species of *Aeromonas* and *Enterobacter* as isolates from the air-held samples were incubated under MA. Eighty percent of the anaerobic isolates were of the *Enterobacter* species. MA isolates incubated in air consisted of *Aeromonas*-like and *Lactobacillus* spp. and those incubated under MA or anaerobic conditions consisted of only *Lactobacillus* spp. It is noteworthy that no fillets tested positive for *Vibrio parahemolyticus, Staphylococcus aureus,* or *Clostridium botulinum* Type E. This study indicates that the media and environment under which isolates are incubated result in a more reliable indicator of the dominating flora.

CLOSTRIDIUM BOTULINUM

Stier et al. (1981) investigated the effect of MAP on fresh king salmon *(Onchorynchus tshawytscha)* skin-on fillets cut into 5-cm^2 blocks and inoculated with Types A and B or E spores (0.1 ml, 3×10^4) and Type E vegetative cells (0.1 ml, 4×10^5) were inoculated onto the surfaces of squares and sandwich packed fillets. The O_2-impermeable packages were flushed with a mixture of 60% CO_2, 25% O_2, and 15% N_2 and stored at either 71.9°F (22.2°C) or 39.9°F (4.4°C). Samples were evaluated for total aerobic Gram-negative (natural spoilage flora) and Gram-positive counts (to evaluate for a shift in microbial flora), toxin production, product appearance, honeycombing, and slime and odor development. The results indicated that all inoculated samples stored at 71.9°F (22.2°C) became toxic within 3 days, the majority by 2 days. Samples stored in either air or under MA at 39.9°F (4.4°C) failed to develop toxin by 57 days although it is considered undesirable to store such MAP foods at temperatures greater than 37.9°F (3.3°C).

Eklund (1982) inoculated salmon with *C. botulinum* Type E spores (10^2, 10^3, or 10^4 per gram) and evaluated toxin production after incubation at either 41°F (5°C) or 50°F (10°C) in air, 60% CO_2, or 90% CO_2. In air, as few as 10^2 spores/g resulted in toxin production; however, 10^3–10^4 spores/g were required under MAP for the same reaction at 50°F (10°C). The air-held samples were considered spoiled at 4 days at 50°F (10°C) but spoilage in the MAP samples was delayed to 10 days. The net result was that MAP inhibited spoilage and masked sensory defects that would normally cause consumer rejection of toxic product. In an accompanying study (Eklund 1982), iced salmon were stored in air for 6 days. When these fish were inoculated with *C. botulinum* Type E and stored at 50°F (10°C) under 60 and 90% CO_2, spoilage occurred prior to toxin production. Fain (1982) conducted similar inoculated pack studies (10^2–10^4 spores/g) and incubated samples under vacuum, 70% CO_2 or 100% CO_2 at either 41°F (5°C) or 50°F (10°C). Similarly, spoilage was detected prior to toxin production.

Cann et al. (1984) evaluated the growth of *C. botulinum* Type E in surface spray and deep-inoculated (10 spores/g) packs of trout and salmon steaks. Samples were incubated at 50, 59, or 68°F (10, 15, or 20°C) under 60% CO_2/40% N_2. Trout developed toxicity between 2 and 4 days at either 59°F (15°C) or 68°F (20°C); and by 8 days at 50°F (10°C) in all deep-inoculated MAP samples. Onset of toxicity in salmon was similar to that in trout; however, toxin development was erratic and of low titer (method not defined). In smoked salmon, toxin development was minimal and occurred at 5–6 days and 10–14 days at 68°F (20°C) and 59°F (15°C), respectively. By 42 days, no toxin was detected in smoked salmon stored at 50°F (10°C).

Post et al. (1985) conducted inoculated pack studies with *C. botulinum* Type E and samples incubated under air, vacuum, N_2, CO_2, and mixtures of N_2, CO_2, and O_2 at 39.2, 46.4, and 78.8°F (4, 8, and 26°C). In concurrence with Eklund (1982), the results indicated that MAP alone was not sufficient to ensure a safe product. It was concluded that a storage temperature of <32°F (<0°C) must be recommended in conjunction with MAP storage of seafoods.

Genigeorgis' group (Garcia and Genigeorgis 1987; Garcia et al. 1987; Ikawa and Genigeorgis 1987) have investigated rockfish *(Sebastes paucispinis)* and salmon *(Oncorhynchus tshawytscha)* in factorial designs including temperature, MAP, inoculum level, and their interactions. Although some results were in agreement with Eklund (1982) and Post et al. (1985), there was generally no toxin detected in samples at 21 days when stored under either 70 or 100% CO_2 at 39.2°F (4°C) or lower prior to the onset of spoilage. The Genigeorgis group has developed a computer-generated model to ensure the safety of MAP from *C. botulinum* outgrowth with an inoculum level of one spore. This modeling is a realistic approach since normally fish flesh would be expected to harbor 1–10 spores/g (Lindsay 1982).

C. BOTULINUM AND VACUUM PACKAGING

Vacuum packaging of seafoods has been controversial since the absence of oxygen and low temperature growth of *C. botulinum* Type E may present a health hazard if not appropriately heat processed after storage and if the product has been abusively handled. The basic concern relates to the accumulation of Type E toxin in the absence of obvious spoilage odors.

Post et al. (1985) inoculated the flesh of cod, whiting, and flounder fillets with 50 spores of *C. botulinum* Type E and packaged each treatment sample under air, vacuum, and mixtures of either 90% CO_2/8% N_2/2% O_2 or 65% CO_2/31% N_2/4% O_2. Samples were incubated at 39.2, 46.4, 53.6, or 78.8°F (4, 8, 12, 26°C) or in cycles of high followed by low temperature storage until there was either definite tissue breakdown and a putrid or trash-like odor (one grade beyond normal consumer rejection), at the point of normal consumer rejection, or Type E toxin was detected.

At both 78.8°F (26°C) and 53.6°F (12°C) all treatments of cod, whiting, and flounder were rejected under normal standards either before or concurrent with toxin detection except for cod stored under 65% CO_2/31% N_2/4% O_2 [78.8°F (26°C)], whiting stored under 90% CO_2/8% N_2/2% O_2 [78.8°F (26°C)], and cod stored under N_2 [53.6°F (12°C)]. At 39.2°F (4°C) storage, all flounder samples were rejected prior to detection of toxin; however, cod and whiting stored under MAP and cod under 100% CO_2 showed detectable toxin between 4 and 9 days before consumer rejection. Of particular interest, cod stored under 100% CO_2 showed toxin between 18 and 21 days but consumer rejection did not occur until 40–53 days. The results of Post et al. (1985) clearly indicated that the production of toxin occurs at different rates under varied holding conditions depending on the species. This provides evidence that MAP could not be given blanket acceptance by the FDA for all seafoods.

Cann et al. (1965) studied the effect of inoculum load, storage temperature, and irradiation on toxin development in a variety of vacuum packaged seafoods. The first trial involved inoculating spores and vegetative cells of *C. botulinum* Type E (5×10^7) into vacuum-packaged herring, kippered herring, smoked haddock, and scallops and incubated the samples at 68°F (20°C) and 50°F (10°C) [except scallops only 68°F (20°C)]. Toxin was detected at 2, 3, 4, and 6 days in herring, kippers, and smoked haddock, respectively at 68°F (20°C). At 50°F (10°C), toxin detection in duplicate samples occurred in herring (12 days) and kippers (23 days) and in only one sample of smoked haddock by 28 days. This demonstrated the net lag in toxin production by reduced temperature.

A second trial involved the mix effects of inoculum (10^1–10^6) × storage temperature 41, 50, or 68°F (5, 10, or 20°C) in vacuum-packaged herring. At 68°F (20°C), toxin was detected at 11 days at the 10^1/ml suspension; at 7 days for suspensions ranging from 10^2 to 10^5/ml, and at 4 days at 10^6/ml inoculum load. At 50°F (10°C) storage, toxin was detected in one sample at 7 to 15 days but not through the duration of the study in the 10^1/ml load; toxin production was delayed to 15 days at the 10^2/ml level and to 11 days at the 10^3–10^5/ml treatments. At 41°F (5°C), no toxin was detected by 36 days at the 10^1/ml level; in only one sample by 15 days at the 10^2 level and in only one sample by 32 days at the 10^3–10^5/ml levels. These results clearly demonstrate the relationship between the dose and storage temperature. Although the 41°F (5°C) storage temperature generally indicated a measure of safety at a low *C. botulinum* Type E load, recommendations for safety must be based on the assumption that the initial load is much higher.

In the third trial, herring, cod, and plaice were vacuum packaged and inoculated with *C. botulinum* Type E at levels of 10^4, 10^5, and 10^6/ml and irradiated at 0.3 Mrad with a Co^{60} source prior to incubation at 50°F (10°C). Control packs were inoculated with spores at the 10^1, 10^2, and 10^3/ml levels and incubated at 50°F (10°C). The results indicated that 0.1% of the spores survived irradiation and toxin was detected by 17 days. Nonirradiated plaice showed no detectable toxin by 28 days; nonirradiated cod showed toxin in only one sample inoculated at 10^3 by 10 days and nonirradiated herring showed toxin at 24 days in only one sample of the 10^1 and 10^2 inoculum loads and was confirmed at the 10^3 load at 24 days. These results indicate that elimination of competitive flora stimulates sporulation of *C. botulinum* Type E in irradiated fish.

Bremner and Statham (1983a) evaluated the effect of competitive microflora in vacuum-packaged scallops with added lactobacilli. Fresh scallops *(Pecten alba)* were shucked and packaged under vacuum, aerobically, or were inoculated with *Lactobacillus plantarum* (8.7 × 10^5 cfu/g) and vacuum packaged. All treatment samples were stored at 39.2°F (4°C). The results indicated that the spoilage pattern of vacuum-packaged scallops with or without lactobacilli spoiled at similar rates. It is interesting to note that the spoilage flora shifted to *Vibrio* spp. (95–97.5%) by 15 days in both vacuum-stored treatments. In air, *Alteromonas* constituted 92.5% of the genera and *Vibrio* 7.5%. It appeared that *Vibrio,* unlike *Alteromonas,* were tolerant to the antibacterial effects of the lactic organisms.

Bremner and Statham (1983b) extended their research efforts to investigate the effects of 0.1% potassium sorbate in conjunction with vacuum packaging. The results indicated that the aerobically held and vacuum packaged scallops were unacceptable by 6 days and had counts of 10^8 cfu/g and 10^7 cfu/g, respectively, and were primarily *Vibrio.* The vacuum-packaged samples had counts of only 10^3 cfu/g by 6 days and reached 10^5 cfu/g by 22 days. The microflora of the sorbate treated scallops shifted to *Vibrio* (25%), *Aeromonas* (25%), *Pseudomonads* (35%), *Acinetobacter* (5%), and enterics (10%). The sorbate-treated samples were judged as acceptable as frozen controls after 28 days. There were no assays for *C. botulinum* Type E growth or toxin, although sorbate is a known inhibitor of this organism.

Miller and Brown (1984) evaluated the effectiveness of potassium sorbate plus chlortetracycline (CTC), ethylenediamine tetraacetic acid (EDTA) plus CTC, and a control dip of distilled, deionized water (DDW) followed by vacuum packaging and storage at 35.6°F (2°C) on the stability of rockfish *(Sebastes* spp.) fillets. The control counts reached 10^8 cfu/cm^2 by 7 days whereas the CTC/ EDTA treatment sample was 1 log unit lower. By 14 days, the sorbate/CTC treatment remained significantly ($p < 0.05$) less than the other treatments. Selectivity for CTC to inhibit TMAO reducing organisms was demonstrated by the slow development of TMA. Suppression of *C. botulinum* Type E was presumed to occur in both CTC treatments due to the oxidation–reduction potential (Eh) of +197 (sorbate plus CTC) and +139 mV (EDTA plus CTC) as compared with +1.0 mV in the control (DDW dip and vacuum packaging) by 21 days.

Eklund et al. (1988), however, investigated the effectiveness of pasteurization on hot smoked salmon spiked with *C. botulinum* Types B and E. Smoked salmon was injected with 10^6 spores followed by pasteurization at 185, 192, and 197.9°F (85, 88.9, 92.2°C) for 85, 65, and 55 minutes, respectively. No Type E toxin was detected after 21 days at 77°F (25°C). Longer times (175, 85, and 65 min) were required at 185, 192, and 197.9°F (85, 88.9, and 92.2°C) to prevent the toxin production form nonproteolytic Type B. For prevention of toxin production during 150-day storage at 77°F (25°C); a pasteurization of 197.9°F

(92.2°C) for 65 minutes was indicated. Since this process could not be guaranteed to inhibit heat-resistant *C. botulinum* Group I and other spore-formers, the authors stated that these pasteurized products should be labelled like smoked fishery products and that storage below 37.9°F (3.3°C) is recommended.

To date, in the U.S., vacuum packaging of refrigerated seafoods is indicated only for hot smoked products with a low A_w and when held at less than 37.9°F (3.3°C). Frozen, vacuum-packaged seafoods do have superior shelf life (Josephson et al. 1985; Yu et al. 1973); and this is the second accepted practice.

Many results associating toxin production under MAP have been conducted by mouse lethality testing. Solberg et al. (1985) investigated the effect of toxic reactions to mice mediated by inherent seafood microflora. Fresh cod (not specified) were stored under 90% CO_2/8% N_2/2% O_2 at 46.4°F (8°C) for 8 days. The predominant microflora were identified and individual endotoxins tested by mouse bioassay. It was determined that Gram-negative endotoxins from *Pseudomonas* spp. resulted in effects ranging from temporary respiratory distress to eye damage to rapid death. Although lactics often become the predominant microflora in MAP systems, Oberlender et al. (1983), for example, demonstrated that 11% of the microflora isolated from swordfish incubated under 100% CO_2 at 35.6°F (2°C) after 22 days were *Pseudomonas* spp. The results of Solberg et al. (1985) indicated that false-positive reactions attributed to *C. botulinum* toxins could be nullified by treatment of MAP assay samples with bovine serum. This may, in part, explain why there have been some erratic and, perhaps, false-positive test results.

MAP STORAGE FOLLOWED BY AIR HOLDING

Wang and Ogrydziak (1986) studied the effects of MAP (80% CO_2/20% air) or air alone on Pacific rockcod (*Sebastes* spp.) at 39.2°F (4°C). A third treatment involved transfer of the MAP samples to air storage for an additional 0–6 days. Although, it was determined that air treatments were spoiled at 7 days, the MAP samples were considered acceptable at 14 days. Transfer of samples from MAP to air storage at either 7 or 14 days resulted in logarithmic bacterial growth which corresponded with a rapid decline of tissue CO_2 concentration. It was again confirmed that pH remained static during MAP while it increased under air storage. Initially, the dominant organisms were *Pseudomonas* spp. (27%), *Acinetobacter* spp. (15%), and *Micrococcus* spp. (14%). *Alteromonas* and *Lactobacillus* accounted for only 6% and 4%, respectively, of the initial flora. After 7 days of storage, *Pseudomonads* composed 60% of the population and the *Alteromonas* and *Lactobacillus* had decreased to 2% and 1%, respectively, of the flora. Spoilage of these fillets was evident. After 4 and 14 days of MAP storage, the *Alteromonas* isolates increased to 23 and 30%; *Lactobacillus* increased to 24 and 36%; *Pseudomonas* had decreased to 10 and 0%; and *Brochothrix thermo-sphacta* remained constant at 11%, respectively. Transfer of the MAP samples to air storage after either 7 or 14 days resulted in an increase of *Pseudomonas; Alteromonas* and *Lactobacillus* decreased while the APC increased to 10^9 cfu/g within 6 days of the environmental change.

MAP IN COMBINATION WITH OTHER PRESERVATIVES

Fey and Regenstein (1982) evaluated the effects of MAP on red hake, chinook, and sockeye salmon. Three-day-old gutted fish were stored at 32–33.8°F (0–1°C) or 42.8°F (6°C) (abusive storage) either in bulk (red hake) or individually (salmon) at ratios of 20:1 or 7:1 gas to fish volumes, respectively. The gas mixtures consisted of air (control) or mixtures of CO_2, O_2, and N_2. The sockeye salmon were evaluated by a factorial design to investigate the mixed effects of MAP (CO_2, O_2, and CO), an antioxidant dip (0.2% wt/vol sodium erythorbate; 0.2% wt/vol citric acid; 0.5% wt/vol sodium chloride solution), and 1% wt/vol potassium sorbate ice.

The results indicated that in the red hake control, the standard plate counts at 5 days were 5×10^3 and rose to 1×10^8 cfu/g by 26 days. For the same time periods under 20% CO_2/21% O_2/59% N_2 (treatment 2), the standard plate counts were 1×10^2 and 6×10^5 cfu/g; for 60% CO_2/21% O_2/19% N_2 (treatment 3) they were 2×10^3 and 2×10^2 cfu/g; and under 60% CO_2/5% O_2/35% N_2 (treatment 4) they were 3×10^2 and 2×10^2 cfu/g, respectively. Psychrotrophic counts over the same time period for the control rose from 5×10^3 to 1×10^8 cfu/g between 5 and 26 days and 1×10^2 to 1×10^6 cfu/g (treatment 2); from 1×10^3 to 8×10^4 cfu/g (treatment 3); and 1×10^2 to 1×10^4 cfu/g (treatment 4). Psychrotrophic counts of chinook salmon (individually packaged) at 23 days were 4 log units greater than those of red hake (bulk packaged). In the factorial evaluation of gas mixtures, antioxidant dip, and potassium sorbate ice, the psychrotrophic counts were held to 10^5 by 17 days in the presence of 60% CO_2/5% O_2/1% CO/34% N_2 plus antioxidant and ice containing 1% potassium sorbate. The remaining experimental treatments had psychrotrophic counts of 10^6–10^7 cfu/g and the control 10^{10} cfu/g in the same time period.

The TMA values for red hake were lowest after 27 days in the presence of reduced O_2 (1,260 μmol TMA/100 g wet fish versus 1,530–1,580 μmol TMA/100 g wet fish) for the remaining treatments. Chinook salmon stored in 1% potassium sorbate ice and MAP was held to 80 μmol by 25 days or about one-fourth that of the next lowest value and one-tenth than of salmon stored at the abusive temperature, 42.8°F (6°C).

Sockeye salmon were treated with 60% CO_2/5% O_2/34–35% N_2/0–1% CO and with or without ice, 1% potassium sorbate ice, or an antioxidant dip. The log psychrotrophic counts after 17 days ranged from 1×10^5 to 5×10^7 cfu/g. Appearance of those fish treated with both the antioxidant dip and potassium sorbate ice was judged most desirable and by sensory evaluation (no CO) and had the greatest score for overall acceptability although not significantly different from the remaining treatments. For red hake and chinook salmon, the combination of 1% potassium sorbate ice and MAP resulted in the highest acceptability values. The combination of two bacteriostatic agents had a net result of reducing odor development, TMA, and maintaining flavor albeit panelists may have determined desirability by worse-case scenario comparisons.

Finne (1982) determined that sorbate did not enhance the preservation effects of MAP when applied to red hake. This conclusion was supported by Barnett et al. (1987) in studies of trout treated with a 2.3% potassium sorbate dip and held under MAP.

CARBONIC ACID DIPS AND MAP

As previously indicated, one of the proposed mechanisms attributed to the effectiveness of CO_2 MAP storage is the hydration of CO_2 in tissue fluids to undissociated carbonic acid (Barnett et al. 1971; Daniels et al. 1985; Golding 1945; Mitsuda et al. 1980; Ogrydziak and Brown 1982; Parkin and Brown 1982; Sears and Eisenberg 1961). In order to evaluate carbonic acid dips, Daniels et al. (1986) used fresh, skinless cod *(Gadus morhua)* fillets. Fillets were either dipped in (1) a carbonic acid (pH 4.6) bath for 5–10 minutes; or (2) distilled water at 35.6°F (2°C); or (3) were untreated. A final study was used to compare a 98% CO_2 atmosphere with the carbonic acid dip. Each of the three treatments were packed in either a high or low oxygen permeable retail tray or were stored in 4.5-kg lots (wholesale pack) at 35.6°F (2°C). The results indicated that bacterial growth was suppressed in the high permeability retail packs (10^6 cfu/cm^2) by 24 days while the wholesale pack had 10^8 cfu/cm^2 in the same time period. In comparison, the low permeability film resulted in significantly lowest 10^4 cfu/cm^2 growth ($p < 0.05$) by 24 days. Although water dipping resulted in a one log reduction of microbial growth, the carbonic acid dip resulted in a 10–12 day extension of shelf life. Storage of fillets in either a 98% CO_2 MAP or after a carbonic acid dip demonstrated a similar lag in logarithmic phase microbial growth. This indicated that the carbonic acid dip could be more immediately bacteriostatic than reliance on the relatively slow hydration of CO_2 in situ. Headspace analysis of package gases indicated that in the low oxygen permeable film, 25% CO_2 was rapidly achieved whereas only 6% CO_2 developed in the high permeability film. Each high and low permeability package contained 1–3% O_2. This may indicate that a low oxygen environment conducive to the outgrowth of *C. botulinum* could develop and this area requires further research.

BIOCHEMICAL INDICATORS

Although results of some sensory evaluation panels have indicated that an undesirable, tough texture may develop during MAP; after 14 days storage no significant difference in salt extractable proteins was seen in rockcod (*Sebastes* spp.). Nonetheless, Morey et al. (1982) determined that rockcod fillets stored under 80% CO_2/20% O_2 at 39.2°F (4°C) had 10% less protein digestibility when stored in air than MAP stored fillets. The C-PER (computerized-protein efficiency ratio) was similar in both treatments after storage.

Lindsay et al. (1986) reviewed chemical and biochemical indices to assess the quality of MAP fish. It was concluded that oxidized sulfur compounds developed at low levels in MAP stored fish and their presence in aerobically stored fish would be attributable to the dominating microflora. Reduced-sulfur volatiles (H_2S and CH_3SH) would, however, be possible indices to assess the later shelf life of air-held and MAP samples. Alcohols such as 1-propanol have been detected in MAP samples and are presumed to be metabolites of the altered array of microflora. It was also reported that evaluation of hypoxanthine in MAP stored fish has met with limited success. TMA and ammonia are similarly poor indicators since their development parallels microbial growth. More research in the area of biochemical indicators is needed.

SUMMARY AND CONCLUSIONS

Temperature control is probably the most critical consideration in MAP. Since it is known that *C. botulinum* can grow at temperatures exceeding 37.9°F (3.3°C), recommendations are to store MAP products at temperatures <33.8°F (<1°C). Abusive handling for even a few hours can result in the outgrowth of *C. botulinum* spores and for this reason strict temperature control must be followed.

Since bacterial growth is inhibited during MAP and there is a shift in characteristic spoilage microflora, odors normally associated with spoilage do not usually develop and may create a false confidence in product safety. More research into the biochemical indicators is needed to develop criteria for acceptability of MAP seafoods. In addition, such criteria may differ from one species to another since aerobic shelf life and spoilage patterns can vary widely.

Although the U.S. Food and Drug Administration does not currently endorse modified atmosphere storage of seafoods, Cann (1984) has indicated that MAP is a safe packaging method that is accepted in some European countries. It is important to emphasize that accurate temperature control is essential. MAP cannot be condoned as a technology to extend shelf life for marginally acceptable product.

Specifically, Cann (1984) has recommended that for white fish, scampi, shrimp, and scallops a 40% CO_2/30% N_2/30% O_2 environment provides the best results. For the fatty fish, e.g., salmon, trout, herring, mackerel and smoked fish products, a 60% CO_2/40% N_2 atmosphere is most desirable. If a greening of smoked fish occurs, which may depend on the strength of the smoke cure, the atmosphere recommended for leaner species is advised. It is indicated that white fish fillets packed 3:1 gas to fillet will keep up to 50% longer than when either unwrapped or under vacuum. Raw, shell-on scampi and shrimp keep up to 30% longer at 32°F (0°C) in MAP than other types of pack; onset of black spot may also be inhibited.

ACKNOWLEDGMENT

This work was funded in part by the Agricultural Experiment Station, College of Agriculture and Life Sciences, Virginia Polytechnic Institute State University, Blacksburg, Virginia.

REFERENCES

Aickin, C. C., and Thomas, R. C. 1975. Micro-electrode measurement of the internal pH of crab muscle fiber. *J. Physiol.* **52**:803–807.

Barnett, H. J., Nelson, R. W., Hunter, P. J., and Groninger, H. 1971. Studies on the use of carbon dioxide in refrigerated brine for preservation of whole fish. *Fish. Bull.* **69**:433–441.

Barnett, H. J., Conrad, J. W., and Nelson, R. W. 1987. Use of laminated high and low density polyethylene flexible packaging to store trout *(Salmo gairdneri)* in a modified atmosphere. *J. Food Protect.* **50**:645–651.

Becker, Z. E. 1933. A comparison between the action of carbonic acid and other acids upon the living cell. *Protoplasma* **25**:161–175.

Bremner, H. A. and Statham, J. A. 1983a. Spoilage of vacuum-packaged chill-stored scallops with added lactobacilli. *Food Technol. Aust.* **35**:284–287.

Bremner, H. A. and Statham, J. A. 1983b. Effect of potassium sorbate on refrigerated storage of vacuum packaged scallops. *J. Food Sci.* **48**:1042–1047.

Brinkman, R., Margaria, R., and Roughton, F. J. W. 1933. The kinetics of the carbon dioxide-carbonic acid reaction. *Proc. R. Soc. Lon.* **232**:65–73.

Brown, W. D., Albright, M., Watts, D. A., Heyer, B., Spruce, B., and Price, R. J. 1980. Modified atmosphere storage of rockfish *(Sebastes miniatus)* and silver salmon *(Oncorhynchus kisutch)*. *J. Food Sci.* **45**:93–96.

Callow, E. H. 1932. Gas storage of pork and bacon. Part I—Preliminary experiments. *J. Soc. Chem. Indust.* **51**:116T–119T.

Cann, D. C. 1984. Packing fish in a modified atmosphere. Torry Advisory Note No. 88.

Cann, D. C., Wilson, B. B., Hobbs, G., and Shewan, J. M. 1965. The growth and toxin production of *Clostridium botulinum* Type E in certain vacuum packed fish. *J. Appl. Bacteriol.* **28**:3, 431–436.

Cann, D. C., Houston, N. C., Taylor, L. Y., Smith, G. L., Thomas, A. B., and Craig, A. 1984. Studies of Salmonids packed and stored under a modified atmosphere. Torry Research Station. Ministry of Agriculture, Fisheries and Food. Aberdeen, Scotland, April, 1984, 53 pp.

Castelli, A., Littaru, G. P., and Barbesi, G. 1969. Effect of pH and CO_2 concentration changes on lipids and fatty acids of *Saccharomyces cerevisiae*. *Arch. Mikrobiol.* **66**:34–39.

Coyne, F. P. 1932. The effect of carbon dioxide on bacterial growth with special reference to the preservation of fish. I. *J. Soc. Chem. Indust.* **51**:119T–121T.

Coyne, F. P. 1933. The effect of carbon dioxide on bacterial growth with special reference to the preservation of fish. II. *J. Soc. Chem. Indust.* **52**:19T–24T.

Daniels, J. A., Krishnamurthi, R., and Rizvi, S. S. H. 1985. A review of effects of carbon dioxide on microbial growth and food quality. *J. Food Protect.* **48**:532–537.

Daniels, J. A., Krishnamurthi, R., and Rizvi, S. S. H. 1986. Effects of carbonic acid dips and packaging films on the shelflife of fresh fish fillets. *J. Food Sci.* **51**:929–931.

Delente, J., Akin, C., Krabbe, E., and Ladenburg, K. 1969. Fluid dynamics of anaerobic fermentation. *Biotechnol. and Bioeng.* **11**:631–646.

Eklund, M. W., Peterson, M. E., Paranjpye, R., and Pelroy, G. A. 1988. Feasibility of a heat-pasteurization process for the inactivation of nonproteolytic *Clostridium botulinum* Types B and E in vacuum-packaged, hot-process (smoked) fish. *J. Food Protect.* **51**:9, 720–726.

Eklund, M. W. 1982. Effect of CO_2-modified atmospheres and vacuum packaging on *Clostridium botulinum* and spoilage organisms of fishery products. In: *Proceedings of the First National Conference on Seafood Packaging and Shipping*, November 15–17, 1982, Washington, DC, pp. 298–331.

Elsden, S. R. 1938. The effect of carbon dioxide on the production of succinic acid by *Bact. coli commune*. *Biochem. J.* **32**:187–193.

Fain, A. R. 1982. *C. botulinum* in fresh fish, incidence and packaging studies. In: *Proceedings of the First National Conference on Seafood Packaging and Shipping*, November 15–17, 1982, Washington, DC, pp. 359–373.

Fey, M. S., and Regenstein, J. 1982. Extending shelf-life of fresh wet red hake and salmon using CO_2-O_2 modified atmosphere and potassium sorbate ice at 1°C. *J. Food Sci.* **47**:1048–1054.

Finne, G. 1982. Research update on controlled and modified atmosphere packaging. In: *Proceedings of the First National Conference on Seafood Packaging and Shipping*, November 15–17, 1982, Washington DC, pp. 374–394.

Foster, J. W. and Davis, J. B. 1949. Carbon dioxide inhibition of anaerobic fumarate formation in molds *Rhizopus nigricans*. *Arch. Biochem.* **21**:135–142.

Fox, D. L. 1981. Presumed carbaminoprotein equilibria and free energy exchanges in reversible carbon dioxide narcosis of protoplasm. *J. Theoret. Biol.* **90**: 441–443.

Garcia, G., and Genigeorgis, C. 1987. Quantitative evaluation of *Clostridium botulinum* nonproteolytic types B, E, and F growth risk in fresh salmon tissue homogenates stored under modified atmospheres. *J. Food Protect.* **50**:390–397.

Garcia, G., Genigeorgis, C., and Lindroth, S. 1987. Risk of growth and toxin production by *Clostridium botulinum* nonproteolytic types B, E, and F in salmon fillets stored under modified atmospheres at low and abused temperatures. *J. Food Protect.* **50**:330–336.

Gill, C. O., and Tan, K. H. 1979. Effect of carbon dioxide on the growth of *Pseudomonas flourescens*. *Appl. Environ. Microbiol.* **38**:237–240.

Golding, N. S. 1945. The gas requirements of molds. IV. A preliminary interpretation of the growth rates of four common mold cultures on the basis of absorbed gases. *J. Dairy Sci.* **28**:737–750.

Gray, R. J. H., Hoover, D. G., and Muir, A. M. 1983. Attenuation of microbial growth on modified atmosphere-packaged fish. *J. Food Protect.* **46**:610–613.

Haard, N. F., and Lee, Y. Z. 1982. Hypobaric storage of Altantic salmon in a carbon dioxide atmosphere. *Can. Inst. Food Sci. Technol. J.* **15**:68–71.

Ikawa, J. Y., and Genigeorgis, C. 1987. Probability of growth and toxin production by nonproteolytic *Clostridium botulinum* in rockfish fillets stored under modified atmospheres. *Int. J. Food Microbiol.* **4**:167–181.

Josephson, D. B., Lindsay, R. C., and Stuiber, D. A. 1985. Effect of handling and packaging on the quality of frozen whitefish. *J. Food Sci.* **50**:1–4.

King, A. D., and Nagel, C. W. 1967. Growth inhibition of a Pseudomonas by carbon dioxide. *J. Food Sci.* **32**:575–579.

King, A. D., and Nagel, C. W. 1975. Influence of carbon dioxide upon the metabolism of *Pseudomonas aereuginosa. J. Food Sci.* **40**:362–366.

Knoche, W. 1980. Chemical reactions of CO_2 in water. In: Bauer, C., Gros, G., and Bartels, H. (eds.) *Biophysics and Physiology of Carbon Dioxide* Springer-Verlag, Berlin, pp. 3–11.

Krebs, H. A., and Roughton, R. J. W. 1948. Carbonic anhydrase as a tool in studying the mechanisms of reactions involving H_2CO3, CO_2 or HCO_3. *Biochem. J.* **43**:550–555.

Krough, A. 1919. The rate of diffusion of gases through animal tissues with some remarks on the coefficient of invasion. *J. Physiol.* **52**:391.

Lannelongue, M., Finne, G., Hanna, M. O., Nickelson, R., and Vanderzant, G. 1982. Storage characreristics of brown shrimp *(Penaeus aztecus)* stored in retail packages containing CO_2-enriched atmospheres. *J. Food Sci.* **47**:911–913, 923.

Ledward, D. A. 1970. Metmyoglobin formation on beef stored in carbon dioxide enriched and oxygen depleted atmospheres. *J. Food Sci.* **35**:33–37.

Lindsay, R. C. 1982. Sorbate and botulism safety aspects for modified atmosphere packaging. In: *Proceedings of the First National Conference on Seafood Packaging and Shipping*. Washington, DC, November 15–17, 1982, pp. 332–358.

Lindsay, R. C., Josephson, D. B., and Olafsdottir, G. 1986. Chemical and biochemical indices for assessing the quality of fish packaged in controlled atmospheres. In: Kramer, D. E., and Liston, T. (eds.) *Seafood Quality Determination,* Elsevier Science Publishers, New York, pp. 221–234.

Miller, S. A., and Brown, W. D. 1984. Effectiveness of chlortetracycline in combination with potassium sorbate or tetrasodium ethylenediaminetetraacetate for preservation of vacuum packed rockfish fillets. *J. Food Sci.* **49**:188–191.

Mitsuda, H. K., Nakajima, K., Mizuno, H., and Kawai, F. 1980. Use of sodium chloride for extending the shelflife of fish fillets. *J. Food Sci.* **45**:661–666.

Mitz, M. A. 1957. The stability of proteins in the presence of carbon dioxide. *Biochim. Biophys. Acta* **25**:426.

Mitz, M. A. 1978. Carbon dioxide as a reagent for proteins. *Enzyme Engin.* **3**:235–239.

Mokhele, K., Johnson, A. R., Barrett, E., and Ogrydziak, D. M. 1983. Microbiological analysis or rock cod (*Sebastes* spp.) stored under elevated carbon dioxide atmospheres. *Appl. Environ. Microbiol.* **45**:878–883.

Molin, G., and Stenstrom, I. M. 1984. Effect of temperature on the microbial flora of herring fillets stored in air or carbon dioxide. *J. Appl. Bacteriol.* **56**:275–282.

Molin, G., Stenstrom, I., and Ternstrom, A. 1983. The microbial flora of herring fillets after storage in carbon dioxide, nitrogen or air at 2°C. *J. Appl. Bacteriol.* **55**:49–56.

Morey, K. S., Satterlee, L. D., and Brown, W. D. 1982. Protein quality of fish in modified atmospheres as predicted by the C-PER assay. *J. Food Sci.* **47**:1399–1400, 1409.

Oberlender, V., Hanna, M. O., Miget, R., Vanderzant, C., and Finne, G. 1983. Storage characteristics of fresh swordfish steaks stored in carbon dioxide-enriched controlled (flow-through) atmospheres. *J. Food Protect.* **46**:434–440.

Ogrydziak, D. M., and Brown, W. D. 1982. Temperature effects in modified-atmosphere storage of seafoods. *Food Technol.* **36**:86–96.

Parkin, K. L., and Brown, W. D. 1982. Preservation of Seafood with modified atmospheres. In: Martin, R. E., Flick, G. J., Hebard, C. E., and Ward, D. R. (eds.) *Chemistry and Biochemistry of Marine Food Products.* AVI Publishing, Westport, Connecticut, pp. 453–465.

Post, L. S., Lee, D. A., Solberg, M., Furgang, D., Specchio, J., and Graham, C. 1985. Development of botulinal toxin and sensory deterioration during storage of vacuum and modified atmosphere packaged fish fillets. *J. Food Sci.* **50**:990–996.

Price, R. J., and Lee, J. S. 1970. Inhibition of *Pseudomonas* species by hydrogen peroxide producing lactobacilli. *J. Milk Food Technol.* **33,** 13–18.

Quinn, E. L., and Jones, C. L. 1936. *Carbon Dioxide.* Reinhold, New York.

Reppond, K. D., and Collins, J. 1983. Pacific cod *(Gadus macrocephalus):* change in sensory and chemical properties when held in ice and in CO_2 modified refrigerated seawater. *J. Food Sci.* **48**:1552–1553.

Reppond, K. D., Collins, J., and Markey, D. 1985. Walleye pollock *(Theragra chalcogramma):* changes in quality when held in ice, slush-ice, refrigerated seawater, and CO_2-modified refrigerated seawater then stored as blocks of fillets at −18°C. *J. Food Sci.* **50**:985–989, 996.

Richter, E. R., and Banwart, G. J. 1983. Microbiological and sensory evaluation of fresh fish packaged in carbon dioxide for retail outlets in the midwest. *J. Food Protect.* **46**:245–247.

Scott, D. N., Fletcher, G. C., and Hogg, M. G. 1986. Storage of snapper fillets in modified atmospheres at −1°C. *Food Technol. Aust.* **38**:234–238.

Sears, D. F., and Eisenberg, R. M. 1961. A model representing a physiological role of carbon dioxide at the cell membrane. *J. Gen. Physiol.* **44**:869–887.

Silliker, J. H., Woodruff, R. E., Lugg, J. R., Wolfe, S. K., and Brown, W. D. 1977. Preservation of refrigerated meats with controlled atmospheres: treatment and post-treatment effects of carbon dioxide on pork and beef. *Meat Sci.* **1**:195–204.

Smikahl, J. R. 1985. Action of carbon dioxide on myoglobin and mitochondrial enzymes in beef heart extracts. M. S. Thesis. University of California, Davis.

Solberg, M., Post, L. S., Furgang, D., and Graham, C. 1985. Bovine serum eliminates rapid nonspecific toxic reactions during bioassay of stored fish for *Clostridium botulinum* toxin. *Appl. Environ. Microbiol.* **49**:644–649.

Stansby, M. E. 1963. Analytical methods. In: Stansby, M. E. (ed.) *Industrial Fishery Technology.* Robert E. Krieger, Huntington, New York, pp. 363–373.

Stenstrom, I. 1985. Microbial flora of cod fillets *(Gadus morhua)* stored at 2°C in different mixtures of carbon dioxide and nitrogen/oxygen. *J. Food Protect.* **48**:585–589.

Stier, R. F., Bell, L., Ito, K. A., Shafer, B. D., Brown, D. A., Seeger, M. L., Allen, B. H., Porcuna, M. N., and Lerke, P. A. 1981. Effect of modified atmosphere storage on *C. botulinum* toxigenesis and the spoilage microflora of salmon fillets. *J. Food Sci.* **46**:1639–1642.

Villemure, G. and Simard, R. E. 1988. Bulk storage of gutted cod *(Gadus morhua)* under carbon dioxide atmosphere in insulated polyethylene containers. *Microbiol. Aliments Nutr.* **6**:181–184.

Wang, M. Y., and Ogrydziak, D. M. 1986. Residual effect of storage in an elevated carbon dioxide atmosphere on the microbial flora of rock cod (*Sebastes* spp.). *Appl. Environ. Microbiol.* **52**:727–732.

Watts, D. A., and Brown, W. D. 1982. Histamine formation in abusively stored Pacific mackerel: effect of CO_2-modified atmosphere. *J. Food Sci.* **47**:1386–1387.

Wolf, S. K. 1980. Use of CO and CO_2 enriched atmospheres for meats, fish, and produce. *Food Technol.* **34**:55–58.

Yu, T. C., Sinnhuber, R. O., and Crawford, D. L. 1973. Effect of packaging on shelf-life of frozen silver salmon steaks. *J. Food Sci.* **38**:1197–1199.

16

Shellfish Depuration

Gary P. Richards

Pollution of the marine environment has compromised the quality and safety of many seafoods, particularly inshore molluscan shellfish. With population centers increasing along the coast, problems with oyster, clam, and mussel contamination are likely to increase. Autochthonous marine bacteria, like the vibrios, also pose a threat to humans who consume raw or undercooked shellfish.

Molluscan shellfish are a gastronomic anomaly for they are one of the few animals frequently eaten raw and alive. Cooked shellfish are intrinsically safer to eat, but cooking renders them unacceptable to some consumers. Although steaming will inactivate contaminating microorganisms, shellfish are seldom steamed for adequate periods to ensure the total inactivation of all potential human pathogens.

An alternate means of shellfish cleansing is through a process known as controlled purification, more commonly called depuration. Depuration is a dynamic process whereby shellfish are allowed to purge contaminants in tanks of clean seawater. Commercial depuration facilities offer some hope of reducing the levels of human pathogens in molluscan shellfish.

HISTORY OF DEPURATION

Shellfish-associated typhoid fever and other illnesses were attributed to the consumption of polluted shellfish as far back as the 1800s (Belding and Lane 1909; Herdman and Boyce 1899; Marvel 1902). Scientists at the turn of the century struggled with methods to eliminate sewage-borne bacteria from shellfish. Attempts to relay shellfish from polluted areas to pristine waters became less feasible as populations expanded and pollution encroached into clean areas. Early scientists examined the use of "degorgeoirs" or disgorging tanks (Herdman and Boyce 1899; Herdman and Scott 1896) which were the predecessors of current depuration tanks. Many of the early depuration studies focused on the disinfection of shellfish in chlorinated seawater (Carmelia 1921; Dodgson 1928; Wells 1916, 1928). Subsequent work on ultraviolet light (Hill et al. 1969b;

Wood 1961) and ozone (Voille 1929) disinfection systems for seawater led to more acceptable techniques which are commonly used today.

The emphasis toward depuration research and the development of commercial depuration plants resulted from a progression of shellfish-associated illnesses among molluscan shellfish consumers. As far back as 1839, oysters were banned in New York City at certain times of the year, in part because of an increased threat of cholera. Early outbreaks of shellfish-associated typhoid fever demonstrated a need for remedial measures to reduce typhoid and other illnesses. Uncertainties surrounding shellfish safety led some processors to experiment with commercial plants to purge environmental pollutants from the shellfish. Early commercial depuration facilities were built in England in 1916 (Dodgson 1936) and in the U.S. (Dodgson 1928) and other countries shortly thereafter. Perhaps the most notable facility in the U.S. is the plant at Newburyport, Massachusetts, which has depurated clams for well over 50 years. Over the years, these and other plants have proven the utility of depuration in reducing bacterial indicator organisms.

Depuration systems are now in operation in the U.S., the U.K., Australia, France, Italy, Japan, Portugal, Spain, New Zealand, and other countries. Depuration is not practiced extensively in the U.S. This is in contrast to the U.K., Australia, Spain, and other countries where depuration is required by regulation. Political, economic, and social pressures have molded a spectrum of strategies by which countries safeguard the shellfish consumer. For the above-mentioned countries, these safeguards include depuration.

There are perhaps many reasons for failure of the U.S. industry to practice depuration extensively. One reason relates to the formation of the National Shellfish Sanitation Program (Frost 1925) after an epidemic of shellfish-associated typhoid fever in the U.S. during the 1920s (Lumsden 1925). The program instituted shellfish growing water criteria based on the levels of coliform bacteria in the water. Only seawater meeting certain coliform limits would be opened for shellfish harvesting; more polluted areas would remain closed. The abundant supply of shellfish from clean water negated the need for depuration for many years. The costs associated with depuration were also a major deterrent to the full commercial utilization of the process.

New interest in shellfish depuration has developed for several reasons. Continually degrading water quality has reduced the acreage of clean and productive shellfish growing areas. Those beds remaining open are suffering from the effects of overfishing and increasing shellfish mortalities from the intrusion of shellfish pathogens. Increased reports of shellfish-borne outbreaks of gastroenteritis, particularly in recent years (Richards 1985, 1987), have focused public attention on the safety of shellfish and have stressed even further the need for depuration. Clean shellfish-growing waters have been unavailable for many years in most European countries where depuration, cooking, or importation remain the only viable options for continued shellfish commerce.

PRACTICAL APPLICATIONS OF DEPURATION

Removal of Bacteria

With the exception of vibrios, shellfish moderately contaminated with most bacterial indicators and pathogens can be adequately depurated within 72 hours. The presence of trace amounts of bacterial pathogens in depurated shellfish does not generally confer illness to consumers because threshold levels required to cause illness are seldom reached. Proper refrigeration of shellstock should be ensured both before and after depuration; storage at elevated temperatures may permit the outgrowth of bacterial pathogens and may render depurated products hazardous. Shellfish quality may also deteriorate in inadequately refrigerated shellstock and shucked products.

Even though depuration causes a reduction in total bacterial counts in shellfish tissues, studies indicate that there remains a persistent microflora for which depuration appears ineffective (Colwell and Liston 1960; Ledo et al. 1983; Son and Fleet 1980; Souness and Fleet 1979; Souness et al. 1979; Vasconcelos and Lee 1972). One-log reductions in aerobic plate counts (APCs) can be achieved readily in most instances; however, reductions to fewer than 10^4 organisms per gram are seldom seen for some shellfish species, perhaps because of the presence of an indigenous gut flora (Colwell and Liston 1960; Son and Fleet 1980; Vasconcelos and Lee 1972). Spoilage microorganisms still present in shellfish after depuration can best be managed by the use of proper refrigeration techniques.

Bacteria of the genus *Vibrio* are indigenous to the marine environment and pose new problems for the depuration plant operator. In fact, vibrios appear to persist within shellfish tissues even during extended periods of depuration. Oysters naturally contaminated with *Vibrio parahaemolyticus* showed no significant difference ($p > 0.1$) in mean counts between nondepurated and depurated oysters (Eyles and Davey 1984). Shellfish naturally contaminated with *Vibrio vulnificus* and non-01 *Vibrio cholerae* appear to depurate more slowly than laboratory-contaminated shellfish (Steslow et al. 1987). Possible mechanisms for this phenomenon are discussed in the section "Sequestration of Microbes in Shellfish," page 402. Even though laboratory-inoculated shellfish tended to depurate more rapidly than naturally contaminated shellfish in some studies, oysters required 16 days to depurate laboratory-acquired *V. vulnificus* to nondetectable levels (Kelly and Dinuzzo 1985).

Vibrio parahaemolyticus and *Vibrio harveyi* were also studied in a laboratory uptake and depuration system (Greenberg et al. 1982). Clams depurated vibrios more slowly than they depurated *E. coli*. The vibrio levels oscillated throughout the 72-hour depuration period with counts actually increasing after 6 hours of depuration. Increases in counts of both *V. parahaemolyticus* and *V. harveyi* may signify growth of these bacteria within the shellfish (Greenberg et al. 1982).

Colonization of vibrios within shellfish tissues would limit the practicality of using conventional depuration as a means of eliminating vibrios. Utilization of water disinfection procedures that are unconventional in the U.S., i.e. ozone or activated oxygen may enhance depuration effectiveness for vibrios and other persistent microorganisms. (See sections "Ozone," page 410, and "Activated Oxygen," page 411.) Like many other bacteria, *Vibrio* outgrowth in improperly refrigerated shellfish meats may lead to outbreaks of illness.

Removal of Viruses

A number of studies have been conducted on the depuration of viruses from shellfish, particularly from the Eastern oyster, *Crassostrea virginica*. Most studies are inconclusive, having been conducted before quantitative analytical methods were available for the detection and enumeration of enteric viruses in shellfish tissues. In fact, there are still no standard methods for virus extraction from shellfish. Nevertheless, there is strong evidence showing that moderate levels of poliovirus depurate within 3 days in hard shell clams *(Mercenaria mercenaria)* and soft shell clams *(Mya arenaria)* (Durgin et al. 1981; Liu et al. 1967a,b), Pacific oysters *(Crassostrea gigas)* (Hoff and Becker 1969), and Eastern oysters (Akin et al. 1966; Durgin et al. 1981; Hamblet et al. 1969; Liu 1968; Meinhold and Sobsey 1982; Mitchell et al. 1966; Powers and Collins 1989; Sobsey et al. 1987). Longer periods were required for the depuration of Manila clams *(Tapes japonica, Tapes philippinarum)* and Olympia oysters *(Ostrea lurida)* (Hoff and Becker 1969). The failure of some shellfish to pump during depuration trials resulted in high poliovirus counts in some depurated products (Metcalf et al. 1979, 1980).

Results of several studies indicate that the various enteroviruses depurate at different rates even in the same shellfish species. Although inconclusive, evidence points to rapid depuration of poliovirus and slow depuration of Coxsackievirus Type B4, hepatitis A virus, and coliphage. Further research is needed to define optimal conditions for virus depletion in depurating shellfish.

Recent studies showed a rapid reduction of poliovirus in laboratory-contaminated Eastern oysters (Sobsey 1990, Control of Hepatitis A Virus Contamination of Shellfish by Depuration, unpublished contract report). This study concluded that poliovirus reductions were more rapid than *E. coli,* fecal streptococci, phage MS-2, and hepatitis A virus during depuration at several temperatures and salinities. Other research on environmentally contaminated green lipped mussels *(Perna canaliculus)* showed poliovirus presence on the day of collection, but not after depuration for 3 days (Lewis et al. 1986). This same study demonstrated high levels of Coxsackie B4 in mussels at the time of collection and persisting at high levels throughout 8 days of depuration.

Hepatitis A virus appears to be particularly resistant to the depuration process. Eastern oysters contaminated with cell-culture adapted strains of hepatitis A virus in a laboratory uptake tank showed persistence of the virus throughout the 5-day depuration period (Sobsey 1990, Control of Hepatitis A Virus Contamination of Shellfish by Depuration, unpublished contract report). Over 40% of the hepatitis remained after depuration for 5 days at temperatures of 53.6°F (12°C) and 64.4°F (18°C) and 2.7% remained at 77°F (25°C). At 77°F (25°C) over 9% of the initial hepatitis A virus was present after depuration at salinities of 8, 18, and 28 ppt. In contrast, < 1% of the *E. coli* and poliovirus remained after depuration for 5 days at 18 and 28 ppt salinity.

Bacteriophage reductions were also prolonged during depuration experiments. Coliphage S-13 were reduced to negligible levels in hard clams after 72 hours of depuration at 75.2°F (24°C), but were reduced only one log unit in 6 days at 60.8°F (16°C) (Canzonier 1971). A wild-type coliphage depurated more slowly than *E. coli* in a commercial depuration facility (Powers and Collins 1986).

Scotti et al. (1983) exposed oysters to *E. coli* and cricket paralysis virus, a picornavirus of insects. The *E. coli* were reduced over three orders of magnitude after only 3 days. No appreciable reductions in the virus counts occurred even after 9 days.

Removal of Toxins, Metals, Pesticides, and Organic Contaminants

No review of depuration would be complete without at least a passing mention of depuration in the context of nonmicrobial contaminants. Toxins (Blogoslawski and Stewart 1983; Blogoslawski et al. 1979), heavy metals (Denton and Burdon-Jones 1981; Okazaki and Panietz 1981), petroleum hydrocarbons (Fossato and Canzonier 1976; Jackim and Wilson 1977), and radionuclides (Clifton et al. 1983; Dahlgaard 1981) are difficult to eliminate; they either do not depurate using current procedures or they depurate so slowly that commercial steps to purify them would be uneconomical. Some of these contaminants may be purged over extended periods, thus making long-term relaying the only possible solution for afflicted shellstock. The economic disadvantages of extended relay have prevented the implementation of commercial ventures to purify nonmicrobially contaminated shellstock.

Improvement in Organoleptic Qualities

Preservation of organoleptic attributes is important to both the shellfish dealer and consumer. The primary intent of depuration is the removal of microbial contaminants. As microbes are reduced, spoilage microorganisms are reduced as well. This may lead to extended shelf life of the shucked product. In addition, depuration allows sand and grit to be purged from the shellfish gut, thus rendering them more palatable to some consumers.

Organoleptic qualities can be enhanced by depurating shellfish in slightly higher salinity seawater. The increased salt picked up by the shellfish reportedly enhances their flavor. The practice of depurating shellfish in higher salinity seawater must be monitored closely to ensure that the effectiveness of the depuration process is not hampered (see section "Salinity," page 414).

PHYSIOLOGY OF SHELLFISH CONTAMINATION AND ELIMINATION

Site of Contaminant Collection

There are perhaps several routes for microbial entry into shellfish. The principal route is through normal feeding processes where mucus is secreted by the shellfish. Sheaths or strands of mucus then entrap food and contaminants which are swept into the mouth by ciliary action.

The association of microbes with the mucus may be by ionic phenomenon, direct chemical reactions, Van der Waal and H^+ ion bonding, or by simple occlusion. Poliovirus association with the mucus appears to be primarily by ionic bonding of viral particles to the sulfate radicals on the mucopolysaccharide moiety of the mucus (Di Girolamo et al. 1977). Changes in salinity and pH affected virus attachment to the mucus as would be expected with ionic bonding (Di Girolamo et al. 1977).

Particulate materials in the water column may have microbial contaminants entraped within or adsorbed to their surfaces. As shellfish filter these food particles out of the water, contaminants may be passively transported to the gut. The size of the particles and the degree of particle digestion within the gut may affect the efficiency of contaminant uptake by gut mucus. Particle- or feces-associated viruses were more efficiently concentrated in oysters than laboratory grown or purified virus preparations that were not associated with large particles (Durgin et al. 1981; Hoff and Becker 1969; Meinhold and Sobsey 1982). It seems likely that microbes not associated with particles may be too small to be readily filtered by the shellfish.

Shortly after shellfish become contaminated with bacteria and viruses, they display high numbers of these contaminants within their feces. The mechanism by which microbes associate with fecal materials appears to involve physical entrapment rather than ionic bonding (Timoney and Abston 1984). Many of the bacteria and viruses acquired by shellfish are rapidly expelled via the feces during normal shellfish activities.

It is well established that bacterial and viral accumulation within molluscan shellfish tissues occurs primarily in the hepatopancreas (Di Girolamo et al. 1975; Durgin et al. 1981; Liu et al. 1966a, 1967a; Metcalf et al. 1979, 1980; Rodrick et al. 1987) and digestive diverticula (Cabelli and Heffernan 1970a) as well as in the siphon of clams (Cabelli and Heffernan 1970a; Durgin 1981; Metcalf et al. 1979, 1980). Low levels of *E coli* and *Salmonella typhimurium* were detected in the mantle, adductor muscle, gills, pericardium, and hemolymph of the soft shell clam (Metcalf et al. 1979, 1980). Several other studies show low levels of bacteria and viruses in the mantle, gills, palps, stomach, esophagus, hemolymph, and mussel tissues (see Table 16.1).

Migration of Microbes Within Shellfish

The digestive gland/hepatopancreas, the location where most microbes are concentrated, contains many dead-end tubules lined with digestive cells. Food vacuoles are found within the digestive cells. Phagocytosis of food and nutrients by the digestive cells may be accompanied by phagocytosis, pinocytosis, or encapsulation of bacteria, viruses, and other contaminants (Moore and Gelder 1983).

Shellfish contain multiple mechanisms to carry nutrients to their tissues and to inactivate or dispel contaminants. Hemocytes within the blood isolate foreign materials from the shellfish by phagocytosis, pinocytosis, or encapsulation depending on the particle's size (Feng 1965, 1967; Tripp 1960, 1961, 1970). Foreign materials incorporated into the hemocytes via phagocytosis are either degraded or remain isolated within phagosomes (Moore and Gelder 1983). Digested bacteria and other materials can be converted to glycogen within the hemocytes (Cheng and Rudo 1976). Nondigested materials can be eliminated by the migration of leukocytes through the epithelium of the gut or through the inner or outer surfaces of the mantle (Tripp 1960). Alternatively, phagocytized microorganisms may undergo vacuolization and subsequent degeneration (Tripp 1960).

Table 16.1 Depuration Studies Showing Low Levels of Bacteria and Virus Accumulation in Specific Tissues of Shellfish

Tissue	*References*
Mantle, gills, palps, stomach, esophagus	Meinhold and Sobsey 1982 Metcalf and Stiles 1965 Metcalf et al. 1979, 1980
Hemolymph	Di Girolamo et al. 1975 Hoff et al. 1966 Liu et al. 1966a, 1966b Metcalf et al. 1979, 1980
Muscle tissues	Meinhold and Sobsey 1982 Metcalf et al. 1979, 1980
Pericardium, adductor muscle	Metcalf et al. 1979, 1980

Studies addressing the migration and viability of microorganisms through shellfish tissues have been performed. Tripp (1960) showed that bacterial and yeast cells were cleared from oysters in a similar manner. After intracardial injection of *Bacillus mycoides,* large numbers of bacteria were phagocytized within 1 hour with the resulting transport of the bacteria from the blood to the adjacent tissues within a few hours. *Saccharomyces cerevisiae* occluded main arteries immediately after intracardial injection into oysters. Within 24 hours, 95% of the yeast cells were phagocytized by leukocytes and the blood vessels were no longer occluded. Most bacteria and yeasts were removed from the oyster tissues by migration of the leukocytes to the gut epithelium, diverticula, palp, and mantle, or through the inner and outer surfaces of the mantle. This clearance process was essentially completed in 24–48 hours for vegetative *B. cereus* cells, but required several weeks for *B. cereus* spores. At day 12, only a few yeast cells were present in the oyster and these persisted for several weeks (Tripp 1960). There was no evidence of microbial replication throughout this study.

More recently Capers et al. (1990) found that naturally accumulated *V. vulnificus* was located primarily in the digestive tract, followed by the adductor muscle, mantle, gills, and hemolymph of Eastern oysters. Depuration changed the distribution of *V. vulnificus* over time with increasing levels occurring in the adductor muscle, mantle, and gills. Total *V. vulnificus* counts did not significantly decrease in whole oysters during depuration (Capers et al. 1990).

Studies on the distribution of radiolabeled viruses in the Pacific oyster showed the presence of viruses at relatively high levels in the epithelium surrounding the tubules of the digestive diverticula (Fig. 16.1) and along the wall of the mid gut (Fig. 16.2) (Hay and Scotti 1986). Considerably fewer viruses were observed in the surrounding connective tissues (Fig. 16.2). Hay and Scotti (1986) did not detect labeled viruses in the gill, mantle, muscle, gonads, or labial palps after laboratory uptake of labeled viruses. Likewise, Tripp (1960) did not detect appreciable levels of bacteria or yeasts in the epithelium of the gills or kidney, the pericardium, or the gonaducts.

Hay and Scotti (1986) showed that viruses were concentrated in the mucus found throughout the digestive tract of oysters. Although the mechanisms of enteric virus migration through shellfish tissues are uncertain, it appears likely that viruses are phagocytized along with food, nutrients, and bacteria by hemocytes along the digestive system. Hemocytes spread throughout the circulatory system depositing the viruses to internal blood sinuses, where the viruses may be further spread throughout internal tissues. Recent research supports the concept of virus accumulation along blood sinuses (Hay and Scotti 1986).

Sequestration of Microbes in Shellfish

The extent of microbial sequestration within shellfish tissues is uncertain. Cellular clearance mechanisms responsible for packaging contaminants in vacuoles or

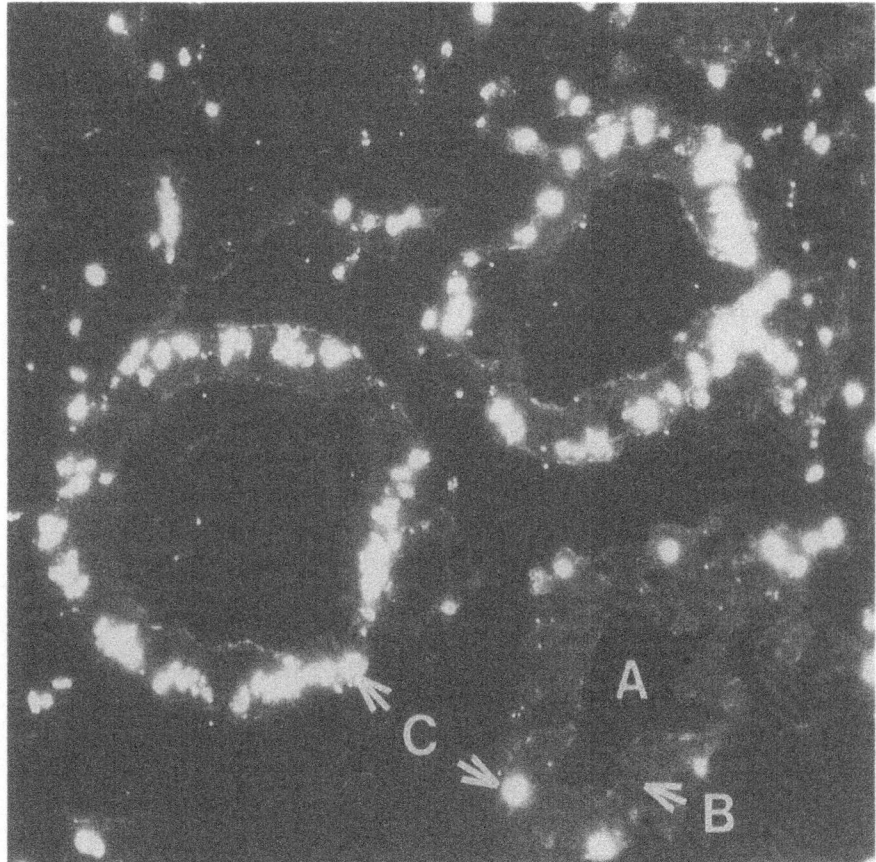

Figure 16.1 Cross section of the digestive diverticula of a Pacific oyster held in water containing radiolabeled virus for 12 hours. The presence of label in the epithelial cells is clearly visible. A, lumen; B, epithelial cells; C, radioactive label. Photograph courtesy of Brenda Hay.

hemocytes may play a role in protecting some types of contaminating microorganisms.

Studies suggest that large numbers of laboratory acquired vibrios may be readily purged because they become entrapped within feces and migrate through the digestive tract. However, the closer association of some of the vibrios with cells of the hepatopancreas and digestive diverticula may permit *Vibrio* colonization within the cells and produce persistently infected shellfish (Capers et al. 1990; Eyles and Davey 1984; Greenburg et al. 1982). Some vibrios are natural fish pathogens that have virulence factors to weaken shellfish defense mechanisms. They may invade tissues, rendering depuration ineffective. Vibrios contain antiphagocytic surface antigens which, in one study, prevented phagocy-

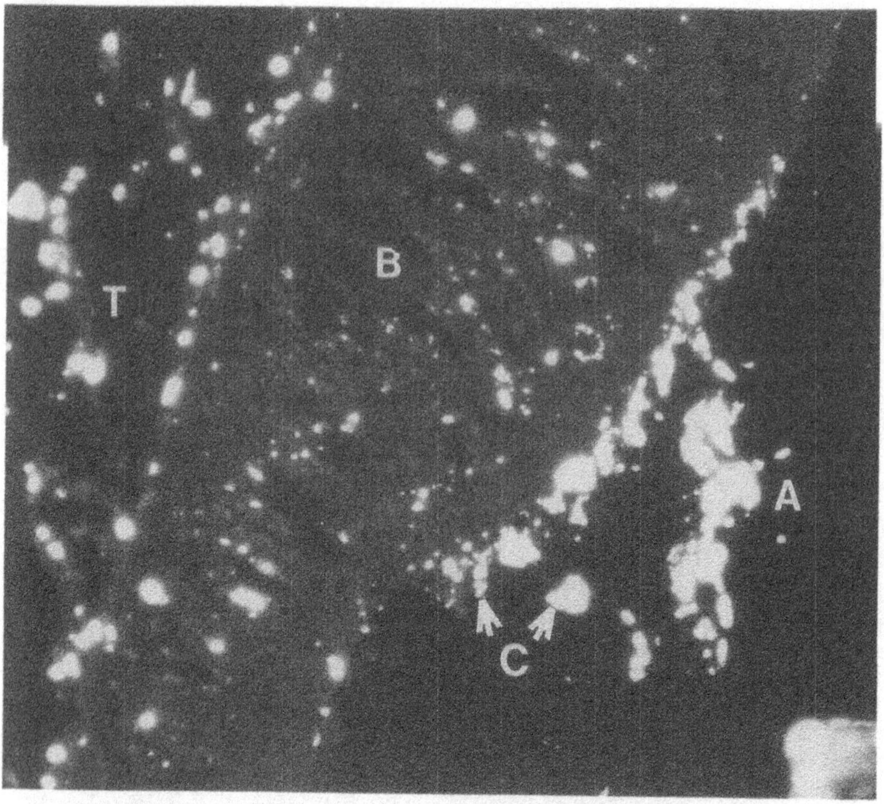

Figure 16.2 Section through the wall of the mid-gut of a Pacific oyster having been held for 12 hours in water containing radiolabeled virus. A, lumen; B, epithelium; C, radioactive label; T, connective tissue. (Reprinted from: Hay B., and Scotti, P. 1986. Evidence for intracellular absorption of virus by the Pacific oyster, *Crassostrea gigas. N.Z. J. Marine Freshwater Res.* **20**:655–659, with permission.)

tosis of *V. vulnificus* in human polymorphonuclear leukocytes (Kreger et al. 1981). The presence of antiphagocytic surface antigens may explain some of the reasons for *Vibrio* persistence in shellfish.

Bombardment of shellfish with vibrios naturally present in seawater may gradually lead to closer association and colonization of some of the vibrios in the intestinal tissues. Such relationships may occur slowly. The time required for vibrios to become closely associated with or to colonize intestinal tissues may not be sufficient in the laboratory challenge of shellfish with vibrios. This may explain why laboratory contaminated shellfish appear to depurate vibrios more readily than naturally contaminated shellfish.

Vibrios are not the only microorganisms that persist in shellfish; certain aerobic bacteria also persist during depuration (Ledo et al. 1983; Son and Fleet 1980; Souness and Fleet 1979; Souness et al. 1979). APCs decreased one log unit in studies on Sydney rock oysters, but seldom below 10^4 per gram of tissue (Son and Fleet 1980). Ayres (1975) showed that purified shellfish contained ca. 5×10^5 bacteria per gram of tissue. These bacteria represented the natural shellfish flora which persisted throughout depuration.

Viruses seated deep within shellfish tissues or in the hemolymph may persist for periods comparable to those observed for viruses in seawater (Canzonier 1971). Coliphage S-13 and *S. aureus* phage P-80 (Canzonier 1971), hepatitis A virus (Sobsey et al. 1987), and Coxsackie B4 virus (Lewis et al. 1986) have been shown to persist within shellfish. The mechanism for their persistence in shellfish undergoing depuration is uncertain.

Clearance of viruses may be less assured in shellfish because of virus size, shape, or antigenic composition. Vacuolization of viruses could actually protect them for extended periods. Bacteria, yeasts, and other respiring microorganisms seem less likely to survive vacuolization or encapsulation because of anticipated inhibition of required metabolic functions. Viruses lack metabolic functions outside a susceptible host and appear to be preserved within the shellfish. Proper refrigeration of shellstock may enhance the preservation of undesirable viral contaminants.

REGULATIONS GOVERNING SHELLFISH DEPURATION

United States

Shellfish purification in the U.S. follows the guidelines provided under the National Shellfish Sanitation Program (U.S. Public Health Service 1986, 1987) and is monitored and enforced by state and federal agencies. Under these guidelines, new depuration plants must undergo a process verification phase during which the entire operation is closely monitored to ensure adequate physiologic activity of the shellfish in the depuration system. A sampling plan and end product standards are also provided under the guidelines of the National Shellfish Sanitation Program. End product standards for soft clams restrict fecal coliforms to a most probable number (MPN) of 170/100 g or to an arithmetic mean of 125/100 g (if samples are analyzed in duplicate). End product standards for hard clams and Eastern oysters limit fecal coliform MPNs to not greater than 100/100 g or to an arithmetic mean of duplicate samples not to exceed 75/100 g (U.S. Public Health Service 1987). Each batch of depurated shellfish must comply with these standards before it can be released to the marketplace. To increase the likelihood of successfully depurating shellfish to these standards, the U.S., unlike most other countries, imposes limits on the amount of contaminants present in waters from which shellfish may be harvested. This is accomplished by classifying shellfish growing waters according to their pollution levels, through sanitary surveys. Shellfish from "approved" waters (containing a mean MPN of ≤70 total coliforms or ≤14 fecal coliforms/100 ml) may be sold without depurating them; however, shellfish from "restricted" waters (containing a mean MPN of ≤700 total coliforms or ≤88 fecal coliforms/100 ml) must be depurated (U.S. Public Health Service 1986). Shellfish in prohibited waters (containing an MPN of >700 total coliforms or >88 fecal coliforms/100 ml) may not be harvested or depurated in the U.S. Adherence to these water quality criteria will almost guarantee compliance with end product standards in properly functioning depuration plants.

The overall plant's performance is based on yet another set of fecal coliform criteria. Fecal coliform levels of processed products are compared after the depuration of every 10 batches of shellfish. A plant is in compliance if the fecal coliform levels show a geometric mean of 50/100g for soft shell clams, and 20/100g for hard shell clams and Eastern oysters. Upper 10% limits of 130/100g of soft shell clams and 70/100g of hard shell clams and oysters also apply. The upper limit means that not more than 10% of the values may exceed that MPN value. Criteria to evaluate overall plant performance were designed to alert plant management of problems as they develop in the depuration process and to ensure timely remedial action.

Other Countries

General guidance and regulations are provided for the depuration of shellfish in many countries including Australia, France, England, Italy, Spain, Scotland, and Ireland. Depuration in Australia is compulsory for the Sydney rock oyster which is the principal commercial species. The states of Tasmania and Victoria produce Pacific oysters commercially, but are not required to depurate this species if it meets the bacteriological criteria of the National Health and Medical Research Council (1987). Depuration standards in Australia include requirements on water quality, condition and handling of shellstock, loading densities, equipment construction and sanitation, and record keeping (National Health and Medical Research Council 1987). The minimum depuration period for the Sydney rock oyster is 36 hours.

Depuration throughout the U.K., France, Spain, and Italy is mandatory for all molluscan shellfish. Current regulations in the U.K. are embodied in the Public Health (Shellfish) Regulations, 1934 and 1948, of England and Wales which authorize local authorities the statutory power to regulate shellfish (West 1986; West et al. 1985). Enforcement of hygenic practices during depuration is guided by the Food and Drug Act (1955) and the Food Hygiene (General) Regulations 1970 (West et al. 1985). More detailed operating criteria are developed for each depuration facility based on plant location, size, depuration system, and type of shellfish to be depurated. Similar legal requirements apply in Northern Ireland, but not in Scotland where shellfish growing waters are much cleaner (West et al. 1985).

DEPURATION PLANT DESIGN

Physical Location of Plant

Site selection for the construction of a depuration facility is an important consideration. The availability of a clean seawater source for use in the depuration tanks must be found. Plants built along coastal waters classified as "Approved" are ideal in that the water may be obtained and used directly in the depuration process. This permits more flexibility in determining whether a recirculated or flow-through seawater system is installed.

Plants not adjacent to clean seawater sources must truck or pipe clean seawater into the facility. Piping may not be practical because of the long distances between the plant and a clean seawater source. Seawater wells may be an acceptable water source in some locations (MacMillan and Redman 1971). Trucking seawater into plants is practiced, but will increase the costs of processing. Plants using trucked water usually have recirculating depuration systems to minimize water consumption.

The availability of shellfish, distance to market, and costs of harvesting and transport must also be considered during site selection.

Plant Construction

As with any seafood processing plant, certain measures must be taken to minimize the possibility of product contamination or recontamination. Proper selection of construction materials will facilitate plant sanitation. The interior of the plants, including product contact surfaces, should be constructed of corrosion-resistant, nonporous materials such as stainless steel, tile, fiberglass, and plastics. Epoxy-painted tanks of concrete or wood may be acceptable in some applications. Wood should be excluded whenever possible, because wood absorbs potential contaminants and is nearly impossible to adequately disinfect. Wood should always be the last choice for depuration tank construction or for shellfish-holding baskets, racks, or trays.

A minimum of three rooms should be built for (a) handling, culling, and refrigerated storage of raw materials; (b) the depuration tanks, pumps, etc.; and for (c) finished product storage. Separate refrigerated storage areas should be available for raw materials and depurated shellstock to reduce the likelihood of inadvertent commingling or cross contamination. Further compartmentalization of portions of the building may simplify sanitation or reduce maintenance. For

Figure 16.3   Example of a large, open-air depuration plant in Florida.

instance, ultraviolet water treatment systems may be segregated from the main depuration area to reduce water contact, corrosion, and possible damage to the light source. Open-air (wall-less) depuration plants are in use in some parts of the U.S. (Fig. 16.3).

Pumps and plumbing fixtures should be installed in locations that will not impair plant sanitation efforts. It may be practical to exclude pumps from wet areas to prevent pump damage and preclude electric shock hazards.

Ventilation, drains and gutters, and vermin control measures should be considered in the early planning stages. Without adequate ventilation, condensate often forms on ceilings of depuration plants and may contaminate shellfish as it drips into depuration tanks or onto shellstock. Insect and rodent control may pose a particular problem to the shellfish processor and adequate measures to exclude their entry to the plant must be considered. Premises should drain well and be kept free of stored equipment, litter, refuse, uncut weeds, and grass. Further guidance on plant construction and design may be found in Title 50 of the U.S. Code of Federal Regulations (Regulations Governing Processed Fishery Products, Chapter 2, Subchapter G, Section 260.98 through 260.102) and in papers by Furfari (1966, 1976) and Nielson et al. (1978).

Waste Management and Sanitation

Plant design can facilitate the elimination of shellfish wastes. Depuration tanks should have a sloped bottom so that shellfish feces and pseudofeces will migrate away from the shellfish during depuration. A sloped bottom also allows wastes to be rinsed more easily from the tank. Trays or baskets used for holding shellfish in the depuration tanks should be designed to allow feces and pseudofeces to fall freely from the tray during depuration. This will help eliminate recontamination of the shellfish.

Dead shellfish or debris, which is culled out before and after depuration, must be quickly removed from the premises. Waste removal will reduce odors, which can attract insects, rodents, and other animals, and will reduce the likelihood of product recontamination.

Commercially supplied disinfectants should be used according to label instructions to ensure adequate contact time with contaminated surfaces. Proper rinsing is necessary to remove residues that might prove toxic to shellfish. For this reason, tanks, pipes, and pumps should be thoroughly rinsed of any chemical sanitizers. Steam or hot chemical disinfection systems may be used to enhance plant sanitation efforts.

WATER DISINFECTION PROCEDURES

Chlorine/Hypochlorite Disinfection

Chlorine is the oldest disinfection procedure for depuration waters (Carmelia 1921; Wells 1916, 1928). Its effectiveness in depuration processes is only marginal due to its extreme toxicity to shellfish. Low levels of chlorine impair shellfish pumping (Dodgson 1928; Galtsoff 1946) and hamper shellfish depuration. Chlorinated waters can be dechlorinated using sodium thiosulfate followed by vigorous aeration or passage through activated charcoal before addition to depuration tanks (Canzonier 1982). There is some evidence that sodium thiosulfate itself may impede shellfish pumping (Kelly 1961a). Water disinfection by means of chlorine gas or liquid hypochlorite solutions is not recommended unless large water reservoirs are available for storage and dechlorination purposes. Hypochlorite solutions are used for water disinfection at the mussel depuration facility at Conway, North Wales (West et al. 1985).

Iodophors

Only scant information is available on the use of iodophors for depuration water. Italy appears to be one of the few countries where iodophor disinfection of depuration water has been practiced commercially (Canzonier 1982, 1984; Fleet 1978). According to one report, rapid bacterial reductions occurred in depuration tank water containing 0.1–0.4 mg iodophor/L (Fleet 1978). There were no reported adverse effects on shellfish feeding activity or on edibility characteristics for shellfish subjected to iodophor-disinfected water (Fleet 1978).

Ultraviolet Light

Ultraviolet light irradiation of seawater is the most common system for the disinfection of depuration waters in the U.S. and the U.K. Ultraviolet (UV) light effectively reduces bacteria and viruses (Chang et al. 1985; Fogh 1955; Hill et al. 1969b, 1970). Numerous studies have demonstrated the effectiveness of UV light in reducing microbes in shellfish during depuration (Romagosa Vila 1957; Vasconcelos and Lee 1972; Wood 1961). Disinfection of seawater is effective provided the water is clear and the exposure time is adequate. High turbidity (Hill et al. 1967; Kelly 1961b), dinoflagellate blooms (Nielson et al. 1978), or the white milky gametes produced during shellfish spawning (Buisson et al. 1981; Furfari 1966) can impair light penetration and reduce the effectiveness of UV light disinfection of seawater. UV light does not produce compounds in seawater which would affect shellfish physiology. This is one of the advantages of this disinfection technique. On the other hand, UV light irradiation of seawater does not inactivate sequestered or endogenous microbial contaminants. Inactivation only occurs when the light is in direct contact with microbes in the seawater. Properly operating UV lamps are crucial; lamps should be checked and replaced as needed to ensure peak effeciency.

Some studies on the effectiveness of UV light irradiation of depuration tank waters showed a time-dependent increase in the numbers of UV-resistant bacteria in the process water (Souness and Fleet 1979). The extent to which resistant bacteria replicate in tank water is unclear. As nonresistant microbes die off, there may be less competition and a greater likelihood of replication among the UV-resistant strains.

Ozone

Ozone is a toxic gas that has been used extensively in France, Australia, and elsewhere for the disinfection of depuration waters (Fauvel et al. 1982; Institute of Maritime Fisheries 1972; Voille 1929). Ozone has the capacity to destroy bacteria and viruses (Katzenelson et al. 1974), but is toxic to shellfish. To prevent shellfish mortality or inactivity during depuration, ozone must be removed from seawater before entering the depuration tanks. This can be accomplished by vigorous aeration of the water.

The mechanism of ozone inactivation of poliovirus appears to be associated with damage to the viral RNA, rather than by alterations of the polypeptide chains of the viral capsid (Roy et al. 1981). Viral inactivation occurred after exposure to a residual ozone concentration <0.3 mg/L for up to 2 minutes.

Ozonation may offer some advantages over UV light irradiation techniques. Very low amounts of residual ozone in the depuration waters may enhance the elimination of vibrios and other potential pathogens that are associated with the gut flora of shellfish. Some scientists believe that residual ozone in the water could also reduce spoilage organisms in the gut and confer a longer shelf life to depurated products. Studies to address these possibilities are underway. Currently, the U.S. Food and Drug Administration does not sanction the use of ozone for depuration processes.

Activated Oxygen

Activated oxygen, also known as UV-generated ozone or commercially as Photozone (Water Management Inc., Engelwood, CO), is a relatively new technology that is undergoing evaluation for possible use in commercial depuration facilities. It is produced from the photooxidation of oxygen by UV light at a wavelength of 180 nm. The UV lamp is made up of two electrodes encased in a quartz housing that encloses a special mixture of gases. It converts oxygen molecules into ionized gas or plasma known as activated oxygen. Activated oxygen consists of several oxygen species: the hydroxyl radical (OH), atomic oxygen (O), ozone (O_3-), hydrogen peroxide (H_2O_2), and hydrogen dioxide (HO_2) (personal communication, Water Management Inc., Englewood, CO). An advantage of activated oxygen over conventional ozone is its reported nontoxicity to fish (personal communication, Water Management Inc.). A number of organic compounds may form in seawater as a consequence of treatment with activated oxygen. The effects of these compounds on the depuration process or the health of the consumer have not been determined.

SYSTEM PARAMETERS

Temperature

Maintaining the appropriate water temperature is a critical factor for any successful depuration process. Each shellfish species will pump within a range of temperatures which are often substantially different among the species. Shellfish obtained from naturally cold waters will obviously have a lower optimum depuration temperature than shellfish that are accustomed to warmer conditions. The geographical location and evolutionary history of the shellfish will also affect the temperatures for most efficient depuration.

Depuration studies have been conducted to determine the effects of water temperature on the reduction of bacteria and viruses from several shellfish species. The depuration of *E. coli* was extensively studied in the hard shell clam (Cabelli and Heffernan 1971; Greenberg et al. 1982; Heffernan and Cabelli 1970), soft shell clam (Arcisz and Kelly 1955; Cabelli and Heffernan 1970b); and the Sydney Rock oyster (Rowse and Fleet 1984; Souness and Fleet 1979). Optimal depuration temperatures for *E. coli* in the various shellfish species are shown in Table 16.2.

Another study (Beck et al. 1966) reported seasonal differences in the

Table 16.2 Optimal Depuration Temperature for the Elimination of Bacteria and Viruses from Shellfish

Contaminant	Shellfish Species[a]	Optimal Depuration Temperatures Among Ranges Tested	References
E. coli	Hard clam	68°F (20°C) [50–59°F (10–15°C) more protracted)]	Cabelli and Heffernan 1971; Heffernan and Cabelli 1970
	Soft clam	53.6–59°F (12–15°C) [46.4°–60.8°F (8–16°C) good, 35.6°F (2°C) poor]	Liu et al. 1966, 1967a
	Sydney Rock oyster	64.4–71.6°F (18–22°C) [55.4–62.6°F (13–17°C) and 75.2–80.6°F (24–27°C) more protracted]	Arcisz and Kelly 1955; Cabelli and Heffernan 1970b; Rowse and Fleet 1984
	Sydney Rock oyster	64.4–68°F (18–20°C) [55.4–62.6°F (13–17°C) and 80.6–84.2°F (27–29°C) more protracted]	Souness and Fleet 1979
Fecal coliforms	Eastern oyster	81.5°F (27.5°C) [75.9–83.6°F (24.4–28.7°C) good, 61.3–65.8°F (16.3–18.8°C), and 67.1°F (19.5°C) more protracted]	Presnell et al. 1969
Total coliforms	Soft clam	53.6–55.4°F (12–13°C) [46.4–60.8°F (8–16°C) good, 35.6°F (2°C) poor]	Arcisz and Kelly 1955; Cabelli and Heffernan 1970b
	Eastern oyster	81.5°F (27.5°C) [75.9–83.6°F (24.4–28.7°C) good, 61.3–65.8°F (16.3–18.8°C), and 67.1°F (19.5°C) more protracted]	Presnell et al. 1969
	Sydney Rock oyster	64.4–68°F (18–20°C) [80.6–84.2°F (27–29°C) more protracted]	Souness and Fleet 1979

Organism	Shellfish species	Temperature	Reference
Salmonella sp.	Soft clam	55.4°F (13°C)	Arcisz and Kelly 1955
	Sydney Rock oyster	64.4–71.6°F (18–22°C) [80.6–84.2°F (27–29°C) more protracted]	Rowse and Fleet 1984
Vibrio sp.	Hard clam	59°F (15°C) [64.4°F and 77°F (18 and 25°C) more protracted]	Greenberg et al. 1982
APC[b]	Sydney Rock oyster	64.4–68°F (18–20°C) [80.6–84.2°F (27–29°C) more protracted]	Souness and Fleet 1979
Coliphage	Hard clam	75.2°F (24°C) [60.8°F (16°C) more protracted]	Canzonier 1971
Polio type 1	Hard clam	68°F (20°C) [50–59°F (10–15°C) more protracted]	Liu et al. 1967a
	Hard clam	66.2°F (19°C) [44.6–46°F (7–8°C) and 55.4–57.2°F (13–14°C) more protracted]	Liu et al. 1967b
	Eastern oyster	82.4°F (28°C) [42.8°F (6°C) more protracted]	Meinhold and Sobsey 1982
	Eastern oyster	Good at 53.6, 62.6, and 73.4°F (12, 17, and 23°C)	Sobsey et al. 1987
Polio type 3	Hard clam	64.4–68°F (18–20°C) [57.3°F (14°C) slow, 41–42.8°F (5–6°C) very slow]	Lui et al. 1967b
	Hard clam	68°F (20°C) [50–59°F (10–15°C) more protracted]	Liu et al. 1967a
Coxsackie B4	Hard clam	68°F (20°C) [59°F (15°C) more protracted]	Liu et al. 1967b
	Hard clam	68°F (20°C) [50–59°F (10–15°C) more protracted]	Liu et al. 1967a
Hepatitis A	Eastern oyster	Poor at 53.6, 62.6, and 73.4°F (12, 17, and 23°C)	Sobsey et al. 1987

[a]Shellfish species: hard clam, *Mercenaria mercenaria*; soft clam, *Mya arenaria*; Eastern oyster, *Crassostrea virginica*; Sydney Rock oyster, *Crassostrea commercialis*

[b]Aerobic plate count microorganisms

elimination of *E. coli* from Olympia and Pacific oysters. Generally speaking, depuration proceeded more slowly during the winter months when water temperatures were low.

Table 16.2 lists a number of temperatures and temperature ranges that were found most effective (optimal) in eliminating specific microbes during the particular studies. Gaps and overlaps exist in the temperatures used among the investigators, making direct comparisons difficult; however, it appears that temperatures representative of normal summertime water temperatures for a particular area may be within the optimal range for the depuration of indigenous shellfish. Shellstock should not be placed in depuration tanks containing waters which deviate substantially in temperature from the natural waters unless they are acclimated to the new conditions. Timing of the depuration process should begin after the shellfish are acclimated to the tank water as indicated by active shellfish pumping.

Salinity

Another critical parameter for the successful depuration of shellfish involves the salinity of the seawater. Salinities affect physiological processes of shellfish including the pumping rates. At very low salinities, pumping may be inhibited. Higher salinities appear to enhance depuration effectiveness as seen in Table 16.3. Most shellfish species appear to depurate well at salinities around 30 ppt. Shellfish should be acclimated to process waters which deviate appreciably from the indigenous water salinities. This period of acclimation should be extended until shellfish begin active pumping. It is recommended that shellfish depuration waters do not deviate in salinity by more than 20% from harvest water salinities (U.S. Public Health Service 1987). In one study, salinities 50–60% below the levels from which hard shell clams were harvested completely stopped the depuration process (Liu et al. 1967b).

Dissolved Oxygen

Dissolved oxygen (DO) levels must be adequate in depuration tank water for shellfish to remain physiologically active. Recommendations in the U.S. called for DO levels of 5 mg/L (minimum) to saturation; however a minimum DO of 50% saturation is now in effect (U.S. Public Health Service 1987). Dissolved oxygen levels are also determined in terms of percent oxygen saturation of the seawater in the U.K., Australia, and Italy (Ayres 1978; West 1986; National Health and Medical Research Council 1987; Gazzetta Ufficiale Della Repubblica Italiana 1978). Percent saturation may be a more valid measure of DO than mg/L because it takes into account the effects of water temperature and salinity on the water's oxygen-holding capacity. Percent saturation in seawater may be calculated using a nomograph or from solubility tables.

Table 16.3 Optimal Salinities for the Elimination of Bacteria and Viruses from Shellfish

Contaminant	Shellfish Species[a]	Optimal Depuration Salinities Among Ranges Tested	References
E. coli	Hard clam	20–30 ppt (15 ppt more protracted)	Heffernan and Cabelli 1970
	Sydney Rock oyster	23–36 and 43–47 ppt (16–20 ppt poor)	Rowse and Fleet 1984
ET Coliforms[b]	Soft clam	25–30 ppt (20 ppt good, 15 ppt poor, 5 ppt very poor)	Cabelli and Heffernan 1970b
Fecal Coliforms	Eastern oyster	24.8–25.5 ppt (16.6–18.0 ppt good, 9.3–11.8 ppt fair, 6.3–6.9 ppt poor)	Presnell et al. 1969
Total Coliforms	Eastern oyster	24.8–25.5 ppt (16.6–18.0 ppt good, 9.3–11.8 ppt fair, 6.3–6.9 ppt poor)	Presnell et al. 1969
Salmonella charity	Sydney Rock oyster	23–36 and 43–47 ppt (16–20 ppt poor)	Rowse and Fleet 1984
Polio type 1	Hard clam	31 ppt (23–28 ppt slow, 17–21 ppt very slow)	Liu et al. 1967b
	Eastern oyster	8, 18, and 28 ppt all good	Sobsey et al. 1987
Hepatitis A	Eastern oyster	28 ppt (8 and 18 ppt slow)	Sobsey et al. 1987

[a]Shellfish species: hard clam, *Mercenaria mercenaria*; soft clam, *Mya arenaria*; Eastern oyster, *Crassostrea virginica*; Sydney Rock oyster, *Crassostrea commercialis*

[b]ET Coliforms, elevated temperature coliforms

Figure 16.4 Photograph of a depuration tank aerated with jets of water.

The level of DO is also affected by the surface area of the tank, flow rates, shellfish loading densities, shellfish activity, and aeration. Dissolved oxygen levels are frequently maintained by spraying seawater over the surface of the tank as seen in Figure 16.4. Aeration should not resuspend feces and pseudofeces from the bottom, or cause excessive turbulence with might disturb shellfish and affect pumping rates.

Supersaturation of the water with gases must be avoided. In a supersaturated system, gas bubbles form in the water and on water-contact surfaces including the shellfish's gills. These bubbles may severely impair shellfish respiration and may cause increased shellfish mortality. Decreasing the water temperature during periods of saturation will accentuate the problem.

Turbidity

Early studies suggested that turbidities around 25 Jackson Turbidity Units (JTU) enhanced bacterial depuration in hard shell clams (Heffernan and Cabelli 1970, 1971), but not in Eastern oysters (Presnell et al. 1969). Pumping rates may be increased for some species by elevations in turbidity. Oysters depurated *E. coli* at similar rates at average water turbidities of 8.8, 19.3, 22.6, and 69.7 JTU (Presnell et al. 1969). Total and fecal coliform reductions in Manila clams were similar at turbidities of 0, 20, 43, and 61 JTU (Vasconcelos 1969).

Poliovirus reductions were also similar in oysters depurated with total solids between 8 and 80 mg/L of seawater (Hamblet et al. 1969; Hill et al. 1969b). Oyster depuration was efficient for the removal of total and fecal coliforms with 1 and 100mg/L of total solids in the process water (Haven et al. 1978).

Very high turbidities may inhibit shellfish pumping (Loosanoff 1961) and may interfere with UV light transmission through seawater. Both of these factors can reduce depuration efficiency. Furfari (1966) recommended that turbidities not exceed 20 JTU for all process seawater. This limit has been generally accepted in depuration plants throughout the U.S.

Shellfish spawning increases turbidity and has been thought to reduce depuration efficiency. A recent study by Power and Collins (1989) showed that spawning reduced *E. coli* depuration efficiency, but not the rate of polio 1 or coliphage Al-5a reduction in mussels.

Flow Rate

Flow rates affect dissolved oxygen levels in seawater, the rate of water disinfection, and hence the rate of shellfish depuration. For hard shell clams, a flow rate of 13 ml per minute per clam provided optimal depuration of *E. coli* (Heffernan and Cabelli 1970). In the same study, the elimination of *E. coli* was prolonged at flow rates of 3, 7, and 27 ml per minute per clam. Another study showed effective depuration of soft shell clams at flow rates of 3–24 ml per minute per clam (4.5–36.3 L per minute per bushel) although at 3 ml per minute per clam, mortalities increased (Cabelli and Heffernan 1970b). Additional studies demonstrated reduced depuration at flow rates <1.1 L per minute per bushel (Goggins 1964; Goggins et al. 1964).

Manila clams effectively purged *E. coli* at flow rates of 20 and 30 ml per minute per clam (Vasconcelos 1971). At lower flow rates, clams exhibited siphon hyperextension and slow retraction of siphons when touched. Eastern oysters had >98% reduction of total and fecal coliforms at flow rates of 0.5, 1, 3, and 5 L per oyster per hour (Presnell et al. 1969). A minimum flow rate of 1 gallon per minute per standard U.S. bushel (3.79 L per minute per bushel) of shellfish is recommended (Furfari 1966; U.S. Public Health Service 1987) for any of the commercial shellfish species depurated in the United States.

Souness and Fleet (1979) compared the number of times water was cycled through UV disinfection units and found that one cycle of tank water per hour gave inadequate depuration, two cycles per hour enhanced depuration rates, whereas three cycles per hour provided optimal depuration efficiencies in their system. Because each depuration system is different, the required flow rates and number of cycles through the UV system must be determined for each facility.

Shellfish Loading Rates

The number or volume of shellfish placed into a depuration tank must be controlled if microbial loads are to be effectively reduced. The minimum recommended volume of process water is 8 ft^3 (0.233 m^3) per bushel of Eastern oysters or hard shell clams and 5 ft^3 (0.14 m^3) per bushel for soft shell clams (U.S. Public Health Service 1987).

Figure 16.5   Depuration tank loaded with soft shell clams.

Shellfish depth within depuration racks must be controlled to ensure adequate system hydraulics and to enhance shellfish pumping rates. Deep layers may isolate shellfish in the center of the layer, prevent proper water passage, and inhibit pumping. Shell opening occurs by passive release of the adductor muscle and by a slight opening force exerted by the elastic hinge ligament. Shellfish near the bottom of a rack may exhibit impaired pumping because of the weight of shellfish above them and their inability to open their shells. To avoid these problems, shellstock should be placed in shallow layers according to the following recommendations: soft clams should be no more than 8 inches (20.3 cm) deep, hard clams and Eastern oysters not greater than 3 inches (7.6 cm) deep (U.S. Public Health Service 1987), and mussels from 6 to 7 inches (15–18 cm) deep (Canzonier 1984). In England, recommended shellfish depths are 3.1 inches (8 cm) for hard shell clams and mussels, and single overlapping layers for oysters (primarily the European flat and Pacific oysters) (West 1986). As a rule-of-thumb, light-shelled species may be stacked higher than heavier-shelled animals. Figure 16.5 shows a depuration tank fully loaded with soft shell clams.

Recirculating vs. Flow-Through Systems

Shellfish may effectively depurate microbial contaminants in either recirculated or flow-through systems. Flow-through systems are preferable if the facility is close to a clean seawater source. Simple UV treatment of incoming water provides further assurances of acceptable water quality. Because depuration waters from flow-through systems are relatively clean from the standpoint of total and settleable solids, and chemical contaminants, their release into the estuaries should be of little environmental concern. As a precaution, effluent should be UV-treated to prevent the introduction of potential pathogens into shellfish-growing waters.

There is generally more concern over the disposal of waste water from recirculating depuration systems. The state or federal agencies having jurisdiction over effluent dumping into the ocean (usually the state's department of environmental protection) will issue permits for effluent discharge after obtaining information on water quality.

Duration of Depuration

The duration of depuration is dependent on many factors, e.g., the contaminant load, physiology of the shellfish, design of the depuration system, effectiveness of water disinfection, water parameters (temperature, salinity, dissolved oxygen, flow rates, etc.), shellfish loading rates, and more. Many studies have shown the effective depuration of bacterial indicator organisms within 48 hours in properly controlled depuration systems. Enteric viruses and both anthropogenic (hospital/septic wastes) and naturally occurring vibrios may require longer depuration periods. Longer depuration intervals should be recommended for shellfish obtained from waters likely to contain low level hospital discharge or septic wastes. Shellfish from waters more heavily polluted with these types of waste may not be appropriate source material for depuration even though the waters may be classified as restricted based on bacteriological criteria.

PLANT OPERATIONS

There are numerous aspects to plant operation that are critical for successful depuration. The first is shellfish harvesting and transport. Harvested shellstock should be culled to eliminate gross debris and empty shells. Shellstock should be stored on board the vessel under sanitary conditions and protected from excessive heat or cold. A roof of canvas or other suitable material will protect shellstock from the hot summer sun and the droppings from seabirds. Shellstock should only be stored for as long as necessary on board the boat. In the event of delays in off-loading, measures should be taken to ensure the product will remain in good condition. The transport vehicle should be maintained in a sanitary manner.

Shellstock may be processed immediately at the plant or placed in a cooler. Some shellfish experts believe that refrigerated storage for several days will enhance the initial pumping rates of shellfish once they are placed into depuration tanks and that this practice may enhance depuration efforts. Just before depuration, it is important to wash shellstock to remove mud and debris, which may otherwise foul the depuration system. Shellstock should again be culled to remove dead or moribund animals and empty shells. These steps will reduce the initial burden of microbes in the system.

The introduction of shellfish into baskets or racks and into the tanks must be done carefully to prevent shell breakage and shock to the animals. A period may be required for the shellfish to acclimate to its new environment especially if the shellfish are placed in water that deviates appreciably in temperature or salinity from harvest area water. Timing of the depuration process should begin only after there is reasonable evidence that the shellfish are pumping, e.g., appearance of gaped shells, extension of siphons, presence of feces or pseudofeces, clearance of particulates in the water.

System monitoring should be performed regularly. Temperature, salinity, dissolved oxygen, flow rates, shellfish appearance, water disinfection equipment, and residual levels of chemical disinfectants (if used) must be monitored at least daily and preferably several times a day. A log book should be maintained showing the date, time, and readings for the above.

Finished product should be closely monitored for indicator organisms by the appropriate state or federal regulatory agency to ensure compliance with applicable laws and regulations. This monitoring activity may be reduced after the effectiveness of the process has been verified by numerous successful runs. Only then should the sampling and assay of product be reduced.

Depurated product must be tagged and then stored in a different cooler than the incoming unprocessed materials. It should be shipped or further processed as soon as possible. Accurate records must be maintained on the distribution of the product. Record keeping, to maintain accountability for the product, is important in the event of a recall or problem.

Plant sanitation is critically important. Thorough cleansing of tanks, plumbing fixtures, racks, floors, walls, etc. will lessen the likelihood of problems arising during the next production run. Pipes, pumps, and tanks should be flushed with solutions containing sodium hypochlorite or other suitable disinfectants and rinsed thoroughly. Figure 16.6 shows a large, multi-tank depuration facility and some of the many plumbing fixtures which should be disinfected and flushed after each use.

CURRENT AND FUTURE RESEARCH

Considerable depuration research was conducted during the 1960s, but diminished during the 1970s and early 1980s. Only recently has there been a resurgence in depuration research due to worsening pollution of coastal waters, diminishing stocks of shellfish from clean waters, and increasing consumer

Figure 16.6 Photograph of a multitank depuration facility in Massachusetts.

awareness concerning outbreaks of shellfish-borne illness. The feasibility of depuration in regard to contemporary problems needs to be assessed.

Improved analytical tools are available to isolate and identify pathogens in shellfish. The recent recognition of *V. vulnificus* as a potentially lethal human pathogen, particularly to the immunocompromised, raw oyster eater, and outbreaks of enteric virus illness associated with raw or lightly steamed oysters and clams (Richards, 1985, 1987) have raised public apprehension concerning the safety of molluscan shellfish and have had negative economic effects on the seafood industry. These problems have contributed to a renewed interest in depuration research.

Government-funded research is underway at universities and government research laboratories to address a host of depuration and safety issues. University contractors are evaluating the effectiveness of depuration in eliminating polio, hepatitis A and other viruses, vibrios, and indicator bacteria from shellfish. The relationships between pathogens and conventional indicators (such as *E. coli,* fecal coliforms, and enterococci) are being studied. The U.S. Public Health Service, through many of its Food and Drug Administration Laboratories, and the National Marine Fisheries Service are working together to develop improved analytical procedures and to evaluate depuration parameters for hepatitis A and

Norwalk-like viruses, vibrios, and other emerging pathogens. Gene probe technologies are advancing and will likely resolve some of the remaining analytical deficiencies. The above agencies are also evaluating system parameters and water disinfection procedures necessary to achieve optimal depuration of pathogens.

Epidemiological studies are addressing the incidence of illness among shellfish eaters. Current and future studies will evaluate and compare illness rates for a host of bacterial and viral pathogens and indicators in individuals consuming shellfish harvested from approved waters. A past study provided information on health risks associated with the consumption of depurated oysters (Grohmann et al. 1981). Clearly, there is need for additional epidemiological evaluations of depurated products.

Additional studies must be conducted on basic shellfish physiology. Information on the mode of contaminant uptake, tissue distribution, and sequestering and elimination mechanisms will provide insight to improve depuration systems. Furthering our knowledge of basic physiological processes and improving depuration systems based on this information will enhance shellfish quality and safety by reducing spoilage microorganisms and pathogens.

Future researchers should recognize past problems and limitations that may affect depuration effectiveness and interpretation of results. The practice of contaminating shellfish with viruses and bacteria in the laboratory may not represent the natural mechanisms by which shellfish become contaminated in the environment. Likewise, shellfish may not accumulate, transfer, or sequester contaminants the same way in laboratory systems. If the mode of uptake is not the same, then we must also question whether the elimination of contaminants is the same in the laboratory as it is in nature. Work on vibrios indicates that the mode of contamination is critical; laboratory-contaminated shellfish depurate vibrios rapidly whereas shellstock naturally contaminated with vibrios appear resistant to depuration.

Standardization of depuration parameters remains a crucial factor in order for scientists to compare research results. Many previous studies lacked vital information on the temperatures, salinities, and other system parameters, making comparisons of data impossible. Fecal coliforms or *E. coli* should be included in all laboratory uptake and depuration studies to serve as a benchmark: to indicate the effectiveness of the depuration system in terms of conventional indicators and to facilitate data comparisons. Likewise, it may be prudent to include poliovirus as a benchmark in depuration studies involving other viruses.

One additional control treatment, not often seen in the depuration literature, involves the use of dry controls. Dry controls consist of shellfish that are not placed in the depuration system, but are placed in a cooler while the bulk of the lot is depurated. Bacteriological and virological analyses of the dry controls should be conducted in parallel with the depurated product. Evidence points to the possibility that substantial reductions in some pathogens, especially the vibrios, may occur during dry storage at refrigeration temperatures. Without dry controls, one might conclude that microbial reductions are a result of depuration, when in fact, they may be by natural clearance mechanisms occurring in non-depurated shellstock as well.

REFERENCES

Akin, E. W., Hamblet, F. E., and Hill, W. F., Jr. 1966. Accumulation and depuration of poliovirus by individual oysters. Gulf Coast Shellfish Sanitation Research Ctr., Dauphin Island, Alabama. Technical memorandum GCSSRC-FY66-5, 5 pp.

Arcisz, W., and Kelly, C. B. 1955. Self-purification of the soft clam, *Mya arenaria*. *Publ. Hlth. Rep.* **70**:605–614.

Ayres, P. A. 1975. The quantitative bacteriology of some commercial bivalve shellfish entering British markets. *J. Hyg. Cambr.* **74**:431–440.

Ayres, P. A. 1978. Shellfish purification in installations using ultraviolet light. Ministry of Agriculture, Fisheries and Food. Fish Research Laboratory leaflet No. 43. Lowestoft, United Kingdom.

Beck W. J., Kelly, C. B., Hoff, J. C., and Presnell, M. W. 1966. Bacterial depuration studies on West Coast shellfish. Presented at National Conference on Depuration, Kingston, Rhode Island.

Belding, D. L., and Lane, F. C. 1909. The shellfisheries of Massachusetts: their present condition and extent. In: *A Report Upon the Mollusk Fisheries of Massachusetts*. Wright & Potter, Boston.

Blogoslawski, W. J., and Stewart, M. E. 1983. Depuration and public health. *J. World Maricult. Soc.* **14**:535–545.

Blogoslawski, W. J., Stewart, M. E., Hurst, J. W., Jr., and Kern, F. G., III. 1979. Ozone disinfection of paralytic shellfish poison in the softshell clam *(Mya arenaria)*. *Toxicon* **17**:650–654.

Buisson, D. H., Fletcher, G. C., and Begg, C. W. 1981. Bacterial depuration of the Pacific oyster *(Crassostrea gigas)* in New Zealand. *N.Z. J. Sci.* **24**:253–262.

Cabelli, V. J., and Heffernan, W. P. 1970a. Accumulation of *Escherichia coli* by the northern quahaug. *Appl. Microbiol.* **19**:239–244.

Cabelli, V. J., and Heffernan, W. P. 1970b. Elimination of bacteria by the soft shell clam, *Mya arenaria*. *J. Fish. Res. Bd. Can.* **27**:1579–1587.

Cabelli, V. J., and Heffernan, W. P. 1971. Seasonal factors relevant to coliform levels in northern quahaugs. *Proc. Natl. Shellfish Assoc.* **61**:95–101.

Canzonier, W. J. 1971. Accumulation and elimination of coliphage S-13 by the hard clam, *Mercenaria mercenaria. Appl. Microbiol.* **21**:1024–1031.

Canzonier, W. J. 1982. Depuration of bivalve mollusks—what it can and cannot accomplish and some practical aspects of plant design and operation. In: *Proceedings of the International Seminar on Management of Shellfish Resources*. Irish Marine Farmers Association, Tralee, Ireland.

Canzonier, W. J. 1984. Technical aspects of bivalve depuration plant operation: pipes, pumps and petri plates. In: O'Sullivan A. J. (ed.) *Mussel Bound. Proceedings of the International Shellfish Seminar*. Environmental Management Services. Bantry, Ireland, pp. 68–96.

Capers, G. M., Tamplin, M. L., Martin, A. L., and Hopkins, L. H. 1990. Distribution and retention of *Vibrio vulnificus* in tissues of the Eastern oyster, *Crassostrea virginica. Abstr. Annu. Meeting Am. Soc. Microbiol,* p. 305.

Carmelia, F. A. 1921. Hypochlorite process of oyster purification. *Publ. Hlth. Ser. Rep.* pp. 876–883.

Chang, J. C. H., Ossoff, S. F., Lobe, D. C., Dorfman, M. H., Dumais, C. M., Qualls, R. G., Johnson, J. B. 1985. U.V. inactivation of pathogenic and indicator microorganisms. *Appl. Environ. Microbiol.* **49:**1361–1365.

Cheng, T. C., and Rudo, B. M. 1976. Distribution of glycogen resulting from the degradation of ^{14}C-labeled bacteria in the American oyster, *Crassostrea virginica. J. Invert. Pathol.* **27:**259–262.

Clifton, R. J., Stevens, H. E., and Hamilton, E. I. 1983. Concentration and depuration of some radionuclides present in a chronically exposed population of mussels *(Mytilus edulis). Marine Ecol. Prog. Ser.* **11:**245–256.

Colwell, R. R., and Liston, J. 1960. Microbiology of shellfish. Bacteriological study of the natural flora of Pacific oysters *(Crassostrea gigas). Appl. Microbiol.* **8:**104–109.

Dahlgaard, H. 1981. Loss of ^{51}Cr, ^{54}Mn, ^{57}Co, ^{59}Fe, ^{65}Zn, ^{134}Cs by the mussel *Mytilus edulis.* In: *International Symposium on the Impacts of Radionuclide Releases into the Marine Environment.* IAEA, Vienna, pp. 361–370.

Denton, G. R. W., and Burdon-Jones, C. 1981. Influence of temperature and salinity on the uptake, distribution and depuration of mercury, cadmium and lead by the black-lip oyster *Saccostrea echinata. Marine Biol.* **64:**317–326.

Di Girolamo, R., Liston, J., and Matches, J. 1975. Uptake and elimination of poliovirus by West Coast oysters. *Appl. Microbiol.* **29:**260–264.

Di Girolamo, R., Liston, J., and Matches, J. 1977. Ionic bonding, the mechanism of viral uptake by shellfish mucus. *Appl. Environ. Microbiol.* **33:**19–25.

Dodgson, R. W. 1928. Report on mussel purification. Ministry of Agriculture and Fisheries. *Fish. Invest. Ser. II.* **10:**1–436.

Dodgson, R. W. 1936. Shellfish and the public health. *Br. Med. J.* **2:**169–173.

Durgin, O. B., Metcalf, T. G., Moulton, E. R., and Hurst, J. W., Jr. 1981. Viral monitoring of commercial shellfish. Contract no. 223-78-2228, U.S. Food and Drug Admin., Washington, DC, 52 pp.

Eyles, M. J., and Davey, G. R. 1984. Microbiology of commercial depuration of the Sydney rock oyster, *Crassostrea commercialis. J. Food Protect.* **47:**703–706.

Fauvel, Y., Pons, G., and Legeron, J. P. 1982. Ozonation de l'eau de mer et epuration des coquillages. *Sci. Peche Nantes* **320:**1–16.

Feng, S. Y. 1965. Pinocytosis of proteins by oyster leucocytes. *Biol. Bull.* **128:**95–105.

Feng, S. Y. 1967. Responses of molluscs to foreign bodies, with special reference to the oyster. *Fed. Proc.* **26:**1685–1692.

Fleet, G. H. 1978. Oyster depuration—a review. *Food Technol. Aust.* **30:**444–454.

Fogh, J. 1955. Ultraviolet light inactivation of poliomyelitis virus. *Proc. Soc. Exp. Biol. Med.* **89:**464–465.

Fossato, V. U., and Canzonier, W. J. 1976. Hydrocarbon uptake and loss by the mussel *Mytilus edulis. Marine Biol.* **36:**243–250.

Frost, H. W. 1925. Report of committee on the sanitary control of the shellfish industry in the United States. *Publ. Hlth. Rep. (Suppl.)* **53:**1–17.

Furfari, S. A. 1966. Depuration plant design. Public Health Service Publ. No. 999-FP-7. U.S. Dept. of Health, Education and Welfare, Washington, DC, 119 pp.

Furfari, S. A. 1976. Shellfish purification: a review of current technology. FAO Technical Conference on Aquaculture, Publ. FIR:AQ/Conf/76/R.11. Kyoto, Japan, 16 pp.

Galtsoff, P. S. 1946. Reaction of oysters to chlorination. U.S. Fish and Wildlife Serv. Res. Rept. 11, U.S. Government Printing Office, Washington, DC.

Gazzetta Ufficiale Della Repubblica Italiana. 1978. Supplemento ordinario alla *Gazzetta Ufficiale* **125:**12–14.

Goggins, P. L. 1964. Depuration in Maine. In: Houser, L. S. (ed.) *Proceedings of the Fifth National Shellfish Sanitation Workshop* U.S. Food and Drug Administration, Washington, DC, pp. 78–92.

Goggins, P. L., Hurst, J. W., and Mooney, P. B. 1964. Soft clam depuration studies. Laboratory studies on shellfish purification. Maine Department of Sea and Shore Fisheries, Augusta, Maine, pp. 19–35ff.

Greenberg, E. P., Dubois, M., and Palhof, B. 1982. The survival of marine vibrios in *Mercenaria mercenaria*, the hardshell clam. *J. Food Safety* **4**:113–123.

Grohmann, G. S., Murphy, A. M., Christopher, P. J., Auty, G., and Greenberg, H. B. 1981. Norwalk virus gastroenteritis in volunteers consuming depurated oysters. *Aust. J. Exp. Biol. Med. Sci.* **59**:219–228.

Hamblet, F. E., Hill, W. F., Jr., Akin, E. W., and Benton, W. H. 1969. Oysters and human viruses: effects of seawater turbidity on poliovirus uptake and elimination. *Am. J. Epidemiol.* **89**:562–571.

Haven, D. S., Perkins, F. O., Morales-Alamo, R., and Rhodes, M. W. 1978. Bacterial depuration by the Eastern Oyster *(Crassostrea virginica)* under controlled conditions. Vol. 1. Biological and technical studies. Special Scientific Report No. 88. Virginia Institute of Marine Science, Gloucester Point, Virginia, p. 64ff.

Hay, B., and Scotti, P. 1986. Evidence for intracellular adsorption of virus by the Pacific oyster, *Crassostrea gigas*. *N.Z. J. Marine Freshwater Res.* **20**:655–659.

Heffernan, W. P., and Cabelli, V. J. 1970. Elimination of bacteria by the northern quahaug *(Mercenaria mercenaria)*: environmental parameters significant to the process. *J. Fish Res. Bd. Can.* **27**:1569–1577.

Heffernan, W. P., and Cabelli, V. J. 1971. The elimination of bacteria by the northern quahaug: variability in the response of individual animals to the development of criteria. *Proc. Natl. Shellfish Assoc.* **61**:102–108.

Herdman, W. A., and Boyce, R. 1899. Oysters and disease. An account of certain observations upon the normal and pathological histology and bacteriology of the oyster and other shellfish. Lancashire Sea-Fisheries Memoir No. 1, London, pp. 35–40.

Herdman, W. A., and Scott, A. 1896. Report on the investigations carried on in 1895 in connection with the Lancashire Sea-Fisheries Laboratory at the University College, Liverpool. *Proc. Trans. Liverpool Biol. Soc.* **10**:103–174.

Hill W. F., Hamblet, F. E., and Akin, E. W. 1967. Survival of poliovirus in flowing turbid seawater treated with ultraviolet light. *Appl. Microbiol.* **15**:533–536.

Hill, W. F., Jr., Akin, E. W., Hamblet, F. E., and Benton, W. H. 1969a. Poliovirus uptake and elimination by the American oyster, *Crassostrea virginica. Proc. Natl. Shellfisheries Assoc.* **60**:5.

Hill, W. F., Jr., Hamblet, F. E., and Benton, W. H. 1969b. Inactivation of poliovirus type 1 by the Kelly-Purdy ultraviolet seawater treatment unit. *Appl. Microbiol.* **17**:1–6.

Hill, W. F. Jr., Hamblet, F. E., Benton, W. H., and Akin, E. W. 1970. Ultraviolet devitalization of eight selected enteric viruses in estuarine water. *Appl. Microbiol.* **19**:805–812.

Hoff, J. C., and Becker, R. C. 1969. The accumulation and elimination of crude and clarified poliovirus suspensions by shellfish. *Am. J. Epidemiol.* **90**:53–61.

Hoff, J. C., Jakubowski, W., and Beck, W. J. 1966. Studies on bacteriophage accumulation and elimination by the Pacific oyster (Crassostrea gigas). In: Beck W. J., and Hoff J. C. (eds.) *1965 Proceedings of the Northwest Shellfish Sanitation Research Planning Conference* U.S. Public. Health Service Publ. No. 999-FP-6, Washington, DC, pp. 74–90.

Institute of Maritime Fisheries. 1972. Use of ozone in sea water for cleansing shellfish. *Effluent Water Treatment J.* **12**:260–262.

Jackim, E., and Wilson L, 1977. Benzo (a) pyrene accumulation and depuration in the soft-shell clam *(Mya arenaria)*. In: Wilt D.S. (ed.) *Proceedings of the Tenth National Shellfish Sanitation Workshop*. Hunt Valley, Maryland, pp. 91–94.

Katzenelson, E., Kletter, B., and Shuval, H. I. 1974. Inactivation kinetics of viruses and bacteria by use of ozone. *J. Am. Water Works Assoc.* **66:**725–729.

Kelly, C. B. 1961a. Accumulation of bacteria by the Pacific and Olympia Oysters. Paper read at Shellfish Sanitation Research Conference, Purdy, Washington.

Kelly, C. B. 1961b. Disinfection of seawater by ultraviolet radiation. *Am. J. Pub. Hlth.* **51:**1670–1680.

Kelly, M. T., and Dinuzzo, A. 1985. Uptake and clearance of *Vibrio vulnificus* from Gulf Coast oysters *(Crassostrea virginica)*. *Appl. Environ. Microbiol.* **50:**1548–1549.

Kreger. A., DeChatelet, L., and Shirley, P. 1981. Interaction of *Vibrio vulnificus* with polymorphonuclear leukocytes: association of virulence with resistance to phagocytosis. *J. Infect. Dis.* **144:**244–248.

Ledo, A., Gonzalez, E., Barja, J. L., and Toranzo, A. E. 1983. Effect of depuration systems on the reduction of bacteriological indicators in cultured mussels *(Mytilus edulis* Linnaeus). *J. Shellfish Res.* **3:**59–64.

Lewis, G., Loutit, M. W., and Austin, F. J. 1986. Enteroviruses in mussels and marine sediments and depuration of naturally accumulated viruses by green lipped mussels *(Perna canaliculus)*. *N.Z. J. Marine Freshwater Res.* **20:**431–437.

Liu, O. C. 1968. Appraisal and planning of virus research program. Northeast Shellfish Sanitation Research Center, U.S. Public Health Service, Narrangansett, Rhode Island, 38 pp.

Liu, O. C., Seraichekas, H. R., and Murphy, B. L. 1966a. Fate of poliovirus in northern quahaugs. *Proc. Soc. Exp. Biol. Med.* **121:**601–607.

Liu, O. C., Seraichekas, H. R., and Murphy, B. L. 1966b. Viral pollution in shellfish. 1. Some basic facts of uptake. *Proc. Soc. Exp. Biol. Med.* **123:**481–487.

Liu, O. C., Seraichekas, H. R., and Murphy, B. L. 1967a. Viral pollution and self-cleansing mechanisms of hard clams. In: Berg G. (ed.) *Transmission of Viruses by the Water Route*. Interscience Publishers, New York, pp. 419–437.

Liu, O. C., Seraichekas, H. R., and Murphy, B. L. 1967b. Viral depuration of the northern quahaug. *Appl. Microbiol.* **15:**307–315.

Loosanoff, V. L. 1961. Effects of turbidities on some larval and adult bivalves In: *Proceedings of the Gulf and Caribean Fish Institute, Session 14,* pp. 80–95.

Lumsden, L. L., Hasseltine, H. E., Leak, J. P., and Veldee, M. V. 1925. A typhoid fever epidemic caused by oyster-borne infection. *Pub. Hlth. Rep. (Suppl.)* **50:**1–102.

MacMillan, R. B., and Redman, J. H. 1971. Hard clam cleansing in New York. *Commercial Fish. Rev.* **33:**25–33.

Marvel, P. 1902. Report of the New Jersey State Board of Health.

Meinhold, A. F., and Sobsey, M. D. 1982. The uptake, elimination and tissue distribution of poliovirus in the American oyster, *Crassostrea virginica. Abstr. Annu. Meeting Am. Soc. Microbiol,* p. 181.

Metcalf, T. G., and Stiles, W. C. 1965. The accumulation of enteric viruses by the American oyster, *Crassostrea virginica. J. Infect. Dis.* **115:**68–76.

Metcalf, T. G., Mullin B., Eckerson, D., Moulton, E., and Larkin, E. P. 1979. Bioaccumulation and depuration of enteroviruses by the soft-shelled clam, *Mya arenaria. Appl. Environ. Microbiol.* **38:**275–282.

Metcalf, T. G., Eckerson, D., Moulton, E., and Larkin, E. P. 1980. Uptake and depletion of particulate-associated polioviruses by the soft shell clam. *J. Food Protect.* **43:**86–88.

Mitchell, J. R., Presnell, M. W., Akin, E. W., Cummins, J. M., and Liu, O. C. 1966. Accumulation and elimination of poliovirus by the Eastern oyster. *Am. J. Epidemiol.* **84**:40–50.

Moore, C. A., and Gelder, S. R. 1983. The role of the "blunt" granules in the hemocytes of *Mercenaria mercenaria* following phagocytosis. *J. Invert. Pathol.* **41**:369–377.

National Health and Medical Research Council. 1987. Code of hygenic practice for oysters and mussels for sale for human consumption. Australian Government Publishing Service, Canberra, Australia, 20 pp.

Nielson, B. J., Haven, D. S., Perkins, F. O., Morales-Alamo, R., and Rhodes, M. W. 1978. Bacterial depuration by the American oyster *(Crassostrea virginica)* under controlled conditions. Vol. 2. Practical considerations and plant design. Special Scientific Report No. 88. Virginia Institute of Marine Science, Gloucester Point, Virginia, 48 pp.

Okazaki, R. K., and Panietz, M. H. 1981. Depuration of twelve trace metals in tissues of the oysters *Crassostrea gigas* and *C. virginica*. *Marine Biol.* **63**:113–120.

Power, U., and Collins, J. K. 1986. Evaluation of depuration as a means of rendering shellfish free from viral pathogens and bacterial indicators. *Irish J. Food Sci. Technol.* **10**:159.

Power, U. F., and Collins, J. K. 1989. Differential depuration of poliovirus, *Escherichia coli*, and a coliphage by the common mussel, *Mytilus edulis*. *Appl. Environ. Microbiol.* **55**:1386–1390.

Presnell, M. W., Cummins, J. M., and Miescier, J. J. 1969. Influence of selected environmental factors on the elimination of bacteria by the Eastern oyster, *Crassostrea virginica*. In: Hammerstrom R. J., and Hill, W. F., Jr. (eds.) *Proceedings of the Gulf and South Atlantic States Shellfish Sanitation Research Conference.* Environmental Health Series, U.S. Public Health Service Publ. No. 999-UIH-9, pp. 47–65.

Richards, G. P. 1985. Outbreaks of shellfish-associated enteric virus illness in the United States: requisite for development of viral guidelines. *J. Food Protect.* **48**:815–823.

Richards, G. P. 1987. Shellfish-associated enteric virus illness in the United States, 1934–1984. *Estuaries* **10**:84–85.

Richards, G. P. 1988. Microbial purification of shellfish: a review of depuration and relaying. *J. Food Protect.* **51**:218–251.

Rodrick, G. E., Schneider, K. R., Steslow, F. A., and Blake, N. J. 1987. Uptake, fate and elimination by shellfish in a laboratory depuration system. *Proc. Oceans 87* **5**:1752–1756.

Romagosa Vila, J. A. 1957. Los rayos ultravioletas en el saneamiento de los moluscos. *An. Bromatol. Tomo* **IX**:401–404.

Rowse, A. J., and Fleet, G. H. 1984. Effects of water temperature and salinity on elimination of *Salmonella charity* and *Escherichia coli* from Sydney rock oysters *(Crassostrea commercialis)*. *Appl. Environ. Microbiol.* **48**:1061–1063.

Roy, D., Wong, P. K. Y., Engelbrecht, R. S., and Chian, E. S. K. 1981. Mechanism of enteroviral inactivation by ozone. *Appl. Environ. Microbiol.* **41**:718–723.

Scotti, P. D., Fletcher, G. C., Buisson, D. H., and Fredericksen, S. 1983. Virus depuration in the Pacific oyster *(Crassostrea gigas)* in New Zealand. *N.Z. J. Sci.* **26**:9–13.

Sobsey, M. D., Davis, A. L., and Rullman, V. A. 1987. Persistence of hepatitis A virus and other viruses in depurated Eastern oysters. *Proc. Oceans 87* **5**:1740–1745.

Son, N. T., and Fleet, G. H. 1980. Behavior of pathogenic bacteria in the oyster, *Crassostrea commercialis,* during depuration, re-laying, and storage. *Appl. Environ. Microbiol.* **40**:994–1002.

Souness, R. A., and Fleet, G. H. 1979. Depuration of the Sydney rock oyster, *Crassostrea commercialis*. *Food Technol. Aust.* **31**:397–404.

Souness, R., Bowery, R. G., and Fleet, G. H. 1979. Commercial depuration of the Sydney rock oyster, *Crassostrea commercialis. Food Technol. Aust.* **31**:531–537.

Steslow, F. A., Schneider, K. R., Sierra, F. J., and Rodrick, G. E. 1987. Ultraviolet light depuration of *Vibrio cholerae* and *Vibrio vulnificus* from Florida oysters. *Abstr. Annu. Meeting Am. Soc. Microbiol.* p. 292.

Timoney, J. F., and Abston, A. 1984. Accumulation and elimination of *Escherichia coli* and *Salmonella typhimurium* by hard clams in an *in vitro* system. *Appl. Environ. Microbiol.* **47**:986–988.

Tripp, M. R. 1960. Mechanisms of removal of injected microorganisms from the American oyster, *Crassostrea virginica* (Gmelin). *Biol. Bull.* **119**:273–282.

Tripp, M. R. 1961. The fate of foreign materials experimentally introduced into the snail, *Australorbis glabratus. J. Parasitol.* **47**:745–751.

Tripp, M. R. 1970. Defense mechanisms of molluscs. *J. Reticuloendothel. Soc.* **7**:173–182.

U.S. Public Health Service. 1986. National Shellfish Sanitation Program manual of operations. Part 1. Sanitation of shellfish growing areas. U.S. Public Health Service, Washington, DC.

U.S. Public Health Service. 1987. National Shellfish Sanitation Program manual of operations. Part II. Sanitation of the harvesting, processing and distribution of shellfish. 1987 revision. U.S. Public Health Service, Washington, DC.

Vasconcelos, G. J. 1969. The effect of turbidity on bacterial purification of Manila clams *(Tapes japonica).* Shellfish Sanitation Technical Report no. NWWHL-71-3, U.S. Public Health Service, Washington, DC, 8 pp.

Vasconcelos, G. J., 1971. The effect of various flow rates on the elimination of bacteria by Manila clams. In: Hoff J. C., and Beck W. J. (eds.) *1967 Proceedings of the Northwest Shellfish Planning Conference*, Environmental Protection Agency, Washington, DC. pp. 25–35.

Vasconcelos, G. J., and Lee J. S. 1972. Microbial flora of Pacific oysters *(Crassostrea gigas)* subjected to ultraviolet-irradiated seawater. *Appl. Microbiol.* **23**:11–16.

Voille, H. 1929. De la stérilisation de l'eau de mer par ozone: applicationes de cette méthode pour le purification des coquillages contaminés. *Rev. Hyg. Méd. Prévent.* **51**:42–46.

Wells, W. F. 1916. Artificial purification of oysters. *Publ. Hlth. Rep.* **31**:1848–1852.

Wells, W. F. 1928. Chlorination as a factor of safety in shellfish production. *Am. J. Pub. Hlth.* **19**:72–77.

West, P. A. 1986. Hazard analysis critical control point (HACCP) concept: application to bivalve shellfish purification systems. *J. R. Soc. Hlth.* **4**:133–140.

West, P. A., Wood, P. C., and Jacob, M. 1985. Control of food poisoning risks associated with shellfish. *J. R. Soc. Hlth.* **1**:15–21.

Wood, P. C. 1961. The principles of water sterilisation by ultra-violet light, and their application in the purification of oysters. Ministry of Agriculture, Fisheries and Food. *Fish. Invest. Ser. II* **23**:1–47.

17 Irradiation

Robert M. Grodner and
Linda S. Andrews

Food irradiation is a process in which food products are exposed to ionizing radiation to improve the shelf life or wholesomeness of that product. Radiation is applied as gamma emission from isotopes of Co^{60} and Cs^{137} or as x-rays, or accelerated electrons from machine sources (Ley 1983). Ionizing radiation may damage DNA at the cellular level, thus interfering with normal biochemical processes. Genetic damage caused by exposure to radiation can kill or damage molds, bacteria, and other microorganisms that cause food spoilage or disease. The degree of damage to the cell is dependent on the dose, type of ionizing radiation, and the physical qualities of the subject cell. The principal advantage of radiation over other preservation techniques relates to the destruction of harmful microorganisms, with little change in temperature. Thus, the wholesomeness, integrity, and sensory quality of the original food product are maintained (Kamplemacher 1983).

HISTORICAL ASPECTS

Use of ionizing radiation as a food preservative is not a new concept. In 1921, a researcher was issued a patent on an irradiation technique that could destroy *Trichinella* in pork (Schwartz 1921). The first use of irradiation by x-rays to destroy pathogenic bacteria was attempted by French researchers in the 1930s (Goresline 1982). Immediately after World War II, improved ionizing radiation sources became available and a wide range of food products were subjected to and shown to be preservable by short-term exposure to irradiation sources. For example, x-rays were used to sterilize milk products (Gaden et al. 1951) and to improve the bacterial quality of vegetables and seafoods (Proctor and Goldblith 1951). It soon became clear that the ability of x-rays to penetrate a food product was limited and it would be necessary to develop improved irradiation sources.

As isotopes became available, new radiation sources were introduced. Currently over 30 nations worldwide use ionizing radiation as a means to preserve or extend the shelf life of a variety of food products. However, approval by the U.S. Food and Drug Administration for use of ionizing radiation on food products in the U.S. has been slow to develop. In 1958, Congress officially classified irradiation as a food additive. Subsequently any food product undergoing irradiation treatment must be tested for wholesomeness. In addition, animal feeding studies have to be conducted to demonstrate the absence of toxic effects of the radiation and of any breakdown products that might be produced during the irradiation process. In more than 30 years of intense studies there has been no confirmed evidence of toxic substances produced in low-dose irradiated food products (IFT 1983). Legislation has been introduced to the U.S. Congress, which would remove ionizing radiation from the food additive list and categorize it as a form of thermal processing. However, these suggested changes have been slow to be approved, apparently due to concern by consumer groups and their lobbying activities in regard to food irradiation and radioactivity in general. Consumer acceptance will come only as people are educated to the beneficial effects of this preservation technique in the absence of undue risk.

UTILIZATION OF IRRADIATION IN THE SEAFOOD INDUSTRY

Research conducted on seafood products since the mid-1950s has shown ionizing radiation to have a positive effect on maintaining the quality and freshness of seafoods. The advantages of using ionizing radiation as a processing method are twofold: first it will reduce or eliminate 90–95% of the microorganisms responsible for spoilage and subsequently will extend the fresh storage shelf life (Novak and Liuzzo 1966, 1967, 1968). Fresh shrimp held on ice normally maintain good quality for up to approximately 7 days. By irradiating the shrimp at low dosage (<150 Krad), it is possible to extend the fresh quality for an additional 7–10 days. Also, this same low dose of irradiation has the ability to reduce and/or eliminate specific pathogenic microorganisms. For example, *Vibrio* species tested to date in seafood, including *Vibrio cholera, V. vulnificus,* and *V. parahaemolyticus,* have been eliminated immediately after treatment with a 100 Krad dosage or less of gamma-rays depending on the specific seafood tested.

TYPES OF IRRADIATION USED IN SEAFOODS

Irradiation of seafood can utilize ionizing radiation from a variety of sources. The sources that have proven to be the most effective in eliminating microorganisms, while maintaining the integrity and quality of seafood, use x-rays, gamma-rays, and electron beam accelerators. "In order to be useful in the treatment of foods, radiation must have the capability of penetrating into the depth of foods" (Urbain 1986). These three sources have the capability of penetrating into as well as through most food products. The least expensive of these methods appears to be the utilization of gamma-rays emitted from decaying Co^{60} or Cs^{137} sources. The amount of time a food product is exposed to the radiation source dictates the level or dosage that is received by that product. The dosage rate is equal to the calculated gamma source emission rate in rads/min $\times$ the length of exposure time.

When attempting to control microorganisms in foods, particularly seafoods, the time of exposure or the specific irradiation dose is very important. Microorganisms respond differently depending on specific cellular characteristics. Consequently, three degrees or levels of control have been identified: radapperization, radurization, and radicidation.

When attempting to eliminate or inactivate all spoilage microorganisms present in the seafood, a high dose of irradation is necessary. This high dose is termed *radappertization* because the effect simulates the traditional thermal canning process as developed by "Appert." "Properly packaged radappertized seafoods will keep indefinitely without refrigeration. The dose necessary to secure sterility is the highest employed in food irradiation systems, and can be as large as about 50 kGy (5,000 Krad or 5 Mrad)" (Urbain 1986). However, at this dose the original characteristic flavor and texture of the product will not be maintained.

Radurization, a treatment with an effect similar to that of heat pasteurization, inactivates 90–95% of the spoilage microorganisms. This reduction in microorganisms extends the shelf life of iced and/or refrigerated fresh seafood. However, if the product is held too long spoilage will occur. The dose level most appropriate for radurization is considered to be that $<$10 kGy (1,000 Krad) and particularly in the range of 1–5 kGy (100–500 Krad). Dose levels above 5 kGy (500 Krad) may interfere with the other sensory qualities of seafood products.

Inactivation of pathogenic non-spore-forming bacteria sometimes presents special problems. Non-spore–forming pathogenic bacteria that have been identified in seafood products include *Salmonella, Shigella, Vibrionacae,* and *Staphylococcus* (see related chapters). *Vibrios* can be eliminated with a relatively low dose (often $<$50,000 rads) of irradiation; however, a higher radiation dose has been found necessary to eliminate or inactivate *Salmonella.* Thus a third term, *radicidation,* exists that takes into account the variation in microbial response. The most effective radicidation dose is in the range between 5 and 8 kGy (500–800 Krad) depending on the seafood and the degree of contamination. At this dose level, each type of seafood would require extensive study to ensure specific pathogens are destroyed while still maintaining the flavor, texture, and wholesomeness of the product.

IRRADIATION EFFECT
ON MICROORGANISMS IN SEAFOOD

Ionizing radiation destroys most microorganisms by producing lethal damage to nuclear DNA. Species of microorganisms differ in their resistance to ionizing radiation. Sensitivity variations may even occur among strains of the same species. Gram-negative bacteria are generally considered more sensitive than Gram-positive species; consequently many of the typical seafood spoilage bacteria are among the least resistant. Each species of bacteria, as well as the particular type of seafood substrate with which one is concerned, should be examined on an individual basis. For example, there is great variation among the different *Salmonella* serotypes in different seafoods. Also, traditional indicators may not work in irradiated food. *Escherichia coli* is far more sensitive than many salmonellae serotypes and enteric viruses; as a result, *E. coli* which is normally useful as an indicator of fecal contamination in fish and shellfish would not be a good indicator of possible *Salmonella* contamination in irradiated seafoods. Gram-positive bacteria such as *Staphylococcus aureus, Micrococci, Bacilli,* and *Clostridium* are among the more irradiation-resistant bacteria. Viruses, in general, are extremely resistant to irradiation. Fish parasites also require a fairly high dosage of irradiation to be inactivated.

Matsuyama et al. (1960) reported that the lethality for vegetative cells of *E. coli* was increased by the presence of NaCl during the irradiation treatment. He speculated that the release of chlorine ions during the irradiation process was responsible for the increased lethality.

Spoilage Organisms

Seafood quality is determined by subjective sensory judgment of the consumer. In addition, microbiologists often use aerobic plate counts (APCs) as an objective measurement. This method, although imprecise, can relate to sensory quality. The microbial flora of freshly caught fish and shellfish naturally reflects that of their environment. The predominant bacterial flora of freshly caught fish or shellfish is the *Pseudomonads* groups, which frequently compose 60% of the total flora. *Achromobacter (Acinetobacter* and *Moraxella)* is usually the second largest group, varying from 14 to 40% (Schewan and Georgala 1957). As spoilage proceeds, the *Pseudomonads* group increases to as much as 90% of the total flora. Other microbial species implicated in spoilage of fresh caught marine fish include *Micrococcus, Flavobacterium, Crytophaga* species with *Corynebacterium, Vibrios, Bacillus, Proteus,* and yeasts (Josephson and Peterson 1983).

In ice-stored fish and shellfish, gamma irradiation below 200 Krad has increased shelf life approximately 7–10 days over the normal shelf life of the specific species involved. This is mainly due to reduction of the original number of microorganisms, especially the spoilage type, immediately after low-dose exposure to gamma irradiation. These results are well documented and can be found in the following specific examples: Fillets of fish such as Bombay duck, rohu, cod, haddock, mullet, and catfish when irradiated with a 100 Krad dosage of gamma irradiation and stored on ice, have been found to maintain good quality with a shelf life extension of 7–15 days beyond that of unirradiated samples (Baldrati et al. 1974; Bhattacharya et al. 1978; Karnop and Antonacopoulos 1977; Oosterhuis 1976; Przybylski et al. 1989; Savagaon 1975). Shark and ray fillets steam cooked after irradiation (100–500 Krad) and irradiated raw (100 Krad) were evaluated for up to 40 days. Analysis of irradiated raw samples showed that irradiation did not extend the shelf life beyond the 10 days observed for control samples due to changes in quality as a result of formation of ammonia. Combined steaming and irradiation extended shelf life to 30–35 days (Ghadi and Lewis 1978). Whole redfish and haddock, irradiated at 100 Krad with an on-board x-ray facility, showed no marked sensory differences between irradiated and unirradiated samples during the first 16 days of iced storage (Ehlermann and Reinacher 1979). However, after 16 days the irradiated samples occasionally were scored higher than the unirradiated ones for sensory quality. The microbial quality of the irraditated samples was improved, with a marked reduction in spoilage microorganisms. Another study comparing cooking and irradiation processing of fresh horse mackerel fillets demonstrated that the combined effect of 200 Krad and 5 minutes of steam cooking at 176–185°F (80–85°C) increased the shelf life to 5–8 weeks compared to only 2 weeks for the uncooked irradiated samples and only 3 weeks for the cooked unirradiated samples. Increasing the cooking time to 15 minutes extended the shelf life for unirradiated samples to 4 weeks, and that of the irradiated samples to 11 weeks (Sasayama and Amano 1970). Irradiation of fresh mackerel at 100, 150, and 250 Krad dosage of gamma irradiation maintained fresh quality for 8, 14, and 35 days respectively, beyond the storage life of unirradiated mackerel (Baldrati et al. 1978; Bhattachryya et al. 1979). Fresh whole and filleted hake were gamma irradiated at 71 Krad and monitored for organoleptic quality by an experienced taste panel. This level of irradiation achieved a 6-day shelf life extension in both forms of the hake. Improvements were atrributed to marked reductions in the levels of *Pseudomonas* and other spoilage bacteria (Avery and Lamprecht 1988). Another study, comparing irradiation dosages at different storage temperatures, found that the minimum radiation dosage necessary for effective radiopasteurization was 400 Krad for samples stored at 39.2°F (4°C) and 250 Krad for samples stored at 19.4°F (−7°C). Fresh quality was maintained for a total of 14 days for the samples stored at 19.4°F (−7°C) and 21 days for samples stored at 39.2°F (4°C). the psychrotrophic bacteria showed a marked increase in the samples stored at the lower temperature, reducing the storage life by 7 days (Ola 1976).

Gamma irradiation of fresh crab meat and shrimp at 100, 150, or 250 Krad demonstrated that the storage life of these products was significantly extended when followed with refrigerated storage (Houwing 1976; Loaharanu 1973).

Total bacterial counts of $<5 \times 10^5$ after 60 days of storage were reported in fresh and frozen grass shrimp when processed with 450 Krad of gamma irradiation (Yeh and Hau 1988) and maintained under refrigerated storage.

Pathogenic Microorganisms

Research with gamma irradiation has primarily focused on low-dose pasteurization of fish and shellfish. This was an historical approach fostered by the U.S. Atomic Energy Commission, which chose to work primarily with pasteurization dose levels on fresh seafood products. The principal reason that pasteurization levels were chosen was that dosages close to or above 1 Mrad will definitely affect the original sensory and physical qualities of seafood which would remove them from the fresh seafood market. Advantages of low-dose pasteurization included control or elimination of many pathogens and parasites as well as increase the shelf life of these fresh seafoods for at least 1 week over the normal shelf life.

Bacteria

Pathogenic bacteria fine their way into seafood products in different ways. The pathogens of primary concern are bacteria that are found naturally in the marine environment. These would include the various *Vibrio* species, *Aeromonas*, and *Clostridium botulinum* Type E. In general, *Vibrio* species are relatively sensitive to low-dose gamma irradiation. When sterile and nonsterile fresh Gulf shrimp were stored for 21 days at 32°F (0°C) and 39.2°F (4°C) following inoculation with 1×10^7 colony forming units (cfu)/g of *V. cholerae* there was a 5 log cycle reduction in the number of cfu/g immediately after irradiation at 50 Krad at both temperatures. However, no *V. cholerae* were recovered immediately after or subsequent to irradiation at 100 Krad (Hinton 1983). Blue crab meat treated in the same way, except given a dosage of 25, 50, and 100 Krad, demonstrated a 5 log and a 3 log reduction in cfu/g, at 32°F (0°C) and 39.2°F (4°C) respectively, immediately after irradiation with 25 Krad. With the more effective dosages of 50 and 100 Krad, no *V. cholerae* were recovered in the crab meat immediately after irradiation (Grodner and Hinton 1987). Also, oyster meats treated in a similar manner demonstrated the same 5 log cycle reduction in cfu/g after treatment with 25 Krad and no *V. cholerae* were recovered in the oyster meat immediately following treatment with a 50 or 100 Krad dosage of gamma irradiation (Grodner and Hinton 1987). Furthermore, the effect of low-dose gamma irradiatin on *V. cholerae* in sterile and nonsterile fresh crayfish tail meat stored for 21 days at 32°F (0°C) and 39.2°F (4°C) inoculated with 1×10^7 cfu/g of *V. cholerae* was studied. In sterile and nonsterile fresh crayfish meat there was a 6 log cycle reduction and 1.5 log cycle reduction, respectively in cfu/g immediately after irradiation at 25 Krad on day 0 at 32°F (0°C). There was a 6 log cycle and a 3.5 log cycle reduction in cfu/g, respectively, immediately after irradiation at 25 Krad on day 0 at 39.2°F (4°C). No *V. cholerae* were recovered immediately after irradiation at 100 Krad at either temperature (Grodner and Hinton 1988).

Other *Vibrios* have been shown to be even more sensitive to low-dose irradiation than *V. cholera*. Twenty-seven strains of *V. parahaemolyticus* suspended in seawater were irradiated with 0–40 Krad. All strains were reduced 4–7 log cycles from the original 10^7 cfu/ml. In fish homogenate made up with seawater the organisms were more resistant and were reduced only 2.7–6.1 log cycles. In most cases complete destruction was obtained with 30–40 Krad (Matches and Liston 1971). Fresh Gulf shrimp inoculated with 10^7 cfu/g of *V. parahemolyticus* (serotype 05:17) strain 116 were reduced to 0 cfu/g immediately following a 30 Krad dose of gamma irradiation. At lower dosages of 5, 10, 15, and 25 Krad, respectively, there was survival of 10^3 cfu/g of *Vibrio* for up to 3 weeks when ice stored at 32°F (0°C). Fresh Gulf oysters demonstrated a similar response, with 0 cfu/g recovered after a 30 Krad dosage and survival of 10^3 cfu/g *V. parahemolyticus* for 1 week at the lower doses (Lewis 1986).

Vibrio vulnificus, which has been implicated in wound infections and intestinal infections of persons with compromised immune systems, has also been studied. Fresh Gulf shrimp and crabmeat were inoculated with 10^6 cfu/g of *V. vulnificus* strain 1008H and irradiated with 15, 25, and 35 Krad, respectively. The cfu/g were reduced to 0 immediately with the 35 Krad dose and reduced by approximately 4 log cycles when treated with 15 and 25 Krad doses but they survived in shrimp stored on ice for 7 days and in crabmeat 21 days before being reduced to 0 at the lower doses (Watson 1987). A study by Palumbo et al. (1986) examined the radiation resistance of *Aeromonas hydrophila,* a psychrotropic pathogen of emerging importance. The results of the study indicated that a pasteurizing dose of ionizing radiation of 150 Krad is sufficient to destroy *Aeromonas hydrophila* in concentrations at or below 10^5 cfu/g when present in retail fresh fish such as bluefish.

Clostridium botulinum Type E has always been a potential problem for the seafood processors because it is found naturally in the coastal environment and can produce botulinum toxin under ideal anaerobic storage conditions. Spores of this bacterium inoculated at 10^3 and 10^4 spores/g into fresh Gulf coast shrimp and irradiated at a dose of 150 Krad produced no toxin over the 31-day iced [32°F (0°C)] storage period. However, when the same inoculation treatment and irradiation at 150, 200, 300 and 500 Krad dosages were given to shrimp stored at 42.8°F (6°C), toxin was found to be produced in all samples after 7 days except in those treated with the 500 Krad, when toxin production occurred only after 30 days storage. Botulinum spores in oysters were only slightly more sensitive with the toxin recovered only after 30 days in iced samples treated with irradiation dosages of 150–500 Krad (Jimes 1967).

Seafood may become exposed to some human intestinal pathogenic bacteria due to the presence of sewage waste water in their growing environment. Of greatest concern here are species of *Salmonella, E. coli, Streptococcus faecalis,* and *Shigella.* Only a few strains of *Salomonella* have been treated and studied with gamma irradiation to date. *Salmonella typhimurium,* when inoculated into Louisiana Gulf oysters and then stored on ice up to 14 days, was not recovered immediately following 200 Krad gamma irradiation (Richard 1966).

Underdal and Rossebo (1972) recommended a dosage of 1.3 Mrad (1,300 Krad) when attempting to reduce *Salmonella senftenberg* in Norwegian fish meal by 10^7 cfu/g even though the initial number of viable *Salmonella* seldom exceed 1 cfu/g of commercial fish meal. *Escherichia coli,* the most common bacteria of the human gut, is often used as an indicator for the possible presence of human intestinal pathogens. Shrimp and oysters inoculated with up to 10^4 cfu/g of *E. coli* and gamma irradiated with up to 200 Krad dosage demonstrated that this bacterium had a 5% survival rate at 100 Krad and a 0.1% surviva rate at 200 Krad (Lee 1966). It would appear that a >200 Krad dosage would be required for the complete (zero) elimination of *E. coli* in these seafood products. *Streptococcus faecalis* also appears to require a dosage of irradiation >200 Krad. Gulf shrimp and oysters were treated with 100 and 200 Krad dosages of gamma irradiation after inoculation with a 10^6 cfu/g of *Streptococcus faecalis* and stored on ice. The number of cfu/g recovered after 14 days was reduced by 10^4 log cycles following the 100 Krad treatment and reduced by 10^5 log cycles following the 200 Krad treatment (Dietrich 1968).

Another major concern of seafood processing plants is the introduction of bacteria through personnel handling. One of major concern is coagulase-positive *Staphylococcus aureus,* which is capable of producing food poisoning toxins. *Staphylococcus aureus,* coagulase-positive (ATCC 9664), was inoculated into fresh Gulf shrimp, irradiated with 100 Krad of gamma irradiation, and stored on ice for 21 days. Although the cfu counts were reduced by 4 log cycles from 10^5 immediately, coagulase enzyme and *Staphylococcus* counts remained at or below 100 cfu/g over the 21-day storage period on ice (Kendall 1969). Another study has shown that *Staphylococcus* in dried and smoked mackerel required as much as a 500 Krad dosage to be inactivated. The organoleptic evaluations in this study indicated irradiation at 500 Krad level did not affect the sensory quality of these food products (Gonzalez et al. 1981).

Several bacteria that were not previously considered pathogenic are emerging as a possible threat to public health. Preliminary studies of *Listeria monocytogenes* indicate this bacteria to be fairly resistant to irradiation treatment, with viable cells remaining after 200 Krad irradiation following inoculation of crab-meat with 10^7 cells/ml (Juneau 1989). More work is needed on the incidence of *Listeria* and *Aeromonas* in seafood and on the effectiveness of low-dose gamma irradiation on eliminating these potential pathogens.

Viruses

Accumulation of viruses in shellfish is a major concern in the seafood industry, with hepatitis A virus being the virus of greatest concern. Irradiation is not seen, at this time, as a promising method for elimination of these pathogens from shellfish. Very little work has been published on the effects of irradiation on virus particles but what has been published is not encouraging. In a typical study by Girolamo et al. (1969) west coast oysters were contaminated with *Poliovirus* by natural uptake from water or by direct inoculation of the digestive tract. Viruses were accumulated rapidly. Irradiation with 50, 100, 150, 200, 300, and 500 Krad was relatively ineffective as a means of inactivation of viruses in these shellfish. Inactivation of pathogenic viruses in fish and shellfish requires doses that are too high to be usable or even generate interest for this usage by the food industry (Josephson and Peterson 1983; Urbain 1986).

Parasites

Currently there is almost no information on the use of irradiation to inactivate parasites carried by marine and freshwater fish and shellfish. *Anisakis* and related genera are found in fish and have recently become more obvious to the public with the popularity of "sushi." Thus this parasite has been implicated as a human pathogens of recent but real concern. *Anisakis* larvae in lightly salted (5%) herring fillets were only partially inactivated (44% survival) by a dose of 600 Krad. It appears that this dosage or larger would be needed to inactivate this parasite completely. Irradiation at this high dosage produces an unacceptable flavor in the herring (Van Mamersen et al. 1969). Other seafood parasites have not been investigated to any great extent for elimination with gamma irradiation.

THE FUTURE OF IRRADIATION IN THE SEAFOOD INDUSTRY

It certainly appears, to the authors, that low-dose gamma radiation is a viable solution for reducing the possible threat of foodborne illness due to the consumption of fresh seafood. Both gamma-rays from radioactive isotopes and the electron beam have been shown to be very effective in the control of spoilage and other bacteria as well as the problems of microbial pathogens that have always been part of the marine environment. For example, *V. vulnificus,* which has also been present in the environment, has now become a major concern of immunocompromised individuals. There apparently is a great deal of data being generated that would seem to indicate that a problem of chronic exposure to food pathogens is just being recognized. In studies with low-dose gamma irradiation to date, in most situations where seafood microorganisms are involved, irradiation below the 200 Krad dose level has either reduced the bacterial load tremendously or eliminated the bacteria completely. However, the dosage required to control viruses and spores of such organisms as *C. botulinum* change the characteristic quality and texture of the seafood, and therefore cannot be an alternative to the safety of a cooked product.

There is no question, if the U.S.F.D.A. will give permission for low-dose pasteurization with either Co^{60} or Cs^{137} radioisotopes or electron beam, this method of processing, when used properly, can protect the consumer public. In addition, this process maintains the fresh quality and nutritional value, with an increased shelf life of seafood, while being a proven safe and effective treatment. This might be the only method we have at present that will allow pathogen-free fresh raw seafood in the market place, especially for the public consuming raw seafoods such as ceviche, sushi, molluscan shellfish (oysters, clams, and mussels), and other exotic raw foods from the sea.

REFERENCES

Avery, K. J. W., and Lamprecht, A. 1988. The shelf-life extensions of fresh hake through gamma-irradiation. *Food Rev.* **15**(*Suppl*):28–29.

Baldrati, G., Guidi, G., Pirazzoli, P., and Vicini, E. 1974. Use of ionizing radiation for the preservation of fresh fish. Radio pasteurization of striped mullet. *Indust. Cons.* **49**:10–12.

Baldrati, G., Pirazzoli, P., Gola, S., and Ambroggi, F. 1978. Use of ionizing radiation for the preservation of fresh fish: radiopasteurization of Saurels *(Trachurus trachurus)*. *Indust. Cons.* **53**:8–10.

Bhattacharya, S. K., Choudhuri, D. R., and Bose, A. N. 1978. Preservation of Bombay duck *(Harpodon nehereus)* and rohu *(Labeo rohita)* by gamma irradiation. *Fish. Technol.* **15**:21–25.

Bhattacharyya, S. K., Chaudhuri, D. R., and Bose, A. N. 1979. Preservation of Indian mackerel *(R. kanagurta)* by gamma irradiation. *J. Food Sci. Technol. India* **16**:61–63.

Dietrich, M. D. 1968. Destruction of *Streptococcus faecalis* in shellfish with ionizing radiation. M.S. Thesis, Louisiana State University Library, Baton Rouge.

Ehlermann, D. A. E., and Reinacher, E. 1979. Some conclusions from shipboard experiments on the radurization of whole fish in the Federal Republic of Germany. *FSTA Lect.* **11**:4G311.

Gaden, E. L., Hanley, E. J., and Collins, V. P. 1951. Preservation of milk by radiation. *Food Technol.* **5**:506.

Ghadi, S. V., and Lewis, N. F. 1978. Enhancement of refrigerated storage of elasmobranchs by gamma irradiation. *FSTA Lect.* **10**:2G58.

Girolamo, R., Liston, J., and Matches, J. 1969. Uptake, elimination, and effects of irradiation on virus in west coast shellfish. *Bacteriol. Proc.* **1969**:26.

Gonzalez, O. N., Sanchez, F. C., Vista, J. B., and Ronquillo, R. P. 1981. Elimination of the risk resulting from pathogenic microorganisms in some Philippine fishery products by gamma irradiation. *Phillip. J. Sci.* **110**:45–53.

Goresline, H. E. 1982. Historical aspects of the radiation preservation of food. In: Josephson, E. S., and Peterson, M. S. (eds.) *Preservation of Food by Ionizing Radiation.* CRC Press, Boca Raton, Florida.

Grodner, R. M., and Hinton, A., Jr. 1985. Low dose gamma irradiation of *Vibrio cholerae* in oysters *(Crassostrea virginica)*. In: *Proceedings of the 10th Conference Tropical and Subtropical Fisheries Technological Society of America*, **TAMU-SG-86-102:**261–274.

Grodner, R. M., and Hinton, A., Jr. 1987. Low dose gamma irradiation of *Vibrio cholerae* in blue crab *(Callinectes sapidus)*. In: *Proceedings of the 11th Tropical and Subtropical Fisheries Technological Society of America*, **TAMU-SG-86-101:**248–260.

Grodner, R. M., and Hinton, A. Jr. 1988. Gamma irradiation of *Vibrio cholerae* in crayfish *(Procambarus clarkii* Gerard). In: *Proceedings of the 12th Tropical and Subtropical Fisheries Technological Society of America*, **SGR-92:**429-441.

Hinton, A., Jr. 1983. Survival of *Vibrio cholerae* 01, biotype El Tor, Serotype Inaba in seafoods. Ph.D. Dissertation, Louisiana State University Library, Baton Rouge.

Houwing, H. 1976. Irradiation of fishery products. I. shrimps. *Voedingsmiddelentechnol.* **9:**10–15.

IFT Scientific Status Summary, 1983. *Radiation Preservation of Foods*. Institute of Food Technologists, Chicago.

Jimes, S. 1967. *Clostridium botulinum* Type E in gulf shrimp and shucked oysters, and toxin production as affected by irradiation dosage, temperature, storage time, and mixed spore concentration. Ph.D. Dissertation, Louisiana State University Library, Baton Rouge.

Josephson, E. S., and Peterson, M. S. 1983. *Preservation of Food by Ionizing Radiation*. CRC Press, Boca Raton, Florida.

Juneau, B. 1989. M.S. Thesis, Effect of Low Dose Gamma Irradiation on *Listeria monosytogenes* in Blue Crab and Crayfish. Department of Food Science, Louisiana State University, Baton Rouge.

Kamplemacher, E. H. 1983. Irradiation for control of *Salmonella* and other pathogens in poultry and fresh meats. *Food Technol.* **37:**117–119.

Karnop, G., and Antonacopoulos, N. 1977. Effects of X-ray irradiation on the quality of fresh fish in ice. *Dsch. Lebensmit. Rundsch.* **73:**217–222.

Kendall, J. H. 1969. Low-dose gamma irradiation of coagulase positive *Staphylococcus aureus* in Gulf shrimp. M.S. Thesis, Louisiana State University, Baton Rouge.

Lee, E. P. 1966. Destruction of *E. coli* in shellfish by gamma irradiation. M.S. Thesis, Louisiana State University Library, Baton Rouge.

Lewis, T. M. 1986. Destruction of *Vibrio parahaemolyticus* on Gulf Coast shrimp and oysters by gamma irradiation. M.S. Thesis. Louisiana State University Library, Baton Rouge.

Ley, F. J. 1983. New interest in the use of irradiation in the food industry. *Soc. Appl. Bacteriol.* **11:**113.

Loaharanu, P. 1973. Gamma-irradiation of fishery products in Thailand with special reference to their microbiological aspects. In: *International Atomic Energy Agency Conference Proceedings*, pp. 21–32.

Matches, J. R., and Liston, J. 1971. Radiation destruction of *Vibrio parahaemolyticus. J. Food Sci.* **36:**339–340.

Matsuyama, A., Okazawa, Y., Namiki, M., and Sumiki, Y. 1960. Enhancement of radiation lethal effect on microorganisms by sodium chloride treatment during irradiation. *J. Jpn. Radiat. Res. Soc.* **1-2:**98–100.

Novak, A. F., and Liuzzo, J. A. 1966. Radiation pasteurization of shrimp and oysters, Louisiana State University for U.S. Atomic Energy Commission Rep. No. ORO-636, Baton Rouge, Louisiana.

Novak, A. F., and Liuzzo, J. A. 1967. Radiation pasteurization of Gulf shellfish,

Louisiana State University for U.S. Atomic Energy Commission Rep. No. ORO-654, Baton Rouge, Louisiana.

Novak, A. F., and Liuzzo, J. A. 1968. Radiation pasteurization of Gulf shellfish, Louisiana State University for U.S. Atomic Energy Commission Rep. No. ORO-660, Baton Rouge, Louisiana.

Ola, F. N. 1976. Aerobic radiopasteurization of hake *(Merluccius hubbsi)* by gamma-irradiation. *Rev. Latinoam. Microbiol.* **18:**179–183.

Oosterhuis, J. J. 1976. Irradiation of fishery products. II. fish fillets. *Voedingsmiddelentechnol.* **9:**10–16.

Palumbo, S., Jenkins, R. K., Buchanan, R. L., and Thayer D. W. 1986. Determination of irradiation *D*-values for *Aeromonas hydrophila. J. Food Protect.* **49:**189–191.

Proctor, B. E., and Goldblith, S. A., 1951. Food processing with ionizing radiation. *Food Technol.* **5:**376–377.

Przybylski, L., Finerty, M., and Grodner, R. 1989. Extension of shelf-life of fresh channel catfish fillets using modified atmospheric packaging and low dose irradiation. *J. Food Sci.* **54(2):**269–273.

Richard, G. M. 1966. Destruction of *Salmonella typhimurium* in gulf coast oysters with gamma irradiation. M.S. Thesis Louisiana State University Library, Baton Rouge.

Sasayama, S., and Amano, K. 1970. Effect of gamma irradiation on the preservation of cooked fish. II. storage life of cooked and irradiated fillets of horse mackerel at 12–15°C. *Bull. Tokai Reg. Fish. Res. Lab.* **63:**115–122.

Savagon, K. A. 1975. Commercial studies on radiation preservation of fish and shell-fish in Canada. *Fish. Technol.* **12:**160–161.

Schewan, J. M., and Georgala, D. L. 1957. The effect of spoilage and handling on the bacterial flora of fish. *Nutr. Soc. Proc.* **16:**161–163.

Schwartz, B. 1921. Effect of x-rays on *Trichinia. J. Agric. Res.* **20:**845–847.

Underdal, B., and Rossebo, L. 1972. Inactivation of strains of *Salmonella senftenberg* by gamma irradiation in fish meal. *J. Appl. Bacteriol.* **35:**371–377.

Urbain, W. M. 1986. *Food Irradiation.* Academic Press, Orlando, Florida.

Van Mameren, J., Van Spreekens, K. J., Houwing, H., and Mossel, D. A. A. 1969. Effect of irradiation on the keeping quality of packaged cod flesh. In: *Freezing and Irradiation of Fish.* Fishing News Books, London.

Watson, M. 1987. Destruction of *Vibrio vulnificus* in shrimp *(Penaeus setiferus)* and blue crab *(callinectes sapidus)* by gamma irradiation. M.S. Thesis Louisiana State University Library, Baton Rouge.

Yeh, L. T., and Hau, L. B. 1988. Preservation of grass shrimp by low dose radiation. *J. Chin. Agric. Chem. Soc.* **26:**92–102.

Index

The manufacturer's authorised representative in the EU is Springer
Nature Customer Service Centre GmbH, Europaplatz 3, 69115 Heidelberg,
Germany. If you have any concerns regarding our products, please
contact ProductSafety@springernature.com

Printed and bound by CPI Group (UK) Ltd, Croydon, CR0 4YY

23/04/2026

02095625-0002